Quantum Probability
& Related Topics

QP–PQ

Volume VIII

Quantum Probability & Related Topics

World Scientific

Singapore • New Jersey • London • Hong Kong

Published by

World Scientific Publishing Co. Pte. Ltd.

5 Toh Tuck Link, Singapore 596224

USA office: 27 Warren Street, Suite 401-402, Hackensack, NJ 07601

UK office: 57 Shelton Street, Covent Garden, London WC2H 9HE

British Library Cataloguing-in-Publication Data
A catalogue record for this book is available from the British Library.

QUANTUM PROBABILITY AND RELATED TOPICS
Volume VIII

ISBN-13 978-981-02-1140-0
ISBN-10 981-02-1140-6
ISBN-13 978-981-02-1141-7 (pbk)
ISBN-10 981-02-1141-4 (pbk)

CONTENTS

Quantum Probability and Related Topics Vol. VIII (pp. 1–18)

WIENER NOISE VERSUS WIGNER NOISE IN QUANTUM ELECTRODYNAMICS

L. ACCARDI Y.G. LU

Centro V. Volterra
Dipartimento di Matematica — Università di Roma II
Via Orazio Raimondo, 00173 Roma - Italy

ABSTRACT

We show that the investigation of the weak coupling limit (WCL) of the quantum electro-magentic (QEM) field naturally leads to the study of the **free** quantum stochastic calculus on a Hilbert module. A new type of Fock space, the so–called *interacting Fock space* also comes out from the limit.

§1. INTRODUCTION

It is well known that in quantum electro–dynamics (QED), an explicit solution of the equations of motion is not known and, for their study, several types of approximations are introduced. Among these approximations, probably the best known is the so called *dipole approximation* (c.f. [10]) in which the response term which couples matter, represented by an electron in position q, to the k-th mode of the EM-field is assumed to be 1. In the present investigation this approximation will not be assumed.

Consider a *free* particle, called the **system**, and characterized by its Hilbert space $L^2(\mathbb{R}^d)$ and Hamiltonian $H_s := -\Delta/2 = p^2/2$, where Δ is the Laplacian and $p = (p_1, p_2, p_3)$

the momentum operator. This system interacts with a quantum field in Fock representation with Hamiltonian informally written as:

$$H_R := \sum_{k\in \mathbb{Z}^d} |k| a_k^+ a_k \tag{1.1}$$

where to each mode $k \in \mathbb{Z}^d$ it is associated a representation of the CCR with creation and annihilation operators a_k^+, a_k respectively ($a_k = (a_{1,k}, a_{2,k}, a_{3,k})$).

The interaction between the system and the field is (neglicting the $\lambda^2\ A^2(q)$ term):

$$\lambda H_I = \lambda \sum_k e^{ik\cdot q} p \otimes \frac{a_k}{\sqrt{|k|}} + \text{ h.c.} = \lambda(A(q)\cdot p + p\cdot A(q)) \tag{1.2}$$

where $e^{\pm ik\cdot q}$ is the response term; $A(q)$ is the vector potential and the tensor product $p\otimes a_k$ has to be meant in the sense:

$$p\otimes a_k = \sum_{j=1}^{d} p_j \otimes a_{j,k} \tag{1.3}$$

and where the factor λ is a small scalar (coupling constant).

The total Hamiltonian we are going to consider is then $H = H_s + H_R + \lambda H_I$. The most important object for such interacting model is the wave operator at time t which, in the interacting picture, can be written as

$$U_t^{(\lambda)} := e^{it(H_S+H_R)} e^{-it(H_S+H_R+\lambda H_I)} \tag{1.4}$$

$U_t^{(\lambda)}$ is the solution of Schördinger equation:

$$\frac{d}{dt}U_t^{(\lambda)} = -i\lambda H_I(t) U_t^{(\lambda)} \,, \qquad U_0^{(\lambda)} = 1 \tag{1.5}$$

where the *evolved Hamiltonian* $H_I(t)$ is given by

$$H_I(t) := e^{it(H_S+H_R)} H_I e^{-it(H_S+H_R)} \tag{1.6}$$

Replacing the sum in (1.1), (1.2) by a continuous integral, and replacing the factor $\frac{1}{\sqrt{|k|}}$ by a cut-off function $\bar{g}(k)$ ($\bar{z}$ is the complex conjugate of $z \in \mathbb{C}$) one obtains the expression:

$$H_I(t) = i\Big[\int_{\mathbb{R}^d} dk e^{itp^2/2} e^{ik\cdot q} e^{-itp^2/2}(-ip)\otimes a_k^+(S_t g)(k) - \text{h.c.}\Big] \tag{1.7}$$

where g is *good* function (e.g. a Schwartz function) and $\{S_t\}_{t\geq 0}$ is the unitary group on $L^2(\mathbf{R}^d)$ given, in momentum representation, by

$$(S_t g)(k) = e^{it|k|}g(k) \tag{1.8}$$

Using the CCR, one can rewrite in the form (1.7)

$$H_I(t) = i\Big[\int_{\mathbf{R}^d} dk e^{-ik\cdot q}e^{-itk\cdot p}e^{-it|k|^2/2}(-ip)\otimes a_k^+(S_t g)(k) - \text{h.c.}\Big] \tag{1.9}$$

where, here and in the following, integrals should be meant in the weak sense on the domain $\mathcal{S}(R^d)$ of Schwartz functions. Moreover, without loss of generality we can simply forget the factor $e^{-it|k|^2/2}$ by transfering it into S_t. This will be done from now on.

Formally the solution of equation (1.5) is given by the iterated series

$$U_t^{(\lambda)} = \sum_{n=0}^{\infty}(-i)^n\lambda^n\int_0^t dt_1\cdots\int_0^{t_{n-1}} dt_n H_I(t_1)\cdots H_I(t_n) \tag{1.10}$$

Now we renormalize time by the replacement $t \hookrightarrow t/\lambda^2$ and study the limit of matrix elements of the form:

$$< \Phi_\lambda(\xi), U_{t/\lambda^2}^{(\lambda)}\Phi'_\lambda(\eta) > \tag{1.11}$$

($\xi,\eta \in L^2(\mathbf{R}^d)$) as $\lambda \to 0$. Where the Φ_λ are the so-called *collective vectors* and their choice is suggested by the usual perturbation theory. The basic goal of the theory is to prove that, as $\lambda \longrightarrow 0$, the limit (1.11) exists and has the form

$$< \Phi(\xi), U(t)\Phi'(\eta) > \tag{1.12}$$

Where $\Phi(\xi), \Phi'(\eta)$ are vectors in the tensor product of the space of the system with a certain limiting space $\mathcal{H}$, called *the noise space*, whose explicit form has to be determined and the limit *process* $\{U(t)\}_{t\geq 0}$ is unitary on this tensor product space for each $t \geq 0$.

Notice that denoting $\mathcal{K} \subset L^2(\mathbf{R}^d)$ the subspace of the Schwartz functions, for any pair $f, g \in \mathcal{K}$, the condition

$$\int |< f, S_t g >|dt < \infty \tag{1.12}$$

is satisfied. The subspace $\mathcal{K}$ will play an important role in our investigation.

§2. The Collective Vectors

One of the basic steps in our investigation is the choice of the collective vectors which is suggested by first order analysis of the usual perturbation theory.

For each $t \in \mathbf{R}_+$, define

$$A^+(S_tg) := \int_{\mathbf{R}^d} dk e^{-ik\cdot q} e^{itk\cdot p} (S_tg)(k) \otimes a_k^+ \tag{2.1a}$$

and define $A(S_tg)$ as the formal adjoint of $A^+(S_tg)$, i.e.

$$A(S_tg) := \int_{\mathbf{R}^d} dk e^{-itk\cdot p} e^{ik\cdot q} (S_{-t}\bar{g})(k) \otimes a_k \tag{2.1b}$$

Both $A^+(S_tg)$, $A(S_tg)$ live on $L^2(\mathbf{R}^d) \otimes \Gamma(L^2(\mathbf{R}^d))$ and behave like creation and annihilation operators respectively:

$$A(S_tg)[\xi \otimes \Phi(f)] = \int_{\mathbf{R}^d} dk e^{-itk\cdot p} e^{ik\cdot q} \xi \otimes (S_{-t}\bar{g})(k) f(k) \Phi(f) \tag{2.2}$$

where $\Phi(f)$ is the coherent vector with respect to the test function f and the integral in (2.2) is well defined (at least) in the weak topology on $L^2(\mathbf{R}^d)$ and for ξ, g, f sufficiently *good* (e.g. Schwartz) functions. In these notations, we have that

$$H(t) = i\big[A^+(S_tg)(-ip) - h.c.\big] \tag{2.3}$$

and

$$-i\lambda \int_0^t dt_1 H(t_1) = \lambda \int_0^t dt_1 \big[A^+(S_{t_1}g)(-ip) - h.c.\big] = A^+(\lambda \int_0^t dt_1 S_{t_1} g)(-ip) - h.c. \tag{2.4}$$

We shall define *the collective annihilator process* by:

$$A_\lambda(t) := \lambda \int_0^{t/\lambda^2} dt_1 \int_{\mathbf{R}^d} dk e^{-itk\cdot p} e^{ik\cdot q} \otimes \overline{S_tg}(k) a_k; \quad t \in \mathbf{R} \tag{2.5a}$$

and its conjugate, *the collective creator process*, by:

$$A_\lambda^+(t) := \lambda \int_0^{t/\lambda^2} dt_1 \int_{\mathbf{R}^d} dk e^{-ik\cdot q} e^{itk\cdot p} \otimes S_tg(k) a_k^+; \quad t \in \mathbf{R} \tag{2.5b}$$

More generally, we shall define the collective creators and annihilators by:

$$A_\lambda^+(S,T,f) := \lambda \int_{S/\lambda^2}^{T/\lambda^2} dt \int_{\mathbf{R}^d} dk e^{-ik\cdot q} e^{itk\cdot p} S_t f(k) a_k^+ \tag{2.6}$$

The first step of our study is to show that in a certain sense, as $\lambda \longrightarrow 0$,

$$A_\lambda^+(S,T,f) \longrightarrow A^+(S,T,f); \qquad A_\lambda(S,T,f) \longrightarrow A(S,T,f)$$

i.e. the collective operators converge to some kind of creation and annihilation operators $A^+(S,T,f)$ and $A(S,T,f)$ acting on some limit Hilbert space.

The consideration of the two point function is naturally suggested by the fact that the vacuum distribution of the creation and annihilation fields is gaussian.

Consider, for each $s,t \geq 0$,

$$< \xi \otimes \Phi, A_\lambda(t) A_\lambda^+(s) \; \eta \otimes \Phi > \tag{2.7}$$

which is equal to

$$\lambda^2 \int_0^{t/\lambda^2} dt_1 \int_0^{s/\lambda^2} dt_2 \int_{\mathbf{R}^d} dk_1 \int_{\mathbf{R}^d} dk_2 \;\; < 0 | a_{k_1} a_{k_2}^+ | 0 >$$

$$< \xi, e^{-it_1 k_1 \cdot p} e^{ik_1 \cdot q} e^{it|k_1|^2/2} (-ip)(S_{t_1} g)(k_1) e^{-ik_2 \cdot q} e^{itk_2 \cdot p} e^{-it|k_2|^2/2} (-ip)(S_{t_2} g)(k_2) \eta >=$$

$$= \lambda^2 \int_0^{t/\lambda^2} dt_1 \int_0^{s/\lambda^2} dt_2 \int_{\mathbf{R}^d} dk e^{-i(t_2-t_1)|k|^2/2} < (ip)\xi, e^{i(t_2-t_1)k\cdot p}(ip)\eta > \eta \otimes \Phi > \tag{2.8}$$

and goes, as $\lambda \to 0$, to

$$t \wedge s \int_{-\infty}^{\infty} d\tau \int_{\mathbf{R}^d} dk < (ip)\xi, e^{i\tau k \cdot p}(ip)\eta > \bar{g}(k) \; (S_\tau g)(k) \tag{2.9}$$

The above result can be rephrased as follows: the approximate *two point function* (2.7) tends to an object which for some aspects, is very similar to a two point functions.

DEFINITION **(2.1)** A collective number vector is a vector of $L^2(\mathbf{R}^d) \otimes \Gamma(L^2(\mathbf{R}^d))$ obtained by applying a (finite) product of collective creation operators to a vector of the form $\Phi \otimes \xi$ where Φ is the vacuum in $\Gamma(L^2(\mathbf{R}^d))$ and ξ is an arbitrary vector in $(L^2(\mathbf{R}^d))$. Such a vector will have the form

$$A_\lambda^+(S_1, T_1, f_1) \cdot \ldots \;\; A_\lambda^+(S_n, T_n, f_n)\xi \otimes \Phi \tag{2.10}$$

In the following, when no confusion can arise, we shall use the simplified notation

$$\prod_{h=1}^{n} A_{\lambda,h}^+ \xi \otimes \Phi \tag{2.11}$$

to denote a vector of the form (2.10).

From now on we shall find it convenient to operate a change in notations with respect to the previous Sections and to write the state space of the composite system in the form $\Gamma(L^2(\mathbf{R}^d)) \otimes L^2(\mathbf{R}^d)$, rather than $L^2(\mathbf{R}^d) \otimes \Gamma(L^2(R^d))$.

§3 THE BREAKDOWN OF THE GAUSSIANITY AND THE LIMITING FIELDS

In this section we shall show that, by applying to the model our procedure, one breaks the CCR and the gaussianity in the sense that the limiting objects no longer enjoy the properties. The simplest situation illustrating this phenomenon is the **four point function**:

$$< A_{\lambda}^{+}(0,T_1,f_1)A_{\lambda}^{+}(0,T_2,f_2)\Phi \otimes \xi,\ A_{\lambda}^{+}(0,T_1',f_1')A_{\lambda}^{+}(0,T_2',f_2')\Phi \otimes \eta > \tag{3.1}$$

By the definition of collective creation and annihilation operators, (3.1) is equal to

$$\begin{aligned}
&< \lambda^2 \int_0^{T_1/\lambda^2} dt_1 \int_0^{T_2/\lambda^2} dt_2 \int_{\mathbb{R}^d}\int_{\mathbb{R}^d} dk_1 dk_2 e^{-ik_1\cdot q} e^{it_1 k_1\cdot p} e^{-ik_2\cdot q} e^{it_2 k_2\cdot p} \\
&\qquad (S_{t_1} f_1)(k_1)(S_{t_2} f_2)(k_2) a_{k_1}^{+} a_{k_2}^{+} |0> \otimes \xi, \\
&\lambda^2 \int_0^{T_1'/\lambda^2} ds_1 \int_0^{T_2'/\lambda^2} ds_2 \int_{\mathbb{R}^d}\int_{\mathbb{R}^d} e^{-ik_1'\cdot q} e^{is_1 k_1'\cdot p} e^{-ik_2'\cdot q} e^{is_2 k_2\cdot p} \\
&\qquad (S_{s_1} f_1')(k_1')(S_{s_2} f_2')(k_2) a_{k_1'}^{+} a_{k_2'}^{+} |0 \otimes \eta >= \\
&= \lambda^4 \int_0^{T_1/\lambda^2} dt_1 \int_0^{T_2/\lambda^2} dt_2 \int_0^{T_1'/\lambda^2} ds_1 \int_0^{T_2'/\lambda^2} ds_2 \int_{\mathbb{R}^{2d}} dk_1 dk_2 \\
&\qquad [< \xi, e^{-it_2 k_2\cdot p} e^{ik_2\cdot q} e^{-it_1 k_1\cdot p} e^{is_1 k_1\cdot p} e^{-ik_2\cdot q} e^{is_2 k_2\cdot p} \eta > \\
&\qquad \bar{f}_1(k_1)(S_{s_1-t_1} f_1')(k_1)\bar{f}_2(k_2)(S_{s_2-t_2} f_2')(k_2) + \\
&+ < \xi, e^{-it_2 k_2\cdot p} e^{ik_2\cdot q} e^{-it_1 k_1\cdot p} e^{ik_1\cdot q} e^{-ik_2\cdot q} e^{is_1 k_2\cdot p} e^{-ik_1\cdot q} e^{is_2 k_1\cdot p} \eta > \\
&\qquad \bar{f}_1(k_1)(S_{s_2-t_1} f_2')(k_1)\bar{f}_2(k_2)(S_{s_1-t_2} f_1')(k_2)]
\end{aligned} \tag{3.2}$$

By the CCR, one can move the $q-$factors together in the right hand side of (3.2). Thus one can rewrite (3.1) as

$$\begin{aligned}
&\lambda^4 \int_0^{T_1/\lambda^2} dt_1 \int_0^{T_2/\lambda^2} dt_2 \int_0^{T_1/\lambda^2} ds_1 \int_0^{T_2/\lambda^2} ds_2 \int_{\mathbb{R}^{2d}} dk_1 dk_2 \\
&[< \xi, e^{-it_2 k_2\cdot p} \cdot e^{is_2 k_2\cdot p} e^{-it_1 k_1\cdot p} e^{-is_1 k_1\cdot p} e^{i/2(s_1-t_1)k_2\cdot k_1} \eta > \\
&\qquad \bar{f}_1(k_1)(S_{s_1-t_1} f_1)(k_1)\bar{f}_2(k_2)(S_{s_2-t_2} f_2')(k_2) + \\
&+ < \xi, c^{-it_2 k_2\cdot p} e^{is_1 k_2\cdot p} e^{-it_1 k_1\cdot p} e^{-is_2 k_1\cdot p} e^{i/2(s_1-t_1)k_2\cdot k_1} \eta > \\
&\qquad \bar{f}_1(k_1)(S_{s_2-t_1} f_2')(k_1)\bar{f}_2(k_2)(S_{s_1-t_2} f_1')(k_2)]
\end{aligned} \tag{3.3}$$

By the change of variables

$$\tau_1 = \lambda^2 t_1 \ , \tau_2 = \lambda^2 t_2 \ , s_1 - \tau_1/\lambda^2 = u \ , \ s_2 - \tau_2/\lambda^2 = v \tag{3.4}$$

The first term of (3.3) becomes

$$\int_0^{T_1} d\tau_1 \int_0^{T_2} d\tau_2 \int_{-\tau_1/\lambda^2}^{(T_1'-\tau_1)/\lambda^2} du \int_{-\tau_2/\lambda^2}^{(T_2'-\tau_2)/\lambda^2} dv \int_{\mathbb{R}^{2d}} dk_1 dk_2$$

$$< \xi, e^{ivk_2\cdot p} e^{iuk_1\cdot p} \eta > e^{\frac{i}{2}uk_1\cdot k_2} \bar{f}_1(k_1)\bar{f}_2(k_2)(S_u f_1')(k_1)(S_v f_2')(k_2) \tag{3.5}$$

and as λ tends to zero, this goes to

$$< \chi_{[0,T_1]}, \chi_{[0,T_2']} >_{L^2(\mathbb{R})} \cdot \int_{-\infty}^{+\infty} du \int_{-\infty}^{+\infty} dv \int_{\mathbb{R}^{2d}} dk_1 dk_2 \bar{f}_1(k_1)\bar{f}_2(k_2) e^{i/2uk_1\cdot k_2}$$

$$(S_u f_1')(S_v f_2')(k_2) \cdot < \xi, e^{i(uk_1+vk_2)\cdot p}\eta > \tag{3.6}$$

By a similar procedure and the Riemann–Lebesgue lemma, one shows that the second term of (3.3) tends to zero as $\lambda \to 0$.

Since there is no way to express (3.6) as the four point function of some quantum gaussian field, we conclude that: the above calculation shows that even if the limits

$$A_\lambda \longrightarrow A\ , \qquad A_\lambda^+ \longrightarrow A^+$$

are supposed to exist, the limit creation and annihilation operators can **not** be Boson Fock creation and annihilation operators althghout we begin our investigation from Boson type objects.

In order to identify the limit space on which our limit noise lives, one must try to determine the limit

$$\lim_{\lambda\to 0} \langle \Phi_\lambda,\ \Phi_\lambda' \rangle =: \quad < \Phi, \Phi' >$$

of the scalar product of two collective vectors.

According to our choice of the collective vectors this amounts to study the limit, as $\lambda \to 0$, of scalar products of the form:

$$\langle \prod_{h=1}^{n} A_{\lambda,h}^+ \Phi \otimes \xi,\ \prod_{h=1}^{n} A_{\lambda,h}^+ \Phi \otimes \eta \rangle$$

THEOREM**(3.1)**. For any $N, n \in \mathbb{N}$, the limit of the scalar product of two collective number vectors:

$$\lim_{\lambda\to 0} \langle \prod_{h=1}^{N} A_{\lambda,h}^+ \Phi \otimes \xi,\ \prod_{h=1} A_{\lambda,h}^+ \Phi \otimes \eta \rangle \tag{3.7}$$

exists and is equal to:

$$\delta_{N,n} \prod_{h=1}^{n} \langle \chi_{[S_h,T_h]},\ \chi_{[S_h',T_h']} \rangle \cdot \int_{-\infty}^{+\infty} du_1 \ldots \int_{-\infty}^{+\infty} du_n \int_{\mathbb{R}^{nd}} dk_1 \ldots dk_n$$

$$\langle\xi, \prod_{h=1}^{n} e^{iu_h k_h \cdot p}\eta\rangle \prod_{h=1}^{n}(S_{u_h} f'_h)(k_h)\overline{f}_h(k_h) \exp\Big(\frac{i}{2}\sum_{r=1}^{n-1}\sum_{h=r}^{n-1} u_r k_r \cdot k_{h+1}\Big) \tag{3.8}$$

REMARK. We shall not prove this Theorem but only discuss some of its consequences. One can find a proof for a more general case in the following section.

Notice that if the factor

$$\langle\xi, \prod_{h=1}^{n} e^{iu_h k_h \cdot p}\eta\rangle \cdot \exp\Big(\frac{i}{2}\sum_{r=1}^{n-1}\sum_{h=r}^{n-1} u_r k_{h+1} \cdot k_r\Big) \tag{3.9}$$

were absent in (3.8), then the expression (3.8) would coincide with the scalar product

$$\langle\otimes_{j=1}^{n}(\chi_{[S_j,T_j]} \otimes f_j),\ \otimes_{j=1}^{n}(\chi_{[S'_j,T'_j]} \otimes f'_j)\rangle \tag{3.10}$$

defined on the tensor product of n copies of the space:

$$L^2(\mathbf{R}) \otimes \mathcal{K}$$

where $L^2(\mathbf{R})$ has the usual scalar product, while $\mathcal{K} \subset L^2(\mathbf{R}^d)$ has the scalar product

$$(f|g) := \int_{\mathbf{R}} dt\langle f, S_t g\rangle_{L^2(\mathbf{R}^d)} \tag{3.11}$$

Notice that the scalar product (3.10) can also be written in the form

$$\langle\prod_{j=1}^{n} A^+(\chi_{[S_j,T_j]} \otimes f_j)\Psi,\ \prod_{j=1}^{n} A^+(\chi_{[S'_j,T'_j]} \otimes f'_j)\Psi\rangle \tag{3.12}$$

where $A^+(\cdot)$ denotes the creation operator on the full Fock space over $L^2(\mathbf{R}) \otimes \mathcal{K}$. Ψ is the vacuum vector in this space.

Our goal in the following sections will be to write the limit (3.7) in the form (3.12) where the $A^+(\cdot)$ are *some sort of creation operators* acting on *some sort of Fock space* in which Ψ plays the role of vacuum vector.

The obstruction to the use of the usual Fock or full Fock space comes precisely from the scalar factor (3.9).

The scalar factor (3.9) is itself the product of two factors, each of which is related to the other two new features arising from our construction, namely:

(i) the $\xi - \eta$-scalar product is related to the fact that the noise lives on a Hilbert module rather than a Hilbert space.

(ii) The exp-factor is related to the fact that the space (more precisely, the module) over the test function space is not the usual Fock, or full Fock space, but the *interacting Fock space*.

§4 THE ADMISSIBLE PAIR PARTITIONS AND THE LIMITS OF MATRIX ELEMENTS

The present section is devoted to recall some basic properties of the *admissible pair partition* (also cf. [4,5,6]).

To each pair of natural integers (m_h, m_h), we associate the closed interval:

$$[m'_h, m_h] := \{x \in \mathbf{N} : m'_h \le x \le m_h\} \tag{4.1}$$

and we shall say that the two pairs (m'_h, m_h), (m'_k, m_k) are admissible if

$$[m'_h, m_h] \cap [m'_k,\ m_k] = \begin{cases} \emptyset \\ [m'_h, m_h] \\ [m'_k, m_k] \end{cases} \tag{4.2}$$

DEFINITION **(4.1)**. Let n be a natural integer. An admissible pair partition of the set $\{1, 2, \ldots, 2n\}$ is a pair partition

$$(m'_1, m_1), \ldots, (m'_n, m_n) \tag{4.3}$$

such that any two pairs (m'_h, m_h), (m'_k, m_k) of the partition are admissible.

LEMMA **(4.2)**. Given n numbers

$$m_1 < m_2 < \ldots < m_n$$

in the set $\{1, \ldots, 2n\}$. If there exists an admissible pair partition

$$(m'_1, m_1), \ldots, (m'_n, m_n) \tag{4.4}$$

of the set $\{1, \ldots, 2n\}$ such that

$$m'_h < m_h \ ; \qquad h = 1, \ldots, n \tag{4.5}$$

then it is unique. Moreover, for each $h_0 = 1, \ldots, n$ the number m'_{h_0} is uniquely determined by the condition:

$$\begin{aligned} m'_{h_0} := \max\{x \in \{1, \ldots, 2n\} \backslash \{m_h\}_{h=1}^n : x < m_{h_0} | \{x+1, \ldots, m_{h_0} - 1\} \cap \\ \cap \{m_1, \ldots, m_n\}| = |\{x+1, \ldots, m_{h_0} - 1\} \cap (\{1, \ldots, 2n\} \backslash \{m_h\}_{h=1}^n)|\} \end{aligned} \tag{4.6}$$

where, by $|\{\cdots\}|$ we denote the cardinality of set $\{\cdots\}$

In the following we shall use the notation

$$a.p.p.\{1, \ldots, 2n\} \tag{4.7}$$

to denote the family of all the admissible pair partitions of the set $\{1,\ldots,2n\}$.

In order to calculate the limit (1.11), one has to identify the limit of matrix elements of arbitrary products of collective creation and annihilation operators.

Theorem (4.3). For any natural integer n and with the notations:

$$A^0_\lambda := A_\lambda \ ; \qquad A^1_\lambda := A^+_\lambda \tag{4.8}$$

the limit

$$\lim_{\lambda\to 0}\langle \Phi\otimes\xi, \sum_{\varepsilon\in\{0,1\}^{2n}} A_\lambda^{\varepsilon(1)}(S_1,T_1,f_1)\ldots A_\lambda^{\varepsilon(2n)}(S_{2n},T_{2n},f_{2n})\Phi\otimes\eta\rangle \tag{4.9}$$

exists and is equal to:

$$\sum_{\{m'_h,m_h\}_{h=1}^n\in a.p.p.\{1,\ldots,2n\}} \prod_{h=1}^n \langle \chi_{[S_{m_h},T_{m_h}]}, \chi_{[S_{m'_h},T_{m'_h}]}\rangle_{L^2(\mathbb{R})}$$

$$\int_{-\infty}^{+\infty} d u_1\ldots\int_{-\infty}^{+\infty} d u_n \int_{\mathbb{R}^{nd}} d k_1\ldots d k_n \prod_{h=1}^n (S_{u_h} f_{m_h})(k_h)\overline{f}_{m'_h}(k_h)$$

$$\langle\xi, \prod_{h=1}^n e^{iu_hk_h\cdot p}\eta\rangle\cdot\exp\Big(\frac{i}{2}\sum_{h=1}^{n-1}\sum_{r=h+1}^{n} u_hk_h\cdot k_r\chi_{(m'_r,m_r)}(m_h)\Big) \tag{4.10}$$

Without loss of generality we shall assume that $S_h = 0$ for all $h = 1,\cdots,2n$ and we consider the matrix elements of arbitrary products of collective creation and annihilation operators

$$\langle\Phi\otimes\xi,\ A_\lambda(0,T_1,f_1)\ldots A^+_\lambda(0,T'_1,f'_1)\ldots A^+_\lambda(0,T'_n,f'_n),\ldots,A_\lambda(0,T_n,f_n)\Phi\otimes\eta\rangle \tag{4.11}$$

Counting from left to right, we denote

$$1\le m_1 < m_2 < \ldots < m_n \le 2n \tag{4.12}$$

the places where the creation operators appear.

Since Φ is the vacuum vector, the expression (4.11) is clearly equal to zero if either $m_n < 2n$ or $m_1 = 1$.

Using the definition of A_λ and A^+_λ, (4.11) becomes

$$\lambda^{2n}\int_0^{T_1/\lambda^2} d t_1\ldots\int_0^{T_{m_1}/\lambda^2} d t_{m_1}\ldots\int_0^{T_{2n}/\lambda^2} d t_{m_n}\int_{\mathbb{R}^{2nd}} d\vec{k}$$

$$\langle\xi, e^{-itk_1\cdot p}e^{ik_1\cdot q}\ldots e^{-ik_{m_1}q}e^{is_{m_1}k_{m_1}\cdot p}\ldots e^{-it_n\cdot k_n\cdot p}e^{ik_n\cdot q}\ldots e^{-ik_{m_n}\cdot q}e^{itm_nk_{m_n}\cdot p}\eta\rangle$$

$$\prod_{h=1}^{n}(S_{t_{m_h}}f'_{m_h})(k_{m_h})\cdot\prod_{\alpha\in\{1,\dots,2n\}\setminus\{m_1,\dots,m_n\}}\overline{(S_{t_\alpha}f_\alpha)}(k_\alpha)$$

$$\langle 0|a_{k_1}\dots a^+_{k_{m_1}}\dots a_{k_n}\dots a^+_{k'_{m_n}}|0\rangle \tag{4.13}$$

Because of the vacuum expectation value in (4.13) becomes

$$\sum_{\substack{(\overline{m}_1,\dots,\overline{m}_n)=(1,\dots,2n)\setminus\{m_h\}_{h=1}^n\\ \overline{m}_h<m_h,\,h=1,\dots,n}}\prod_{h=1}^{n}\delta_{k_{\overline{m}_h},\,k_{m_h}}$$

$$\lambda^{2n}\int_0^{T_1/\lambda^2}dt_1\dots\int_0^{T_{2n}/\lambda^2}dt_{2n}\int_{\mathbb{R}^{nd}}d\vec{k}\sum_{\substack{(\overline{m}_1,\dots,\overline{m}_n)=(1,\dots,2n)\setminus\{m_n\}_{h=1}^n\\ \overline{m}_h<m_h,\,h=1,\dots,n}}$$

$$\langle\xi,e^{-it_{\overline{m}_1}k_1\cdot p}e^{ik_1\cdot q}\dots e^{-im_1\cdot q}e^{it_{m_1}k_1\cdot p}e^{-it_{\overline{m}_n}k_n\cdot p}e^{ik_n\cdot q}\dots e^{-ik_n\cdot q}e^{it_{m_n}k_n\cdot p}\eta\rangle$$

$$\prod_{h=1}^{n}\overline{f}_{\overline{m}_h}(k_h)(S_{t_{m_h}-t_{\overline{m}_h}}f_{m_h})(k_h) \tag{4.14}$$

LEMMA **(4.4)**. In the limit $\lambda\to 0$ the only terms of the summation (4.14) which give a non zero contribution are those corresponding to the choice of those n-uples $(\overline{m}_1,\dots,\overline{m}_n)$ which satisfy the equation

$$\chi_{[\overline{m}_r+1,2n-1]}(m_h)-\chi_{[\overline{m}_r+1,2n-1]}(\overline{m}_h)$$

$$+\chi_{[m_r+1,2n-1]}(\overline{m}_h)-\chi_{[m_r,2n-1]}(m_h)=0 \tag{4.15}$$

Proof. Remember that the indices $\overline{m}_j$ in (4.14) correspond to annihilators in (4.11) and are associated to exponential factors of the form

$$e^{-it_{\overline{m}_j}k_j\cdot p}e^{ik_j\cdot q} \tag{4.16}$$

in (4.14). While the indices m_j correspond to creators and are associated to exponential factors of the form

$$e^{-ik_j\cdot q}e^{it_{m_j}k_j\cdot p} \tag{4.17}$$

We bring all the factors $\exp(\pm ik_j\cdot q)$ to the right of all the factors $\exp(\pm t_{\alpha_j}k_j\cdot p)$ ($\alpha_j=m_j$ or $\overline{m}_j$) with the exception of the factor $\exp(it_{m_n}k_n\cdot p)$, which remains on the extreme right.

The scalar phase factor which arises when all the permutations have been performed has the form $\exp(i\Sigma)$ where Σ is a real number which can be expressed as the sum of four different types of terms:

$$-\sum_{r=1}^{n-1}\sum_{h=r}^{n-1}t_{m_h}k_r\cdot k_h \tag{4.18a}$$

$$\sum_{h=1}^{n-1}\sum_{r=1}^{n} \chi_{[m_h+1,2n-1]}(\overline{m}_r) \quad t_{\overline{m}_r} k_h \cdot k_r \tag{4.18b}$$

$$\sum_{r=1}^{n-1}\sum_{h=1}^{n} \chi_{[\overline{m}_h+1,2n-1]}(m_r) \quad t_{m_r} k_h \cdot k_r \tag{4.18c}$$

$$-\sum_{r=1}^{n}\sum_{h=1}^{n} \chi_{[\overline{m}_h+1,2n-1]}(\overline{m}_r) \quad t_{\overline{m}_r} k_r \cdot k_h \tag{4.18d}$$

The sum of the four expressions $(4.18a, b, c, d)$ can be rewritten as

$$\sum_{1\le h,r\le n} [\chi_{[m_r+1,2n-1]}(\overline{m}_h)t_{\overline{m}_h} + \chi_{[\overline{m}_r+1,2n-1]}(m_h)t_{m_h}$$

$$-\chi_{[m_r,2n-1]}(m_h)t_{m_h} - \chi_{[\overline{m}_r+1,2n-1]}(\overline{m}_h)t_{\overline{m}_h}]k_h \cdot k_r \tag{4.19}$$

which is equal to

$$\begin{aligned} &\sum_{1\le h,r\le n} k_h \cdot k_r (t_{m_h} - t_{\overline{m}_h})\chi_{[\overline{m}_r+1,2n-1]}(m_h) \\ &+ \sum_{1\le h,r\le n} t_{\overline{m}_h} k_h \cdot k_r (\chi_{[\overline{m}_r+1,2n-1]}(m_h) - \chi_{[\overline{m}_r+1,2n-1]}(\overline{m}_h)) \\ &- \sum_{1\le h,r\le n} (t_{m_h} - t_{\overline{m}_h})k_h \cdot k_r \chi_{[m_r,2n-1]}(m_h) \\ &+ \sum_{1\le h,r\le n} t_{\overline{m}_h} k_n \cdot k_r (\chi_{[m_r+1,2n-1]}(\overline{m}_h) - \chi_{[m_r,2n-1]}(m_h)) \end{aligned}$$

By the Riemann–Lebesgue Lemma, we conclude that, in the limit $\lambda \to 0$, only those terms of the summation (4.14) can survive, for which

$$\sum_{1\le h,r\le n} t_{\overline{m}_h} k_h \cdot k_r [\chi_{[\overline{m}_r+1,2n-1]}(m_h) - \chi_{[\overline{m}_r+1,2n-1]}(\overline{m}_h)$$

$$+ \chi_{[m_r+1,2n-1]}(\overline{m}_h) - \chi_{[m_r,2n-1]}(m_h)] = 0$$

for almost all $(t_1, \ldots, t_{2n}) \in \mathbb{R}^{2n}$ and almost all $(k_1, \ldots, k_n) \in \mathbb{R}^{nd}$ and this condition is equivalent to equation (4.15).

LEMMA **(4.5)**. For any natural integer n and any choice of $1 < m_1 < \ldots < m_n = 2n$ in the set $\{1, \ldots, 2n\}$, the equation (4.15), in the unknowns $\overline{m}_1, \ldots, \overline{m}_n$ has a unique solution satisfying

$$\overline{m}_h < m_h \; ; \qquad h = 1, \ldots, n \tag{4.20}$$

Moreover the solution $(\overline{m}_1, \ldots, \overline{m}_h)$ is characterized by the property that

$$(\overline{m}_1, m_1), \; (\overline{m}_2, m_2), \ldots, (\overline{m}_n, m_n) \tag{4.21}$$

is the unique admissible partition of the set $\{1, \ldots, 2n\}$ associated to the set $\{m_h\}_{h=1}^n$ in the sense specified by Lemma (4.3).

Proof. We distinguish two cases:

1)
$$\chi_{[\overline{m}_r+1,2n-1]}(m_h) = 0$$

In this case, by admissibility one must also have:

$$\chi_{[m_r,2n-1]}(m_h) = 0$$

because $\overline{m}_r < m_r$. Hence

$$\chi_{[m_r+1,2n-1]}(\overline{m}_h) = \chi_{[\overline{m}_r+1,2n-1]}(\overline{m}_h)$$

and therefore

$$\overline{m}_h > \overline{m}_r \Leftrightarrow \overline{m}_h > m_r$$

2)
$$\chi_{[\overline{m}_r+1,2n-1]}(m_h) = 1$$

In this case, if $\chi_{[m_r,2n-1]}(m_h) = 1$, by the same arguments as above, we have

$$\overline{m}_h > \overline{m}_r \Leftrightarrow \overline{m}_h > m_r \qquad \forall h, r = 1, \ldots, n$$

Notice that this is equivalent to say that

$$m_h > \overline{m}_r \Leftrightarrow \overline{m}_h > \overline{m}_r \qquad \forall h, r = 1, \ldots, n$$

and clearly this can be reformulated as

$$m_h > \overline{m}_r \Rightarrow \overline{m}_h > \overline{m}_r \qquad \forall h, r = 1, \ldots, n$$

Now if $\chi_{[m_r,2n-1]}(m_h) = 0$, one must have

$$\chi_{[\overline{m}_r+1,2n-1]}(\overline{m}_h) = 1$$

That is, in any case, $\chi_{[\overline{m}_r+1,2n-1]}(m_h) = 1$ implies that

$$\chi_{[\overline{m}_r+1,2n-1]}(\overline{m}_h) = 1$$

which is equevalent to

$$m_h > \overline{m}_r \Rightarrow \overline{m}_h > \overline{m}_r$$

summing up, we have that either $[\overline{m}_h, m_h] \cap [\overline{m}_r, m_r] = \emptyset$ or $[\overline{m}_h, m_h] \subseteq [\overline{m}_r, m_r]$. Hence the partition (6.14) is admissible.

§5 The Limit Hilbert Modules

In order to interpret the expression (4.10), obtained in Theorem (4.1) for the limit of matrix elements any product of collective creators and annihilators, we can describe in this section the Hilbert module on which the limits of the collective creation and annihilation processes live.

For each $f \in \mathcal{K}$, define

$$\tilde{f}(t) := \int_{\mathbf{R}^d} e^{-ikq} e^{ikp} (S_t f)(k) dk \tag{5.1}$$

then, $\tilde{f}$ is a map from $\mathbf{R}$ to the bounded operators $L^2(\mathbf{R}^d)$. Denote by $\mathcal{P}$ the C^*- algebra generated by $\{e^{ikp}; k \in \mathbf{R}^d\}$; $\mathcal{F}$ the $\mathcal{P}$-right-linear span of $\{\tilde{f};\ f \in \mathcal{K}\}$. Then $\mathcal{F}$ is a $\mathcal{P}$-right module and therefore the algebraic tensor product between $L^2(\mathbf{R})$ and $\mathcal{F}$ (denoted by $L^2(\mathbf{R}) \odot \mathcal{F}$) is a $\mathcal{P}$ (in fact $1 \odot \mathcal{P}$)-right module. On the $\mathcal{P}$-right module $L^2(\mathbf{R}) \odot \mathcal{F}$, we introduce a $\mathcal{P}$-right bi-linear, $\mathcal{P}$- value form:

$$(\cdot|\cdot)\ : L^2(\mathbf{R}) \odot \mathcal{F} \times L^2(\mathbf{R}) \odot \mathcal{F} \longrightarrow \mathcal{P} \tag{5.2}$$

by

$$(\alpha \odot \tilde{f} | \beta \odot \tilde{g}) := < \alpha, \beta >_{L^2(\mathbf{R})} \cdot \int_{\mathbf{R}} du \int_{\mathbf{R}^d} dk e^{-iukp} \bar{f}(k)(S_u g)(k) \tag{5.3}$$

In the following we shall identify with the eqivalence class of f with respect to the bi-linear form given by the right hand of (5.3). Thus $L^2(\mathbf{R}) \odot \mathcal{F}$ becomes a $\mathcal{P}$-pre-Hilbert module. The positivity of the right hand side of (5.3) follows from Theorem (4.1). For each $n \in \mathbf{N}$, we denote by $\left(L^2(\mathbf{R}) \odot \mathcal{F}\right)^{\odot n}$ the algebraic tensor product of n-copies of $L^2(\mathbf{R}) \odot \mathcal{F}$ and define $\mathcal{P}$-right bi-linear, $\mathcal{P}$- valued form:

$$(\cdot|\cdot)\ : \left(L^2(\mathbf{R}) \odot \mathcal{F}\right)^{\odot n} \times \left(L^2(\mathbf{R}) \odot \mathcal{F}\right)^{\odot n} \longrightarrow \mathcal{P} \tag{5.4}$$

by

$$((\alpha_1 \odot \tilde{f}_1) \odot \cdots \odot (\alpha_n \odot \tilde{f}_n) | (\beta_1 \odot \tilde{g}_1) \odot \cdots \odot (\beta_n \odot \tilde{g}_n)) :=$$

$$:= \prod_{h=1}^{n} < \alpha_h, \beta_h >_{L^2(\mathbf{R})} \int_{\mathbf{R}^n} du_1 \cdots du_n \int_{\mathbf{R}^{nd}} dk_1 \cdots dk_n \prod_{h=1}^{n} [e^{-iu_h k_h p} \bar{f}_h(k_h)(S_{u_h} g_h)(k_h)$$

$$\exp\left(\frac{i}{2} \sum_{1 \le r \le h \le n-1} u_r k_r k_{h+1}\right) \tag{5.5}$$

For each $n \in \mathbf{N}$, with the $\mathcal{P}$-right bi-linear, $\mathcal{P}$- valued form given by (5.5), $\left(L^2(\mathbf{R}) \odot \mathcal{F}\right)^{\odot n}$ becomes a $\mathcal{P}$-pre-Hilbert module and the notion $\left(L^2(\mathbf{R}) \odot \mathcal{F}\right)^{\otimes n}$ will be used to denote it.

Since for each $n \in \mathbf{N}$, $(L^2(\mathbf{R}) \odot \mathcal{F})^{\otimes n}$ is a $\mathcal{P}$–pre–Hilbert module, the direct sum $\mathbf{C} \oplus \bigoplus_{n=1}^{\infty}(L^2(\mathbf{R}) \odot \mathcal{F})^{\otimes n}$ makes sense and will be denoted by $\Gamma\Big(L^2(\mathbf{R}) \odot \mathcal{F}\Big)$ and called the Fock module over $L^2(\mathbf{R}) \odot \mathcal{F}$. In this pre–Hilbert module, the vector $\Psi := 1 \oplus 0 \oplus 0 \cdots$ is called the **vacuum vector**. One can easily show that

LEMMA **(5.1)**. The number vector subset

$$\Gamma := \Big\{A^+(\alpha_1 \odot \tilde{f}_1) \cdots A^+(\alpha_n \odot \tilde{f}_n)\Phi;\ n \in \mathbf{N},\ \alpha_j \in L^2(\mathbf{R}),\ \tilde{f}_j \in \mathcal{F},\ j = 1, \cdots, n\Big\} \quad (5.6)$$

is a $\mathcal{P}$–total subset of $\Gamma\Big(L^2(\mathbf{R}) \odot \mathcal{F}\Big)$.

DEFINITION **(5.2)** For each element of $L^2(\mathbf{R}) \odot \mathcal{F}$, the creator with respect to this element, denoted by $A^+(\cdot)$, is defined on the $\mathcal{P}$–right linear span of Γ by $\mathcal{P}$–right linearity and

$$A^+(\alpha \odot \tilde{f})\Big[(\alpha_1 \odot \tilde{f}_1) \odot \cdots \odot (\alpha_n \odot \tilde{f}_n)\Psi\Big] := (\alpha \odot \tilde{f}) \odot (\alpha_1 \odot \tilde{f}_1) \odot \cdots \odot (\alpha_n \odot \tilde{f}_n)\Phi \quad (5.7)$$

where $n \in \mathbf{N}$, α, $\alpha_j \in L^2(\mathbf{R})$, $\tilde{f}$, $\tilde{f}_j \in \mathcal{F}$, $j = 1, \cdots, n$. The formal adjoint is called annihilator and denoted by $A(\cdot)$.

REMARK. In general, $A^+(\alpha_1 \odot \tilde{f}_1)A^+(\alpha_2 \odot \tilde{f}_2)$ is not necessarily equal to $A^+(\alpha_2 \odot \tilde{f}_2)A^+(\alpha_1 \odot \tilde{f}_1)$.

DEFINITION **(5.3)** For each $k_0 \in \mathbf{R}^d$, e^{ik_0p} is defined as the operator on $\Gamma\Big(L^2(\mathbf{R}) \odot \mathcal{F}\Big) \otimes \mathcal{H}_0$: for all $\alpha \in L^2(\mathbf{R})$, $\tilde{f} \in \mathcal{F}$

$$e^{ik_0p}A^+(\alpha \odot \tilde{f}) := A^+(\alpha \odot \widetilde{e^{\frac{i}{2}k_0p}f})e^{ik_0p} \quad (5.8)$$

Notice that the above definition implies the identity: $\Big(e^{ik_0p}\Big)^* = e^{-ik_0p}$

THEOREM **(5.4)** For each $n \in \mathbf{N}$, $\alpha, \alpha_j \in L^2(\mathbf{R})$, $\tilde{f}, \tilde{f}_j \in \mathcal{F}$ $(j = 1, \cdots, n)$

$$A(\alpha \odot \tilde{f})A^+(\alpha_1 \odot \tilde{f}_1) \cdots A^+(\alpha_n \odot \tilde{f}_n)\Psi \quad (5.9)$$

is equal to

$$(\alpha \odot \tilde{f}|\alpha_1 \odot \tilde{f}_1)A^+(\alpha_2 \odot \tilde{f}_2) \cdots A^+(\alpha_n \odot \tilde{f}_n)\Psi \quad (5.10)$$

Proof. For each $n \in \mathbf{N}$, $\alpha, \alpha_j \in L^2(\mathbf{R})$, $\tilde{f}, \tilde{f}_j \in \mathcal{F}$ $(j = 1, \cdots, n)$, $\beta_k \in L^2(\mathbf{R}), \tilde{g}_k \in \mathcal{F}$ $(k = 0, 1, \cdots, n)$, since $A = (A^+)^+$, we have

$$< A^+(\alpha_1 \odot \tilde{f}_1) \cdots A^+(\alpha_n \odot \tilde{f}_n)\Phi, A(\alpha \odot \tilde{f})A^+(\beta_0 \odot \tilde{g}_0)A^+(\beta_1 \odot \tilde{g}_1) \cdots A^+(\beta_n \odot \tilde{g}_n)\Phi >=$$

$$=< A^+(\alpha \odot \tilde{f})A^+(\alpha_1 \odot \tilde{f}_1)\cdots A^+(\alpha_n \odot \tilde{f}_n)\Phi, A^+(\beta_0 \odot \tilde{g}_0)A^+(\beta_1 \odot \tilde{g}_1)\cdots A^+(\beta_n \odot \tilde{g}_n)\Phi > \tag{5.11}$$

Denote $\alpha := \alpha_0, \tilde{f} := \tilde{f}_0$ By definition one knows that (5.11) is equal to

$$\prod_{h=0}^{n}\langle\alpha_h,\ \beta_h\rangle \cdot \int_{-\infty}^{+\infty} \mathrm{d}\,u_0 \ldots \int_{-\infty}^{+\infty} \mathrm{d}\,u_n \int_{\mathbf{R}^{(n+1)d}} \mathrm{d}\,k_0 \ldots \mathrm{d}\,k_n$$

$$\prod_{h=0}^{n} e^{iu_h k_h \cdot p} \prod_{h=0}^{n}(S_{u_h} g_h)(k_h)\overline{f}_h(k_h) \cdot \exp\Big(\frac{i}{2}\sum_{r=0}^{n-1}\sum_{h=r}^{n-1} u_r k_r \cdot k_{h+1}\Big) \tag{5.12}$$

Rewriting the expression (5.12) in the form:

$$\prod_{h=1}^{n}\langle\alpha_h,\ \beta_h\rangle \cdot \int_{-\infty}^{+\infty} \mathrm{d}\,u_1 \ldots \int_{-\infty}^{+\infty} \mathrm{d}\,u_n \int_{\mathbf{R}^{nd}} \mathrm{d}\,k_1 \ldots \mathrm{d}\,k_n$$

$$\prod_{h=1}^{n} e^{iu_h k_h \cdot p} \prod_{h=1}^{n}(S_{u_h} g_h)(k_h)\overline{f}_h(k_h) \cdot \exp\Big(\frac{i}{2}\sum_{r=1}^{n-1}\sum_{h=r}^{n-1} u_r k_r \cdot k_{h+1}\Big)$$

$$\langle\alpha,\ \beta_0\rangle \cdot \int_{-\infty}^{+\infty} \mathrm{d}\,u_0 \int_{\mathbf{R}^d} \mathrm{d}\,k_0\ e^{iu_0 k_0 \cdot p} \cdot (S_{u_0} g_0)(k_0)\overline{f}(k_0) \cdot \exp\Big(\frac{i}{2}\sum_{h=1}^{n} u_0 k_0 \cdot k_h\Big) \tag{5.13}$$

one finishes the proof.

Now we compute the matrix elements of products of creators and annihilators on the limit Hilbert module, which we interpreted as limit noise space. For simplicity we shall not distinguish f from $\tilde{f}$. It is necessary to evaluate

$$< \Psi, A^{\varepsilon(1)}(\alpha_1 \odot f_1)\cdots A^{\varepsilon(2n)}(\alpha_{2n} \odot f_{2n})\Psi > \tag{5.14}$$

where, $n \in \mathbf{N},\ \alpha_j \in L^2(\mathbf{R}), f_j \in \mathcal{F}\ (j = 1,\cdots,2n),\ \varepsilon \in \{0,1\}^{2n}$ and

$$A^0 := A, \qquad A^1 := A^+$$

It is sufficcient to consider the case:

$$\varepsilon(1) = 0\ , \qquad \varepsilon(2n) = 1 \tag{5.15}$$

LEMMA **(5.5)** The matrix element (5.14) is not equal to zero only if

$$\sum_{h=1}^{2n}\varepsilon(h) = n \tag{5.16}$$

Proof. The Lemma can be easily verified if $n = 1$. Suppose by induction that $\sum_{h=1}^{2n} \varepsilon(h) \neq n$ implies that (5.14) is equal to zero, and consider

$$< \Psi, A^{\varepsilon(1)}(\alpha_1 \odot f_1) \cdots A^{\varepsilon(2(n+1))}(\alpha_{n+1} \odot f_{2(n+1)})\Psi > \tag{5.17}$$

Denote

$$h := \min\{x \in \{1, \cdots, 2n\};\ \varepsilon(x) = 1\} \tag{5.18}$$

the position where the first creator is. By Theorem (5.4), (5.17) is equal to

$$< \Phi, A^{\varepsilon(1)}(\alpha_1 \odot f_1) \cdots A^{\varepsilon(h-2)}(\alpha_{h-2} \odot f_{h-2})(\alpha_{h-1} \odot f_{h-1}|\alpha_h \odot f_h)$$

$$A^{\varepsilon(h+1)}(\alpha_{h+1} \odot f_{h+1}) \cdots A^{\varepsilon(2(n+1))}(\alpha_{n+1} \odot f_{2(n+1)})\Phi > \tag{5.19}$$

By applying Lemma (5.3), one can move the inner product $(\alpha_{h-1} \odot f_{h-1}|\alpha_h \odot f_h)$ in (5.19) out from the inner product. Since in order to produce this inner product we have used one creator and one annihilator, it follows that

$$\sum_{r=1}^{2(n+1)} \varepsilon(r) \neq n+1 \iff \sum_{1 \leq r \leq 2(n+1),\ r \notin \{h-1,h\}} \varepsilon(r) \neq n \tag{5.20}$$

By the induction assumption the proof is completed.

The same technique can be used to prove the following

THEOREM **(5.6)** Denote

$$\{m_h\}_{h=1}^n := \{r \in \{1, \cdots, 2n\};\ \varepsilon(r) = 1\}, \qquad 1 < m_1 < \cdots < m_n = 2n \tag{5.21}$$

The inner product (5.14) is equal to zero if $\{m_h\}_{h=1}^n$ does not define an admissble pair partition of $\{1, \cdots, 2n\}$; if it does (then Lemma (4.3) implies that is unique), (5.14) is equal to

$$\prod_{h=1}^n \langle \alpha_{m_h}, \alpha_{m'_h} \rangle_{L^2(\mathbb{R})}$$

$$\int_{-\infty}^{+\infty} du_1 \ldots \int_{-\infty}^{+\infty} du_n \int_{\mathbb{R}^{nd}} dk_1 \ldots dk_n \prod_{h=1}^n (S_{u_h} f_{m_h})(k_h) \overline{f}_{m'_h}(k_h)$$

$$\langle \xi, \prod_{h=1}^n e^{iu_h k_h \cdot p} \eta \rangle \cdot \exp\Big(\frac{i}{2} \sum_{h=1}^{n-1} \sum_{r=h+1}^{n} u_h k_h \cdot k_r \chi_{(m'_r, m_r)}(m_h)\Big) \tag{5.22}$$

where, $\{m'_h, m_h\}_{h=1}^n$ is the unique admissible pair partition of $\{1, \ldots, 2n\}$ determinered by $\{m_h\}_{h=1}^n$.

§6 THE LIMIT STOCHASTIC PROCESS

In this section we state, without proof, our main result on the WCL problem for the quantum EM-field.

THEOREM **(6.1)** For each $t \geq 0$, $\{S_h, T_h\}_{h=1}^{N}$, $\{S'_h, T'_h\}_{h=1}^{N'} \subset \mathbf{R}$, $\{f_h\}_{h=1}^{N}$, $\{f'_h\}_{h=1}^{N'} \subset K$ and $\xi, \eta \in L^2(\mathbf{R}^d)$, the limit

$$< \prod_{h=1}^{N} A_\lambda^+(S_h, T_h, f_h)\Phi \otimes \xi, U_{t/\lambda^2} \prod_{h=1}^{N'} A_\lambda^+(S'_h, T'_h, f'_h)\Phi \otimes \eta > \tag{6.1}$$

exsits and is equal to the solution of the quantum stochastic differential equation with respect to the free Brownian Motion:

$$U(t) = 1 + \int_0^t \Big(dA_s^+(g)(-ip) - (-ip)^+ dA_s(g) - (-ip)^+(g|g)_-(-ip)ds\Big)U(s) \tag{6.2}$$

On the $\mathcal{P}$-full Fock module described in Section 5. Where, the half-inner product $(\cdot|\cdot))_-$ is defined by

$$(f|g)_- := \int_{-\infty}^{0} dt \int_{\mathbf{R}^d} dk e^{-itkp} \bar{f}(k)(S_t g)(k) \tag{6.3}$$

The proof that the solution of the quantum stochastic differential equation (6.2) is effectively a unitary operator and in fact the very meaning of this equation, depends on the theory of the **free stochastic calculus over a Hilbert module**, which has been recently developed in [7].

REFERENCES

[1] L. Accardi, R. Alicki, A. Frigerio, Y.G. Lu: An invitation to the weak coupling and low density limits. *Quantum Probability and Related Topics, QP VI*, 3–61.

[2] L. Accardi, Y.G. Lu: On the weak coupling limit for quantum electrodynamics. *Prob. Meth. in Math. Phys.*, 16–29.

[3] L. Accardi, Y.G. Lu: From the weak coupling limit to a new type of quantum calculus. *Quantum Probability and Related Topics, QP VII*, 1–14.

[4] M. Bozejko, R. Speicher: An example of a generalized Brownian Motion, to appear in Commun. Math. Phys.

[5] M. Bozejko, R. Speicher: An example of a generalized Brownian Motion II. *Quantum Probability and Related Topics, QP VII*, 67–78.

[6] R. Speicher: A new example of "independence" and "white niose" Probab. Th. Rel. Fields, **84** (1990), 141–159.

[7] Y.G. Lu: Free stochastic calculus on Hilbert module, preprint.

Quantum Probability and Related Topics Vol. VIII (pp. 19–35)

QUANTUM STOCHASTIC FLOWS ON MANIFOLDS I

by

David Applebaum

Department of Mathematics, Statistics and Operational Research
Nottingham Polytechnic, Burton Street, Nottingham, NG7 4BU
England

Abstract

An example of a quantum stochastic flow on a manifold is constructed in closed form out of two commuting smooth vector fields and a single canonical Brownian motion. This flow is shown to be of the class whose structure maps do not all preserve the initial algebra.

1. Introduction

In two earlier papers ([App 1,2]), the author has considered the problem of understanding the theory of stochastic flows of diffeomorphisms of manifolds driven by classical Brownian motion from the point of view of quantum stochastic calculus. The main result of this analysis is that these flows can be implemented spatially by a unitary operator valued stochastic process which is precisely the "stochastic pullback" of the flow to the Hilbert space of random half-densities on the manifold with finite second moment. Moreover this unitary process satisfies a stochastic differential equation of the type introduced in [Hu Pa 1,2] with unbounded coefficients which are determined by the vector fields which drive the flow.

In the present paper we aim to "quantise" this theory to obtain a genuine quantum stochastic flow on the algebra $\mathcal{A}$ of smooth functions (of compact support) on the manifold which cannot be reduced to a Brownian flow of the type described above. Our main focus is on the simplest possible example of such a flow which we are able to construct explicitly from two commuting Brownian flows by use of the duality transform and second quantisation maps as is described in §4 below. The main features of these flows are that

(a) They are driven by complex conjugate pairs rather than real vector fields thus breaking the symmetry between the annihilation and creation processes which we find in the classical case.

(b) The Lindblad generator of the flow does not preserve the algebra $\mathcal{A}$ so that we obtain a further class of examples of the $\mathcal{A}$-$\mathfrak{M}$ flows described in [App 3, 4].

At the end of §4, we make some remarks about the physical interpretation of these flows which we feel may be relevant to the study of quantum fields on curved space-times. The flow describes the annihilation and creation of particles on the manifold whose positions and momenta are quantised along distinct families of integral curves. It is feasible that a physical realisation of such flows as a collective description of interacting bose fields in the weak coupling limit may be achieved using the techniques of [AFL].

Notation

We will use the same notation as in [App 2]. Note in particular, for a test function f, the use of a(f), $a(f)^\dagger$ and W(f) for annihilation, creation and Weyl operators (respectively) and $\psi(f)$ for exponential vectors. $\mathcal{E}$ will denote the dense subspace of boson Fock space comprising finite linear combinations of exponential vectors. We will use Einstein summation convention where appropriate Some knowledge of elementary quantum stochastic calculus as described in e.g. [HuPa 2] will be assumed.

Addendum to [App 2]

Formula (5.14) on page 114 is valid only when all the covariant derivatives are mutually commuting.

Acknowledgement

My thanks to Kalyan Sinha for an extremely valuable discussion during the train journey from Hausach to Frankfurt.

2. Deterministic Flows ([AMR])

Let M be a smooth, orientable manifold modelled on $\mathbb{R}^d$ which is equipped with a volume form μ_V and let μ be the corresponding Borel measure on M. We denote by $C^\infty(M)$ (respectively, $C_K^\infty(M)$) the algebra of smooth real-valued functions on M (respectively, of compact support).

Let Y be a complete, smooth vector field on M so that the differential equation

$$\frac{dc}{dt} = Y(c(t)) \quad ; \qquad c(0) = x \qquad \ldots(2.1)$$

has a unique solution for all $x \in M$ and for all $t \in \mathbb{R}$. We then obtain an autonomous flow of diffeomorphisms $\xi = \{\xi_t, t \in \mathbb{R}\}$ of M by the prescription

$$\xi_t(x) = c_x(t) \qquad \ldots(2.2)$$

where c_x is the solution of (2.1). Note that for each $s,t \in \mathbb{R}$ we have

$$\xi_{s+t} = \xi_s \circ \xi_t$$

and

$$\xi_t^{-1} = \xi_{-t} \qquad \ldots(2.3)$$

so that ξ is a one-parameter subgroup of Diff(M).
The induced flow of automorphisms of $C^\infty(M)$,$\{j_t, t \in \mathbb{R}\}$ is given by

$$j_t(f) = f \circ \xi_t \qquad \ldots(2.4)$$

for $f \in C^\infty(M)$.
We denote by $\mathfrak{H}_0$ the complex, separable Hilbert space of half - densities on M so that $\psi \in \mathfrak{H}_0 \Rightarrow \psi = g\mu_V^{1/2}$ for some $g \in L^2(M,\mu)$.

Note that the inner product in $\mathfrak{H}_0$ is given by

$$\langle g_1\mu_V^{1/2}, g_2\mu_V^{1/2} \rangle = \int \overline{g_1}\, g_2 \, d\mu \qquad \ldots(2.5)$$

for $g_1, g_2 \in L^2(M,\mu)$.

Let $\mathfrak{D}_0$ denote the dense linear manifold in $\mathfrak{H}_0$ given by $\{\psi \in \mathfrak{H}_0 ; \psi = f\mu_V^{1/2}$ for some $f \in C_K^\infty(M)\}$. We obtain a strongly continuous, one parameter group of unitary operators in $\mathfrak{H}_0$, $V = (V(t), t \in \mathbb{R})$ by continuous extension of the prescription

$$V(t)\psi = j_t(f)\,(\xi_t^*(\mu_V))^{1/2} \qquad \ldots(2.6)$$

for $\psi = f\mu_V^{1/2} \in \mathfrak{D}_0$ where ξ_t^* is the pullback of the flow to the bundle of d-forms.
Note that for $f \in C_K^\infty(M)$, we then have

$$j_t(f) = V(t)\, f\, V(t)^{-1} \qquad \ldots(2.7)$$

Strong differentiation in (2.6) shows that the infinitesimal generator of $V = (V(t), t \in \mathbb{R})$ is $-iT_Y$ where T_Y is the unique skew-adjoint extension of the operator which acts on $\mathfrak{D}_0$ as $Y + \frac{1}{2}\mathrm{div}_\mu(Y)$ where $\mathrm{div}_\mu(Y)$ is the divergence (with respect to μ_V) of the vector field Y. Hence we may write, for $t \in \mathbb{R}$,

$$V(t) = \exp(tT_Y) \qquad \ldots(2.8)$$

3 Brownian Flows ([App 1])

Let $B = (B(t), t \in \mathbb{R}^+)$ denote canonical one-dimensional Brownian motion with the underlying probability space $(\Omega,\mathfrak{F},P)$ so that for $\omega \in \Omega$, $t \in \mathbb{R}^+$, we have

$$B(t)\omega = \omega(t) \qquad \ldots(3.1)$$

Now for each $s,t \in \mathbb{R}^+$ with $s \le t$ define $\Phi_{s,t} : M \times \Omega \to M$ by

$$\Phi_{s,t}(x,\omega) = \xi_{\omega(t)-\omega(s)}(x) \qquad \ldots(3.2)$$

for $x \in M$, $\omega \in \Omega$, then $\Phi \ (\Phi_{s,t},\ s,t \in \mathbb{R}^+,\ s \le t)$ is a stochastic flow of diffeomorphisms of M, in fact it is the simplest example of a Brownian flow in the sense of [Kun]. Our argument will lose none of its generality if we restrict our attention to maps of the form $\Phi_{0,t}$ which we will denote by Φ_t.(Note that these do not satisfy a group law as in (2.3))

For each $t \in \mathbb{R}^+$, define $J_t : C_K^\infty(M) \to L^\infty(M \times \Omega, \mu \times P)$ by

$$J_t(f) = f \circ \Phi_t \qquad \ldots(3.3)$$

for each $f \in C_K^\infty(M)$, then it follows by (3.1) and Itô's formula that

$$dJ_t(f) = J_t(Y(f))\, dB(t) + \frac{1}{2} J_t(Y^2(f))\, dt \qquad \ldots(3.4)$$

so Φ satisfies the stochastic differential equation

$$d\Phi_t = Y(\Phi_t) \circ dB(t) \qquad \ldots(3.5)$$

where $\circ$ here denotes the Stratonovitch differential.

Let $\mathfrak{h}$ denote the Hilbert space $L^2(\Omega,\mathfrak{F},P;\mathfrak{h}_0)$ and $\mathfrak{D}$ be the dense linear manifold in $\mathfrak{h}$ comprising square integrable maps from Ω into $\mathfrak{D}_0$ We obtain a family of unitary operators $U = (U(t),\ t \in \mathbb{R}^+)$ in $\mathfrak{h}$ by continuous extension of the prescription

$$(U(t)\ \Psi)(\omega) = V(\omega(t))\ \Psi(\omega) \qquad \ldots(3.6)$$

for $\Psi \in \mathfrak{D}$, $\omega \in \Omega$. Hence we see that for each $t \in \mathbb{R}^+$, $f \in C_K^\infty(M)$ we have

$$J_t(f) = U(t)\, f\, U(t)^{-1} \qquad \ldots(3.7)$$

(c.f. (2.7)).

For every $\Psi \in \mathfrak{D}$, we have $\Psi(\omega) = f(\omega)\mu^{1/2}$ for each $\omega \in \Omega$, where f is a square integrable map from Ω into $C_K^\infty(M)$ Hence by (2.6) and (3.3) we find that

$$(U(t)\Psi)(\omega) = J_t(f(\omega))(\omega)(\Phi_t^*(\mu_Y^{1/2}))(\omega) \qquad \ldots(3.8)$$

for each $\Psi \in \mathfrak{D}$, $\omega \in \Omega$ (c.f. (2.6)).

A further application of Itô's formula then shows that $U(t)$ satisfies the operator valued stochastic differential equation

$$dU(t) = U(t)(T_Y\, dB(t) + \tfrac{1}{2}\, T_Y^2\, dt) \qquad \ldots(3.9)$$

(see [Sau] for an alternative approach).

Using the canonical isomorphism between $L^2(\Omega,\mathfrak{F},P;\mathfrak{h}_0)$ and $\mathfrak{h}_0 \otimes L^2(\Omega,\mathfrak{F},P)$ to identify the two spaces it is not difficult to verify that for each $t \in \mathbb{R}^+$,

$$U(t) = \exp(T_Y \otimes B(t)) \qquad \ldots(3.10)$$

where $B(t)$ is here acting as a self-adjoint multiplication operator in $L^2(\Omega,\mathfrak{F},P)$. (Here we have abused notation to the extent of identifying the essentially skew-adjoint operator $T_Y \otimes B(t)$ with its closure).

We close this section by noting the following result from [Kun] which we will use in section 5 below.

Let $Y_0,\ldots,Y_n$ be complete, smooth vector fields on M which generate a finite - dimensional Lie algebra, then the unique solution of the stochastic differential equation given below yields a stochastic flow of diffeomorphisms on M

$$d\Phi_t \qquad Y_j(\Phi_t) \circ dB^j + Y_0(\Phi_t)\, dt \qquad \ldots(3.11)$$

where $(B^j,\ 1 < j < n)$ are independent Brownian motions.
If we now define unitary operators $(U(t), t \in \mathbb{R}^+)$ as in (3.8), it follows that these are a solution of

$$dU(t) \qquad U(t)\left[T_{Y_j}\, dB^j + \left(T_{Y_0} + \frac{1}{2}\sum_{j=1}^{n} T_{Y_j}^{\,2} \right) dt \right] \qquad \ldots(3.12)$$

4. Quantum Stochastic Flows

Let Y_1 and Y_2 be complete smooth vector fields on M for which

$$[\, Y_1 \,,\, Y_2 \,] = 0 \qquad \ldots(4.1)$$

Let T_{Y_j} $(j \quad 1,2)$ denote the corresponding skew-adjoint operators in $\mathfrak{h}_0$ defined as in § 2.

<u>Lemma 1</u> T_{Y_1} and T_{Y_2} commute

Proof Let $(\xi_t^1,\ t \in \mathbb{R})$ and $(\xi_t^2,\ t \in \mathbb{R})$ be the one parameter groups of diffeomorphisms generated by Y_1 and Y_2 respectively. By (4.1) it follows that these commute and hence so also do their pullbacks to d-forms. Then by (2.6) the one-parameter unitary groups associated to these flows will also commute and the required result follows □

Now let D denote the duality transform from $L^2(\Omega, \mathfrak{F}, P)$ onto $\Gamma(L^2(\mathbb{R}^+))$ which maps exponential martingales to exponential vectors. From now on we identify $\mathfrak{h}$ with its image under the isomorphism $I \otimes D$.
Note that for each $t \in \mathbb{R}^+$

$$D\, B(t)\, D^{-1} \qquad Q(t) = \frac{A(t) + A(t)^\dagger}{\sqrt{2}} \qquad \ldots(4.2)$$

Let $\Gamma(iI)$ denote the second quantisation of the operator of multiplication by i in $L^2(\mathbb{R}^+)$ acting in $\Gamma(L^2(\mathbb{R}^+))$ and note that

$$\Gamma(iI)\, Q(t)\, \Gamma(-iI) = P(t) = \frac{-i(A(t) - A(t)^\dagger)}{\sqrt{2}} \qquad \ldots(4.3)$$

We define unitary processes $(\mathcal{U}(t),\ t \in \mathbb{R}^+)$ and $(\mathcal{B}(t),\ t \in \mathbb{R}^+)$ by

$$\mathcal{U}(t) = (I \otimes D)\, \exp(T_{Y_1} \otimes B(t))\, (I \otimes D^{-1})$$

so by (4.2)

$$\mathcal{U}(t) = \exp(T_{Y_1} \otimes Q(t)) \qquad \ldots(4.4)$$

and $$\mathcal{B}(t) = (I \otimes \Gamma(iI)D)\, \exp(T_{Y_2} \otimes B(t))\, (I \otimes D\Gamma(-iI))$$

so by (4.2) and (4.3)

$$\mathcal{B}(t) = \exp(T_{Y_2} \otimes P(t)) \qquad \ldots(4.5)$$

Using the fact that for $s,\ t \in \mathbb{R}^+$

$$[\, Q(s)\ ,\ P(t)\,] = i \min(s,t) I$$

together with the result of lemma (4.1) we find that on the dense domain $\mathcal{D}_0 \underline{\otimes} \mathcal{E}$ in $\mathfrak{h}$, we have the commutation relation

$$[\, T_{Y_1} \otimes Q(s)\ ,\ T_{Y_2} \otimes P(t)\,] = i \min(s,t)\, (T_{Y_1} T_{Y_2} \otimes I) \qquad \ldots(4.6)$$

Now by (4.4), (4.5), (4.6) and the Campbell-Baker-Hausdorff formula we obtain for each $t \in \mathbb{R}^+$,

$$\mathcal{U}(t)\, \mathcal{B}(t) = \mathcal{B}(t)\, \mathcal{U}(t)\, \exp(it\, T_{Y_1} T_{Y_2} \otimes I) \qquad \ldots(4.7)$$

Define a unitary process $\mathfrak{W} = (\mathfrak{W}(t),\ t \in \mathbb{R}^+)$ in $\mathfrak{h}$ by

$$\mathfrak{W}(t) = \mathfrak{U}(t)\ \mathfrak{V}(t)\ \exp(-\tfrac{1}{2}\ it\ T_{Y_1}T_{Y_2} \otimes I) \qquad \ldots(4.8)$$

<u>Theorem 2</u> For each $t \in \mathbb{R}^+$,

$$\mathfrak{W}(t) = \exp(\ (T_{Y_1} \otimes Q(t)) + (T_{Y_2} \otimes P(t))) \qquad \ldots(4.9)$$

Proof Define strongly continuous one-parameter unitary groups in $\mathfrak{h}$, $(\mathfrak{U}_\tau(t),\ \tau \in \mathbb{R})$ and $(\mathfrak{V}_\tau(t),\ \tau \in \mathbb{R})$ by
$\mathfrak{U}_\tau(t) = \exp(\tau\ T_{Y_1} \otimes Q(t))$ and $\mathfrak{V}_\tau(t) = \exp(\tau\ T_{Y_2} \otimes P(t))$.
Now define a family of unitary operators $(\mathfrak{W}_\tau(t),\ \tau \in \mathbb{R})$ by

$$\mathfrak{W}_\tau(t) = \mathfrak{U}_\tau(t)\ \mathfrak{V}_\tau(t)\ \exp(-\tfrac{1}{2}\ i\tau^2 t\ T_{Y_1}T_{Y_2} \otimes I) \qquad \ldots(4.10)$$

Straightforward algebraic manipulation using (4.7) establishes the group property

$$\mathfrak{W}_{\sigma+\tau}(t) = \mathfrak{W}_\sigma(t)\mathfrak{W}_\tau(t)$$

and strong continuity of the map $\tau \to \mathfrak{W}_\tau(t)$ is deduced by a standard argument. Hence $(\mathfrak{W}_\tau(t),\ \tau \in \mathbb{R})$ is a strongly continuous one-parameter unitary group. We compute its infinitesimal generator by strong differentiation in (4.10) to find

$$\mathfrak{W}_\tau(t) = \exp(\tau\ [(T_{Y_1} \otimes Q(t)) + (T_{Y_2} \otimes P(t))])$$

and the required result is obtained when we put $\tau = 1$ □

We now introduce the complex conjugate pair of vector fields

$$Z = \frac{Y_1 + iY_2}{\sqrt{2}} \qquad \text{and} \qquad \bar{Z} = \frac{Y_1 - iY_2}{\sqrt{2}}$$

and observe that by linearity we have

$$T_Z \;=\; Z + \tfrac{1}{2}\,\mathrm{div}_{\mu}(Z) \qquad \frac{T_{Y_1} + iT_{Y_2}}{\sqrt{2}}$$

and
$$T_Z^{\dagger} \qquad - T_{\bar{Z}}$$

By (4.2) and (4.3) we can rewrite (4.9) as

$$\mathfrak{B}(t) \;=\; \exp(\;(T_Z \otimes A(t)^{\dagger}) \qquad (T_Z^{\dagger} \otimes A(t))) \qquad \ldots.(4.12)$$

It follows by lemma 1 that there exists a self-adjoint operator X on $\mathfrak{H}_0$ and two Borel functions $\alpha, \beta : \mathbb{R} \to \mathbb{R}$ such that $T_{Y_1} \quad i\alpha(X)$ and $T_{Y_2} = i\beta(X)$. Writing the spectral decomposition of X as

$\int_{-\infty}^{\infty} \lambda \; dE(\lambda)$, we see that

$$T_Z \;=\; \int_{-\infty}^{\infty} z(\lambda)\; dE(\lambda) \quad \text{and} \quad T_Z^{\dagger} \;=\; -\int_{-\infty}^{\infty} \overline{z(\lambda)}\; dE(\lambda) \qquad \ldots(4.13)$$

$$\text{where} \qquad z(\lambda) \;=\; \frac{\beta(\lambda) - i\,\alpha(\lambda)}{\sqrt{2}}$$

Combining (4.12) and (4.13), we find that for each $t \in \mathbb{R}^{+}$,

$$\mathfrak{B}(t) \qquad \int_{-\infty}^{\infty} \exp(\; z(\lambda)\; A(t)^{\dagger} \quad \overline{z(\lambda)}\; A(t)\;) \otimes dE(\lambda)$$

$$\int_{-\infty}^{\infty} W_{\lambda}(t) \otimes dE(\lambda) \qquad \ldots(4.14)$$

where for each $\lambda \in \mathbb{R}$, $W_\lambda = (W_\lambda(t),\ t \in \mathbb{R}^+)$ is a Weyl process i.e. a family of Weyl operators in $\Gamma(L^2(\mathbb{R}^+))$ defined by

$$W_\lambda(t) \quad = \quad W(z(\lambda)\ \chi_{[0,t)}) \qquad \ldots(4.15)$$

It is shown in [HuPa 1], that each W_λ satisfies the quantum s.d.e.

$$dW_\lambda \qquad W_\lambda(\ z(\lambda)\ dA^\dagger \quad \overline{z(\lambda)}\ dA \quad \tfrac{1}{2}|z(\lambda)|^2\ dt) \qquad \ldots(4.16)$$

with initial condition $W_\lambda(0) \quad I$.

Now consider again the process $\mathfrak{B}$ defined by (4.9).

<u>Theorem 3</u> $\mathfrak{B}$ satisfies the quantum s.d.e.

$$d\mathfrak{B} \quad \mathfrak{B}\ (T_z\ dA^\dagger - T_z^\dagger\ dA \quad \frac{1}{2}\ T_z^\dagger\ T_z\ dt) \qquad \ldots(4.17)$$

with initial condition $\mathfrak{B}(0) = I$

Proof Let f and g be locally bounded functions in $L^2(\mathbb{R}^+)$, $u \in \mathfrak{D}_0$, $v \in \mathfrak{D}_0$, then for each $t \in \mathbb{R}^+$, by (4.14), we obtain

$$\langle\ u \otimes \psi(f),\ \mathfrak{B}(t)(v \otimes \psi(g))\ \rangle \ = \int_{-\infty}^{\infty} \langle\ \psi(f),\ W_\lambda(t)\ \psi(g)\rangle\ d\langle u,\ E(\lambda)\ v\rangle$$

The map $t \to \langle\ u \otimes \psi(f),\ \mathfrak{B}(t)(v \otimes \psi(g))\ \rangle$ is then clearly differentiable and using (4.16), we find that

$$\frac{d}{dt}\ \langle\ u \otimes \psi(f),\ \mathfrak{B}(t)(v \otimes \psi(g))\ \rangle$$

$$= \int_{-\infty}^{\infty} \langle\psi(f), W_\lambda(t)[\ \overline{f(t)}\ z(\lambda) \quad g(t)\ \overline{z(\lambda)} - \tfrac{1}{2}|z(\lambda)|^2]\ \psi(g)\rangle\ d\langle u, E(\lambda)v\rangle$$

$$\langle\, u \otimes \psi(f),\ \mathfrak{B}(t)[\ \overline{f(t)}\ T_z - g(t)\ T_z^\dagger - \tfrac{1}{2}\ T_z^\dagger\ T_z\ ]\ v \otimes \psi(g)\rangle$$

by the spectral theorem. The result now follows □

We observe that each $\mathfrak{B}(t)$ is a generalised Weyl operator in the sense of [HuPa 1].

Using the fact that for each $t \in \mathbb{R}^+$,

$\mathfrak{B}(t)^\star \ = \ \exp(\ T_z^\dagger \otimes A(t) \quad T_z \otimes A(t)^\dagger)$, a similar argument to that of theorem 3 shows that

$$d\mathfrak{B}^\star \qquad \mathfrak{B}^\star (T_z^\dagger\ dA - T_z\ dA^\dagger \quad \tfrac{1}{2}\ T_z^\dagger\ T_z\ dt)$$

$$= \ (T_z^\dagger\ dA - T_z\ dA^\dagger \quad \tfrac{1}{2}\ T_z^\dagger\ T_z\ dt)\ \mathfrak{B}^\star \qquad \ldots(4.18)$$

where we have used the fact that for each $t \in \mathbb{R}^+$

$$[\ \mathfrak{B}(t)^\star\ ,\ T_z\ ] \qquad [\ \mathfrak{B}(t)^\star\ ,\ T_z^\dagger\ ] \qquad 0 \text{ on } \mathfrak{D}_0 \underline{\otimes}\ \mathfrak{E}.$$

Now define a spatial quantum flow $j = (j(t),\ t \in \mathbb{R}^+)$ on the algebra $\mathcal{A} \quad C_K^\infty(M)$ by

$$j(t)(f) \ = \ \mathfrak{B}(t)\ f\ \mathfrak{B}(t)^\star \qquad \ldots(4.19)$$

Let $\mathfrak{M}$ denote the two-sided $\mathcal{A}$ module generated by $\mathcal{A}$, T_z, $T_z^\dagger$ and $\frac{1}{2}\ T_z^\dagger\ T_z$.

<u>Theorem 4</u> j is an $(\mathcal{A},\mathfrak{M})$ flow wherein for each $f \in \mathcal{A}$, $t \in \mathbb{R}^+$

$$dj_t(f) \qquad j_t(Z(f))dA^\dagger + j_t(\bar{Z}(f))dA + j_t(\tau(f))dt \qquad \ldots(4.20)$$

with
$$\tau(f) \qquad \tfrac{1}{2}\ (T_z^\dagger\ T_z\ f \quad 2\ T_z^\dagger\ f\ T_z + f\ T_z^\dagger\ T_z) \qquad \ldots(4.21)$$

Proof For $u \in \mathfrak{D}_0, v \in \mathfrak{D}_0$ and g,h locally bounded functions in $L^2(\mathbb{R}^+)$, consider

$$\frac{d}{dt} \langle u \otimes \psi(g), j_t(f)(v \otimes \psi(h))\rangle$$

and argue as in [HuPa 2], [App 2,3] and theorem 3 above □

Observe that the flows of classical type driven by $\mathfrak{U}$ and $\mathfrak{B}$ are obtained by taking Y_2 and Y_1 (respectively) to be zero.

We close this section by investigating the cohomological structure of the map $\tau\ \mathcal{A} \to \mathfrak{M}$. This task is facilitated if we rewrite τ in terms of the skew adjoint operators T_{Y_1} and T_{Y_2}. After some rearrangement we find that (c.f. [App 4]

$$\tau = \tau_1 + \tau_2$$

where for each $f \in \mathcal{A}$,

$$\tau_1(f) = \frac{1}{4}\left(Y_1^2(f) + Y_2^2(f)\right) \qquad \ldots(4.22)$$

$$\text{and} \qquad \tau_2(f) = -\frac{1}{2}\, i \left(T_{Y_1}\, f\, T_{Y_2} - T_{Y_2}\, f\, T_{Y_1}\right) \qquad \ldots(4.23)$$

Clearly τ_2 is the obstruction to j being an Evans-Hudson flow.

Let $\mathfrak{b}$ denote the Hochschild coboundary operator for the complex comprising $\mathfrak{M}$ valued chains on $\mathcal{A}$. We obtain the following

<u>Theorem 5</u> For each $f,g \in \mathcal{A}$,

$$(\mathfrak{b}\tau_2)(f,g) = \frac{1}{2}\, i(df \wedge dg)(Y_1\ ,\ Y_2) \qquad \ldots(4.24)$$

Proof A straightforward but tedious computation establishes the following simpler form for τ_2,

$$\tau_2(f) \quad = \quad \frac{1}{2}\, i\, (Y_1(f)\, T_{Y_2} - Y_2(f)\, T_{Y_1}) \qquad \ldots(4.25)$$

where we have used the fact (easily deduced from lemma 1) that

$$Y_1(\mathrm{div}_\mu(Y_2)) \quad = \quad Y_2(\mathrm{div}_\mu(Y_1))$$

Now compute $(b\tau_2)(f,g) \quad \tau_2(f)g - \tau_2(fg) + f\tau_2(g)$ using (4.25), and the result follows immediately □

Notes

1) The argument described above clearly generalises to the construction of spatial quantum flows driven by an equation of the type (4.17) where the operator T_Z is replaced by an arbitrary normal operator L in any complex separable Hilbert space $\mathfrak{h}_0$ wherein L and L^* share a common invariant domain. Given a * - algebra $\mathcal{A} \subseteq B(\mathfrak{h}_0)$ which also leaves this domain invariant we would again obtain an example of an $\mathcal{A}$-$\mathfrak{M}$ flow.

2) An alternative direction for generalising these ideas is to replace $L^2(\mathbb{R}^+)$ by a complex separable Hilbert space $\mathcal{H}$ and the indicator functions $\chi_{[0,t)}$ in (4.3) and (4.4) by an arbitrary state vector $g \in \mathcal{H}$.

Writing $$a^\dagger(Z,g) \quad = \quad T_Z \otimes a(g)^\dagger \qquad \ldots(4.26)$$

we see that we have the commutation relations

$$[\ a(Z,f)\ ,\ a^\dagger(Z,g)\ ] \qquad \langle f,g\rangle\, T_Z\, T_Z^\dagger \otimes I \qquad \ldots(4.27)$$

for $f,g \in \mathcal{H}$, which suggests a physical interpretation of $a^\dagger(Z,g)$ as an operator which creates a particle in the state g which is constrained in such a way that successive position measurements can only be made along the integral curves of Y_1 and successive momentum measurements can only be made along the integral curves of Y_2.

5 Some Remarks About the General Case

Let $Y_1^0, Y_2^0, \ldots, Y_1^n, Y_2^n$ be $2n + 2$ complete, smooth vector fields on M and let $\mathcal{L}$ denote the Lie algebra which they generate. The most general quantum stochastic flow on a manifold driven by a finite number of annihilation and creation processes has the infinitesimal form

$$dj_t(f) \quad j_t(Z^j(f))dA_j^\dagger + j_t(\bar{Z}^j(f))dA_j + j_t(Z^0(f) + \tau(f))dt \quad \ldots(5.1)$$

where $f \in \mathcal{A}$, $t \in \mathbb{R}^+$ and each $Z^j \quad Y_1^j + Y_2^j$ for $0 \le j \le n$. Formally we may obtain j as a spatial quantum flow driven by the unitary process whose infinitesimal form is

$$d\mathfrak{B} \quad \mathfrak{B}(T_Z^j \, dA_j^\dagger + T_{\bar{Z}}^j \, dA_j + \left(T_Z^0 + \frac{1}{2}\sum_{j=1}^{n} T_{\bar{Z}}^j \, T_Z^j\right)dt \quad \ldots(5.2)$$

so that τ in (5.1) is a superposition of terms of the form given by (4.21).

In the case where $\mathcal{L}$ is commutative we may directly construct $\mathfrak{B}$ as a finite product of unitaries of the form (4.12). In the general case, we can only prove existence and unitarity of solutions to (5.2) if the coefficients $\{T_Z^j,\ 0 < j < n\}$ satisfy an analyticity condition of the type discussed in [Fag].

We would like to find some more natural condition (from a geometrical point of view) for the existence of quantum stochasic flows on manifolds. To this end let us suppose that the Lie algebra $\mathcal{L}$ is finite dimensional. By Kunita's result discussed in §3 and using the n-dimensional versions of (4.2) and (4.3), we can assert the existence of unitary processes satisfying the equations

$$d\mathfrak{U} \quad \mathfrak{U}(T_{Y_1^j} \, dQ_j + \left(T_{Y_1^0} + \frac{1}{2}\sum_{j=1}^{n}\left(T_{Y_1^j}\right)^2\right)dt \;) \quad \ldots(5.3)$$

and

$$d\mathfrak{S} = \mathfrak{S}(T_{Y_2^j} \, dP_j + \left(T_{Y_2^0} + \frac{1}{2}\sum_{j=1}^{n}\left(T_{Y_2^j}\right)^2\right)dt \;) \quad \ldots(5.4)$$

Now suppose there exists a unitary process $\mathfrak{B}^{\mathfrak{U}}$ satisfying the equation

$$d\mathfrak{B}^{\mathfrak{U}} = \mathfrak{B}^{\mathfrak{U}}\left(\mathfrak{U}\, T_{Y_2^j}\, \mathfrak{U}^* dP_j + \left(\mathfrak{U}\, T_{Y_2^o}\, \mathfrak{U}^* + \frac{1}{2}\sum_{j=1}^{n} \mathfrak{U}\left(T_{Y_2^j}\right)^2 \mathfrak{U}^*\right) dt\right) \qquad \ldots(5.5)$$

then we have (c.f. [Ev Hu]) that $\mathfrak{B}$ is given by the prescription

$$\mathfrak{B}(t) \quad \mathfrak{B}^{\mathfrak{U}}(t)\ \mathfrak{U}(t) \qquad \ldots(5.6$$

for each $t \in \mathbb{R}^+$. A proof of the existence and unitarity of $\mathfrak{B}^{\mathfrak{U}}$ will be postponed to a future article.

References

[AMR] R.Abraham, J.E.Marsden, T.Ratiu, Manifolds, Tensor Analysis and Applications, Addison-Wesley (1983), Springer-Verlag (1988)

[AFL] L.Accardi, A.Frigerio, Y.G.Lue, The Weak Coupling Limit as a Quantum Functional Central Limit, Commun.Math.Phys. 131, 537-70 (1990)

[App 1] D.Applebaum, An Operator Theoretic Approach to Stochastic Flows on Manifolds, to appear in Séminaire de Probabilité

[App 2] D.Applebaum, Towards a Quantum Theory of Classical Diffusions on Riemannian Manifolds, Quantum Probability and Applications 6, 93-111, World Scientific (1991)

[App 3] D.Applebaum On a Class of Stochastic Flows Driven by Quantum Brownian Motion, to appear in Journal of Theoretical Probability

[App 4] D.Applebaum, Unitary Evolutions and Horizontal Lifts in Quantum Stochastic Calculus, Commun.Math.Phys. 140, 63-80 (1991)

[EvHu] M.P.Evans, R.L.Hudson, Perturbations of Quantum Diffusions, J.London Math.Soc. 41, 373-84 (1990)

[Fag] F.Fagnola, On Quantum Stochastic Differential Equations With Unbounded Coefficients Probab.Th.Rel.Fields 86, 501-16 (1990)

[HuPa 1] R.L.Hudson, K.R.Parthasarathy, Construction of Quantum Diffusions, Quantum Probability and Applications 1, 173-94, Springer LNM 1055, (1984)

[HuPa 2] R.L.Hudson, K.R.Parthasarathy, Quantum Itô's Formula and Stochastic Evolution, Commun.Math.Phys. 93, 301-23, (1984)

[Kun] H.Kunita, Stochastic Flows and Stochastic Differential Equations, C.U.P. (1990)

[Sau] J.L.Sauvageot, From Classical Geometry to Quantum Stochastic Flows : An Example, to appear in Quantum Probability and Applications 7 (World Scientific)

Quantum Probability and Related Topics Vol. VIII (pp. 37–46)

CHARACTERIZATIONS OF SOME OPERATORS ON FOCK SPACE

Stéphane ATTAL

Université Louis Pasteur
Département de mathématiques
7, rue René Descartes
67084 Strasbourg Cedex, France

I Introduction

On the boson Fock space Φ we study operators which commute with all the projections (also called conditional expectations) $\mathbb{E}_t$ from Φ onto the Fock space $\Phi_{t]}$ over $L^2([0,t])$.

We first show that a bounded operator has this property if and only if it can be represented as a quantum stochastic integral with respect to the number process. We also characterize the Maassen-Meyer kernel of such an operator if it admits one. We finish on an example, the Wiener space endomorphisms which respect the natural filtration.

II Notations

Let Φ, $\Phi_{t]}$ and $\Phi_{[t}$ be respectively the boson Fock spaces over $L^2(\mathbb{R}^+)$, $L^2([0,t])$ and $L^2(]t,+\infty[)$, let $\mathbb{E}_t$ be the projection from Φ onto $\Phi_{t]}$. We will often interpret Φ as $L^2(\Omega,\mathcal{F},\mathcal{P})$ where $(\Omega,\mathcal{F},\mathcal{P})$ is the canonical Wiener space; Brownian motion on Ω will be denoted W and its natural filtration $(\mathcal{F}_t)_{t\geq 0}$. One has the following interpretations $\Phi_{t]} = L^2(\Omega,\mathcal{F}_t,P)$ and $\mathbb{E}_t = \mathbb{E}[\,\cdot\,/\mathcal{F}_t]$, $t\in\mathbb{R}^+$.

We denote by $\mathbf{1}$ the vacuum element of Φ, and by I the identity operator on Φ. For f in $L^2(\mathbb{R}^+)$ we define

$$\begin{cases} f_{t]} \stackrel{\triangle}{=} f\mathbb{1}_{[0,t]}, \quad f_{[t} \stackrel{\triangle}{=} f\mathbb{1}_{]t,+\infty[} \\ \varepsilon(f), \text{ the associated coherent vector.} \end{cases}$$

Let $L^2_{lb}(\mathbb{R}^+)$ be the space of locally bounded elements of $L^2(\mathbb{R}^+)$. Let $\mathcal{E}$ (resp. $\mathcal{E}_{lb}$) be the linear space generated by the vectors $\varepsilon(f)$ when f ranges over $L^2(\mathbb{R}^+)$ (resp. $L^2_{lb}(\mathbb{R}^+)$).

The annihilation, creation and number processes defined in Hudson & Parthasarathy [1], will respectively be denoted A, $A^\dagger$ and Λ.

In Hudson & Parthasarathy [1] and Meyer [4] it is proved that if K, L and H are adapted processes, defined on $\mathcal{E}$, which verify

$$\text{(II.1)} \qquad \int_0^t \Big(\|K_s\varepsilon(f_{s]})\|^2 + \|L_s\varepsilon(f_{s]})\|^2 + |f(s)|^2\|H_s\varepsilon(f_{s]})\|^2\Big)ds < \infty$$

for all t in $\mathbb{R}^+$ and f in $L^2(\mathbb{R}^+)$, then the stochastic integral

$$I_t \triangleq \int_0^t K_s\,dA_s + \int_0^t L_s\,dA_s^\dagger + \int_0^t H_s\,d\Lambda_s \quad , \; t \in \mathbb{R}^+,$$

is well defined as an adapted process of operators on $\mathcal{E}$ which verifies for all t and f,

$$\text{(II.2)} \quad I_t\varepsilon(f_{t]}) = \int_0^t f(s)I_s\varepsilon(f_{s]})\,dW_s + \int_0^t f(s)K_s\varepsilon(f_{s]})\,ds + \int_0^t L_s\varepsilon(f_{s]})\,dW_s + \\ + \int_0^t f(s)H_s\varepsilon(f_{s]})\,dW_s.$$

In the following, each time that one of these stochastic integrals appears, we implicitly suppose that the processes K, L and H verify (II.1).

Let F be any element of Φ. By the classical previsible representation theorem, F can be written

$$F = \mathbb{E}[F] + \int_0^\infty G_s\,dW_s,$$

where G is a previsible process. For every t, let $F_t \triangleq \mathbb{E}_tF$. If F belongs to $\mathcal{E}$, equation (II.2) proves that, for every t,

$$\text{(II.3)} \qquad I_tF_t = \int_0^t I_sG_s\,dW_s + \int_0^t K_sG_s\,ds + \int_0^t L_sF_s\,dW_s + \int_0^t H_sG_s\,dW_s.$$

If, furthermore, the operators I_t, K_t, L_t, H_t, $t \in \mathbb{R}^+$, are bounded and extended to all Φ, if their extensions verify (II.3) for every F in Φ, we will say that

$$I_t = \int_0^t K_s\,dA_s + \int_0^t L_s\,dA_s^\dagger + \int_0^t H_s\,d\Lambda_s$$

on all Φ.

In the part dealing with Maassen-Meyer kernels we use the classical short notations of Maassen [2] and Meyer [3]. Let $\mathcal{P}$ be the set of finite subsets of $\mathbb{R}^+$ Let $\mathfrak{P}^3$ be the set of elements (U, V, W) in $\mathcal{P}^3$ such that U, V and W are two by two disjoint. We will use the sign sum to denote a union of disjoint elements of $\mathcal{P}$.

Let us recall that a kernel operator is an operator on Φ of the form

$$K = \int_{\mathfrak{P}^3} \widehat{K}(U, V, W)\,dA_U^\dagger\,d\Lambda_V\,dA_W$$

where the set function $\widehat{K}$ satisfies

i) a compact time support condition: $\widehat{K}(U,V,W) = 0$ unless U, V and W are contained in some bounded interval $[0,T]$

ii) a majoration of the form

$$|\widehat{K}(U,V,W)| \leq M^{|U+V+W|}.$$

Such a kernel acts on test-functions, that is elements F of Φ with a chaotic expansion,

$$F = \int_{A\in\mathcal{P}} \widehat{F}(A)\, dW_A,$$

verifying a compact time support condition $\big(\widehat{F}(A) = 0$ unless A is included in some $[0,T]\big)$ and a domination condition $\big(|\widehat{F}(A)| \leq M^{|A|}\big)$.

In the following we always denote by $\widehat{T}$ the kernel of an operator T (if it admits one); in the same way, we also denote by $\widehat{F}(A)$ the coefficients of the chaotic expansion of a vector F of Φ.

Then one knows that KF is also a test-function with chaotic expansion given by the following expression

$$\text{(II.4)} \qquad \widehat{KF}(A) = \sum_{U+V+W=A} \int_{N\in\mathcal{P}} \widehat{K}(U,V,N)\widehat{F}(N+V+W)\, dN.$$

Finally, for P in $\mathcal{P}$ we define $\vee P \triangleq \max P$ (with $\vee\emptyset \triangleq 0$).

III Integral representation

As announced in the introduction we consider a bounded linear operator T from Φ into itself which verifies $\mathbb{E}_t T = T\,\mathbb{E}_t$, for all t in $\mathbb{R}^+$.

An example is $T = \mathbb{E}_A$, where A is a Borel subset of $\mathbb{R}^+$ and $\mathbb{E}_A$ the projection on Φ_A (the Fock space over $L^2(A)$); another example (the starting point of this work) will be given in section **V**. The first result is a complete characterization of these operators.

Theorem III.1 - *Let T be a bounded operator on Φ . The following two assertions are equivalent.*

i) $\mathbb{E}_t T = T\,\mathbb{E}_t$, *for all t in $\mathbb{R}^+$.*

ii) $T1$ *belongs to $\mathbb{C}$ and there exists a bounded adapted process H such that*

$$T = T1\, I + \int_0^\infty H_s\, d\Lambda_s, \quad \textit{on all } \Phi.$$

Proof

i)$\Rightarrow$ ii)

Let $(M_t)_{t\geq 0}$ be the (bounded) martingale associated to T, that is

$$M_t = (\mathbb{E}_t T)_{|\Phi_{t]}} \otimes I_{|\Phi_{[t}} = (T\,\mathbb{E}_t)_{|\Phi_{t]}} \otimes I_{|\Phi_{[t}} = T_{|\Phi_{t]}} \otimes I_{|\Phi_{[t}}.$$

For all $a < t$ and all u in $\Phi_{a]}$, we have $||(M_t - M_a)u||^2 = ||Tu - Tu||^2 = 0$. But, since each $\mathbb{E}_t$ is self adjoint, if T verifies condition i) its adjoint T^* verifies i) too. So if M_t^* denotes the adjoint of M_t, the process M^* is also the martingale associated to T^*. Hence, for every $a < t$ and every u in $\Phi_{a]}$, we also have $||(M_t^* - M_a^*)u||^2 = 0$. This implies that M is a regular martingale with respect to the null measure, in the sense of Parthasarathy & Sinha [6]. Their theorem 3.10 shows that, for all t,

$$M_t = M_0 + \int_0^t H_s\, d\Lambda_s \quad \text{on } \mathcal{E}_{lb}, \tag{III.2}$$

for a bounded adapted process H (the terms $K\,dA$ and $L\,dA^\dagger$ in the decomposition vanish since they can be estimated in terms of the measure).

By definition $M_0 u$ belongs to $\Phi_{0]} \simeq \mathbb{C}\mathbf{1}$, for all u in Φ. So we get $M_0 = T\mathbf{1}\, I$.

It is easy to verify that M_t converges strongly to T when t tends to $+\infty$. In other words,

$$T = T\mathbf{1}\, I + \int_0^\infty H_s\, d\Lambda(s),$$

on $\mathcal{E}_{lb}$. That is, for every $F \in \mathcal{E}_{lb}$, $F \triangleq \mathbb{E}[F] + \int_0^\infty G_s\, dW_s$, we have

$$\begin{aligned} TF &= T\mathbf{1}F + \int_0^\infty (M_s - T\mathbf{1}\, I)\, G_s\, dW_s + \int_0^\infty H_s G_s\, dW_s \\ &= T\mathbf{1}\, \mathbb{E}[F] + \int_0^\infty M_s\, G_s\, dW_s + \int_0^\infty H_s G_s\, dW_s. \end{aligned}$$

In order to finish proving ii), we are going to show that this equality is valid for every $F \in \Phi$.

It is proved in Parthasarathy & Sinha [6] that, for every $s \in \mathbb{R}^+$,

$$H(s) = L_s U_s^{-1} - M_s,$$

where,

i) the L_s, $s \in \mathbb{R}^+$, are bounded operators such that, in our case,

$$||L_s||^2 \leq 2\exp(s)||M_s||^2,$$

ii) for every $s \in \mathbb{R}^+$, $\exp(-s/2)U_s$ is an unitary operator.

Let $F = \mathbb{E}[F] + \int_0^\infty G_s\,dW_s$ be any element of Φ. Let

$$F^n = \mathbb{E}[F^n] + \int_0^\infty G_s^n\,dW_s,\ n \in \mathbb{N},$$

be a sequence in $\mathcal{E}_{lb}$ converging to F. Then

$$\begin{aligned}
&||TF - \Big(T\mathbf{1}\mathbb{E}[F] + \int_0^\infty M_s G_s\,dW_s + \int_0^\infty H_s G_s\,dW_s\Big)||^2 \\
&\leq 2\Big(||T(F-F^n)||^2 + ||T\mathbf{1}(\mathbb{E}[F] - \mathbb{E}[F^n]) + \int_0^\infty M_s(G_s - G_s^n)\,dW_s + \\
&\qquad\qquad + \int_0^\infty H_s(G_s - G_s^n)\,dW_s||^2\Big) \\
&\leq 8\Big(||T||^2||F-F^n||^2 + |T\mathbf{1}|^2||F-F^n||^2 + \int_0^\infty ||M_s||^2||G_s - G_s^n||^2\,ds + \\
&\qquad\qquad + \int_0^\infty ||L_sU_s^{-1}(G_s - G_s^n)||^2\,ds + \int_0^\infty ||M_s||^2||G_s - G_s^n||^2\,ds\Big).
\end{aligned}$$

As it is easy to verify, for every $s \in \mathbb{R}^+$, we have $||M_s|| = ||T||$, so

$$\begin{aligned}
&||TF - \Big(T\mathbf{1}F + \int_0^\infty M_s G_s\,dW_s + \int_0^\infty H_s G_s\,dW_s\Big)||^2 \\
&\leq 8\Big(2||T||^2||F-F^n||^2 + 2||T||^2\int_0^\infty ||(G_s - G_s^n)||^2\,ds + \\
&\qquad\qquad + \int_0^\infty 2\exp(s)||M_s||^2||U_s^{-1}(G_s - G_s^n)||^2\,ds\Big) \\
&\leq 8\Big(6||T||^2||F-F^n||^2 + 2||T||^2\Big(\int_0^\infty \exp(s)\exp(-2s)||G_s - G_s^n||^2\,ds\Big). \\
&\leq 8\Big(6||T||^2||F-F^n||^2 + 2||T||^2\Big(\int_0^\infty ||G_s - G_s^n||^2\,ds\Big).
\end{aligned}$$

So, this expression converges to 0 as n tends to $+\infty$. That is,

$$T = T\mathbf{1}\,I + \int_0^\infty H_s\,d\Lambda_s,$$

on all Φ.

ii)⇒i)
If T satisfies ii), by (II.2), we have, for every f in $L^2(\mathbb{R}^+)$ and every t in $\mathbb{R}^+$,

$$T\varepsilon(f_{t]}) = T\mathbf{1}\,\varepsilon(f_{t]}) + \int_0^t f(s)\Big(\int_0^s H_u\,d\Lambda_u\Big)\varepsilon(f_{s]})\,dW_s + \int_0^t f(s)H_s\varepsilon(f_{s]})\,dW_s.$$

So $(T\varepsilon(f_{t]}))_{t\geq 0}$ is a square integrable martingale which converges to $T\varepsilon(f)$. Therefore $T\varepsilon(f_{t]}) = \mathbb{E}_t\, T\varepsilon(f)$, for every t. Of course, one also has $T\varepsilon(f_{t]}) = T\,\mathbb{E}_t\varepsilon(f)$. So i) is verified on the dense subset $\mathcal{E}$, hence everywhere.

∎

In section **V** we will need to characterize when such operators are isometries.

Proposition IV.3 – *Let T be an operator on Φ such that $T\,\mathbb{E}_t = \mathbb{E}_t\,T$, for all t in $\mathbb{R}^+$. Using the same notations as in theorem* III.1, *the following two assertions are equivalent.*

i) *The operator T is an isometry*

ii) *The process $L_t \triangleq H_t + M_t$, $t \in \mathbb{R}^+$, is a process of (adapted) isometries.*

Proof

We know from theorem III.1, that for each $t \in \mathbb{R}^+$, $dM_t = H_t\, d\Lambda_t$. So, $dM_t^* = H_t^*\, d\Lambda_t$ and, by the quantum Itô formula,

$$d(M_t^*\, M_t) = (H_t^*\, M_t + M_t^*\, H_t + H_t^*\, H_t)\, d\Lambda_t.$$

We easily deduce that the process M is a process of isometries if and only if M_0 is an isometry and, for every t,

$$H_t^*\, M_t + M_t^*\, H_t + H_t^*\, H_t = 0$$

$$\text{i.e. } L_t^*\, L_t = M_t^*\, M_t.$$

Hence, L_t is an isometry for every t if and only if M_t is an isometry for every t. But, as

$$\langle\, M_t\,\varepsilon(g)\,,\, M_t\,\varepsilon(f)\,\rangle = \langle\, H_t\,\varepsilon(g_{t]})\,,\, H_t\,\varepsilon(f_{t]})\,\rangle\,\langle\, \varepsilon(g_{[t})\,,\, \varepsilon(f_{[t})\,\rangle,$$

for every $f, g \in L^2(\mathbb{R}^+)$, it is easy to verify that T is an isometry if and only if M_t is an isometry for every t.

∎

IV Maassen-Meyer kernels

With theorem III.1, we know that all the operators which commute with the $\mathbb{E}_t$ admit an integral representation. We are now going to characterize those which admit a Maassen-Meyer kernel.

Theorem IV.1 – *Let T be an operator on Φ admitting a Maassen-Meyer kernel.*

i) *All the spaces* $\operatorname{Im}\mathbb{E}_t$ *are stable under T if and only if $\widehat{T}(U,V,W) = 0$ for every (U,V,W) in $\mathfrak{P}^3$ such that $\vee(U+V+W)$ belongs to U.*

ii) *All the spaces* Ker $\mathbb{E}_t$ *are stable under* T *if and only if* $\widehat{T}(U,V,W) = 0$ *for every* (U,V,W) *in* $\mathfrak{P}^3$ *such that* $\vee(U+V+W)$ *belongs to* W.

iii) *The operator* T *commutes with all the* $\mathbb{E}_t$ *if and only if* $\widehat{T}(U,V,W) = 0$ *for every* (U,V,W) *in* $\mathfrak{P}^3$ *such that* $\vee(U+V+W)$ *does not belong to* V.

Proof

i)

Since each $\mathbb{E}_t$ is a projection, the space Im $\mathbb{E}_t$ is stable under T if and only if $\mathbb{E}_t T \mathbb{E}_t = T \mathbb{E}_t$. But one has, for every F in Φ, t in $\mathbb{R}^+$, $\widehat{\mathbb{E}_t F}(U) = \widehat{F}(U)\mathbb{1}_{U\subset[0,t]}$. So, by (II.4), for every test-functions F and G and every t in $\mathbb{R}^+$, an easy calculation gives

$$\begin{aligned}
&\langle G, (T\mathbb{E}_t - \mathbb{E}_t T \mathbb{E}_t)F\rangle = \\
&= \int \sum_{U+V+W=A} \widehat{T}(U,V,N)\widehat{F}(N+V+W)\widehat{G}(A)\mathbb{1}_{U\not\subset[0,t]}\mathbb{1}_{N+V+W\subset[0,t]}\, dN\, dA \\
&= \int \widehat{T}(U,V,N)\widehat{F}(N+V+W)\widehat{G}(U+V+W)\mathbb{1}_{U\not\subset[0,t]}\mathbb{1}_{N+V+W\subset[0,t]}\, dN\, dU\, dV\, dW \\
&= \int \sum_{V+W=L} \widehat{T}(U,V,N)\widehat{F}(N+L)\widehat{G}(U+L)\mathbb{1}_{U\not\subset[0,t]}\mathbb{1}_{N+L\subset[0,t]}\, dN\, dU\, dL.
\end{aligned}$$

So, if we set, for every (U,L,N) in $\mathfrak{P}^3$,

$$\Psi(U,L,N) \triangleq \sum_{V+W=L} \widehat{T}(U,V,N)\mathbb{1}_{U\not\subset[0,t]}\mathbb{1}_{N+L\subset[0,t]},$$

we have

$$\langle G, (T\mathbb{E}_t - \mathbb{E}_t T \mathbb{E}_t)F\rangle = \int \Psi(U,L,N)\widehat{F}(N+L)\widehat{G}(U+L)\, dU\, dN\, dL.$$

This must be equal to 0, for every t in $\mathbb{R}^+$ and F, G test functions. But, in Meyer [5], it is proved that the right member of previous equality vanishes, for every F, G test-functions, if and only if Ψ vanishes. Using the Mœbius inversion formula,

$$\widehat{T}(U,L,N)\mathbb{1}_{U\not\subset[0,t]}\mathbb{1}_{N+L\subset[0,t]} = \sum_{V\subset L} \Psi(U,V,N)(-1)^{|L\setminus V|},$$

we obtain that the space Im $\mathbb{E}_t$ is stable under T if and only if $\widehat{T}(U,L,N) = 0$ for every (U,L,N) in $\mathfrak{P}^3$ such that $\vee U \geq t$ and $\vee(N+L) < t$. Since this is valid for every t, we have proved i).

ii)

Since each $\mathbb{E}_t$ is a projection, the space Ker $\mathbb{E}_t$ is stable under T if and only if $\mathbb{E}_t T \mathbb{E}_t = \mathbb{E}_t T$. Making the same calculation as in part i), we obtain that this condition is equivalent to $\widehat{T}(U,L,N)=0$ for every (U,L,N) in $\mathfrak{P}^3$ such that there exists a t in $\mathbb{R}^+$ satisfying $\vee L \geq t$ and $\vee(U+N) < t$. That is $\widehat{T}(U,L,N)=0$ for every (U,L,N) in $\mathfrak{P}^3$ such that $\vee(U+N+L)$ belongs to L.

iii)

One knows that an operator T commutes with a projection if and only if T preserves the image and the kernel of this projection. So we may conclude iii) from i) and ii).

■

We can also make the following easy remark.

Proposition IV.2 – *Let T be an operator on Φ admitting a Maassen-Meyer kernel. If T commutes with all the $\mathbb{E}_t$, then each of the operators H_t, $t \in \mathbb{R}^+$, given in theorem* III.1, *admits a kernel $\widehat{H_t}$, given by*

$$\widehat{H_t}(A,B,C) = \begin{cases} \widehat{T}(A,B+\{t\},C) & \textit{if } (A,B,C) \neq (\emptyset,\emptyset,\emptyset) \textit{ and } A+B+C \subset [0,t], \\ 0 & \textit{otherwise.} \end{cases}$$

■

V Application to Wiener space endomorphisms

We are going to see a non-trivial example of operators commuting with all the conditional expectations $\mathbb{E}_t$.

Let $\widetilde{T}$ be an endomorphism of $(\Omega,\mathcal{F},\mathcal{P})$, that is a measure-preserving transformation of the Wiener space. In other words, $\widetilde{T}$ transforms the Brownian motion $(W_t)_{t\geq 0}$ into another Brownian motion $(\widetilde{W_t})_{t\geq 0}$.

Among these endomorphisms, we consider those such that $(\widetilde{W_t})_{t\geq 0}$ is $(\mathcal{F}_t)_{t\geq 0}$-adapted. An example is the well known Levy transformation

$$\widetilde{W_t} \triangleq \int_0^t \operatorname{sign}(W_s)\,dW_s = |W_t| - L_t, \quad t \in \mathbb{R}^+,$$

where $(L_t)_{t\geq 0}$ is the local time of W in 0. More generaly, since the process $\widetilde{W}$ is a locally square integrable martingale for $(\mathcal{F}_t)_{t\geq 0}$, one has, for every t in $\mathbb{R}^+$,

$$\widetilde{W_t} = \int_0^t G_s\,dW_s,$$

for a $(\mathcal{F}_t)_{t\geq 0}$-previsible process G and $\widetilde{W}$ is a Brownian motion if and only if $G_s^2 = 1$ for almost all s in $\mathbb{R}^+$.

As is well known, to study endomorphisms of measure spaces is equivalent to studying the associated isometry T of $L^2(\Omega)$ defined by $TF = F \circ \widetilde{T}$. In our case we have the following result.

Lemma V.1 - *Let T be an isometry of Φ associated to an endomorphism $\widetilde{T}$ of $(\Omega, \mathcal{F}, \mathcal{P})$, let $\widetilde{W}$ denote the image of W under T. Then $\widetilde{W}$ is adapted to the filtration of W if and only if T commutes with all the $\mathbb{E}_t$, t in $\mathbb{R}^+$.*

Proof

Let $(\widetilde{\mathcal{F}}_t)_{t\geq 0}$ be the natural filtration of $\widetilde{W}$, if $\widetilde{W}$ is $(\mathcal{F}_t)_{t\geq 0}$-adapted then $\widetilde{\mathcal{F}}_t$ is included in $\mathcal{F}_t$, for all t in $\mathbb{R}^+$. But one always has that if $(X_t)_{t\geq 0}$ is a martingale for $(\mathcal{F}_t)_{t\geq 0}$, then $(TX_t)_{t\geq 0}$ is a martingale for $(\widetilde{\mathcal{F}}_t)_{t\geq 0}$; so,

$$TX_t = C + \int_0^t P_s \, d\widetilde{W}_s, \quad t \in \mathbb{R}^+,$$

for a $(\widetilde{\mathcal{F}}_t)_{t\geq 0}$-previsible process P. If $\widetilde{W}$ is $(\mathcal{F}_t)_{t\geq 0}$-adapted, then P is a $\mathcal{F}_t$-previsible process and

$$TX_t = C + \int_0^t P_s G_s \, dW_s, \quad t \in \mathbb{R}^+,$$

so $(TX_t)_{t\geq 0}$ is a martingale for $(\mathcal{F}_t)_{t\geq 0}$.
If one applies this latter result to the square integrable $\mathcal{F}$-martingale $(\varepsilon(f_{t]}))_{t\geq 0}$, for a f in $L^2(\mathbb{R}^+)$, then the $\mathcal{F}$-martingale $(T\varepsilon(f_{t]}))_{t\geq 0}$ converges to $T\varepsilon(f)$, so, as in the end of the proof of theorem III.1, T commutes with all the $\mathbb{E}_t$ on $\mathcal{E}$, hence everywhere.

In order to prove the converse, let us suppose that T commutes with all the $\mathbb{E}_t$. Then we have, for every t, $\widetilde{W}_t = TW_t = T\,\mathbb{E}_t W_t = \mathbb{E}_t TW_t = \mathbb{E}_t \widetilde{W}_t$. So $\widetilde{W}_t$ is $\mathcal{F}_t$-measurable, for every t.

■

Using lemma V.1, theorem III.1 and proposition III.3, we have finally proved the following theorem.

Theorem V.2 - *Let T be an isometry of Φ associated to an endomorphism of the Wiener space which respects the natural filtration of the Brownian motion W. Then T can be written as*

$$T = I + \int_0^\infty H_s \, d\Lambda_s.$$

If we denote, for each t, $M_t \triangleq I + \int_0^t H_s \, d\Lambda_s$ then $M_t + H_t$ is an isometry.

■

The author thanks Martin Lindsay and the referee for their useful remarks.

References

[1] R.L. Hudson & K.R. Parthasarathy : Quantum Itô's formula and stochastic evolutions. *Comm. in Math. Phys.* 93, p 301-323, 1984.

[2] H. Maassen : Quantum Markov processes on Fock space described by integral kernels. *Quantum Prob. & Appl. II.*, p 361-374, L.N.M. 1136 Springer 1985.

[3] P.A. Meyer : Eléments de probabilités quantiques I–V. *Séminaire de probabilités XX*, p 186-312, 1986.

[4] P.A. Meyer : Quelques remarques au sujet du calcul stochastique sur l'espace de Fock. *Séminaire de probabilités XX*, p 321-330, 1986.

[5] P.A. Meyer : Eléments de probabilités quantiques VI. *Séminaire de probabilités XXI*, p 34-49, 1987.

[6] K.R. Parthasarathy & K.B. Sinha : Representation of bounded quantum martingales in Fock space. *Journ. Funct. Anal.* 67 (1986), p 126-151.

Quantum Probability and Related Topics Vol. VIII (pp. 47–69)

CHARACTERIZATIONS OF OPERATORS COMMUTING WITH SOME CONDITIONAL EXPECTATIONS ON MULTIPLE FOCK SPACES

Stéphane ATTAL

Université Louis Pasteur
Département de mathématiques
7, rue René Descartes
67084 Strasbourg Cedex, France

I Introduction

In order to study endomorphisms of the Wiener space which respect the natural filtration $(\mathcal{F}_t)_{t\geq 0}$, Attal [1] gives a characterization of the bounded operators on simple Fock space which commute with all the conditionnal expectations $\mathbb{E}[\,\cdot\,/\mathcal{F}_t]$, $t \in \mathbb{R}^+$.

We first extend this result to the case of multiple Fock space, showing that these operators are exactely those which can be written as a sum of quantum stochastic integrals with respect to the conservation processes.

Trying to investigate this result we also characterize, among operators admitting an integral or a Maassen-kernel representation, those which respect the range, those which respect the kernel of these projections, and those which commute with them.

In multiple Fock spaces, the projections $\mathbb{E}[\,\cdot\,/\mathcal{F}_t]$, $t \in \mathbb{R}^+$, are only particular elements of a set of natural conditional expectations. Following the same scheme as previously, we give analogue characterizations for this bigger family.

II Notations

Let $\mathcal{G}$ be a separable Hilbert space with a fixed orthonormal basis $(e_\alpha)_{\alpha\in\mathcal{N}}$, where $\mathcal{N}$ is an at most countable set of indices which does not contain 0. For every $t \in \mathbb{R}^+$, let Φ, $\Phi_{t]}$ and $\Phi_{[t}$ be respectively the boson Fock spaces over $L^2(\mathbb{R}^+;\mathcal{G})$, $L^2([0,t];\mathcal{G})$ and $L^2(]t,+\infty[;\mathcal{G})$, $\mathbb{E}_t$ the projection from Φ onto $\Phi_{t]}$ and $\mathbb{E}_{[t}$ the projection from Φ onto $\Phi_{[t}$.

We interprete Φ as $L^2(\Omega,\mathcal{F},\mathcal{P})$ where $(\Omega,\mathcal{F},\mathcal{P})$ is the canonical Wiener space; Brownian motion on Ω is denoted $(W^\alpha_\cdot)_{\alpha\in\mathcal{N}}$ and, for every $t \in \mathbb{R}^+$, dW^0_t will denote dt. The natural filtration of W will be denoted $(\mathcal{F}_t)_{t\geq 0}$. One has the following interpretations :

$$\begin{cases} \Phi_{t]} = L^2(\Omega, \mathcal{F}_t, P), & t \in \mathbb{R}^+ \\ \mathbb{E}_t = \mathbb{E}[\,\cdot\,/\mathcal{F}_t], & t \in \mathbb{R}^+. \end{cases}$$

The vacuum element of Φ will be denoted $\mathbf{1}$ and the identity operator on Φ will be denoted I.

For every f in $L^2(\mathbb{R}^+;\mathcal{G})$, we define

$$\begin{cases} f^\beta, \text{ its } \beta\text{-th component in the orthonormal basis } (e_\alpha)_{\alpha\in\mathcal{N}},\ \beta\in\mathcal{N} \\ f^0, \text{ the constant function } \mathbf{1} \\ f_{t]} \triangleq f\mathbb{1}_{[0,t]}, \quad t\in\mathbb{R}^+ \\ f_{[t} \triangleq f\mathbb{1}_{]t,+\infty[}, \quad t\in\mathbb{R}^+ \\ \varepsilon(f), \text{ the associated coherent vector.} \end{cases}$$

Let $L^2_{lb}(\mathbb{R}^+;\mathcal{G})$ be the space of locally bounded elements of $L^2(\mathbb{R}^+;\mathcal{G})$. Let $\mathcal{E}$ (resp. $\mathcal{E}_{lb}$) be the linear space generated by the vectors $\varepsilon(f)$ when f ranges over $L^2(\mathbb{R}^+;\mathcal{G})$ (resp. $L^2_{lb}(\mathbb{R}^+;\mathcal{G})$).

The creation, annihilation and conservation families defined in Parthasarathy & Sinha [10], will be respectively denoted $(a_0^\alpha)_{\alpha\in\mathcal{N}}$, $(a_\alpha^0)_{\alpha\in\mathcal{N}}$, and $(a_\beta^\alpha)_{\alpha,\beta\in\mathcal{N}}$.

We will always use the following convention : the letters at the beginning of the Greek alphabet α,β,γ will be understood to be running over $\mathcal{N}$, those at the end of the alphabet ρ,σ,τ will run over $\overline{\mathcal{N}} \triangleq \mathcal{N}\cup\{0\}$, but the pairs (ρ,σ) will be understood to run over $\mathcal{M} \triangleq \overline{\mathcal{N}}^2 \setminus \{(0,0)\}$ only.

In Parthasarathy & Sinha [10] and Meyer [7] it is proved that if $(H_\sigma^\rho)_{\rho,\sigma}$ is a family of adapted processes of operators defined on $\mathcal{E}$ which verifies

$$\text{(II.1)}\quad \int_0^t \sum_\alpha \Big(\|H_0^\alpha(s)\varepsilon(f_{s]})\|^2 + \|H_\alpha^0(s)\varepsilon(f_{s]})\|^2 + \|\sum_\beta f^\beta(s) H_\beta^\alpha(s)\varepsilon(f_{s]})\|^2 \Big)\, ds < \infty$$

for all t in $\mathbb{R}^+$ and f in $L^2(\mathbb{R}^+;\mathcal{G})$, then the quantum stochastic integrals

$$I_t \triangleq \sum_{\rho,\sigma} \int_0^t H_\sigma^\rho(s)\, da_\sigma^\rho(s) \quad , \ t\in\mathbb{R}^+,$$

are well defined as adapted processes of operators on $\mathcal{E}$ which verify, for all t and all f,

$$\text{(II.2)}\qquad I_t\varepsilon(f_{t]}) = \sum_\alpha \int_0^t f^\alpha(s) I_s\, \varepsilon(f_{s]})\, dW_s^\alpha + \sum_{\rho,\sigma} \int_0^t f^\sigma(s) H_\sigma^\rho(s)\varepsilon(f_{s]})\, dW_s^\rho.$$

In the following, each time that one of these quantum stochastic integrals appears, we implicitely suppose that the processes $(H_\sigma^\rho)_{\rho,\sigma}$ verify (II.1).

Let F be any element of Φ, then F can be written

$$F = \mathbb{E}[F] + \sum_{\alpha\in\mathcal{N}} \int_0^\infty G_s^\alpha \, dW_s^\alpha,$$

where G is a previsible process. Let $F_t \triangleq \mathbb{E}_t F$, $t \in \mathbb{R}^+$ If F belongs to $\mathcal{E}$, equation (II.2) proves that

$$\text{(II.3)} \qquad I_t F_t = \sum_\alpha \int_0^t I_s \, G_s^\alpha \, dW_s^\alpha + \sum_{\rho,\sigma} \int_0^t H_\sigma^\rho(s) G_s^\sigma \, dW_s^\rho,$$

where G_s^0 denotes F_s, $s \in \mathbb{R}^+$.

If, furthermore, the operators I_t, $t \in \mathbb{R}^+$ and $H_\sigma^\rho(t)$, $t \in \mathbb{R}^+$, $(\rho, \sigma) \in \mathcal{M}$, are bounded and extended to all Φ, if their extensions verify (II.3) for every F in Φ, we say that $I_t = \sum\limits_{\rho,\sigma} \int_0^t H_\sigma^\rho(s) \, da_\sigma^\rho(s)$ on all Φ.

In the part dealing with Maassen kernels, we suppose that the dimension of $\mathcal{G}$ is finite. One must notice that we use the terminology "Maassen kernels" to denote, in multiplicity 1, the three arguments kernels, called Maassen-Meyer kernels (cf Maassen [6] and Meyer [8]), and, in finite multiplicity, their natural extension, described in Dermoune [3]. We always use their common "short notation".

Let $\mathcal{P}$ be the set of finite subsets of $\mathbb{R}^+$ Let $\mathfrak{P}(\mathcal{M})$ be the set of elements $A = (A_\sigma^\rho)_{\rho,\sigma}$ of $\mathcal{P}^{\mathcal{M}}$ such that A_σ^ρ and $A_{\sigma'}^{\rho'}$ are disjoint for every $(\rho, \sigma) \neq (\rho', \sigma')$. In the same way, let $\mathfrak{P}(\mathcal{N})$ be the set of elements $U = (U_\alpha)_\alpha$ of $\mathcal{P}^{\mathcal{N}}$ such that U_α and $U_{\alpha'}$ are disjoint for every $\alpha \neq \alpha'$. We will use the sign sum to denote an union of disjoint elements of $\mathcal{P}$.

For every $A \in \mathfrak{P}(\mathcal{M})$, for every α in $\mathcal{N}$, we denote

$$\begin{cases} A^\alpha \triangleq (A_\sigma^\alpha)_\sigma \text{ and } A_\alpha \triangleq (A_\alpha^\rho)_\rho \\ \Sigma^\alpha A \triangleq \sum_\sigma A_\sigma^\alpha \text{ and } \Sigma_\alpha A \triangleq \sum_\rho A_\alpha^\rho \end{cases}$$

We also denote $\Sigma^\cdot$ (resp. $\Sigma_\cdot$) the mapping from $\mathfrak{P}(\mathcal{M})$ to $\mathfrak{P}(\mathcal{N})$ such that $\Sigma^\cdot A \triangleq (\Sigma^\alpha A)_\alpha$ (resp. $\Sigma_\cdot A \triangleq (\Sigma_\alpha A)_\alpha$).

If A and N belong respectively to $\mathfrak{P}(\mathcal{M})$ and $\mathfrak{P}(\mathcal{N})$, then $A^\circ N$ denotes the element of $\mathfrak{P}(\mathcal{M})$ such that

$$(A^\circ N)_\sigma^\rho = \begin{cases} A_\sigma^\rho \text{ if } \rho \neq 0 \\ N_\sigma \text{ if } \rho = 0 \end{cases}$$

and $A^{:}N$ denotes the element of $\mathfrak{P}(\mathcal{M})$ such that

$$(A^{:}N)_\sigma^\rho = \begin{cases} A_\sigma^\rho \text{ if } \rho \neq \sigma \\ N_\rho \text{ if } \rho = \sigma. \end{cases}$$

Finally, for U and V in $\mathfrak{P}(\mathcal{N})$, U and V disjoint, $U+V$ denotes the element of $\mathfrak{P}(\mathcal{N})$ such that $(U+V)_\alpha = U_\alpha + V_\alpha$ for all α in $\mathcal{N}$.

Let us recall that a Maassen kernel is an operator on Φ of the form

$$K = \int_{A\in\mathfrak{P}(\mathcal{M})} \widehat{K}(A) \prod_{\rho,\sigma} da_\sigma^\rho(A_\sigma^\rho)$$

where the set function $\widehat{K}$ verifies

i) a compact time support condition: $\widehat{K}(A)=0$ unless all the components A_σ^ρ of A are contained in some bounded interval $[0,T]$,

ii) a majoration of the form

$$|\widehat{K}(A)| \le M^{\sum_{\rho,\sigma}|A_\sigma^\rho|}.$$

Such a kernel acts on test-functions, that is, elements F of Φ with a chaotic expansion,

$$F = \int_{U\in\mathfrak{P}(\mathcal{N})} \widehat{F}(U) \prod_\alpha dW_{U_\alpha}^\alpha,$$

verifying a compact time support condition $(\widehat{F}(U)=0$ unless each U_α is included in some $[0,T])$ and a domination condition $(|\widehat{F}(U)| \le M^{\sum_\alpha |U_\alpha|})$.

In the following we will always denote $\widehat{T}$ the kernel of an operator T (if it admits one); in the same way, we will also denote $\widehat{F}(U)$, $U\in\mathfrak{P}(\mathcal{N})$, the coefficients of the chaotic expansion of a vector F of Φ.

Then one knows that KF is also a test-function with chaotic expansion given by the following expression, where $G(U)$ denotes the set of elements A in $\mathfrak{P}(\mathcal{M})$ such that, for all α in $\mathcal{N}$, $\Sigma^\alpha A + A_\alpha^0 = U_\alpha$,

$$\text{(II.4)} \qquad \widehat{KF}(U) = \sum_{A\in G(U)} \int_{N\in\mathfrak{P}(\mathcal{N})} \widehat{K}(A^\circ N)\widehat{F}(N+\Sigma.A) \prod_\alpha dN_\alpha$$

(The corresponding formula given in Dermoune [3] p 400 contains a misprint.) Finally, for P in $\mathcal{P}$ we define $\vee P \triangleq \max P$, with $\vee\emptyset \triangleq 0$, and if A belongs to $\mathfrak{P}(\mathcal{M})$, $\vee A \triangleq \vee(\sum_{\rho,\sigma} A_\sigma^\rho)$.

III Characterizations of integral representations

As announced in the introduction we first consider a bounded linear operator T from Φ into itself which verifies $\mathbb{E}_t T = T\,\mathbb{E}_t$, for every $t\in\mathbb{R}^+$.

An example is $T = \mathbb{E}_A$, where A is any Borel subset of $\mathbb{R}^+$ and $\mathbb{E}_A$ the projection on Φ_A (the Fock space over $L^2(A;\mathcal{G})$); another example is the isometries associated to Wiener space endomorphisms which respect the natural filtration $(\mathcal{F}_t)_{t\geq 0}$, studied in Attal [1].

The first result is a complete characterization of these operators.

Theorem III.1 – *Let T be a bounded operator on Φ. The following two assertions are equivalent :*

i) $\mathbb{E}_t T = T\,\mathbb{E}_t$, *for all t in $\mathbb{R}^+$*

ii) *$T\mathbf{1}$ belongs to $\mathcal{G}$ and there exists a family of bounded processes $(H^\alpha_\beta)_{\alpha,\beta}$ such that*

$$T = T\mathbf{1}\,I + \sum_{\alpha,\beta}\int_0^\infty H^\alpha_\beta(s)\,da^\alpha_\beta(s)$$

on all Φ.

Remark

Because of our conventions, the operators a^α_β are nor creation, nor annihilation, but only conservation processes.

Proof

i)$\Rightarrow$ ii)

Let $(M_t)_{t\geq 0}$ be the (bounded) martingale associated to T, that is,

$$M_t \triangleq (\mathbb{E}_t\,T)_{|\Phi_{t]}} \otimes I_{|\Phi_{[t}} = (T\,\mathbb{E}_t)_{|\Phi_{t]}} \otimes I_{|\Phi_{[t}} = T_{|\Phi_{t]}} \otimes I_{|\Phi_{[t}}.$$

For every $a < t$, u in $\Phi_{a]}$, we have $||(M_t - M_a)u||^2 = ||Tu - Tu||^2 = 0$. But, since each $\mathbb{E}_t$ is self adjoint, if T verifies condition i), its adjoint T^* verifies i) too. So, if M_t^* denotes the adjoint of M_t, the process M^* is also the martingale associated to T^*. Hence, for every $a < t$ and every u in $\Phi_{a]}$, we also have $||(M_t^* - M_a^*)u||^2 = 0$. This implies that M is a regular martingale with respect to the null measure, in the sense of Parthasarathy & Sinha [10]. Their theorem 3.8 shows that, for all t,

$$M_t = M_0 + \sum_{\alpha,\beta}\int_0^t H^\alpha_\beta(s)\,da^\alpha_\beta(s) \ \text{ on } \mathcal{E}_{lb}, \tag{III.2}$$

for a family of bounded adapted processes $(H^\alpha_\beta)_{\alpha,\beta}$ (the terms $K^0_\alpha\,da^0_\alpha$ and $L^\alpha_0\,da^\alpha_0$, $\alpha \in \mathcal{N}$, in the decomposition vanish since they can be estimated in terms of the measure).

We have $T\mathbf{1} = T\,\mathbb{E}_0\mathbf{1} = \mathbb{E}_0\,T\mathbf{1} \in \Phi_{0]} \simeq \mathcal{G}$, so we get $M_0 = T\mathbf{1}\,I$.

It is easy to verify that M_t converges strongly to T when t tends to $+\infty$. In other words,

$$T = T1\,I + \sum_{\alpha,\beta} \int_0^\infty H^\alpha_\beta(s)\,da^\alpha_\beta(s),$$

on $\mathcal{E}_{lb}$. That is, for every $F \in \mathcal{E}_{lb}$, $F \triangleq I\!E[F] + \sum_\alpha \int_0^\infty G^\alpha_s\,dW^\alpha_s$, we have

$$\begin{aligned} TF &= T1F + \sum_\alpha \int_0^\infty (M_s - T1\,I)\,G^\alpha_s\,dW^\alpha_s + \sum_{\alpha,\beta} \int_0^\infty H^\alpha_\beta(s) G^\beta_s\,dW^\alpha_s \\ &= T1\,I\!E[F] + \sum_\alpha \int_0^\infty M_s\,G^\alpha_s\,dW^\alpha_s + \sum_{\alpha,\beta} \int_0^\infty H^\alpha_\beta(s) G^\beta_s\,dW^\alpha_s. \end{aligned}$$

In order to finish proving ii), we are going to show that this equality is valid for every $F \in \Phi$.

It is proved in Parthasarathy & Sinha [10] that, for every $\alpha, \beta \in \mathcal{N}$, $s \in I\!R^+$,

$$H^\alpha_\beta(s) = L^\alpha_\beta(s)(U^\alpha_s)^{-1} - \delta_{\alpha\beta} M_s,$$

where,

i) the $L^\alpha_\beta(s)$, $\alpha, \beta \in \mathcal{N}$, $s \in I\!R^+$, are bounded operators such that, in our case, for every $u \in \Phi$,

$$\sum_\beta \|L^\alpha_\beta(s)\,u\|^2 \leq 4\exp(s)\|M_s\|^2\,\|u\|^2,$$

ii) for every $\alpha \in \mathcal{N}$, $s \in I\!R^+$, $\exp(-s/2)U^\alpha_s$ is an unitary operator,

iii) $\delta_{\alpha\beta}$ is the Kronecker symbol.

Let $F = I\!E[F] + \sum_\alpha \int_0^\infty G^\alpha_s\,dW^\alpha_s$ be any element of Φ. Let

$$F^n = I\!E[F^n] + \sum_\alpha \int_0^\infty (G^n_s)^\alpha\,dW^\alpha_s,\ n \in I\!N,$$

be a sequence in $\mathcal{E}_{lb}$ converging to F. Then

$$\begin{aligned} &\Big\|TF - \Big(T1I\!E[F] + \sum_\alpha \int_0^\infty M_s\,G^\alpha_s\,dW^\alpha_s + \sum_{\alpha,\beta}\int_0^\infty H^\alpha_\beta(s) G^\beta_s\,dW^\alpha_s\Big)\Big\|^2 \\ &\leq 2\Big(\|T(F - F^n)\|^2 + \|T1(I\!E[F] - I\!E[F^n]) + \sum_\alpha \int_0^\infty M_s(G^\alpha_s - (G^n_s)^\alpha)\,dW^\alpha_s + \\ &\qquad + \sum_{\alpha,\beta} \int_0^\infty H^\alpha_\beta(s)(G^\beta_s - (G^n_s)^\beta)\,dW^\alpha_s\|^2\Big) \end{aligned}$$

$$\leq 8\Big(||T||^2||F-F^n||^2+|T\mathbf{1}|^2||F-F^n||^2+\sum_\alpha\int_0^\infty ||M_s||^2||G_s^\alpha-(G_s^n)^\alpha||^2\,ds+$$

$$+\sum_{\alpha,\beta}\int_0^\infty ||L_\beta^\alpha(s)(U_s^\alpha)^{-1}(G_s^\alpha-(G_s^n)^\alpha)||^2\,ds+\sum_\alpha\int_0^\infty ||M_s||^2||G_s^\alpha-(G_s^n)^\alpha||^2\,ds\Big).$$

As it is easy to verify, for every $s\in I\!\!R^+$, we have $||M_s||=||T||$, so

$$||TF-\Big(T\mathbf{1}F+\sum_\alpha\int_0^\infty M_s\,G_s^\alpha\,dW_s^\alpha+\sum_{\alpha,\beta}\int_0^\infty H_\beta^\alpha(s)G_s^\beta\,dW_s^\alpha\Big)||^2$$

$$\leq 8\Big(2||T||^2||F-F^n||^2+2||T||^2\sum_\alpha\int_0^\infty ||(G_s^\alpha-(G_s^n)^\alpha)||^2\,ds+$$

$$+\sum_\alpha\int_0^\infty 4\exp(s)||M_s||^2||(U_s^\alpha)^{-1}(G_s^\alpha-(G_s^n)^\alpha)||^2\,ds\Big)$$

$$\leq 8\Big(6||T||^2||F-F^n||^2+4||T||^2\Big(\sum_\alpha\int_0^\infty \exp(s)\exp(-2s)||G_s^\alpha-(G_s^n)^\alpha||^2\,ds\Big).$$

$$\leq 8\Big(6||T||^2||F-F^n||^2+4||T||^2\Big(\sum_\alpha\int_0^\infty ||G_s^\alpha-(G_s^n)^\alpha||^2\,ds\Big).$$

So, this expression converges to 0 as n tends to $+\infty$. That is,

$$T=T\mathbf{1}\,I+\sum_{\alpha,\beta}\int_0^\infty H_\beta^\alpha(s)\,da_\beta^\alpha(s),$$

on all Φ.

ii)$\Rightarrow$i)
If T verifies ii), if

$$T_t\triangleq T\mathbf{1}\,I+\sum_{\alpha,\beta}\int_0^t H_\beta^\alpha(s)\,da_\beta^\alpha(s),\ t\in I\!\!R^+,$$

then by (II.2), we have, for every $f\in L^2(I\!\!R^+;\mathcal{G})$ and every $t\in I\!\!R^+$,

$$T\varepsilon(f_{t]})=T\mathbf{1}\,\varepsilon(f_{t]})+\sum_\alpha\int_0^t f^\alpha(s)(T_s-T\mathbf{1}\,I)\varepsilon(f_{s]})\,dW_s^\alpha+$$

$$+\sum_{\alpha,\beta}\int_0^t f^\beta(s)H_\beta^\alpha(s)\varepsilon(f_{s]})\,dW_s^\alpha.$$

So $(T\varepsilon(f_{t]}))_{t\geq 0}$ is a $\mathcal{G}$-valued square integrable martingale which converges to $T\varepsilon(f)$, therefore $T\varepsilon(f_{t]})=I\!\!E_t\,T\varepsilon(f)$, for every t.

Of course one also has $T\varepsilon(f_{t]}) = T\,\mathbb{E}_t\varepsilon(f)$, so i) is verified on the dense subset $\mathcal{E}$, hence everywhere. ■

Theorem III.1 proves that if a bounded operator T commutes with all the $\mathbb{E}_t$, then it has an integral representation and this represntation contains only conservation terms. But, if we study operators for which we already know that they have an integral representation, then we can be much more precise.

Theorem III.2 – *Let T be an operator on Φ, defined on $\mathcal{E}$. Let us suppose that T admits an integral representation,*

$$T = \sum_{\rho,\sigma} \int_0^\infty H_\sigma^\rho(s)\, da_\sigma^\rho(s),$$

with the operators $H_\sigma^\rho(s)$ closable, for almost all s.

i) *All the spaces $\operatorname{Im}\mathbb{E}_t$ are stable under T if and only if all the processes (H_0^α), $\alpha \in \mathcal{N}$, in the representation of T, vanish.*

ii) *All the spaces $\operatorname{Ker}\mathbb{E}_t$ are stable under T if and only if all the processes (H_α^0), $\alpha \in \mathcal{N}$, in the representation of T, vanish.*

iii) *The operator T commutes with all the $\mathbb{E}_t$ if and only if all the processes (H_α^0) and (H_0^α), $\alpha \in \mathcal{N}$, in the representation of T, vanish.*

Proof

i)

Since $\mathbb{E}_t$ is a projection, the space $\operatorname{Im}\mathbb{E}_t$ is stable under T if and only if $T\,\mathbb{E}_t = \mathbb{E}_t T\,\mathbb{E}_t$.

Let, for every $t \in \mathbb{R}^+$, $T_t \triangleq \sum_{\rho,\sigma} \int_0^t H_\sigma^\rho(s)\, da_\sigma^\rho(s)$. For every $f \in L^2(\mathbb{R}^+;\mathcal{G})$, one has

$$T\varepsilon(f) = \sum_\alpha \int_0^\infty f^\alpha(s) T_s\, \varepsilon(f_{s]})\, dW_s^\alpha + \sum_{\rho,\sigma} \int_0^\infty f^\sigma(s) H_\sigma^\rho(s) \varepsilon(f_{s]})\, dW_s^\rho,$$

so

$$T\,\mathbb{E}_t\, \varepsilon(f) = T\varepsilon(f_{t]}) = \sum_\alpha \int_0^t f^\alpha(s) T_s \varepsilon(f_{s]})\, dW_s^\alpha + \sum_{\rho,\beta} \int_0^t f^\beta(s) H_\beta^\rho(s) \varepsilon(f_{s]})\, dW_s^\rho + \\ + \sum_\alpha \int_0^\infty H_0^\alpha(s) \varepsilon(f_{s\wedge t]})\, dW_s^\alpha$$

and

$$\mathbb{E}_t T\,\mathbb{E}_t\, \varepsilon(f) = \sum_\alpha \int_0^t f^\alpha(s) T_s\, \varepsilon(f_{s]})\, dW_s^\alpha + \sum_{\rho,\sigma} \int_0^t f^\sigma(s) H_\sigma^\rho(s) \varepsilon(f_{s]})\, dW_s^\rho.$$

Finally

$$(T\,\mathbb{E}_t - \mathbb{E}_t T\,\mathbb{E}_t)\varepsilon(f) = \sum_\alpha \int_t^\infty H_0^\alpha(s)\varepsilon(f_{t]})\,dW_s^\alpha. \tag{III.3}$$

If T preserves the range of each $\mathbb{E}_t$, this must be 0.

So, by uniqueness of previsible representations, for every $\alpha \in \mathcal{N}$, every $f \in L^2(\mathbb{R}^+;\mathcal{G})$, every $t \in \mathbb{R}^+$, there exists a set $P(\alpha,f,t) \subset [t,+\infty[$, with null Lebesgue measure, such that

$$\forall s \in P(\alpha,f,t)^c,\ s \geq t,\ \ H_0^\alpha(s)\varepsilon(f_{t]}) = 0.$$

Enlarging $P(\alpha,f,t)$ with a null set, we may suppose $H_0^\alpha(s)$ closable for each s in $P(\alpha,f,t)^c$.

Let $P(\alpha,f)^c \triangleq \bigcap_{t\in\mathbb{Q}^+} P(\alpha,f,t)^c$, then

$$\forall\alpha \in \mathcal{N},\ \forall f \in L^2(\mathbb{R}^+;\mathcal{G}),\ \forall s \in P(\alpha,f)^c,\ \forall t \in \mathbb{Q}^+,\ t \leq s,\ H_0^\alpha(s)\varepsilon(f_{t]}) = 0.$$

Let $\alpha \in \mathcal{N}$ and $f \in L^2(\mathbb{R}^+;\mathcal{G})$ be fixed and $s \in P(\alpha,f)^c$. Let $(t_n)_{n\in\mathbb{N}}$ be a sequence in $\mathbb{Q}^+$ such that t_n tends to s, as n goes to $+\infty$, and such that $t_n \leq s$, $n \in \mathbb{N}$. Then, for any $n \in \mathbb{N}$, $H_0^\alpha(s)\varepsilon(f_{t_n]}) = 0$. So, by closability, we also have $H_0^\alpha(s)\varepsilon(f_{s]}) = 0$.

Let $\mathcal{H}$ be any countable dense subset of $L^2(\mathbb{R}^+;\mathcal{G})$, let $P(\alpha)^c \triangleq \bigcap_{f\in\mathcal{H}} P(\alpha,f)^c$.

We have

$$\forall\alpha \in \mathcal{N},\ \forall s \in P(\alpha)^c,\ \forall f \in \mathcal{H},\ \ H_0^\alpha(s)\varepsilon(f_{s]}) = 0.$$

By totality of the set $\{\varepsilon(f_{s]});\ f \in \mathcal{H}\}$ in $\Phi_{s]}$ and by closability of $H_0^\alpha(s)$ we conclude that, for every $\alpha \in \mathcal{N}$, for almost all s, $H_0^\alpha(s)$ vanishes on $\mathcal{E}$. That is, each process H_0^α, $\alpha \in \mathcal{N}$, vanishes.

The converse is easy to prove. Indeed, if in the representation of T all the processes $H_0^\alpha(s)$ vanish, equality (III.3) shows then that $T\,\mathbb{E}_t = \mathbb{E}_t T\,\mathbb{E}_t$ is true for every t and on $\mathcal{E}$.

ii)

Since $\mathbb{E}_t$ is a projection, the space $\operatorname{Ker}\mathbb{E}_t$ is stable under T if and only if $\mathbb{E}_t T = \mathbb{E}_t T\,\mathbb{E}_t$.

For every $f \in L^2(\mathbb{R}^+;\mathcal{G})$, $t \in \mathbb{R}^+$, one has

$$\mathbb{E}_t T\varepsilon(f) = \sum_\alpha \int_0^t f^\alpha(s)T_s\varepsilon(f_{s]})\,dW_s^\alpha + \sum_{\alpha,\sigma}\int_0^t f^\sigma(s)H_\sigma^\alpha(s)\varepsilon(f_{s]})\,dW_s^\alpha + $$
$$+\sum_\beta \int_0^\infty f^\beta(s)\mathbb{E}_t H_\beta^0(s)\varepsilon(f_{s]})\,ds.$$

So

$$(\text{III.4}) \qquad (\mathbb{E}_t T - \mathbb{E}_t T \mathbb{E}_t)\varepsilon(f) = \sum_\beta \int_t^\infty f^\beta(s) \mathbb{E}_t H_\beta^0(s)\varepsilon(f_{s]})\, ds.$$

If T preserves each $\operatorname{Ker} \mathbb{E}_t$, this must vanish for every $t \in \mathbb{R}^+$ and $f \in L^2(\mathbb{R}^+; \mathcal{G})$. Hence, for every f and every t, there exists a null-Lebesgue measured set $P(f,t) \subset [t, +\infty[$, such that

$$\forall s \in P(f,t)^c,\ s \geq t,\ \sum_\beta f(s) \mathbb{E}_t H_\beta^0(s)\varepsilon(f_{s]}) = 0.$$

Let $P(f)^c \triangleq \bigcap_{t \in \mathbb{Q}^+} P(f,t)^c$, $f \in L^2(\mathbb{R}^+; \mathcal{G})$. We then have

$$\forall f \in L^2(\mathbb{R}^+; \mathcal{G}),\ \forall s \in P(f)^c,\ \forall t \in \mathbb{Q}^+,\ t \leq s,\ \ \mathbb{E}_t\Big(\sum_\beta f^\beta(s) H_\beta^0(s)\varepsilon(f_{s]})\Big) = 0.$$

For any $f \in L^2(\mathbb{R}^+; \mathcal{G})$ and $s \in P(f)^c$, let $(t_n)_{n \in \mathbb{N}}$ be a sequence in $\mathbb{Q}^+$ converging to s, with $t_n \leq s$ for all n. Then $\mathbb{E}_{t_n}(\sum_\beta f^\beta(s) H_\beta^0(s)\varepsilon(f_{s]})) = 0$ for every n and converges to $\mathbb{E}_s(\sum_\beta f^\beta(s) H_\beta^0(s)\varepsilon(f_{s]})) = \sum_\beta f^\beta(s) H_\beta^0(s)\varepsilon(f_{s]})$.

So, by closability of the $H_\beta^0(s)$, we have

$$\forall f \in L^2(\mathbb{R}^+; \mathcal{G}),\ \forall s \in P(f)^c,\ \ \sum_\beta f^\beta(s) H_\beta^0(s)\varepsilon(f_{s]}) = 0.$$

That is, by (II.2) the operator $\sum_\beta \int_0^\infty H_\beta^0(s)\, da_\beta^0(s)$ vanishes on $\mathcal{E}$. Using theorem III.3.1 of Attal [2], which proves that an operator of the form $\sum_{\rho,\sigma} \int_0^\infty H_\sigma^\rho(s)\, da_\sigma^\rho(s)$, with the $H_\sigma^\rho(s)$ closable, vanish if and only if each H_σ^ρ vanishes, we conclude that the H_β^0, $\beta \in \mathcal{N}$, vanish on $\mathcal{E}$, hence on their domain.

The converse is proved in the same way as in i).

iii)

Since $\mathbb{E}_t$ is a projection, T commutes with $\mathbb{E}_t$ if and only if T preserves $\operatorname{Im} \mathbb{E}_t$ and $\operatorname{Ker} \mathbb{E}_t$. So we easily conclude from i) and ii). ∎

Remark : In this theorem the closability condition is necessary. Indeed, let, for example, $H_{\beta_0}^0$ be any counter-example to the uniqueness of integral representations of operators (cf Lindsay [5], Attal [2]). That is, a process of operators which is not the null process but such that $\int_0^\infty H_{\beta_0}^0(s)\, da_{\beta_0}^0(s)$ vanishes on its domain (from the previously quoted references, this implies that the $H_{\beta_0}^0(s)$, $s \in \mathbb{R}^+$, are not almost all closable). Then, it is easy to verify, from the proof of the theorem, that any operator T, of the form

$$T = \sum_{\alpha,\beta} \int_0^\infty H_\beta^\alpha(s)\, da_\beta^\alpha(s) + \int_0^\infty H_{\beta_0}^0(s)\, da_{\beta_0}^0(s)$$

commutes with all the $\mathbb{E}_t$, $t \in \mathbb{R}^+$, although $H_{\beta_0}^0$ does not vanish.

For every α in $\mathcal{N}$, let Φ^α and $\Phi_{t]}^\alpha$ be respectively the boson Fock spaces over $L^2(\mathbb{R}^+; \mathbb{C}e_\alpha)$ and $L^2([0,t]; \mathbb{C}e_\alpha)$, let $\mathbb{E}_t^\alpha$ be the projection from Φ^α onto $\Phi_{t]}^\alpha$, then one has $\Phi \simeq \underset{\alpha\in\mathcal{N}}{\otimes} \Phi^\alpha$, $\Phi_{t]} \simeq \underset{\alpha\in\mathcal{N}}{\otimes} \Phi_{t]}^\alpha$ and $\mathbb{E}_t = \underset{\alpha\in\mathcal{N}}{\otimes} \mathbb{E}_t^\alpha$. So the $\mathbb{E}_t$ are particular elements of the projection family $\{\mathbb{E}_{(t_\alpha)_\alpha}$ from Φ onto $\underset{\alpha\in\mathcal{N}}{\otimes} \Phi_{t_\alpha]}^\alpha$; $(t_\alpha)_\alpha \in (\overline{\mathbb{R}^+})^{\mathcal{N}}\}$ (with the convention $\Phi_{+\infty]}^\alpha \stackrel{\Delta}{=} \Phi^\alpha$).

Our aim is to find characterisations similar to theorems III.1 and III.2 for the family $(\mathbb{E}_{(t_\alpha)_\alpha})_{(t_\alpha)_\alpha \in (\overline{\mathbb{R}^+})^{\mathcal{N}}}$. In the following, $(\overline{\mathbb{R}^+})^{\mathcal{N}}$ will be denoted $\mathfrak{R}$. For every $t \in \mathbb{R}^+$ and every $\alpha \in \mathcal{N}$, the operator $I \otimes \ldots \otimes \mathbb{E}_t^\alpha \otimes I \otimes \ldots$ will also be denoted, without any ambiguity, $\mathbb{E}_t^\alpha$.

Lemma III.5 – *Let T be an operator from $\mathcal{E}$ onto Φ.*

i) *The spaces* $\operatorname{Im} \mathbb{E}_{(t_\alpha)_\alpha}$, $(t_\alpha)_\alpha \in \mathfrak{R}$, *are stable under T if (and only if) all the spaces* $\operatorname{Im} \mathbb{E}_t^\alpha$, $\alpha \in \mathcal{N}$, $t \in \mathbb{R}^+$, *are stable under T.*

ii) *The spaces* $\operatorname{Ker} \mathbb{E}_{(t_\alpha)_\alpha}$, $(t_\alpha)_\alpha \in \mathfrak{R}$, *are stable under T if (and only if) all the spaces* $\operatorname{Ker} \mathbb{E}_t^\alpha$, $\alpha \in \mathcal{N}$, $t \in \mathbb{R}^+$, *are stable under T.*

iii) *The operator T commutes with all the* $\mathbb{E}_{(t_\alpha)_\alpha}$, $(t_\alpha)_\alpha \in \mathfrak{R}$, *if (and only if) it commutes with all the* $\mathbb{E}_t^\alpha$, $\alpha \in \mathcal{N}$, $t \in \mathbb{R}^+$.

Proof

One remarks that for every $(t_\alpha)_\alpha \in \mathfrak{R}$, $\mathbb{E}_{(t_\alpha)_\alpha} = \prod_\alpha \mathbb{E}_{t_\alpha}^\alpha$. One remarks also easily that all the $\mathbb{E}_t^\alpha$ commute with each other.

i)

Let us suppose that T preserves all the spaces $\operatorname{Im} \mathbb{E}_t^\alpha$, that is $T\,\mathbb{E}_t^\alpha = \mathbb{E}_t^\alpha\, T\, \mathbb{E}_t^\alpha$ for all $\alpha \in \mathcal{N}$, $t \in \mathbb{R}^+$. Then

$$\begin{aligned} T\,\mathbb{E}_{(t_\alpha)_\alpha} &= T \prod_\alpha \mathbb{E}_{t_\alpha}^\alpha = T\,\mathbb{E}_{t_{\alpha_1}}^{\alpha_1} \prod_{\alpha\neq\alpha_1} \mathbb{E}_{t_\alpha}^\alpha = \mathbb{E}_{t_{\alpha_1}}^{\alpha_1}\, T\,\mathbb{E}_{t_{\alpha_1}}^{\alpha_1} \prod_{\alpha\neq\alpha_1} \mathbb{E}_{t_\alpha}^\alpha = \\ &= \mathbb{E}_{t_{\alpha_1}}^{\alpha_1}\, T\Big(\prod_{\alpha\neq\alpha_1} \mathbb{E}_{t_\alpha}^\alpha\Big)\mathbb{E}_{t_{\alpha_1}}^{\alpha_1} = \cdots = \prod_\alpha \mathbb{E}_{t_\alpha}^\alpha\, T \prod_\alpha \mathbb{E}_{t_\alpha}^\alpha = \\ &= \mathbb{E}_{(t_\alpha)_\alpha}\, T\,\mathbb{E}_{(t_\alpha)_\alpha}. \end{aligned}$$

ii) is obtained with the same kind of calculation and iii) is then trivial. ∎

So, in order to achieve our goal, we will establish an analogue of theorem III.2 for any family $(\mathbb{E}_t^\alpha)_{t\geq 0}$, $\alpha \in \mathcal{N}$. We first need some simple lemmas.

Lemma III.6 – *Let T be a linear mapping from $\mathcal{E}$ to Φ. Let us suppose that T admits an integral representation, $T = \sum_{\rho,\sigma} \int_0^\infty H_\sigma^\rho(s)\, da_\sigma^\rho(s)$. For every $t \in \mathbb{R}^+$, let $T_t \triangleq \sum_{\rho,\sigma} \int_0^t H_\sigma^\rho(s)\, da_\sigma^\rho(s)$. Let α_0 be fixed in $\mathcal{N}$.*

i) *We have $T\,\mathbb{E}_t^{\alpha_0} = \mathbb{E}_t^{\alpha_0}\, T\, \mathbb{E}_t^{\alpha_0}$, for every $t \in \mathbb{R}^+$, if and only if $T_s\, \mathbb{E}_t^{\alpha_0} = \mathbb{E}_t^{\alpha_0}\, T_s\, \mathbb{E}_t^{\alpha_0}$, for every s and $t \in \mathbb{R}^+$.*

ii) *We have $\mathbb{E}_t^{\alpha_0}\, T = \mathbb{E}_t^{\alpha_0}\, T\, \mathbb{E}_t^{\alpha_0}$, for every $t \in \mathbb{R}^+$, if and only if $\mathbb{E}_t^{\alpha_0}\, T_s = \mathbb{E}_t^{\alpha_0}\, T_s\, \mathbb{E}_t^{\alpha_0}$, for every s and $t \in \mathbb{R}^+$.*

iii) *The operator T commutes with all the $\mathbb{E}_t^{\alpha_0}$, $t \in \mathbb{R}^+$ if and only if each T_s $s \in \mathbb{R}^+$, commutes with all the $\mathbb{E}_t^{\alpha_0}$.*

Proof

We just prove i), as ii) is obtained by the same kind of calculation and iii) is trivial from i) and ii).

Let us suppose that, on $\mathcal{E}$, $T\,\mathbb{E}_t^{\alpha_0} = \mathbb{E}_t^{\alpha_0}\, T\, \mathbb{E}_t^{\alpha_0}$ for every $t \in \mathbb{R}^+$. Notice that, by (II.2) we always have, for every $f, g \in L^2(\mathbb{R}^+; \mathcal{G})$, $s \in \mathbb{R}^+$,

$$\begin{aligned}\langle \varepsilon(g),\, T_s\, \varepsilon(f) \rangle &= \langle \varepsilon(g_{s]}),\, T\varepsilon(f_{s]}) \rangle \langle \varepsilon(g_{[s}),\, \varepsilon(f_{[s}) \rangle \\ &= \langle \mathbb{E}_s\, \varepsilon(g),\, T\, \mathbb{E}_s\, \varepsilon(f) \rangle \langle \mathbb{E}_{[s}\, \varepsilon(g),\, \mathbb{E}_{[s}\, \varepsilon(f) \rangle .\end{aligned}$$

As $\mathbb{E}_t^{\alpha_0}\varepsilon(f)$ can be written $\varepsilon(h)$, where $h^\alpha = f^\alpha$ if $\alpha \neq \alpha_0$ and $h^{\alpha_0} = f_{t]}^{\alpha_0}$, we have

$$\begin{aligned}\langle \varepsilon(g), T_s \mathbb{E}_t^{\alpha_0}\varepsilon(f) \rangle &= \langle \mathbb{E}_s\, \varepsilon(g),\, T\, \mathbb{E}_s\, \mathbb{E}_t^{\alpha_0}\varepsilon(f) \rangle \langle \mathbb{E}_{[s}\, \varepsilon(g),\, \mathbb{E}_{[s}\, \mathbb{E}_t^{\alpha_0}\varepsilon(f) \rangle \\ &= \langle \mathbb{E}_s\, \varepsilon(g),\, T\, \mathbb{E}_t^{\alpha_0} \mathbb{E}_s\, \varepsilon(f) \rangle \langle \mathbb{E}_{[s}\, \varepsilon(g),\, \mathbb{E}_t^{\alpha_0} \mathbb{E}_{[s}\, \varepsilon(f) \rangle \\ &= \langle \mathbb{E}_s\, \varepsilon(g),\, \mathbb{E}_t^{\alpha_0}\, T\, \mathbb{E}_t^{\alpha_0} \mathbb{E}_s\, \varepsilon(f) \rangle \langle \mathbb{E}_t^{\alpha_0} \mathbb{E}_{[s}\, \varepsilon(g),\, \mathbb{E}_t^{\alpha_0} \mathbb{E}_{[s}\, \varepsilon(f) \rangle \\ &= \langle \mathbb{E}_t^{\alpha_0} \mathbb{E}_s\, \varepsilon(g),\, T\, \mathbb{E}_t^{\alpha_0} \mathbb{E}_s\, \varepsilon(f) \rangle \langle \mathbb{E}_{[s}\, \mathbb{E}_t^{\alpha_0}\varepsilon(g),\, \mathbb{E}_{[s}\, \mathbb{E}_t^{\alpha_0}\varepsilon(f) \rangle \\ &= \langle \mathbb{E}_s\, \mathbb{E}_t^{\alpha_0}\varepsilon(g),\, T\, \mathbb{E}_s \mathbb{E}_t^{\alpha_0}\varepsilon(f) \rangle \langle \mathbb{E}_{[s}\, \mathbb{E}_t^{\alpha_0}\varepsilon(g),\, \mathbb{E}_{[s}\, \mathbb{E}_t^{\alpha_0}\varepsilon(f) \rangle \\ &= \langle \mathbb{E}_t^{\alpha_0}\varepsilon(g),\, T_s\, \mathbb{E}_t^{\alpha_0}\varepsilon(f) \rangle \\ &= \langle \varepsilon(g),\, \mathbb{E}_t^{\alpha_0}\, T_s\, \mathbb{E}_t^{\alpha_0}\varepsilon(f) \rangle .\end{aligned}$$

In order to prove the converse, let us suppose that, for every $s \in \mathbb{R}^+$, T_s preserves all the spaces $\operatorname{Im} \mathbb{E}_t^{\alpha_0}$, $t \in \mathbb{R}^+$. Then, for every $F \in \mathcal{E}$,

$$\begin{aligned}\|(T\mathbb{E}_t^{\alpha_0} - \mathbb{E}_t^{\alpha_0} T \mathbb{E}_t^{\alpha_0})F\| &\leq \|(T\mathbb{E}_t^{\alpha_0} - T_s\mathbb{E}_t^{\alpha_0})F\| + \|(T_s\mathbb{E}_t^{\alpha_0} - \mathbb{E}_t^{\alpha_0}\, T\mathbb{E}_t^{\alpha_0})F\| \\ &\leq \|(T - T_s)\mathbb{E}_t^{\alpha_0} F\| + \|\mathbb{E}_t^{\alpha_0}(T_s - T)\mathbb{E}_t^{\alpha_0} F\|.\end{aligned}$$

As T_s converges strongly to T, as s tends to $+\infty$, we conclude. ∎

Let us recall an easy lemma which is included in the proof of Attal [2], theorem III.3.1.

Lemma III.7 – *Let us consider every $f \in L^2(\mathbb{R}^+;\mathcal{G})$ as having their components indexed by $\overline{\mathcal{N}}$, instead of $\mathcal{N}$, by adding the component $f^0 \triangleq \mathbf{1}$. For every $\sigma \in \overline{\mathcal{N}}$, there exists a countable dense subset, $\mathcal{H}^\sigma$, of $L^2(\mathbb{R}^+;\mathcal{G})$ such that*

i) *for every element f of $\mathcal{H}^\sigma$, the component f^σ does not vanish on an interval $I_f \subset \mathbb{R}^+$ and the other components (except f^0) vanish on the same interval,*

ii) *for every $t \in \mathbb{R}^+_*$, the space $\mathcal{H}^\sigma_t \triangleq \{f \in \mathcal{H}^\sigma;\ I_f \ni t\}$ is also dense.* ∎

We now can prove the analogue of theorem III.2 for any family $(\mathbb{E}^\alpha_t)_{t\geq 0}$, $\alpha \in \mathcal{N}$.

Proposition III.8 – *Let T be an operator on Φ, defined on $\mathcal{E}$. Let us suppose that T admits an integral representation,*

$$T = \sum_{\rho,\sigma} \int_0^\infty H^\rho_\sigma(s)\, da^\rho_\sigma(s),$$

with the operators $H^\rho_\sigma(s)$ closable, for almost all s. Let α_0 be fixed in $\mathcal{N}$

i) *All the spaces $\operatorname{Im}\mathbb{E}^{\alpha_0}_t$, $t \in \mathbb{R}^+$, are stable under T if and only if all the processes $(H^{\alpha_0}_\sigma)$, $\sigma \neq \alpha_0$, in the representation of T, vanish and almost all the operators $H^\rho_\sigma(s)$, $\rho \neq \alpha_0$, $\sigma \neq \alpha_0$, $s \in \mathbb{R}^+$, also preserve all the spaces $\operatorname{Im}\mathbb{E}^{\alpha_0}_t$, $t \in \mathbb{R}^+$.*

ii) *All the spaces $\operatorname{Ker}\mathbb{E}^{\alpha_0}_t$, $t \in \mathbb{R}^+$, are stable under T if and only if all the processes $(H^\rho_{\alpha_0})$, $\rho \neq \alpha_0$, in the representation of T, vanish and almost all the operators $H^\rho_\sigma(s)$, $\rho \neq \alpha_0$, $\sigma \neq \alpha_0$, $s \in \mathbb{R}^+$, also preserve all the spaces $\operatorname{Ker}\mathbb{E}^{\alpha_0}_t$, $t \in \mathbb{R}^+$.*

iii) *The operator T commutes with all the $\mathbb{E}^{\alpha_0}_t$, $t \in \mathbb{R}^+$, if and only if all the processes $(H^{\alpha_0}_\sigma)$, $\sigma \neq \alpha_0$, $(H^\rho_{\alpha_0})$, $\rho \neq \alpha_0$, in the representation of T, vanish and almost all the operators $H^\rho_\sigma(s)$, $\rho \neq \alpha_0$, $\sigma \neq \alpha_0$, $s \in \mathbb{R}^+$, also commute with all the $\mathbb{E}^{\alpha_0}_t$, $t \in \mathbb{R}^+$.*

Proof

i)

Since $\mathbb{E}^{\alpha_0}_t$ is a projection, the space $\operatorname{Im}\mathbb{E}^{\alpha_0}_t$ is stable under T if and only if $T\,\mathbb{E}^{\alpha_0}_t = \mathbb{E}^{\alpha_0}_t\, T\, \mathbb{E}^{\alpha_0}_t$.

If $F = \mathbb{E}[F] + \sum_\alpha \int_0^\infty G^\alpha_s\, dW^\alpha_s \in \Phi$, then

$$\mathbb{E}^{\alpha_0}_t F = \mathbb{E}[F] + \sum_{\alpha\neq\alpha_0} \int_0^\infty \mathbb{E}^{\alpha_0}_t G^\alpha_s\, dW^\alpha_s + \int_0^t G^{\alpha_0}_s\, dW^{\alpha_0}_s$$

It is obvious that if $F \in \Phi_{t]}$, $\mathbb{E}_t^{\alpha_0} F = F$. Using these two remarks, we have

$$T\,\mathbb{E}_t^{\alpha_0}\varepsilon(f) = \sum_{\alpha\neq\alpha_0}\int_0^\infty f^\alpha(s)T_s\,\mathbb{E}_t^{\alpha_0}\varepsilon(f_{s]})\,dW_s^\alpha + \int_0^t f^{\alpha_0}(s)T_s\,\varepsilon(f_{s]})\,dW_s^{\alpha_0} +$$
$$+\sum_{\substack{\rho,\sigma\\ \sigma\neq\alpha_0}}\int_0^\infty f^\sigma(s)H_\sigma^\rho(s)\mathbb{E}_t^{\alpha_0}\varepsilon(f_{s]})\,dW_s^\rho + \sum_\rho\int_0^t f^{\alpha_0}(s)H_{\alpha_0}^\rho(s)\varepsilon(f_{s]})\,dW_s^\rho$$

and

$$\mathbb{E}_t^{\alpha_0}T\mathbb{E}_t^{\alpha_0}\varepsilon(f) = \sum_{\alpha\neq\alpha_0}\int_0^\infty f^\alpha(s)\mathbb{E}_t^{\alpha_0}T_s\mathbb{E}_t^{\alpha_0}\varepsilon(f_{s]})\,dW_s^\alpha + \int_0^t f^{\alpha_0}(s)T_s\varepsilon(f_{s]})\,dW_s^{\alpha_0} +$$
$$+\sum_{\substack{\rho\neq\alpha_0\\ \sigma\neq\alpha_0}}\int_0^\infty f^\sigma(s)\mathbb{E}_t^{\alpha_0}H_\sigma^\rho(s)\mathbb{E}_t^{\alpha_0}\varepsilon(f_{s]})\,dW_s^\rho + \sum_{\sigma\neq\alpha_0}\int_0^t f^\sigma(s)H_\sigma^{\alpha_0}(s)\varepsilon(f_{s]})\,dW_s^{\alpha_0} +$$
$$+\sum_\rho\int_0^t f^{\alpha_0}(s)H_{\alpha_0}^\rho(s)\varepsilon(f_{s]})\,dW_s^\rho.$$

Consequently

$$\text{(III.9)}\quad (T\,\mathbb{E}_t^{\alpha_0} - \mathbb{E}_t^{\alpha_0}\,T\,\mathbb{E}_t^{\alpha_0})\varepsilon(f) =$$
$$= \sum_{\alpha\neq\alpha_0}\int_0^\infty f^\alpha(s)(T_s\,\mathbb{E}_t^{\alpha_0} - \mathbb{E}_t^{\alpha_0}\,T_s\,\mathbb{E}_t^{\alpha_0})\varepsilon(f_{s]})\,dW_s^\alpha +$$
$$+\sum_{\substack{\rho\neq\alpha_0\\ \sigma\neq\alpha_0}}\int_0^\infty f^\sigma(s)(H_\sigma^\rho(s)\,\mathbb{E}_t^{\alpha_0} - \mathbb{E}_t^{\alpha_0}\,H_\sigma^\rho(s)\,\mathbb{E}_t^{\alpha_0})\varepsilon(f_{s]})\,dW_s^\rho +$$
$$+\sum_{\sigma\neq\alpha_0}\int_t^\infty f^\sigma(s)H_\sigma^{\alpha_0}\,\mathbb{E}_t^{\alpha_0}\,\varepsilon(f_{s]})\,dW_s^{\alpha_0}.$$

If T preserves all the $\operatorname{Im}\mathbb{E}_t^{\alpha_0}$, $t\in\mathbb{R}^+$, this must vanish for every $t\in\mathbb{R}^+$ and every $f\in L^2(\mathbb{R}^+;\mathcal{G})$. By lemma III.6 i), the first term in the right member vanishes. So, for every $t\in\mathbb{R}^+$, $f\in L^2(\mathbb{R}^+;\mathcal{G})$,

$$\text{(III.10)}\quad \sum_{\rho\neq\alpha_0}\int_0^\infty\Big(\sum_{\sigma\neq\alpha_0}f^\sigma(s)(H_\sigma^\rho(s)\,\mathbb{E}_t^{\alpha_0} - \mathbb{E}_t^{\alpha_0}\,H_\sigma^\rho(s)\,\mathbb{E}_t^{\alpha_0})\varepsilon(f_{s]})\Big)\,dW_s^\rho +$$
$$+\int_t^\infty\Big(\sum_{\sigma\neq\alpha_0}f^\sigma(s)H_\sigma^{\alpha_0}(s)\,\mathbb{E}_t^{\alpha_0}\varepsilon(f_{s]})\Big)\,dW_s^{\alpha_0} = 0.$$

So, for any $\rho\neq\alpha_0$, $f\in L^2(\mathbb{R}^+;\mathcal{G})$, $t\in\mathbb{R}^+$, there exists a null-measured subset of $\mathbb{R}^+$, say $P(\rho,f,t)$, such that

$$\forall s\in P(\rho,f,t)^c,\quad \sum_{\sigma\neq\alpha_0}f^\sigma(s)(H_\sigma^\rho(s)\,\mathbb{E}_t^{\alpha_0} - \mathbb{E}_t^{\alpha_0}\,H_\sigma^\rho(s)\,\mathbb{E}_t^{\alpha_0})\varepsilon(f_{s]}) = 0.$$

Since $H^\rho_\sigma(s)$ is closable,

$$K(\rho,\sigma,s;t) \triangleq H^\rho_\sigma(s)\,\mathbb{E}^{\alpha_0}_t - \mathbb{E}^{\alpha_0}_t\, H^\rho_\sigma(s)\,\mathbb{E}^{\alpha_0}_t = (I-\mathbb{E}^{\alpha_0}_t)H^\rho_\sigma(s)\,\mathbb{E}^{\alpha_0}_t$$

is closable too.

For any $\sigma \neq \alpha_0$, let $P(\rho,t)^c \triangleq \bigcap_{f\in\mathcal{H}^\sigma} P(\rho,f,t)^c$, $\rho\neq\alpha_0$, $t\in\mathbb{R}^+$. Then

$$\forall\rho\neq\alpha_0,\ \forall t\in\mathbb{R}^+,\ \forall s\in P(\rho,t)^c,\ \forall f\in\mathcal{H}^\sigma,\quad \sum_{\sigma\neq\alpha_0} f^\sigma(s)K(\rho,\sigma,s;t)\varepsilon(f_{s]})=0.$$

So

$$\forall\rho\neq\alpha_0,\ \forall t\in\mathbb{R}^+,\ \forall s\in P(\rho,t)^c,\ \forall f\in\mathcal{H}^\sigma_s,\quad K(\rho,\sigma,s;t)\varepsilon(f_{s]})=0.$$

Using totality of the set $\{\varepsilon(f);\ f\in\mathcal{H}^\sigma_s\}$ in Φ (lemma III.7), using closability of the operators $H^\rho_\sigma(s)$, we easily conclude that for every $\rho\neq\alpha_0$, $\sigma\neq\alpha_0$, $t\in\mathbb{R}^+$, almost all $s\in\mathbb{R}^+$, $H^\rho_\sigma(s)$ respects the space $\operatorname{Im}\mathbb{E}^{\alpha_0}_t$.

From (III.10), we also deduce that, for any $f\in L^2(\mathbb{R}^+;\mathcal{G})$, $t\in\mathbb{R}^+$, we have, for almost all $s\geq t$,

$$\sum_{\sigma\neq\alpha_0} f^\sigma(s)H^{\alpha_0}_\sigma(s)\,\mathbb{E}^{\alpha_0}_t\varepsilon(f_{s]})=0.$$

In the same way as previously, using the dense subsets $H^{\alpha_0}_\sigma$, $\sigma\neq\alpha_0$, using that almost all $H^{\alpha_0}_\sigma(s)\,\mathbb{E}^{\alpha_0}_t$ are closable, we deduce that, for every $\sigma\neq\alpha_0$, every $t\in\mathbb{R}^+$, there exists a null-set, say $P(\sigma,t)\subset[t,+\infty[$, such that

$$\forall s\in P(\sigma,t)^c,\ s\geq t,\ \forall f\in L^2(\mathbb{R}^+;\mathcal{G}),\quad H^{\alpha_0}_\sigma(s)\,\mathbb{E}^{\alpha_0}_t\varepsilon(f_{s]})=0.$$

Let $P(\sigma)^c \triangleq \bigcap_{t\in\mathbb{Q}^+} P(\sigma,t)^c$, $\sigma\neq\alpha_0$. For any $s\in P(\sigma)^c$, let $(t_n)_{n\in\mathbb{N}}$ be a sequence in $\mathbb{Q}^+$ which converges to s and such that $t_n\leq s$, $n\in\mathbb{N}$. We then have, for every $n\in\mathbb{N}$, $H^{\alpha_0}_\sigma(s)\,\mathbb{E}^{\alpha_0}_{t_n}\varepsilon(f_{s]})=0$ and $\mathbb{E}^{\alpha_0}_{t_n}\varepsilon(f_{s]})$ converges to $\mathbb{E}^{\alpha_0}_s\varepsilon(f_{s]})=\varepsilon(f_{s]})$. So, by closability, $H^{\alpha_0}_\sigma(s)\varepsilon(f_{s]})=0$.

That is, $H^{\alpha_0}_\sigma$ vanishes for every $\sigma\neq\alpha_0$.

So we have proved half of i).

In order to prove the converse, let us suppose that the $H^{\alpha_0}_\sigma$, $\sigma\neq\alpha_0$ vanish and almost all the $H^\rho_\sigma(s)$, $\rho\neq\alpha_0$, $\sigma\neq\alpha_0$, preserves all the $\operatorname{Im}\mathbb{E}^{\alpha_0}_t$. Then equation (III.9) becomes

$$(T\,\mathbb{E}^{\alpha_0}_t - \mathbb{E}^{\alpha_0}_t\,T\,\mathbb{E}^{\alpha_0}_t)\varepsilon(f) = \sum_{\alpha\neq\alpha_0}\int_0^\infty f^\alpha(s)(T_s\,\mathbb{E}^{\alpha_0}_t - \mathbb{E}^{\alpha_0}_t\,T_s\,\mathbb{E}^{\alpha_0}_t)\varepsilon(f_{s]})\,dW^\alpha_s.$$

For any fixed f in $L^2(\mathbb{R}^+;\mathcal{G})$, let $(X_t)_{t\geq0}$ be the process in Φ defined by

$$X_u \triangleq (T_u\,\mathbb{E}^{\alpha_0}_t - \mathbb{E}^{\alpha_0}_t\,T_u\,\mathbb{E}^{\alpha_0}_t)\varepsilon(f_{u]}),\quad u\in\mathbb{R}^+.$$

We then have

$$\begin{cases} X_u = \sum_{\alpha \neq \alpha_0} \int_0^u f^\alpha(s) X_s \, dW_s^\alpha \\ X_0 = 0. \end{cases}$$

That is, $X_u = 0$ for every u. This achieves the proof of i).

ii)
For every $f \in L^2(\mathbb{R}^+;\mathcal{G})$, $t \in \mathbb{R}^+$, we have

$$\begin{aligned} \mathbb{E}_t^{\alpha_0} T\varepsilon(f) = \sum_{\alpha\neq\alpha_0} \int_0^\infty f^\alpha(s) \mathbb{E}_t^{\alpha_0} T_s \varepsilon(f_{s]}) \, dW_s^\alpha + \int_0^t f^{\alpha_0}(s) T_s \varepsilon(f_{s]}) \, dW_s^{\alpha_0} + \\ + \sum_{\substack{\rho,\sigma \\ \rho\neq\alpha_0}} \int_0^\infty f^\sigma(s) \mathbb{E}_t^{\alpha_0} H_\sigma^\rho(s) \varepsilon(f_{s]}) \, dW_s^\rho + \sum_\sigma \int_0^t f^\sigma(s) H_\sigma^{\alpha_0}(s) \varepsilon(f_{s]}) \, dW_s^{\alpha_0}. \end{aligned}$$

So

$$\begin{aligned} (\mathbb{E}_t^{\alpha_0} T - \mathbb{E}_t^{\alpha_0} T \, \mathbb{E}_t^{\alpha_0})\varepsilon(f) = \sum_{\alpha\neq\alpha_0} \int_0^\infty f^\alpha(s) (\mathbb{E}_t^{\alpha_0} T_s - \mathbb{E}_t^{\alpha_0} T_s \, \mathbb{E}_t^{\alpha_0}) \varepsilon(f_{s]}) \, dW_s^\alpha + \\ + \sum_{\substack{\rho\neq\alpha_0 \\ \sigma\neq\alpha_0}} \int_0^\infty f^\sigma(s) (\mathbb{E}_t^{\alpha_0} H_\sigma^\rho(s) - \mathbb{E}_t^{\alpha_0} H_\sigma^\rho(s) \, \mathbb{E}_t^{\alpha_0}) \varepsilon(f_{s]}) \, dW_s^\rho + \\ + \sum_{\rho\neq\alpha_0} \int_t^\infty f^{\alpha_0}(s) \mathbb{E}_t^{\alpha_0} H_{\alpha_0}^\rho(s) \varepsilon(f_{s]}) \, dW_s^{\alpha_0}. \end{aligned}$$

If T preserves all the spaces $\operatorname{Ker} \mathbb{E}_t^{\alpha_0}$, $t \in \mathbb{R}^+$, by lemma III.6 ii), the latter equality becomes

$$\begin{aligned} \sum_{\rho\neq\alpha_0} \int_0^\infty \sum_{\sigma\neq\alpha_0} f^\sigma(s) (\mathbb{E}_t^{\alpha_0} H_\sigma^\rho(s) - \mathbb{E}_t^{\alpha_0} H_\sigma^\rho(s) \, \mathbb{E}_t^{\alpha_0}) \varepsilon(f_{s]}) \, dW_s^\rho + \\ + \sum_{\rho\neq\alpha_0} \int_t^\infty f^{\alpha_0}(s) \mathbb{E}_t^{\alpha_0} H_{\alpha_0}^\rho(s) \varepsilon(f_{s]}) \, dW_s^\rho = 0. \end{aligned}$$

$$\text{Let } K_\sigma^\rho(s) \triangleq \begin{cases} \mathbb{E}_t^{\alpha_0} H_\sigma^\rho(s) - \mathbb{E}_t^{\alpha_0} H_\sigma^\rho(s) \, \mathbb{E}_t^{\alpha_0} & \text{if } \rho\neq\alpha_0 \text{ and } \sigma\neq\alpha_0, \\ \mathbb{E}_t^{\alpha_0} H_{\alpha_0}^\rho(s) & \text{if } \rho\neq\alpha_0,\ \sigma=\alpha_0 \text{ and } s\geq t, \\ 0 & \text{if } \rho\neq\alpha_0,\ \sigma=\alpha_0 \text{ and } s<t. \end{cases}$$

Then almost all the $K_\sigma^\rho(s)$ are closable and, in the same way as previously, using the dense subsets of lemma III.7, we obtain that every K_σ^ρ, $\rho\neq\alpha_0$, $\sigma\neq\alpha_0$, vanishes.

So, if $\rho\neq\alpha_0$ and $\sigma\neq\alpha_0$, almost all the $H_\sigma^\rho(s)$ preserve all the spaces $\operatorname{Ker} \mathbb{E}_t^{\alpha_0}$.

On the other hand, if $\rho \neq \alpha_0$, for any $f \in L^2(\mathbb{R}^+;\mathcal{G})$, $t \in \mathbb{R}^+$, there exists a null-set, $P(\rho,f,t) \subset [t,+\infty[$, such that

$$\forall s \in P(\rho,f,t)^c,\ s \geq t,\ \ \mathbb{E}_t^{\alpha_0}\, H_{\alpha_0}^{\rho}(s)\varepsilon(f_{s]}) = 0.$$

For any $s \in P(\rho,f)^c \triangleq \bigcap_{t\in\mathbb{Q}^+} P(\rho,f,t)^c$, let $(t_n)_{n\in\mathbb{N}}$ be a sequence in $\mathbb{Q}^+$ converging to s, such that $t_n \leq s$. Then $\mathbb{E}_{t_n}^{\alpha_0}\, H_{\alpha_0}^{\rho}(s)\varepsilon(f_{s]}) = 0$ and converges to $\mathbb{E}_s^{\alpha_0}\, H_{\alpha_0}^{\rho}(s)\varepsilon(f_{s]}) = H_{\alpha_0}^{\rho}(s)\varepsilon(f_{s]})$. So, following the same scheme as in i), we easily deduce that the $H_{\alpha_0}^{\rho}$, $\rho \neq \alpha_0$ vanish.

The proof of the converse is the same as in i).

The equivalence iii) is obvious from i) and ii). ■

Using proposition III.8 and lemma III.5, it is now easy to give characterizations similar to theorem III.2, for the family $(\mathbb{E}_{(t_\alpha)_\alpha})_{(t_\alpha)_\alpha\in\mathfrak{R}}$. For any $\alpha \in \mathcal{N}$, $\mathfrak{R}_\alpha$ denotes the subset of elements $(t_\beta)_\beta$ of $\mathfrak{R}$ such that $t_\alpha = +\infty$.

Theorem III.11 – *Let T be an operator on Φ, defined on $\mathcal{E}$. Let us suppose that T admits an integral representation, $T = \sum_{\rho,\sigma} \int_0^\infty H_\sigma^\rho(s)\, da_\sigma^\rho(s)$, with the operators $H_\sigma^\rho(s)$ closable, for almost all s.*

i) *All the spaces* $\operatorname{Im}\mathbb{E}_{(t_\alpha)_\alpha}$, $(t_\alpha)_\alpha \in \mathfrak{R}$, *are stable under T if and only if*

$$T = \sum_\alpha \int_0^\infty H_\alpha^0(s)\, da_\alpha^0(s) + \sum_\alpha \int_0^\infty H_\alpha^\alpha(s)\, da_\alpha^\alpha(s),$$

and, for each $\alpha \in \mathcal{N}$, almost all the $H_\alpha^0(s)$ and $H_\alpha^\alpha(s)$ respect all the spaces $\operatorname{Im}\mathbb{E}_{(t_\beta)_\beta}$, $(t_\beta)_\beta \in \mathfrak{R}_\alpha$.

ii) *All the spaces* $\operatorname{Ker}\mathbb{E}_{(t_\alpha)_\alpha}$, $(t_\alpha)_\alpha \in \mathfrak{R}$, *are stable under T if and only if*

$$T = \sum_\alpha \int_0^\infty H_0^\alpha(s)\, da_\alpha^0(s) + \sum_\alpha \int_0^\infty H_\alpha^\alpha(s)\, da_\alpha^\alpha(s),$$

and, for each $\alpha \in \mathcal{N}$, almost all the $H_0^\alpha(s)$ and $H_\alpha^\alpha(s)$ respect all the spaces $\operatorname{Ker}\mathbb{E}_{(t_\beta)_\beta}$, $(t_\beta)_\beta \in \mathfrak{R}_\alpha$.

iii) *The operator T commutes with all the* $\mathbb{E}_{(t_\alpha)_\alpha}$, $(t_\alpha)_\alpha \in \mathfrak{R}$ *if and only if*

$$T = \sum_\alpha \int_0^\infty H_\alpha^\alpha(s)\, da_\alpha^\alpha(s),$$

and, for each $\alpha \in \mathcal{N}$, almost all the $H_\alpha^\alpha(s)$ commute with all the $\mathbb{E}_{(t_\beta)_\beta}$, $(t_\beta)_\beta \in \mathfrak{R}_\alpha$. ■

From this theorem and theorem III.1, the following result is obvious.

Theorem III.12 – *Let T be a bounded operator on Φ. The following two assertions are equivalent.*

i) $\mathbb{E}_{(t_\alpha)_\alpha} T = T\,\mathbb{E}_{(t_\alpha)_\alpha}$, *for all* $(t_\alpha)_\alpha \in \mathfrak{R}$.

ii) $T1$ *belongs to* $\mathcal{G}$ *and there exists a family of bounded processes* $(H^\alpha_\alpha)_\alpha$ *such that*

$$T = T1I + \sum_\alpha \int_0^\infty H^\alpha_\alpha(s)\,da^\alpha_\alpha(s),$$

on all Φ and, for each $\alpha \in \mathcal{N}$, almost all the $H^\alpha_\alpha(s)$ commutes with all the $\mathbb{E}_{(t_\beta)_\beta}$, $(t_\beta)_\beta \in \mathfrak{R}_\alpha$. ∎

IV Characterisations of kernel representations

In this section we suppose that the dimension of $\mathcal{G}$ is finite. We follow the same scheme as in section **III** and give the corresponding characterisations for operators admitting a Maassen kernel.

First of all, we have to recall a result, used in the proof of uniqueness of Maassen kernel representations (Attal [2], proposition IV.5).

Lemma IV.1 – *Let Ψ be an element of $L^2(\mathfrak{P}(\mathcal{N}))$ such that, for every test-functions F and G,*

$$\int_{\mathfrak{P}(\mathcal{M})} \Psi(A)\widehat{F}(\Sigma.A)\widehat{G}(\Sigma^{\cdot}A)\prod_{\rho,\sigma} dA^\rho_\sigma = 0.$$

Then Ψ vanishes. ∎

Now, for operators admitting a Maassen kernel representation, we have the following analogue of theorem III.2.

Theorem IV.2 – *Let T be an operator on Φ admitting a Maassen kernel $\widehat{T}$.*

i) *All the spaces* $\operatorname{Im}\mathbb{E}_t$ *are stable under T if and only if $\widehat{T}(A) = 0$ for every A in $\mathfrak{P}(\mathcal{M})$ such that $\vee A$ belongs to A^α_0 for some α.*

ii) *All the spaces* $\operatorname{Ker}\mathbb{E}_t$ *are stable under T if and only if $\widehat{T}(A) = 0$ for every A in $\mathfrak{P}(\mathcal{M})$ such that $\vee A$ belongs to A^0_α for some α.*

iii) *The operator T commutes with all the $\mathbb{E}_t$ if and only if $\widehat{T}(A) = 0$ for every A in $\mathfrak{P}(\mathcal{M})$ such that $\vee A$ belongs to A^α_0 or A^0_α for some α.*

Proof

i)

As in theorem III.2, the space Im $\mathbb{E}_t$ is stable under T if and only if $\mathbb{E}_t\, T\, \mathbb{E}_t = T\, \mathbb{E}_t$. But one has, for every F in Φ, t in $\mathbb{R}^+$, $\widehat{\mathbb{E}_t F}\,(U) = \widehat{F}(U)\mathbb{1}_{U\subset[0,t]}$. So, by (II.4), for every test-functions F and G and every t in $\mathbb{R}^+$, an easy calculation gives

$$\langle G\,,(T\,\mathbb{E}_t - \mathbb{E}_t\, T\,\mathbb{E}_t)F\,\rangle =$$
$$= \int_{\mathfrak{P}(\mathcal{N})^2} \sum_{A\in G(U)} \widehat{T}(A^\circ N)\widehat{F}(N+\Sigma.A)\widehat{G}(U)\mathbb{1}_{\sum_\alpha A_0^\alpha \not\subset[0,t]}\mathbb{1}_{N+\Sigma\, A\subset[0,t]}\prod_\alpha dN_\alpha \prod_\alpha dU_\alpha.$$

$$= \int_{\mathfrak{P}(\mathcal{N})\times\mathfrak{P}(\mathcal{M})} \widehat{T}(A^\circ N)\widehat{F}(N+\Sigma.A)\widehat{G}(\Sigma^{\cdot} A + A^0)\mathbb{1}_{\sum_\alpha A_0^\alpha \not\subset[0,t]}\mathbb{1}_{N+\Sigma\, A\subset[0,t]}$$
$$\prod_\alpha dN_\alpha \prod_{\rho,\sigma} dA_\sigma^\rho.$$

Let, for every α, $B_\alpha^\alpha \triangleq A_\alpha^0 + A_\alpha^\alpha$, we then have

$$\langle G\,,(T\,\mathbb{E}_t - \mathbb{E}_t\, T\,\mathbb{E}_t)F\,\rangle =$$
$$= \int_{\mathfrak{P}(\mathcal{M})} \sum_{\forall\alpha,\, A_\alpha^0 + A_\alpha^\alpha = B_\alpha^\alpha} \widehat{T}(A^\circ N)\widehat{F}(N+\Sigma.A)\widehat{G}(\Sigma^{\cdot} A + A^0)\mathbb{1}_{\sum_\alpha A_0^\alpha \not\subset[0,t]}\mathbb{1}_{N+\Sigma\, A\subset[0,t]}$$
$$\prod_\alpha dN_\alpha \prod_{\substack{\alpha,\sigma\\ \alpha\neq\sigma}} dA_\sigma^\rho \prod_\alpha dB_\alpha^\alpha.$$

If we change the notations, that is if we replace each N_α by B_α^0, each A_σ^α, $\alpha\neq\sigma$, by B_σ^α, each A_α^α by C_α, then $N+\Sigma.A$ becomes $\Sigma.B$, $\Sigma^{\cdot} A + A^0$ becomes $\Sigma^{\cdot} B$, $A^\circ N$ becomes $B^{\cdot}_{\cdot}C$, and the latter equality becomes

$$\langle G\,,(T\,\mathbb{E}_t - \mathbb{E}_t\, T\,\mathbb{E}_t)F\,\rangle =$$
$$= \int_{\mathfrak{P}(\mathcal{M})} \sum_{\forall\alpha,\, C_\alpha\subset B_\alpha^\alpha} \widehat{T}(B^{\cdot}_{\cdot}C)\widehat{F}(\Sigma.B)\widehat{G}(\Sigma^{\cdot} B)\mathbb{1}_{\sum_\alpha B_0^\alpha \not\subset[0,t]}\mathbb{1}_{\Sigma\, B\subset[0,t]}\prod_{\rho,\sigma} dB_\sigma^\rho.$$

For every $t\in\mathbb{R}^+$, $B\in\mathfrak{P}(\mathcal{M})$, let

$$\Psi_t(B) \triangleq \sum_{\forall\alpha,\, C_\alpha\subset B_\alpha^\alpha} \widehat{T}(B^{\cdot}_{\cdot}C)\mathbb{1}_{\sum_\alpha B_0^\alpha \not\subset[0,t]}\mathbb{1}_{\Sigma\, B\subset[0,t]}.$$

We then have

$$\langle G\,,(T\,\mathbb{E}_t - \mathbb{E}_t\, T\,\mathbb{E}_t)F\,\rangle = \int_{\mathfrak{P}(\mathcal{M})} \Psi_t(B)\widehat{F}(\Sigma.B)\widehat{G}(\Sigma^{\cdot} B)\prod_{\rho,\sigma} dB_\sigma^\rho.$$

But this must be equal to 0 for every $t \in I\!\!R^+$, F and G test-functions. So, by lemma IV.1, Ψ_t vanishes for every t. That is, for every $t \in I\!\!R^+$, for every $B \in \mathfrak{P}(\mathcal{M})$ such that $\sum_\alpha B_0^\alpha \not\subset [0,t]$, $\Sigma.B \subset [0,t]$ we have

$$\Phi(B) \triangleq \sum_{\forall \alpha,\, C_\alpha \subset B_\alpha^\alpha} \widehat{T}(B\dot{}C) = 0.$$

But it is easy to verify the following Möbius inversion formula :

$$\widehat{T}(B) = \sum_{\forall \alpha,\, C_\alpha \subset B_\alpha^\alpha} \Phi(B\dot{}C) \prod_\alpha (-1)^{|B_\alpha^\alpha \setminus C_\alpha|}.$$

So, if B is such that $\sum_\alpha B_0^\alpha \not\subset [0,t]$ and $\Sigma.B \subset [0,t]$, we have $B_\alpha^\alpha \subset [0,t]$, for every α. So, if $C \in \mathfrak{P}(\mathcal{N})$ is such that, for every α, $C_\alpha \subset B_\alpha^\alpha$, we have $\sum_\alpha (B\dot{}C) \not\subset [0,t]$ and $\Sigma.(B\dot{}C) \subset [0,t]$. So $\Phi(B\dot{}C) = 0$, so $\widehat{T}(B)$ vanishes for every such B.

It is now obvious to verify, since this is valid for all t, that this $\widehat{T}(A) = 0$ for every A in $\mathfrak{P}(\mathcal{M})$ such that $\vee A$ belongs to a A_0^α.

In order to prove the converse, let us suppose that $\widehat{T}(B)$ vanishes for every B such that $\vee B \in B_0^\alpha$, for a α. If $t \in I\!\!R^+$ is such that $\sum_\alpha B_0^\alpha \not\subset [0,t]$ and $\Sigma.B \subset [0,t]$, we have $\vee B \in \sum_\alpha B_0^\alpha$, so $\widehat{T}(B) = 0$, thus, as previously, for every $C \in \mathfrak{P}(\mathcal{N})$ such that, for every α, $C_\alpha \subset B_\alpha^\alpha$, $\widehat{T}(B\dot{}C) = 0$. That is,

$$\Phi(B) = \sum_{\forall \alpha,\, C_\alpha \subset B_\alpha^\alpha} \widehat{T}(B\dot{}C) = 0.$$

That is, Ψ_t vanishes for every $t \in I\!\!R^+$ That is T preserves all the spaces $\operatorname{Im} I\!\!E_t$.

ii)

As in theorem III.2, the space $\operatorname{Ker} I\!\!E_t$ is stable under T if and only if $I\!\!E_t T = I\!\!E_t T I\!\!E_t$. Making the same calculation as in part i), we obtain that this condition is equivalent to $\widehat{T}(A) = 0$ for every A in $\mathfrak{P}(\mathcal{M})$ such that there exists a t in $I\!\!R^+$ verifying $\sum_\alpha A_\alpha^0 \not\subset [0,t]$, $\Sigma^{\cdot} A \subset [0,t]$. That is, $\widehat{T}(A) = 0$ for every A in $\mathfrak{P}(\mathcal{M})$ such that $\vee A$ belongs to a A_α^0.

iii)

As in theorem III.2, T commutes with all the $I\!\!E_t$ if and only if all the spaces $\operatorname{Im} I\!\!E_t$ and $\operatorname{Ker} I\!\!E_t$ are stable under T, so we conclude from i) and ii). ■

Now, let us give an analogue for the families $(I\!\!E_t^{\alpha_0})_{t \geq 0}$, $\alpha \in \mathcal{N}$, of projections and then deduce the result for the family $(I\!\!E_{(t_\alpha)_\alpha})_{(t_\alpha)_\alpha \in \mathfrak{R}}$.

For any $A \in \mathfrak{P}(\mathcal{M})$ and $\alpha_0 \in \mathcal{N}$, we denote by $A(\alpha_0)$ the set $\sum_{\sigma} A_\sigma^{\alpha_0} + \sum_{\rho \neq \alpha_0} A_{\alpha_0}^\rho$ (that is, the union of all the element of the matrix $(A_\sigma^\rho)_{\rho,\sigma \in \overline{\mathcal{N}}^2}$, $A_0^0 \triangleq \emptyset$, which are in the line or in the column containing $A_{\alpha_0}^{\alpha_0}$).

Proposition IV.3 – *Let T be an operator on Φ admitting a Maassen kernel $\widehat{T}$, let α_0 be fixed in $\mathcal{N}$.*

i) *All the spaces* $\operatorname{Im} \mathbb{E}_t^{\alpha_0}$ *are stable under T if and only if $\widehat{T}(A) = 0$ for every $A \in \mathfrak{P}(\mathcal{M})$ such that $\vee A(\alpha_0)$ belongs to $A_\sigma^{\alpha_0}$ for some $\sigma \neq \alpha_0$.*

ii) *All the spaces* $\operatorname{Ker} \mathbb{E}_t^{\alpha_0}$ *are stable under T if and only if $\widehat{T}(A) = 0$ for every $A \in \mathfrak{P}(\mathcal{M})$ such that $\vee A(\alpha_0)$ belongs to $A_{\alpha_0}^\rho$ for some $\rho \neq \alpha_0$.*

iii) *The operator T commutes with all $\mathbb{E}_t^{\alpha_0}$ if and only if $\widehat{T}(A) = 0$ for every $A \in \mathfrak{P}(\mathcal{M})$ such that $\vee A(\alpha_0)$ does not belong to $A_{\alpha_0}^{\alpha_0}$.*

Proof

For every $F \in \Phi$, $t \in \mathbb{R}^+$ and $U \in \mathfrak{P}(\mathcal{N})$, it is easy to verify that

$$(\mathbb{E}_t^{\alpha_0} F)^\frown(U) = F(U) \mathbb{1}_{U_{\alpha_0} \subset [0,t]}.$$

i)

According to this remark, we have, for every F, G test-functions, $t \in \mathbb{R}^+$,

$$\langle G, (T\,\mathbb{E}_t^{\alpha_0} - \mathbb{E}_t^{\alpha_0}\, T\, \mathbb{E}_t^{\alpha_0})F \rangle =$$
$$= \int_{\mathfrak{P}(\mathcal{N})^2} \sum_{A \in G(U)} \widehat{T}(A^\circ N) \widehat{F}(N + \Sigma.A) \widehat{G}(U) \mathbb{1}_{N_{\alpha_0} + \Sigma_{\alpha_0} A \subset [0,t]} \mathbb{1}_{\sum_{\sigma \neq \alpha_0} A_\sigma^{\alpha_0} \not\subset [0,t]} \prod_\alpha dN_\alpha \prod_\alpha dU_\alpha.$$

So, making the same changement of notations as in theorem IV.2, we obtain

$$\langle G, (T\,\mathbb{E}_t^{\alpha_0} - \mathbb{E}_t^{\alpha_0}\, T\, \mathbb{E}_t^{\alpha_0})F \rangle =$$
$$= \int_{\mathfrak{P}(\mathcal{M})} \sum_{\forall \alpha,\, C_\alpha \subset B_\alpha^\alpha} \widehat{T}(B^\cdot C) \widehat{F}(\Sigma.B) \widehat{G}(\Sigma^\cdot B) \mathbb{1}_{\Sigma_{\alpha_0} B \subset [0,t]} \mathbb{1}_{\sum_{\sigma \neq \alpha_0} B_\sigma^{\alpha_0} \not\subset [0,t]} \prod_{\rho,\sigma} dB_\sigma^\rho.$$

Let, for every $T \in \mathbb{R}^+$, $B \in \mathfrak{P}(\mathcal{M})$,

$$\Psi_t(B) \triangleq \sum_{\forall \alpha,\, C_\alpha \subset B_\alpha^\alpha} \widehat{T}(B^\cdot C) \mathbb{1}_{\Sigma_{\alpha_0} B \subset [0,t]} \mathbb{1}_{\sum_{\sigma \neq \alpha_0} B_\sigma^{\alpha_0} \not\subset [0,t]}.$$

We then have that T preserves all the spaces Ker $\mathbb{E}_t^{\alpha_0}$ if and only if Ψ_t vanishes for every t. Thus, for any $B \in \mathfrak{P}(\mathcal{M})$ such that $\sum_{\sigma\neq\alpha_0} B_\sigma^{\alpha_0} \not\subset [0,t]$ and $\sum_\rho B_{\alpha_0}^\rho \subset [0,t]$, we have

$$\Phi(B) \triangleq \sum_{\forall\alpha,\, C_\alpha \subset B_\alpha^\alpha} \widehat{T}(B{:}C) = 0.$$

But, if B is such as described below, in the same way as in the proof of theorem IV.2, $B{:}C$ also verifies the same conditions. Now, using the Möbius inversion formula we easily conclude that T preserves the space Im $\mathbb{E}_t^{\alpha_0}$ if and only if $\widehat{T}(B) = 0$ for every $B \in \mathfrak{P}(\mathcal{M})$ such that $\sum_{\sigma\neq\alpha_0} B_\sigma^{\alpha_0} \not\subset [0,t]$ and $\sum_\rho B_{\alpha_0}^\rho \subset [0,t]$. Since this must be valid for every t, we finaly have that T preserves all the spaces Im $\mathbb{E}_t^{\alpha_0}$ if and only if $\widehat{T}(B) = 0$ for every $B \in \mathfrak{P}(\mathcal{M})$ such that $\vee \sum_{\sigma\neq\alpha_0} B_\sigma^{\alpha_0} > \vee\sum_\rho B_{\alpha_0}^\rho$. That is i).

ii)

The same kind of calculation shows that the spaces Ker $\mathbb{E}_t^{\alpha_0}$ are stable under T if and only if $\widehat{T}(A) = 0$ for every $A \in \mathfrak{P}(\mathcal{M})$ such that $\sum_{\rho\neq\alpha_0} B_{\alpha_0}^\rho \not\subset [0,t]$ and $\sum_\sigma B_\sigma^{\alpha_0} \subset [0,t]$. Since this must be valid for every $t \in \mathbb{R}^+$, we conclude that this is equivalent to $\widehat{T}(B) = 0$ for every $B \in \mathfrak{P}(\mathcal{M})$ such that $\vee \sum_{\rho\neq\alpha_0} A_{\alpha_0}^\rho > \vee\sum_\sigma A_\sigma^{\alpha_0}$. That is ii).

iii)

From i) and ii), iii) is obvious. ■

Using the same remark as for the proof of theorem III.11, the following result is simply deduced from proposition IV.3.

Theorem IV.4 – *Let T be an operator on Φ admitting a Maassen kernel $\widehat{T}$.*

i) *All the spaces* Im $\mathbb{E}_{(t_\alpha)_\alpha}$ *are stable under T if and only if $\widehat{T}(A) = 0$ for every $A \in \mathfrak{P}(\mathcal{M})$ such that there exists a $\alpha \in \mathcal{N}$ verifying $\vee A(\alpha)$ belongs to A_σ^α for some $\sigma \neq \alpha$.*

ii) *All the spaces* Ker $\mathbb{E}_{(t_\alpha)_\alpha}$ *are stable under T if and only if $\widehat{T}(A) = 0$ for every $A \in \mathfrak{P}(\mathcal{M})$ such that there exists a $\alpha \in \mathcal{N}$ verifying $\vee A(\alpha)$ belongs to A_α^ρ for some $\rho \neq \alpha$.*

iii) *The operator T commutes with all the $\mathbb{E}_{(t_\alpha)_\alpha}$ if and only if $\widehat{T}(A) = 0$ for every $A \in \mathfrak{P}(\mathcal{M})$ such that there exists a $\alpha \in \mathcal{N}$ verifying $\vee A(\alpha)$ does not belong to A_α^α.* ■

References

[1] S. Attal : Characterizations of some operators on Fock space. Preprint. To appear in *Quantum Probability & Appl. VIII.*

[2] S. Attal : Problèmes d'unicité dans les représentations d'opérateurs sur l'espace de Fock. Preprint. To appear in *Séminaire de probabilités XXVI.*

[3] A. Dermoune : Formule de composition pour une classe d'opérateurs. *Séminaire de probabilités XXIV*, p 397-401, 1990.

[4] R.L. Hudson & K.R. Parthasarathy : Quantum Itô's formula and stochastic evolutions. *Comm. Math. Phys.* 93, p 301-323, 1984.

[5] J.M. Lindsay : Independence for quantum stochastic integrators. *Quantum Prob.& Appl. VI*, proceeding of the Trento conference on Quantum Probability, World Scientific 1991, p 325-332.

[6] H. Maassen : Quantum Markov processes on Fock space described by integral kernels. *Quantum Prob. & Appl. II.*, p 361-374, L.N.M. 1136 Springer 1985.

[7] P.A. Meyer : Diffusions quantiques III §1. *Séminaire de probabilités XXIV*, p 384-391, 1990.

[8] P.A. Meyer : Eléments de probabilités quantiques I-V. *Séminaire de probabilités XX*, p 186-312, 1986.

[9] P.A. Meyer : Eléments de probabilités quantiques VI. *Séminaire de probabilités XXI*, p 34-49, 1987.

[10] K.R. Parthasarathy & K.B. Sinha : Representation of a class of quantum martingales II. *Quantum Prob. & Appl. III.*, p 232-250, L.N.M. 1303 Springer 1988.

Quantum Probability and Related Topics Vol. VIII (pp. 71-79)

A note on processes on bialgebras, quantum flows, and convolution semigroups of instruments

A. Barchielli, G. Lupieri

Dipartimento di Fisica dell'Università di Milano, Via Celoria 16, I-20133 Milano, Italy, and Istituto Nazionale di Fisica Nucleare, Sezione di Milano

Abstract: The idea of *measurements continuous in time in quantum mechanics* [7,8] led to the introduction of the notion of convolution semigroup of instruments [9-12] and to the study of dilations of such semigroups obtained by the use of quantum stochastic calculus [13-16]. In [17] we discussed the connections between these notions and that of stochastic process with independent increments over a $*$-bialgebra [18-23]. In the present paper we want to show how our results can be formulated in terms of particular Evans-Hudson quantum flows (EH-flows) [24-33] and of perturbations of them [34].

1 An EH-flow

Convolution semigroups of instruments on groups have been constructed [15,17] by means of two ingredients: a quantum increment process over a $*$-bialgebra, namely the coefficient algebra of the group, and a unitary evolution, both obtained as solutions of certain quantum stochastic differential equations. In this note we show how the increment process corresponds to an EH-flow [24] with structure maps having additional invariance properties; the construction of the semigroup of instruments is then interpreted as a perturbation of the flow (in the sense of [34]) by means of the unitary evolution. We shall follow closely [17], and we shall not repeat all definitions and results.

We consider the coefficient algebra $\mathcal{K}(G)$ of a locally compact maximally almost periodic group G. Let us recall that $\mathcal{K}(G)$ is the involutive algebra of the functions $f(x)$, such that $f(x) = \langle D(x)\xi|\eta\rangle$, for some finite-dimensional unitary representation D of G and some vectors ξ and η in the representation space. $\mathcal{K}(G)$ can be turned in a $*$-bialgebra, with the counit defined by $\delta(f) = f(e)$, where e is the neutral element of the group, and the comultiplication defined by $\Delta(f)(x,y) = f(yx)$.

In the symmetric Fock space Γ over $L^2(\mathbb{R}) \otimes \mathcal{H}$ we introduce the increment process defined by the quantum stochastic differential equation [17,23]

$$dj_{st} = j_{st} * dM_t\,, \qquad j_{ss} = \mathbb{1}\delta\,, \tag{1.1}$$

$$M_t(f) = \Lambda_t\big(\pi(f) - \delta(f)\mathbb{1}\big) + A_t\big(\eta(f^*)\big) + A_t^\dagger\big(\eta(f)\big) + \psi(f)t\mathbb{1}\,, \tag{1.2}$$

where (π, η, ψ) is a generating triple on the $*$-bialgebra $\mathcal{K}(G)$ with respect to the Hilbert space $\mathcal{H}$ ([17], Def. 1.1) and $*$ denotes the convolution between linear maps on $\mathcal{K}(G)$ ([17], eq. (1.3)). Let us recall that π is a $*$-representation of $\mathcal{K}(G)$ on $\mathcal{H}$ (necessarily bounded) and that ψ and η are linear maps from $\mathcal{K}(G)$ into $\mathbb{C}$ and $\mathcal{H}$, respectively, such that

$$\pi(f)\eta(g) = \eta(fg) - \eta(f)\delta(g)\,, \tag{1.3}$$

$$\delta(f)\psi(g) - \psi(fg) + \psi(f)\delta(g) = -\langle\eta(f^*)|\eta(g)\rangle\,. \tag{1.4}$$

Here let us assume that $\mathcal{H}$ *is separable*. Then, the differential equation above can be rewritten by using the notations of [31]. Let us fix a c.o.n.s. $\{\xi_i,\ i = 1, 2, \ldots\}$ in $\mathcal{H}$ and define for $i, j = 1, 2, \ldots$

$$\Lambda_0^i(t) := A_t(\xi_i)\,, \quad \Lambda_i^0(t) := A_t^\dagger(\xi_i)\,, \quad \Lambda_i^j(t) := \Lambda_t(|\xi_i\rangle\langle\xi_j|)\,, \quad \Lambda_0^0(t) := t\,. \tag{1.5}$$

Then, Itô's table becomes $d\Lambda_j^i(t)\, d\Lambda_l^k(t) = \hat\delta_l^i\, d\Lambda_j^k$, where $\hat\delta_l^i = 0$ if $i = 0$ and/or $l = 0$, $\hat\delta_l^i = \delta_l^i$ otherwise.

Now let us introduce the linear maps τ_j^i, $i, j \geq 0$, from $\mathcal{K}(G)$ into $\mathbb{C}$ by

$$\begin{aligned} \tau_0^0(f) &:= \psi(f)\,, \quad \tau_j^i(f) := \langle\xi_i|\,[\pi(f) - \delta(f)\mathbb{1}]\,\xi_j\rangle\,, \quad i, j \geq 1\,, \\ \tau_i^0(f) &:= \langle\eta(f^*)|\xi_i\rangle\,, \quad \tau_0^i(f) := \langle\xi_i|\eta(f)\rangle\,, \quad i \geq 1\,. \end{aligned} \tag{1.6}$$

Then, let us define the *structure maps* $\vartheta_j^i : \mathcal{K}(G) \to \mathcal{K}(G)$ by

$$\vartheta_j^i := \left(\mathrm{Id} \otimes \tau_j^i\right) \circ \Delta\,, \qquad i, j \geq 0\,. \tag{1.7}$$

Finally, with these notations, eqs. (1.1) and (1.2) can be written as

$$dj_{st}(f) = \sum_{i,j\geq 0} j_{st}(\vartheta_j^i(f))\, d\Lambda_i^j(t)\,. \tag{1.8}$$

By using the properties of $*$-bialgebras and the definition of generating triple, one can obtain the following properties of the structure maps: the ϑ_j^i are linear maps from $\mathcal{K}(G)$ into itself such that, for $i, j \geq 0$, $f, g \in \mathcal{K}(G)$,

$$\vartheta_j^i(1) = 0\,, \qquad \vartheta_j^i(f)^* = \vartheta_i^j(f^*)\,, \tag{1.9}$$

$$\vartheta_j^i(fg) = f\vartheta_j^i(g) + \vartheta_j^i(f)g + \sum_{k\geq 1} \vartheta_k^i(f)\vartheta_j^k(g)\,. \tag{1.10}$$

The series in the r.h.s. of (1.10) can be rewritten in terms of a converging numerical series involving the τ_j^i and in this way no topology on the bialgebra is required. Equations (1.9) and (1.10) are just the characterizing properties of the structure maps of EH–flows (see [31], Proposition 28.1). However, in order to satisfy the definition of EH–flows further steps are needed.

Let us denote right translations by R_x, i.e. $R_x(f)(y) = f(yx)$, and interpret $\mathcal{K}(G)$ as a subalgebra of $L^\infty(G,\omega) \subset \mathcal{L}(L^2(G,\omega))$, where ω is a left invariant Haar measure on G. We define a linear map J_{st} from $\mathcal{K}(G)$ into $\mathcal{L}(\Gamma \otimes L^2(G,\omega))$ by

$$J_{st}(f)(x) := j_{st} \circ R_x(f) \,. \tag{1.11}$$

Let us stress that $R_x(\mathcal{K}(G)) = \mathcal{K}(G)$. Then, the differential equation (1.8) with initial condition (1.1) becomes

$$\mathrm{d}J_{st}(f) = \sum_{i,j\geq 0} J_{st}(\vartheta^i_j(f))\, \mathrm{d}\Lambda^j_i(t)\,, \qquad J_{ss}(f) = \mathbb{1} \otimes f\,; \tag{1.12}$$

here $\Lambda^j_i(t)$ is identified with $\Lambda^j_i(t) \otimes \mathbb{1}$. The proof of (1.12) is based on the identities $\Delta \circ R_x = (R_x \otimes \mathrm{Id}) \circ \Delta$ and

$$\vartheta^i_j \circ R_x = R_x \circ \vartheta^i_j\,, \qquad \forall x \in G\,. \tag{1.13}$$

Now, J_{st} is indeed an EH–flow on the algebra $\mathcal{K}(G)$ (cf. [31], eq. (28.1) and Proposition 28.1). Let us stress that if $\mathcal{H}$ is finite–dimensional there is no problem in going from (1.1), (1.2) to (1.12); in the case of an infinite–dimensional Hilbert space $\mathcal{H}$ however, one has to check some regularity conditions on the structure maps, such as the Mohari–Sinha conditions given in [31] Section 28. These conditions assure the strong convergence of the series in (1.10) and the existence of an EH–flow satisfying (1.12).

We have seen that from the generating triple (π, η, ψ) one can define the maps ϑ^i_j satisfying properties (1.9), (1.10) and (1.13). On the contrary let the ϑ^i_j be linear maps satisfying those properties and define $\tau^i_j(f) := \vartheta^i_j(f)(e)$. By (1.13) we have $\vartheta^i_j(f)(x) = [R_x \circ \vartheta^i_j(f)](e) = \vartheta^i_j(R_x(f))(e) = \tau^i_j(R_x(f)) = [(\mathrm{Id} \otimes \tau^i_j) \circ \Delta(f)](x)$, so that ϑ^i_j and τ^i_j are indeed connected by eq. (1.7). Then, by using eqs. (1.6) as definition of (π, η, ψ), one can see that properties (1.9) and (1.10) are equivalent to the defining properties of a generating triple. Summing up, the EH–flow we have constructed is a particular one, because it is defined on the commutative algebra $\mathcal{K}(G)$ and its structure maps satisfy eq. (1.13).

The similarity between eqs. (1.1), (1.2) and EH–flows has already been noticed in [23] p. 573. Let us stress that what we have done also has meaning, at least at a formal level, in the case of a generic $*$-bialgebra provided it is concretely represented as an operator algebra and eq. (1.13) is written as $\vartheta^i_j = [\mathrm{Id} \otimes (\delta \circ \vartheta^i_j)] \circ \Delta$.

2 Measurements continuous in time

The notion of perturbation of an EH–flow, introduced in [34], gives a way of generating new flows from a given one. What we want to stress here is that the construction of continuous measurements given in [13,15,17] reduces essentially to a significant class of such perturbations.

In the hypotheses of Theorem 4.4 of [17] (continuity hypotheses), a unique projection valued measure P_{st} on G with values in $\mathcal{L}(\Gamma)$ exists, such that $j_{st}(f) = \int_G f(x)\, P_{st}(\mathrm{d}x)$. Moreover, P_{st} enjoys the properties given by eqs. (4.9)-(4.11) of [17]. In this case eq. (1.11) becomes

$$J_{st}(f)(x) = \int_G f(yx)\, P_{st}(\mathrm{d}y)\,. \tag{2.1}$$

From here one can extend J_{st} to become a map defined on $L^\infty(G,\omega)$ (see [12] for analogous extensions). This does not mean that one can extend eq. (1.12) to the whole $L^\infty(G,\omega)$ in the general case. The extension of the differential equation (1.12) will be possible only on some class of smooth functions, depending on the concrete form of the structure maps.

Let us introduce now a separable Hilbert space $\mathcal{H}_S$ and some coefficients $R^i_j \in \mathcal{L}(\mathcal{H}_S)$, $i,j \geq 0$, such that

$$R^i_j + R^{j*}_i + \sum_{s\geq 1} R^{s*}_i R^s_j = 0\,, \qquad R^i_j + R^{j*}_i + \sum_{s\geq 1} R^i_s R^{j*}_s = 0\,, \tag{2.2}$$

where the series are strongly convergent. Then, we consider the unitary operator $U(t,s) \in \mathcal{L}(\mathcal{H}_S \otimes \Gamma)$ defined by the Hudson–Parthasarathy equation

$$\mathrm{d}U(t,s) = \sum_{i,j\geq 0} \big(R^i_j \otimes \mathrm{d}\Lambda^j_i(t)\big) U(t,s)\,, \qquad U(s,s) = \mathbb{1}\,. \tag{2.3}$$

The interpretation is that we have a physical system S described in the Hilbert space $\mathcal{H}_S$ and interacting with the "Bose fields" described in the Fock space Γ. The evolution operator of the composed system, in the interaction picture with respect to the "free" dynamics of the fields, is $U(t,s)$.

Let $\Omega \in \Gamma$ be the Fock vacuum and let us define $E_0 : \mathcal{L}(\mathcal{H}_S \otimes \Gamma \otimes L^2(G,\omega)) \to \mathcal{L}(\mathcal{H}_S \otimes L^2(G,\omega))$ by

$$\langle v_1 \otimes u_1 | E_0[A]\, v_2 \otimes u_2\rangle = \langle v_1 \otimes \Omega \otimes u_1 | A\, v_2 \otimes \Omega \otimes u_2\rangle\,, \tag{2.4}$$

$\forall u_1, u_2 \in L^2(G,\omega)$, $\forall v_1, v_2 \in \mathcal{H}_S$, $\forall A \in \mathcal{L}(\mathcal{H}_S \otimes \Gamma \otimes L^2(G,\omega))$, and analogously define $E_0 : \mathcal{L}(\mathcal{H}_S \otimes \Gamma) \to \mathcal{L}(\mathcal{H}_S)$. Then, the equation

$$\int_G f(y)\mathcal{I}_t(\mathrm{d}y)[X] = E_0[U(t,0)^*\, X \otimes j_{0,t}(f)\, U(t,0)]$$
$$\equiv E_0[U(t+s,s)^*\, X \otimes j_{s,t+s}(f)\, U(t+s,s)]\,, \qquad X \in \mathcal{L}(\mathcal{H}_S)\,, \tag{2.5}$$

defines a convolution semigroup of instruments (see [17], Section 4).

Let ϱ be a statistical operator on $\mathcal{H}_S$ ($\varrho \geq 0$, $\mathrm{Tr}\{\varrho\} = 1$) and write $\langle \varrho, X\rangle$ for $\mathrm{Tr}\{\varrho X\}$, $X \in \mathcal{L}(\mathcal{H}_S)$. We define a (classical) stochastic process Y_t with values in G via Kolmogorov theorem and

$$\mathbb{P}[Y_{t_1} Y_{t_0}^{-1} \in B_1, \ldots, Y_{t_n} Y_{t_{n-1}}^{-1} \in B_n | \varrho] :=$$
$$= \langle \varrho, \mathcal{I}_{t_1}(B_1) \circ \mathcal{I}_{t_2 - t_1}(B_2) \circ \ldots \circ \mathcal{I}_{t_n - t_{n-1}}(B_n)[\mathbb{1}]\rangle$$

$$= \langle \varrho, E_0 \Big[U(t_n,0)^* \Big(\mathbb{1} \otimes \big(P_{0,t_1}(B_1) \cdots P_{t_{n-1},t_n}(B_n)\big)\Big) U(t_n,0)\Big]\rangle , \tag{2.6}$$

where $0 \le t_1 \le \ldots \le t_n$ and the B_j's are Borel subsets of G. The property that $\mathcal{I}_t$ is a semigroup under convolution guarantees consistency. The process Y_t is defined up to the initial condition Y_0 and its increments describe the continuous measurement process over the system S. The last line of the previous equation can be interpreted by saying that something is measured on the fields (this measurement is described by the projection valued measures P_{st}) while they are interacting with S and the second line says that the whole procedure is interpreted as an indirect measurement on S.

Another way of characterizing a convolution semigroup of instruments is by introducing its associated semigroup of "probability operators" [11,12], defined by

$$T_t[X \otimes f](x) = \int_G f(yx)\mathcal{I}_t(\mathrm{d}y)[X] , \qquad X \in \mathcal{L}(\mathcal{H}_S) . \tag{2.7}$$

In the definition of T_t one starts with f in some class of continuous functions and then one extends it to $\mathcal{L}(\mathcal{H}_S) \otimes L^\infty(G,\omega)$ [12]. The process Y_t can be equivalently defined by giving the expectation values of functions:

$$\mathbb{E}\big[f_1(Y_{t_1}) f_2(Y_{t_2}) \cdots f_n(Y_{t_n}) \big| Y_0 = g; \varrho\big] = \tag{2.8}$$

$$= \langle \varrho\, , T_{t_1}\big[(\mathbb{1} \otimes f_1) T_{t_2 - t_1}\big[\cdots T_{t_{n-1}-t_{n-2}}\big[(\mathbb{1} \otimes f_{n-1}) T_{t_n - t_{n-1}}\big[\mathbb{1} \otimes f_n\big]\big] \cdots \big]\big](g)\rangle .$$

Equations (2.5) and (2.7) are equivalently written as

$$T_t[X \otimes f] = E_0\big[\tilde{J}_{0t}(X \otimes f)\big] , \tag{2.9}$$

where the map $\tilde{J}_{st} :\ \mathcal{L}(\mathcal{H}_S) \otimes L^\infty(G,\omega) \to \mathcal{L}(\mathcal{H}_S \otimes \Gamma \otimes L^2(G,\omega))$ is defined by

$$\tilde{J}_{st}(X \otimes f) := [U(t,s)^* \otimes \mathbb{1}]\, [X \otimes J_{st}(f)]\, [U(t,s) \otimes \mathbb{1}] . \tag{2.10}$$

The map $\tilde{J}_{st}$ is again an EH–flow, now on a particular noncommutative algebra, defined as a perturbation of a trivial ampliation of the original flow J_{st}. In order to obtain the structure maps of $\tilde{J}_{st}$ we have to differentiate eq. (2.10) and for this, in the general case, f must be a smooth function, say $f \in \mathcal{K}(G)$. Moreover, for simplicity, let us assume here that we have only a finite set of indices ($\mathcal{H}$ finite dimensional). By differentiating eq. (2.10) we obtain

$$\mathrm{d}\tilde{J}_{st}(X \otimes f) = \sum_{i,j \ge 0} \tilde{J}_{st}(\Theta^i_j(X \otimes f))\, \mathrm{d}\Lambda^j_i , \tag{2.11}$$

$$\Theta^i_j(X \otimes f) := \Big(R^{j*}_i X + X R^i_j + \sum_{k \ge 1} R^{k*}_i X R^k_j \Big) \otimes f + X \otimes \vartheta^i_j(f) +$$

$$+ \sum_{k \ge 1} R^{k*}_i X \otimes \vartheta^k_j(f) + \sum_{l \ge 1} X R^l_j \otimes \vartheta^i_l(f) + \sum_{k,l \ge 1} R^{k*}_i X R^l_j \otimes \vartheta^k_l(f) . \tag{2.12}$$

One can check that the Θ^i_j satisfy eqs. (1.9) and (1.10) with obvious changes in the notations; therefore they are again structure maps.

Another way of looking at the flow $\tilde{J}_{st}$ is to relate it to the "output fields" [14,16]. Let us define

$$J_{st}^{\text{out}}(f) := [U(s,0)^* \otimes \mathbb{1}]\, \tilde{J}_{st}(\mathbb{1} \otimes f)\, [U(s,0) \otimes \mathbb{1}]\,; \tag{2.13}$$

then, this new map satisfies the equations

$$\mathrm{d}J_{st}^{\text{out}}(f) = \sum_{i,j\geq 0} J_{st}^{\text{out}}(\vartheta_j^i(f))\, \mathrm{d}\Lambda_i^{j\,\text{out}}(t)\,, \qquad J_{ss}^{\text{out}}(f) = \mathbb{1} \otimes \mathbb{1} \otimes f\,, \tag{2.14}$$

$$\Lambda_i^{j\,\text{out}}(t) := U(t,0)^*\,(\mathbb{1}\otimes\Lambda_i^j(t))U(t,0) \equiv \lim_{T\to+\infty} U(T,0)^*\,(\mathbb{1}\otimes\Lambda_i^j(t))U(T,0)\,. \tag{2.15}$$

Note that with respect to the EH–flow defined in the previous section the structure maps have not been changed, but only the integrators have been modified by a unitary transformation.

The new formulation of continuous measurements in terms of perturbed EH–flows suggests two possible generalizations. For a generic locally compact group G the algebra $\mathcal{K}(G)$ can be too poor, even trivial. In this case one has to work with other classes of functions, which, however, cannot be given a $*$-bialgebra structure. Then, only the approach via EH–flows continues to have meaning. Up to now, convolution semigroups of instruments on these more general groups have been studied by using probability operators [11,12]; eqs. (2.9) and (2.10) allow now to treat this case also by using quantum stochastic calculus.

The second generalization is to consider EH–flows defined by structure maps not satisfying condition (1.13). This allows to consider "operator representations" by commutative EH–flows of a large class of classical processes. Many of these processes have been already studied [31–33,...]. Then, one can think of combining these flows with unitary evolutions $U(t,s)$ (by following more or less what is done in this section) in order to obtain a larger class of stochastic processes. The type of processes obtained in this way should be that defined by eq. (1.2) of [11], which is a generalization of eq. (2.6) of this note.

3 An example

In the theory of EH–flows it is usual to study whether the flow can be implemented by unitary (or isometric) Markovian cocycles (see, for instance, [32], Section 3 and eq. (3.5) in particular). Now we show how this can be done in our case; however, we consider only a particular example chosen in such a way that we have not to dwell on unbounded coefficients.

Let $\tilde{P}$ be a projection valued measure with values in $\mathcal{L}(\mathcal{H})$ and v a vector in $\mathcal{H}$. It easy to check that the following formulae define a generating triple:

$$\pi(f) := \int_G f(y)\tilde{P}(\mathrm{d}y)\,, \quad \eta(f) := [\pi(f) - \delta(f)\mathbb{1}]\,v\,, \quad \psi(f) := \langle v|\eta(f)\rangle\,. \tag{3.1}$$

Then, the definitions of Section 1 give for the structure maps

$$\vartheta^i_j(f)(x) = \int_G [f(yx) - f(x)]\,\langle \xi_i|\tilde{P}(\mathrm{d}y)\xi_j\rangle\,, \quad i,j \geq 0\,, \tag{3.2}$$

where $\xi_0 := v$. The structure maps in this case can be considered as defined on the whole space $L^\infty(G,\omega)$.

Let g_j, $j = 1,2,\ldots$, be a sequence of elements of G and let us choose $\tilde{P}$ of the form

$$\tilde{P}(B) = \sum_{j\geq 1|g_j\in B} |\xi_j\rangle\langle\xi_j|\,. \tag{3.3}$$

Then, eq. (3.2) can be written as

$$\vartheta^i_j(f) = \sum_{k\geq 1} (f_{g_k} - f)\,\langle\xi_i|\xi_k\rangle\langle\xi_k|\xi_j\rangle\,, \quad i,j\geq 0\,, \tag{3.4}$$

where $f_g(x) := f(gx)$. By using eqs. (2.8)–(2.12), we obtain in particular

$$\frac{\mathrm{d}}{\mathrm{d}t}\mathbb{E}[f(Y_t)|Y_0 = g;\varrho] = \tag{3.5}$$

$$= \sum_{k\geq 1}\left\langle \varrho, E_0\left[U(t,0)^*\left(\left(R_0^{k*} + \langle v|\xi_k\rangle\mathbb{1}\right)\left(R_0^k + \langle\xi_k|v\rangle\mathbb{1}\right)\right)\otimes J_{0t}(f_{g_k} - f)(g)\,U(t,0)\right]\right\rangle,$$

which shows that Y_t is a jump process with amplitude of jumps g_k. Note that not all the g_j's have to be distinct and that, if some of the g_j's is equal to e, the corresponding term disappears from (3.5). This example is strongly related to Example 28.12 of [31] and to the counting processes studied in [14].

Let us introduce the following bounded operators on $L^2(G,\omega)$, for $i,j\geq 1$,

$$\begin{aligned} S_0^0 &:= -\tfrac{1}{2}\|v\|^2\mathbb{1}\,, & S_j^i &:= \left(L_{g_j^{-1}} - \mathbb{1}\right)\delta_j^i\,, \\ S_j^0 &:= -\langle v|\xi_j\rangle\mathbb{1}\,, & S_0^i &:= \langle\xi_i|v\rangle L_{g_i^{-1}}\,, \end{aligned} \tag{3.6}$$

where L_x is the left translation on $L^2(G,\omega)$. These operators satisfy

$$S_j^i + S_i^{j*} + \sum_{s\geq 1} S_i^{s*}S_j^s = 0\,, \qquad S_j^i + S_i^{j*} + \sum_{s\geq 1} S_s^i S_s^{j*} = 0\,, \tag{3.7}$$

which are the same as eqs. (2.2), and

$$\vartheta^i_j(f) = fS_j^i + S_i^{j*}f + \sum_{s\geq 1} S_i^{s*}fS_j^s\,. \tag{3.8}$$

Let us consider the operators V_{st} on $\Gamma\otimes L^2(G,\omega)$ defined by

$$\mathrm{d}V_{st} = \sum_{i,j\geq 0}\left[\mathrm{d}\Lambda_i^j(t)\otimes S_j^i\right]V_{st}\,, \qquad V_{ss} = \mathbb{1}\,. \tag{3.9}$$

Equations (3.7) guarantee that the V_{st} are unitary operators and eq. (3.8) guarantees that

$$J_{st}(f) = V_{st}^*\,(\mathbb{1}\otimes f)\,V_{st}\,. \tag{3.10}$$

This can be proved by differentiating both sides of (3.10).

Due to eqs. (2.2), (2.3), (3.7), (3.9), we see that $U(t,s)$ and V_{st} are, from the mathematical point of view, very similar objects. However, from the physical point of view they are completely different: $U(t,s)$ represents the dynamics of the composed system "S + fields", while V_{st} contains the measured observables. This two objects can be combined together by defining

$$\tilde{V}(t,s) := \left[\mathbb{1}_{\mathcal{H}_S} \otimes V_{st}\right] \left[U(t,s) \otimes \mathbb{1}_{L^2(G,\omega)}\right] \tag{3.11}$$

Then, we have that $\tilde{V}(t,s)$ too satisfies an equation of the type of (2.3) and that we can write

$$\tilde{J}_{st}(X \otimes f) = \tilde{V}(t,s)^* \,(X \otimes \mathbb{1} \otimes f)\, \tilde{V}(t,s). \tag{3.12}$$

Finally let us stress that eq. (2.8) can now be written as

$$\mathbb{E}\left[f_1(Y_{t_1}) f_2(Y_{t_2}) \cdots f_n(Y_{t_n}) \middle| Y_0; \varrho\right] =$$

$$= \langle \varrho, E_0[\tilde{V}(t_1,0)^* \tilde{f}_1 \tilde{V}(t_2,t_1)^* \tilde{f}_2 \cdots \tilde{V}(t_n,t_{n-1})^* \tilde{f}_n V(t_n,0)]\rangle\,, \tag{3.13}$$

where $\tilde{f}_j := \mathbb{1} \otimes \mathbb{1} \otimes f_j$.

References

1. L. Accardi, A. Frigerio, V. Gorini (eds.): *Quantum Probability and Applications to the Quantum Theory of Irreversible Processes.* Lect. Notes Math., vol. 1055. Berlin: Springer 1984
2. L. Accardi, W. von Waldenfels (eds.): *Quantum Probability and Applications II.* Lect. Notes Math., vol. 1136. Berlin: Springer 1985
3. L. Accardi, W. von Waldenfels (eds.): *Quantum Probability and Applications III.* Lect. Notes Math., vol. 1303. Berlin: Springer 1988
4. L. Accardi, W. von Waldenfels (eds.): *Quantum Probability and Applications IV.* Lect. Notes Math., vol. 1396. Berlin: Springer 1989
5. L. Accardi, W. von Waldenfels (eds.): *Quantum Probability and Applications V.* Lect. Notes Math., vol. 1442. Berlin: Springer 1990
6. L. Accardi (ed.): *Quantum Probability and Related Topics VI.* Singapore: World Scientific 1991
7. G. M. Prosperi: *The quantum measurement process and the observation of continuous trajectories.* In [1] pp. 301–326
8. M. D. Srinivas: *Quantum theory of continuous measurements.* In [1] pp. 356–364
9. A. S. Holevo: *A noncommutative generalization of conditionally positive definite functions.* In [3] pp. 128–148
10. A. S. Holevo: *Limits theorems for repeated measurements and continuous measurement processes.* In [4] pp. 229–255
11. A. Barchielli: *Some Markov semigroups in quantum probability.* In [5] pp. 86–98
12. A. Barchielli, G. Lupieri: *Convolution semigroups of instruments in quantum probability.* In *Probability Theory and Mathematical Statistics* Vol. I, ed. by B. Grigelionis *et al.* (Vilnius, Mokslas, and VSP, Utrecht, 1990) pp. 78–90.

13. A. Barchielli, G. Lupieri: *Dilations of operation valued stochastic processes.* In [2] pp. 57–66
14. A. Barchielli: *Input and output channels in quantum systems and quantum stochastic differential equations.* In [3] pp. 37–51
15. A. Barchielli, G. Lupieri: *Convolution semigroups in quantum probability and quantum stochastic calculus.* In [4] pp. 107–127
16. A. Barchielli: *Applications of quantum stochastic calculus to quantum optics.* In [6] pp. 111–126
17. A. Barchielli, G. Lupieri: *Semigroups of positive–definite maps on ∗-bialgebras.* Preprint IFUM 399/FT, April 1991. To appear in *Quantum Probability and Related Topics VII*, (World Scientific, Singapore, 1992).
18. W. von Waldenfels: *Ito solution of the linear quantum stochastic differential equation describing light emission and absorption.* In [1] pp. 384–411
19. M. Schürmann: *Positive and conditionally positive linear functionals on coalgebras.* In [2] pp. 475–492
20. P. Glockner, W. von Waldenfels: *The relations of the non–commutative coefficient algebra of the unitary group.* In [4] pp. 182–220
21. M. Schürmann: *Noncommutative stochastic processes with independent and stationary additive increments.* In [4] pp. 339–351
22. M. Schürmann: *Gaussian states on bialgebras.* In [5] pp. 347–367
23. P. Glockner: *Quantum stochastic differential equations on ∗-bialgebras.* Math. Proc. Camb. Phil. Soc. **109** (1991) 571–595
24. M. Evans, R. L. Hudson: *Multidimensional quantum diffusions.* In [3] pp. 69–88
25. R. L. Hudson: *Quantum diffusions on the algebra of all bounded operators on a Hilbert space.* In [4] pp. 256–269
26. L. Accardi, R. L. Hudson: *Quantum stochastic flows and nonabelian cohomology.* In [5] pp. 54–69
27. D. Applebaum: *Quantum diffusions on involutive algebras.* In [5] pp. 70–85
28. R. L. Hudson, P. Shepperson: *Stochastic dilations of quantum dynamical semigroups using one–dimensional quantum stochastic calculus.* In [5] pp. 216–218
29. P. Robinson: *Quantum diffusions on the rotation algebras and the quantum Hall effect.* In [5] pp. 326–333
30. D. Applebaum: *Towards a quantum theory of classical diffusions on Riemannian manifolds.* In [6] pp. 93–109
31. K. R. Parthasarathy: *An Introduction to Quantum Stochastic Calculus.* Indian Statistical Institute, Delhi Centre
32. K. S. Parthasarathy, K. B. Sinha: *Markov chains as Evans–Hudson diffusions in Fock space.* Preprint I.S.I., Delhi Centre.
33. F. Fagnola: *Pure birth and pure death processes as quantum flows in Fock space.* Sankhya, Vol. 53, Series A
34. R. H. Hudson: *Algebraic theory of quantum diffusions.* In A. Truman, I. M. Davies (eds.): *Stochastic Mechanics and Stochastic Processes.* Lect. Notes Math., vol. 1325. Berlin: Springer 1988; pp. 113–124

Quantum Probability and Related Topics Vol. VIII (pp. 81–85)

THE QUANTUM FUNCTIONAL ITO FORMULA HAS THE PSEUDO-POISSON STRUCTURE $df(X) = [f(X+B)-f(X)]^{\mu}_{\nu}dA^{\nu}_{\mu}$

V P Belavkin

Department of Mathematics
University of Nottingham
University Park
Nottingham NG7 2RD
England

Abstract. To replace the false functional generalisation [1] of Hudson and Parthasarathy's Ito formula [2] we exposite a universal pseudo-Poisson one, recently discovered within the $*$-Ito non-commutative algebraic approach in [3]. Our formula unifies the classical Ito formulas for Wiener and Poisson cases, extends [4] the chain rule $df(X_1,\ldots,X_l) = \sum f_i'(x)\, dX_i$ to the case of non-commuting stochastically differentiable operators $X_i(t)$, and also admits the non-adapted generalization [5].

The functional Ito formula gives an extension of the chain rule

$$df(x_t) = f'(x_t)c_t dt, \qquad c_t = \frac{d}{dt}x_t \tag{1}$$

to the non-differentiable but stochastically differentiable trajectories

$$x_t = x_0 + \int_0^t (a_s dn_s + b_s dw_s + c_s ds),$$

defined by an adapted stochastic process $x_t\colon \Omega \to \mathbb{R}$. The forward stochastic differential $dx_t = x_{t+dt} - x_t$ in (1) is different from the backward one $x_t - x_{t-dt}$, and in general has the form

$$dx_t = (z_t - x_t)dn_t + b_t dw_t + c_t dt. \tag{2}$$

Here $dn_t = n_{t+dt} - n_t$ is the increment $dn_t \in \{0,1\}$ of the standard Poisson process n_t, producing the jumps $x_t \to z_t$ of the value $a_t = z_t - x_t$ when $dn_t = 1$, $dw_t = w_{t+dt} - w_t$ is the increment $dw_t \approx \sqrt{dt}$ of the standard Wiener process w_t, producing the diffusion $(dx_t)^2 = b_t^2 dt$ when $dn_t = 0$, a_t and b_t are stochastic adapted processes, called (stochastic) derivatives of x_t with respect to n_t and w_t, and c_t is called the Lebesgue time derivative of x_t. The adapticity of $z_t(\omega)$, $b_t(\omega)$ and $c_t(\omega)$ and hence of the integral (2) means their casuality respectively to $\omega = (w,n)$, i.e. $x_t(\omega)$ might be random only due to the dependence of the presence and the past $w_{t]} = \{w_s \mid s \le t\}$, $n_{t]} = \{n_s \mid s \le t\}$ of the basic stochastic processes $w = \{w_t\}$ and $n = \{n_t\}$, but not on their future $w_{(t} = \{w_s \mid s > t\}$, $n_{(t} = \{n_s \mid s > t\}$.

The classical functional Ito formula [1,2] extended to the general case (2) with the jumps $\Delta x_t \neq 0$ reads

$$df(x_t) = [f(z_t) - f(x_t)]dn_t + f'(x_t)b_t dw_t + (f'(x_t)c_t + \tfrac{1}{2}f''(x_t)b_t^2)dt \tag{3}$$

as it follows from the formal multiplication table

$$dn_t^2 = dn_t, \quad (dw_t)^2 = dt, \quad (dt)^2 = 0 = dn_t dw_t = dw_t dn_t = dw_t dt = dt dw_t.$$

Obviously this formula yields different structures for the Wiener case ($a_t = 0$) and for the Poisson case ($b_t = 0$) which is especially simple if also $c_t = 0$:

$$df(x_t) = [f(x_t + a_t) - f(x_t)]dn_t. \tag{4}$$

The last stochastic differential formula for $x_t = x_0 + \int_0^t a\, dn$ enjoys an elementary sense. Because dn_t has with probability one the only possible non-zero value 1, $dx_t = a_t dn_t$ is equal to a_t, when $dn_t = 1$ and 0, when $dn_t = 0$, and hence the forward differential $df(x_t)$ is either the finite difference $f(x_t + a_t) - f(x_t)$ (if $dn_t = 1$) or zero (if $dn_t = 0$).

The classical Ito formula, originally written for the products $(x_t^* x_t)(\omega) = x_t^*(\omega)x_t(\omega)$ of two stochastic processes x_t, x_t^*, was extended by Hudson and Parthasarathy [2] as $d(X_t^* X_t) = dX_t^* X_t + X_t^* dX_t + dX_t^* dX_t$ to the operator-valued functions X_t, X_t^*, having the quantum stochastic differentials

$$dX_t = (Z_t - X_t)dN_t + B_t^- dA_t + B_t dA_t^+ + C_t dt. \tag{5}$$

Here Z_t, B_t^-, B_t and C_t are the adapted [3] quantum stochastic processes respectively to the canonical number N_t, annihilation A_t, creation A_t^+ and scalar time processes tI with the increments having the non-commutative multiplication table

$$dN_t^2 = dN_t, \quad dA_t dN_t = dA_t, \quad dN_t dA_t^\dagger = dA_t^\dagger, \quad dA_t dA_t^\dagger = dtI, \tag{6}$$

with all the other products being equal zero, in particular $dN_t dA_t = dA_t^+ dN_t = dA_t dA_t = 0$.

The purpose of this paper is to show how the recently discovered [6,7] pseudo-Poisson structure of the non-commutative $*$-Ito algebra and its universal triangular matrix representation in a complex Minkovski space [7,8] enables us to extend the chain rule (1) and the functional Ito formula (3) to the non-commuting stochastic operators as a particular case of the quantum stochastic one [3-5]. In the one-dimensional case, corresponding to (5) the quantum functional Ito formula can be written as

$$df(X_t) = \left(f\begin{bmatrix} X_t & B_t^- & C_t \\ 0 & Z_t & B_t \\ 0 & 0 & X_t \end{bmatrix} - f\begin{bmatrix} X_t & 0 & 0 \\ 0 & X_t & 0 \\ 0 & 0 & X_t \end{bmatrix} \right)_\nu^\mu dA_\mu^\nu(t). \tag{7}$$

Here $f[Z_\nu^\mu]$ denotes the function $f(\mathbf{Z})$ of a triangular block-matrix $\mathbf{Z} = [Z_\nu^\mu] = \boldsymbol{X}+\boldsymbol{B}$,

$$\boldsymbol{X} = \begin{bmatrix} X & 0 & 0 \\ 0 & X & 0 \\ 0 & 0 & X \end{bmatrix}, \quad \boldsymbol{B} = \begin{bmatrix} 0 & \boldsymbol{B}^- & C \\ 0 & Z-X & B \\ 0 & 0 & 0 \end{bmatrix} \Rightarrow f(\mathbf{Z}) = \begin{bmatrix} f(X) & G^- & H \\ 0 & f(Z) & G \\ 0 & 0 & f(X) \end{bmatrix}$$

indexed by $\mu, \nu \in \{-,0,+\}$ with the elements $Z_\nu^\mu = (\mathbf{Z})_\nu^\mu = 0$ if $\mu = +$ or $\nu = -$, $Z_-^- = X = Z_+^+$, $Z_0^0 = Z$, $Z_+^0 = B$, $Z_0^- = B^-$, and $A_0^0 = N$, $A_-^0 = A$, $A_0^+ = A^+$, $A_-^+ = tI$, so the convolution over (μ, ν) can be taken only for $\mu = -,0$ and $\nu = 0,+$:

$$df(X_t) = [f(Z_t) - f(X_t)]dN_t + G_t^- dA_t + G_t dA_t^+ + H_t dt,$$

where G^-, G and H have to be found as the elements $(f[Z_\nu^\mu])_0^-$, $(f[Z_\nu^\mu])_+^0$ and $(f[Z_\nu^\mu])_+^-$.

Within this approach the chain rule (1) as well as Wiener–Ito's formula

$$df(x_t) = f'(x_t)b_t dw_t + \tfrac{1}{2}f''(x_t)b_t^2 dt \tag{8}$$

can be unified with the Poisson one (4). Indeed, the formula (1) corresponding to the case $a_t = 0 = b_t$, can be written in the difference matrix form

$$df(x_t) = \left(f\begin{bmatrix} x_t & c_t \\ 0 & x_t \end{bmatrix} - f\begin{bmatrix} x_t & 0 \\ 0 & x_t \end{bmatrix} \right)_\nu^\mu da_\mu^\nu, \quad a_-^+(t) = t,$$

because the function $f(z)$ of a 2×2 triangular matrix $z = [z_\nu^\mu]_{\nu=-,+}^{\mu=-,+}$ with $z_-^- = x = z_+^+$, $z_+^- = c$, $z_-^+ = 0$ has the triangular form $f(\boldsymbol{x}_t) + \boldsymbol{h}_t$, where $f(\boldsymbol{x}_t) = [f(x_t)\delta_\nu^\mu]_{\nu=-,+}^{\mu=-,+}$, and $\boldsymbol{h} = [h_\nu^\mu]_{\nu=-,+}^{\mu=-,+}$ is nilpotent: $h_+^- = f'(x_t)c_t$, $h_\nu^\mu = 0$ if $(\mu,\nu) \neq (-,+)$.

Similarly to this the formula (8), corresponding to the case $a_t = 0 = c_t$, can be written also as the pseudo-Poisson one (7)

$$df(x) = \left(f\begin{bmatrix} x_t & b_t & 0 \\ 0 & x_t & b_t \\ 0 & 0 & x_t \end{bmatrix} - f\begin{bmatrix} x_t & 0 & 0 \\ 0 & x_t & 0 \\ 0 & 0 & x_t \end{bmatrix} \right)_\nu^\mu da_\mu^\nu \tag{9}$$

with $a_-^0(t) + a_0^+(t) = w_t$, $a_-^+(t) = t$. Here $(\boldsymbol{g})_\nu^\mu$ denote the matrix elements g_ν^μ of the nilpotent matrix

$$g = \begin{bmatrix} g_-^- & g_0^- & g_+^- \\ g_-^0 & g_0^0 & g_+^0 \\ g_-^+ & g_0^+ & g_+^+ \end{bmatrix} = f(x+b) - f(x) = \begin{bmatrix} 0 & g & h \\ 0 & 0 & g \\ 0 & 0 & 0 \end{bmatrix},$$

giving the canonical pseudo-Euclidean representation $g = f'(x)b$, $h = \frac{1}{2}f''(x)b^2$ of the difference of the diagonal matrix-function $f(x) = [f(x)\delta_\nu^\mu]$ and the function $f(z)$ of the triangular matrix $z = x+b$, the nilpotent variation of x:

$$x = \begin{bmatrix} x & 0 & 0 \\ 0 & x & 0 \\ 0 & 0 & x \end{bmatrix}, \quad b = \begin{bmatrix} 0 & b & 0 \\ 0 & 0 & b \\ 0 & 0 & 0 \end{bmatrix} \quad \Rightarrow \quad f(x+b) = \begin{bmatrix} f(x) & f'(x)b & \frac{1}{2}f''(x)b^2 \\ 0 & f(x) & f'(x)b \\ 0 & 0 & f(x) \end{bmatrix}.$$

The summation in (10) can be taken only over the pairs $(\mu, \nu) \in \{(-,0), (0,+), (-,+)\}$, corresponding to the three non-zero elements $g_0^- = f'(x)b$, $g_+^0 = f'(x)b$, $g_+^- = \frac{1}{2}f''(x)b^2$ of the matrix g, and the increments da_0^+, da_-^0 in (10) are appearing only in the sum $da_0^+ + da_-^0 = dw$, $g_+^0 da_0^+ + g_0^- da_-^0 = g(da_0^+ + da_-^0)$, generating the two-dimensional nilpotent Ito algebra with the base dw and dt. So, the quantum Ito formula (7), extended for the multivariable functions f in [4] and being invariant under the non-adapted generalization [5,8] of the quantum stochastic calculus, is bringing also something new for the classical stochastic calculus. Moreover, it can offer a non-commutative chain rule

$$dX_t = C_t dt \Rightarrow df(X_t) = \left(f\begin{bmatrix} X_t & C_t \\ 0 & X_t \end{bmatrix} - f\begin{bmatrix} X_t & 0 \\ 0 & X_t \end{bmatrix}\right)_\nu^\mu da_\mu^\nu = \left(f\begin{bmatrix} X_t & C_t \\ 0 & X_t \end{bmatrix}\right)_+^- dt$$

and a new non-commutative Ito formula

$$dX_t = C_t dt + B_t dw_t + (Z_t - X_t) dn_t \Rightarrow df(X_t) = \left(f\begin{bmatrix} X_t & C_t \\ 0 & X_t \end{bmatrix}\right)_+^- dt +$$

$$+ \sum_{\mu<\nu} \left(f\begin{bmatrix} X_t & B_t & 0 \\ 0 & X_t & B_t \\ 0 & 0 & X_t \end{bmatrix}\right)_\nu^\mu da_\mu^\nu + [f(Z_t) - f(X_t)] dn_t, \tag{10}$$

where $- < 0 < +$, $a_0^+ + a_-^0 = w$, $a_-^+ = t$ for the stochastic calculus of random operators $X_t(\omega)$ with adapted $C_t(\omega)$, $B_t(\omega)$ and $Z_t(\omega)$ in a Hilbert space $\mathcal{H}$. The last formula can be again written as $df(X_t) = [f(X_t + B_t) - f(X_t)]_\nu^\mu da_\mu^\nu$ but in terms of 4×4 block-matrices, indexed by $\{-, 0, 1, +\}$ with $a_1^1 = n$, $B_1^1 = Z - X$, $B_+^0 = B = B_0^-$, $B_+^- = C$ and $B_\mu^\nu = 0$ for all other indices μ, ν.

Acknowledgements

I would like to thank L Accardi, A Barchielli, F Fagnola, A Frigerio, R L Hudson and J M Lindsay for their stimulating discussions, hospitality and encouragement to write this paper.

References

[1] G O S Ekhaguere, The functional Ito formula in quantum stochastic calculus, *J Math Phys* **31** (12) 2921 (1990).

[2] R L Hudson and K R Parthasarathy, Quantum Ito's formula and stochastic evolution, *Comm Math Phys* **93**, 301 (1984).

[3] V P Belavkin, A new form and *-algebraic structure of quantum stochastic integrals in Fock space, Rediconti del Seminario Matematico e Fisico di Milano, LVIII (1988), 117, Pavia 1990.

[4] V P Belavkin, A quantum stochastic calculus in Fock space of input and output non-demolition processes. In: Proc Fifth Quantum Probability Conference, Heidelberg 1988, eds. L Accardi and W von Waldenfels, LNM **1442**, 99, Springer Verlag, Berlin 1990.

[5] V P Belavkin, A quantum non-adapted Ito formula and stochastic analysis in Fock scale, *J Funct Analysis*, **102**, 414 (1991).

[6] V P Belavkin, A pseudo-euclidean representation of conditionally positive maps, *Matemat Zametki*, **49** (6), 135 (1991).

[7] V P Belavkin, Chaotic states and stochastic integration in quantum systems, *Uspechi Matem Nauk*, **1**, 3 (1992).

[8] V P Belavkin, Kernel representation of *-semigroups associated with infinitely divisible states. In: Proc Seventh Quantum Probability Conference, New Delhi, 1990, ed. L Accardi, Quantum Prob and Related Fields, World Scientific, Singapore, 1992.

Quantum Probability and Related Topics Vol. VIII (pp. 87–94)

THE KERNEL OF A FOCK SPACE OPERATOR II

V P Belavkin and J M Lindsay

§0. Introduction

In a previous note Fock space operators, which are given formally by multiple quantum stochastic integrals of the form

$$\iiint_{\Gamma^3} x(\rho,\sigma,\tau)\, dA_\rho^* d\Lambda_\sigma dA_\tau, \tag{0.1}$$

were analysed ([L]). Various senses in which such an operator is determined by its *kernel* x were established. In this note we show how the kernel may be recovered from the operator.

§1. Notations and Preliminaries

Fix a σ-finite, non-atomic, separable measure space $M = (S, \mathcal{F}, m)$ in which each singleton set is measurable, and let $\mathcal{R}$ be a countable generating subring of $\mathcal{F}$ consisting of sets having finite measure. *We shall adopt all the notations of* [L] — in particular, Γ is the finite power set of S: $\{\sigma \subset S: \#\sigma < \infty\}$ and Φ is the natural surjection $\bigcup_{n \geq 0} S^{(n)} \to \Gamma$ taking a point in the Cartesian product to the set consisting of its coordinates. However we shall abbreviate the cumbersome notation Γ_m, for the symmetric measure, to μ.

For a finite collection $\{E_i : i \in \lambda\}$ of mutually disjoint subsets of S, let E_λ denote the following subset of Γ:

$$\Phi\left(\times_{i\in\lambda} E_i\right) \tag{1.1}$$

with the convention $E_\emptyset := \{\emptyset\}$. Thus the elements of E_λ are finite sets σ whose elements may be labeled $\{s_i : i \in \lambda\}$ in such a way that $s_i \in E_i$ for each i. Note that if $\{E_i : i \in \lambda\}$, $\{E_j : j \in \mu\}$ are mutually disjoint, then $\chi_{E_{\lambda\cup\mu}} = \chi_{E_\lambda} \circ \chi_{E_\mu}$, where the *symmetric product* $\circ$ is given by

$$f \circ g(\sigma) = \sum_{\alpha \subset \sigma} f(\alpha) g(\bar{\alpha})$$

— $\bar{\alpha}$ denoting the complement of α in σ: $\sigma \backslash \alpha$.

Let $\mathcal{G}_0$ denote the collection of subsets of Γ of the form (1.1) where each E_i belongs to $\mathcal{R}$, and define

$$\mathcal{D}_0 := \text{ linear span } \left\{ \chi_{E_\lambda} : E_\lambda \in \mathcal{G}_0 \right\}.$$

The notion of *local integrability* we shall use here is the following:

$$L^1_{\text{loc.}}(\Gamma) := \left\{ g \in L^0(\Gamma) : \int_{\Gamma_n(E)} |g(\omega)| \, d\omega < \infty \;\forall n \in \mathbb{N}, E \in \mathcal{R} \right\}. \tag{1.2}$$

The operators we are considering are given by the prescription:

$$(X_3 k)(\sigma) = \int_\Gamma \sum_{|\alpha| = \sigma} x(\alpha_1, \alpha_2, \omega) k(\omega \cup \alpha_2 \cup \alpha_3) \, d\omega$$

(see [L] Definition 2.1) or, equivalently, by

$$\int_\Gamma \sum_{\alpha \subset \sigma} x'(\alpha, \bar{\alpha}, \omega) k(\omega \cup \bar{\alpha}) \, d\omega$$

where

$$x' : (\alpha, \beta, \gamma) \mapsto \sum_{\sigma \subset \beta} x(\alpha, \sigma, \gamma). \tag{1.3}$$

For $E_\lambda \in \mathcal{G}_0$, if $x(\alpha, \beta, \cdot) \chi_{E_\tau}(\cdot) \in L^1$ for a.a. (α, β) and each subset τ of λ, then $X_3 \chi_{E_\lambda}$ is well-defined. This follows from the identity

$$\chi_{E_\lambda}(\omega \cup \gamma) = \sum_{\tau \subset \lambda} \chi_{E_\tau}(\omega) \chi_{E_{\lambda \setminus \tau}}(\gamma),$$

for a.a. $(\omega, \gamma) \in \Gamma^2$ Thus if x is locally integrable in its third argument, then $X_3 k$ is at least well-defined for $k \in \mathcal{D}_0$. Moreover, by [L], Lemma 2.2, null set variations of x and k only result in null set variations of $X_3 k$, so that X_3 may be viewed as a (partially defined) linear operator on $L^0(\Gamma)$.

The *number operator* N is the multiplication operator given by $Nk(\sigma) = \#\sigma \, k(\sigma)$, and for functions $\phi : \mathbb{N} \to \mathbb{C}$, $\phi(N)$ is defined through the functional calculus. Thus the function spaces defined in [L], (2.5), may be described as

$$\mathcal{H}_a = \{k \in L^0(\Gamma) : \|k\|_{(a)} := \|a^{N/2}k\| < \infty\}.$$

(An unqualified norm will always mean the $L^2(\Gamma)$- norm.) The scale of Fock spaces $\{\mathcal{H}_c : c > 0\}$ has been exploited in [B] for analysing a generalisation of (0.1) to operator-valued x.

§2. From kernel to operator

For a measurable function $x : \Gamma^3 \to \mathbb{C}$, $a, b, c > 0$ let

$$x'_{a,b,c} : (\alpha,\beta,\gamma) \mapsto x(\alpha,\beta,\gamma)\Big/\sqrt{a^{\#\alpha}b^{2\#\beta}c^{\#\gamma}}$$

$$\|x'\|_{a,b,c} = \left\{\iint \sup_\beta |x'_{a,b,c}(\alpha,\beta,\gamma)|^2 \,\mathrm{d}\alpha\,\mathrm{d}\gamma\right\}^{\frac{1}{2}}$$

where x is the transform of x given by (1.3). Recall the identity

$$\iiint_{\Gamma^3} h(\alpha\cup\beta)x'(\alpha,\beta,\gamma)k(\beta\cup\gamma)\,\mathrm{d}\alpha\,\mathrm{d}\beta\,\mathrm{d}\gamma = \int_\Gamma h(\sigma)(X_3 k)(\sigma)\,\mathrm{d}\sigma, \tag{2.1}$$

which is valid when either side is defined ([L], (2.4)). In particular a locally integrable kernel always defines a *quadratic form* on $\mathcal{D}_0$. In this section we give conditions for a function x to be the kernel of a reasonable *operator*. The basic estimate is

Lemma 2.1: Let $p > a > 0$, $q > c > 0$, $h, k \in L^0(\Gamma)$ and $x \in L^0(\Gamma^3)$. Then

$$\int_\Gamma |h(\sigma)(X_3 k)(\sigma)|\,\mathrm{d}\sigma \leqslant \|h\|_{(p)}\|k\|_{(q)}\|x'\|_{a,\sqrt{(p-a)(q-c)},c} \tag{2.2}$$

Proof: Let $b = \sqrt{(p-a)(q-c)}$. (2.1), several applications of Minkowski's inequality, Cauchy-Schwarz and the $\oint$-Lemma, give

$$\begin{aligned}
&\int |h(\sigma)(X_3k)(\sigma)|\,\mathrm{d}\sigma\\
&\leqslant \iiint \mathrm{d}\alpha\mathrm{d}\beta\mathrm{d}\gamma\,|h(\alpha\cup\beta)x'(\alpha,\beta,\gamma)k(\beta\cup\gamma)|\\
&= \iiint \mathrm{d}\alpha\,\mathrm{d}\beta\,\mathrm{d}\gamma\,\sqrt{a}^{\#\alpha}\sqrt{p-a}^{\,\#\beta}|h(\alpha\cup\beta)x'_{a,b,c}(\alpha,\beta,\gamma)k(\beta\cup\gamma)|\sqrt{q-c}^{\,\#\beta}\sqrt{c}^{\#\gamma}\\
&\leqslant \left\{\int \mathrm{d}\alpha\,\mathrm{d}\beta\; a^{\#\alpha}(p-a)^{\#\beta}|h(\alpha\cup\beta)|^2\right\}^{1/2}\int \mathrm{d}\gamma\;\sqrt{c}^{\#\gamma}\left\{\iint \mathrm{d}\alpha\,\mathrm{d}\beta\;(q-c)^{\#\beta}|x'_{a,b,c}(\alpha,\beta,\gamma)k(\beta\cup\gamma)|^2\right\}^{1/2}
\end{aligned}$$

$$\leqslant \left\{\int \mathrm{d}\sigma p^{\#\sigma}|h(\sigma)|^2\right\}^{1/2}\int \mathrm{d}\gamma\sqrt{c^{\#\gamma}}\left\{\int \mathrm{d}\beta\,(q-c)^{\#\beta}|k(\beta\cup\gamma)|^2\int \mathrm{d}\alpha\ \sup_{\omega}|x'_{a,b,c}(\alpha,\omega,\gamma)|^2\right\}^{1/2}$$

$$\leqslant \|h\|_{(p)}\left\{\int \mathrm{d}\gamma c^{\#\gamma}\int \mathrm{d}\beta(q-c)^{\#\beta}|k(\beta\cup\gamma)|^2\right\}^{1/2}\|x'\|_{a,b,c}$$

$$= \|h\|_{(p)}\|k\|_{(q)}\|x'\|_{a,b,c}.$$

■

(2.2) may be re-written as an L^2- operator norm estimate:

$$\|(a+b_1)^{-N/2}X_3(b_2+c)^{-N/2}\| \leqslant \|x'\|_{a,\sqrt{b_1b_2},c} \tag{2.3}$$

for $a, b_1, b_2, c > 0$, from which the next result is immediate.

Proposition 2.2: Let $x \in L^0(\Gamma^3)$ satisfy $\|x'\|_{a,b,c} < \infty$, for some $a,b,c > 0$. Then X_3 defines bounded operators from $\mathcal{H}_{(c+be^{-1})}$ to $\mathcal{H}_{(a+be)^{-1}}$, for each $e > 0$, with bound at most $\|x'\|_{a,b,c}$. In particular,

(i) if $a < 1$ then X_3 is densely defined on L^2:

$$\mathrm{Dom}(X_3) \supset \mathcal{H}_{(c+b^2/(1-a))};$$

(ii) if $a < 1, c < 1$ and $\|x'\|_{a,\sqrt{(1-a)(1-c)},c} < \infty$ then X is bounded on L^2.

In the next section we shall merely impose a local integrability condition on the kernel, in the sense of (1.2), so the next result is relevant.

Proposition 2.3: Let $x \in L^0(\Gamma^3)$ satisfy $x'_{a,1,1} \in L^2 \otimes L^2_{\mathrm{loc.}} \otimes L^2_{\mathrm{loc.}}$ for some $a < 1$, then X_3 is densely defined on $L^2(\Gamma)$: $\mathrm{Dom}(X_3) \supset \mathcal{D}_0$.

Proof: If $k \in \mathcal{D}_0$ then, for some $n \in \mathbb{N}$ and $E \in \mathcal{R}$, $|k| \leqslant \|k\|_\infty \chi_n$, where χ_n is the indicator function of $\Gamma_{\leqslant n}(E)$. Thus, for each $g \in L^2(\Gamma)$,

$$\int_\Gamma |g(\sigma)(X_3k)(\sigma)|\mathrm{d}\sigma$$

$$\leqslant \iiint \mathrm{d}\alpha\,\mathrm{d}\beta\,\mathrm{d}\gamma\sqrt{a}^{\#\alpha}\sqrt{1-a}^{\#\beta}|g(\alpha\cup\beta)x'_{a,1,1}(\alpha,\beta,\gamma)k(\beta\cup\gamma)|\sqrt{1-a}^{-\#\beta}$$

$$\leqslant \|g\|\,\|k\|_\infty\int \mathrm{d}\gamma\left\{\iint \mathrm{d}\beta\,\mathrm{d}\alpha|x'_{a,1,1}(\alpha,\beta,\gamma)|^2\chi_n(\beta\cup\gamma)(1-a)^{-\#\beta}\right\}^{1/2}$$

$$\leqslant \|g\|\,\|k\|_\infty C,$$

where C is a constant depending only on n, a, E and an integral of $x'_{a,1,1}$ which is *local* in the second and third argument. Thus $X_3 k \in L^2(\Gamma)$. ■

§3. From operator to kernel

First we relate matrix elements of an integral-sum operator with integrated values of its kernel.

Proposition 3.1: Let $E_\lambda, E_\mu, E_\nu \in \mathcal{G}_0$ and let $x \in L^1_{\text{loc.}}(\Gamma^3)$, then

(i) $X_3 \chi_{E_\nu}$ is well-defined

(ii) $\chi_{E_\lambda}\left(X_3 \chi_{E_\nu}\right)$ is integrable

and, if *all* the sets are mutually disjoint, then

$$\text{(iii)} \int_{E_\lambda \times E_\mu \times E_\nu} x' = \int_{E_{\lambda\cup\mu}} \left(X_3 \chi_{E_{\mu\cup\nu}}\right) - \sum_{\omega \subsetneq \mu} \int_{E_{\lambda\cup\varpi} \times E_\omega \times E_{\varpi\cup\nu}} x'. \tag{3.1}$$

Proof: If x is locally integrable, then so is x', so (i) and (ii) follow from remarks in §1 and the identity (2.1). Mutual disjointness of the sets implies the (a.e.) identity

$$\chi_{E_{\lambda\cup\mu}}(\alpha \cup \beta)\, \chi_{E_{\mu\cup\nu}}(\beta \cup \gamma) = \sum_{\omega \subset \mu} \chi_{E_{\lambda\cup\varpi}}(\alpha)\, \chi_{E_\omega}(\beta)\, \chi_{E_{\varpi\cup\nu}}(\gamma)$$

so the result follows from (2.1). ■

In order to actually *recover* the kernel of the operator it is necessary to make some assumption on the measure space — namely that it supports a Vitali system. We do not wish to make too technical a diversion and therefore settle for a particular type which is easier to visualise.

Definition 3.2: A system of summable sets $\mathfrak{N} = \{F_{i_1,\ldots,i_n} : n \in \mathbb{N}, \boldsymbol{i} \in \mathbb{N}^n\}$ is a *net of partitions* for M if

($\mathfrak{N}$ i) $\{F_{i_1} : i_1 \in \mathbb{N}\}$ is a partition of S and, for each $n \geqslant 2$ and $i_1, \ldots, i_{n-1}$, $\{F_{i_1,\ldots,i_n} : i_n \in \mathbb{N}\}$ is a partition of $F_{i_1,\ldots,i_{n-1}}$;

($\mathfrak{N}$ ii) the linear span of $\{\chi_F : F \in \mathfrak{N}\}$ is dense in $L^1(M)$;

($\mathfrak{R}$ iii) for every null set E of M, and every $\varepsilon > 0$, there is a set F which is a countable union from $\mathfrak{R}$ such that $F \supset E$ and $m(F) < \varepsilon$;

($\mathfrak{R}$ iv) $\lim_{n\to\infty} \sup_{i\in\mathbb{N}^n} m(F_{i_1,\ldots,i_n}) = 0.$

Notation: Given a net of partitions $\mathfrak{R}$, the sets $\{F_i : i \in \mathbb{N}^n\}$ are called *of nth rank* and for $x \in S$, $F_x^{(n)}$ denotes the unique set in $\mathfrak{R}$, of nth rank, which contains x. For $\sigma \in \Gamma$ let $F_\sigma^{(n)}$ be defined as in §1 if $\{F_s^{(n)} : s \in \sigma\}$ are mutually disjoint, and put $F_\sigma^{(n)} = \{\emptyset\}$ otherwise. Notice that, by σ-finiteness and ($\mathfrak{R}$ iv), for almost all $\sigma \in \Gamma \setminus \{\emptyset\}$, $F_\sigma^{(n)}$ eventually *gets off the ground*, and (also by ($\mathfrak{R}$ iv))

$$\mu(F_\sigma^{(n)}) \to 0 \text{ as } n \to \infty \text{ for a.a. } \sigma \neq \emptyset. \tag{3.2}$$

Proposition 3.3: Suppose that $x \in L^1_{\text{loc.}}(\Gamma^3)$ and that the measure space M supports a net of partitions $\mathfrak{R} = \{F_i : i \in \bigcup_n \mathbb{N}^n\}$. Then

$$x'(\rho,\sigma,\tau) = \lim_{n\to\infty} \mu\left(F^{(n)}_{\rho\cup\sigma\cup\tau}\right)^{-1} \int_{F^{(n)}_{\rho\cup\sigma}} (X_3 \chi_{F^{(n)}_{\sigma\cup\tau}})$$

for a.a. $(\rho,\sigma,\tau) \in \Gamma^3$.

Proof: By non-atomicity and the previous remark we may suppose that ρ, σ and τ have no points in common and the sets $F_x^{(n)}$, with x running through $\rho \cup \sigma \cup \tau$, are distinct for sufficiently large n — in particular $\mu^3\left(F_\rho^{(n)} \times F_\sigma^{(n)} \times F_\tau^{(n)}\right) = \mu\left(F^{(n)}_{\rho\cup\sigma\cup\tau}\right)$. By Proposition 3.1 and the Lebesgue–Vitali Theorem or, more particularly, de Possel's theorem (see e.g. [ShG]), the absolute difference between $x'(\rho,\sigma,\tau)$ and the above limit is, for almost all (ρ,σ,τ),

$$\lim_{n\to\infty} \sum_{\omega \subsetneq \sigma} \left\{\mu\left(F_{\bar\omega}^{(n)}\right) / \mu^3(F(n,\rho,\omega,\tau))\right\} \int_{F(n,\rho,\omega,\tau)} x', \tag{3.3}$$

where $F(n,\rho,\omega,\tau) = F^{(n)}_{\rho\cup\bar\omega} \times F^{(n)}_\omega \times F^{(n)}_{\bar\omega\cup\tau}$. By (3.2), for almost all σ, $\mu\left(F_\alpha^{(n)}\right) \to 0$ as $n \to \infty$ for $\alpha \subset \sigma$, $\alpha \neq \emptyset$, and so another application of the Lebesgue–Vitali Theorem shows that the limit (3.3) is zero for almost all (ρ,σ,τ). This completes the proof. ∎

This gives a further uniqueness statement for integral-sum operators (c.f. [L]).

Corollary 3.4: Suppose that the measure space M supports a net of partitions. If $x \in L^1_{\text{loc.}}(\Gamma^3)$ and $X_3 f = 0 \ \forall f \in \mathcal{D}_0$, then $x = 0$ a.e.

One might hope for a converse for Proposition 3.2 of the form:

If $X : \mathcal{D}_0 \to L^0$ is a linear operator and if

$$(\rho, \sigma, \tau) \mapsto \lim_{n \to \infty} \mu\big(F^{(n)}_{\rho \cup \sigma \cup \tau}\big)^{-1} \int_{F^{(n)}_{\rho \cup \sigma}} (X \chi_{F^{(n)}_{\sigma \cup \tau}})$$

exists a.e. and defines a function $x' \in L^1_{\text{loc.}}(\Gamma^3)$ then $X = X_3$.

The following example shows that this is false:

Example 3.5: Let M be the Lebesgue space $\mathbb{R}$, and let X be a second quantised shift thus $Xk(\sigma) = k(\sigma + t)$ where $t \neq 0$ and $\sigma + t := \{s_1 + t, \ldots, s_n + t\}$ when $\sigma = \{s_1, \ldots, s_n\}$. Then X is defined on all of $L^2(\Gamma)$, and leaves each $\mathcal{H}_a$ invariant, but for almost all (ρ, σ, τ), the terms of the sequence $\int_{F^{(n)}_{\rho \cup \sigma}} \big(X \chi_{F^{(n)}_{\sigma \cup \tau}}\big)$ will eventually be zero.

We end by summarising what is required of a converse. For an operator X to be an integral-sum operator with locally integrable kernel, there must exist a sequence of absolutely continuous measures $\big(\Phi^{(n)}$ on $\Gamma \times \Gamma_n \times \Gamma\big)$ such that

$$\Phi^{(0)}(E_\lambda \times \{\emptyset\} \times E_\nu) = \int_{E_\lambda} (X \chi_{E_\nu}) \text{ and, for } n = \#\mu \geqslant 1,$$

$$\Phi^{(n)}(E_\lambda \times E_\mu \times E_\nu) = \int_{E_{\lambda \cup \mu}} (X \chi_{E_{\mu \cup \nu}}) - \sum\nolimits_{\omega \subsetneq \mu} \Phi^{(n-1)}\big(E_{\lambda \cup \bar{\omega}} \times E_\omega \times E_{\bar{\omega} \cup \nu}\big)$$

for $E_\lambda, E_\mu, E_\nu \in \mathcal{G}_0$ and $\{E_i : i \in \lambda \cup \mu \cup \nu\}$ mutually disjoint.

References

[B] V P Belavkin: A quantum non-adapted Itô formula and generalised stochastic integration in Fock space, *J. Funct. Anal.* **102** no.2 (1991) 414-417.

[L] J M Lindsay: The kernel of a Fock space operator I, *Quantum Probability & Related Topics VIII (this volume).*

[ShG] G E Shilov and B L Gurevich: Integral, measure and derivative: a unified approach (transl. R A Silverman) Prentice-Hall, N.J. (1966).

Mathematics Department
University Park
Nottingham NG7 2RD
England

Quantum Probability and Related Topics Vol. VIII (pp. 95–97)

AN APPLICATION OF FIXED POINT THEORY TO QUANTUM SDE

by
Andreas Boukas
Department of Mathematics, American College of Greece
6 Gravias Str., Aghia Paraskevi, Athens, Greece GR 15342

Abstract. The existence and uniqueness of solutions of a certain class of quantum stochastic differential equations in Boson Fock space is established with the use of Tichonov's fixed point theorem.

1. INTRODUCTION

Let H_0 be a separable (initial) Hilbert space, let $\Gamma(L^2([0,\infty),\mathbb{C}))$ be the Fock space over the space of square integrable complex-valued functions on $[0,\infty)$, let $A^+=\{A^+(t)/t\geq 0\}$ and $A=\{A(t)/t\geq 0\}$ be respectively the creation and annihilation processes of [1], and let $A_1,A_2,A_3\in L(H_0\otimes\Gamma(L^2([0,\infty),\mathbb{C})))$, the space of bounded linear operators from $H_0\otimes\Gamma(L^2([0,\infty),\mathbb{C}))$ whose identity is denoted by I. The existence and uniqueness of solutions of the quantum stochastic differential equation (QSDE) $dX(t)=A_1X(t)dA^+(t)+A_2X(t)dA(t)+A_3X(t)dt$ with initial condition $X(0)=I$, was proved in [1] with the use of Picard's method of successive approximations. In an effort to establish the same result via the usually "cleaner" fixed point theory methods we consider adapted processes $X=\{X(t)\}$ (in the sense of [1]) with t ranging over some finite interval and with domain of interest for each $X(t)$, $H_0\otimes\text{span}\{y(f)/f\in B\}$, where B is the unit ball of $L^2([0,\infty),\mathbb{C})$. The initial condition is generalized to $X(t_0)=X_0$ where $t_0\geq 0$ and $X_0\in L(H_0\otimes\Gamma(L^2([0,\infty),\mathbb{C})))$ but we assume, for technical reasons, that $\{A_i=\tilde{A}_i\otimes I/i=1,2,3\}$ is a commuting family with each $\tilde{A}_i\in L(H_0)$ compact and self-adjoint, where I is the identity of $L(\Gamma(L^2([0,\infty),\mathbb{C})))$. The solution obtained lives on a finite interval $[t_0,t_0+\delta]$ but can be continuously extended by reapplying our method to starting time $t_0+\delta$. The fixed point theorem suitable for our class of QSDE is that of Tichonov ([2]) which states that "a compact convex nonempty subset of a locally convex topological vector space has the fixed point property".

2. THE TOPOLOGICAL SET UP

Let $\{A_i=\tilde{A}_i\otimes I\in L(H_0\otimes\Gamma(L^2([0,\infty),\mathbb{C})))/i=1,2,3\}$ be a commuting family with each $\tilde{A}_i$ compact self-adjoint, let $U=\{u_n\}_{n=1}^{\infty}$ be an orthonormal basis of H_0 consisting of common eigenvectors of $\tilde{A}_1,\tilde{A}_2$ and $\tilde{A}_3$ ([3]), and let $D=\{f_n\}_{n=1}^{\infty}$ be a dense subset of B. If $\tau>0$ is arbitrary then for an adapted process $X=\{X(t)/t_0\leq t\leq t_0+\tau\}$ we define for each $n,k,N,K\in\mathbb{N}\overset{\text{def}}{=}\{1,2,3,\dots\}$

$$ {}_{n,N}\rho^{\tau}_{k,K}(X) \stackrel{\text{def}}{=} \sup_{t_0 \le t \le t_0+\tau} |\langle X(t)u_n \otimes y(f_N), u_k \otimes y(f_K)\rangle| \tag{2.1} $$

provided that the right hand side of (2.1) is finite. For i=1,2,3 and $n,k,N,K\in\mathbb{N}$

$$ {}_{n,N}\rho^{\tau}_{k,K}(A_i X) \le \|A_i\| \; {}_{n,N}\rho^{\tau}_{k,K}(X) \tag{2.2} $$

since the norm of a compact self-adjoint operator in a Hilbert space is equal to the supremum of the set of its eigenvalues.

Let $M \stackrel{\text{def}}{=} \sqrt{2}\,(\|A_1\|^2+\|A_2\|^2)^{1/2}+\tau\|A_3\|+\|X_0\|$ and choose $\delta\in(0,\tau]$ such that $\delta M\le 1$. If we define

$$ S \stackrel{\text{def}}{=} \{\text{adapted } X=\{X(t)/t_0\le t\le t_0+\delta\}/t \longrightarrow \langle X(t)u_n\otimes y(f_N),u_k\otimes y(f_K)\rangle \text{ is bounded for all } n,k,N,K\in\mathbb{N}\} \tag{2.3} $$

$$ G \stackrel{\text{def}}{=} \{X\in S/\text{for all } n,k,N,K\in\mathbb{N} \text{ and } t,t'\in[t_0,t_0+\delta],\ {}_{n,N}\rho_{k,K}(X)\le\delta M \text{ and } |\langle (X(t)-X(t'))u_n\otimes y(f_N),u_k\otimes y(f_K)\rangle|\le(M-\|X_0\|)\,|t-t'|^{1/2}\} \tag{2.4} $$

where ${}_{n,N}\rho_{k,K} \stackrel{\text{def}}{=} {}_{n,N}\rho^{\delta}_{k,K}$ as in (2.1), then we can prove the following:

Theorem 2.1. $(S,\{{}_{n,N}\rho_{k,K}\})$ is a metrizable locally convex topological vector space.

Proof. It is easy to see that S is a vector space and that each ${}_{n,N}\rho_{k,K}:S\longrightarrow[0,\infty)$ is a semi-norm. Moreover, if ${}_{n,N}\rho_{k,K}(X)=0$ for all $n,k,N,K\in\mathbb{N}$ then for $u,v\in H_0$ and $f,g\in B$ we may choose $\{f_\lambda\}_{\lambda=1}^{\infty}$, $\{g_\mu\}_{\mu=1}^{\infty}\subset D$ such that $f_\lambda\longrightarrow f$ and $g_\mu\longrightarrow g$ in the L^2-sense, as $\lambda,\mu\longrightarrow\infty$, to conclude that for every $t\in[t_0,t_0+\delta]$

$$ \langle X(t)u\otimes y(f),v\otimes y(g)\rangle=\sum_{n,k=1}^{\infty}\langle u,u_n\rangle\langle u_k,v\rangle \lim_{\lambda,\mu}\langle X(t)u_n\otimes y(f_\lambda),u_k\otimes y(g_\mu)\rangle=0 $$

Thus $X(t)=0$ on $H_0\otimes\text{span}\{y(f)/f\in B\}$ for all $t\in[t_0,t_0+\delta]$ and so $(S,\{{}_{n,N}\rho_{k,K}\})$ is a semi-normed vector space which equipped with the topology induced by the countably many seminorms ${}_{n,N}\rho_{k,K}$ becomes a metrizable locally convex topological vector space whose topology can therefore be described by sequential convergence.

Theorem 2.2. G is a compact convex nonempty subset of S.

Proof. The zero process is in G so $G\neq\emptyset$. By the triangle inequality G is convex. If $\{X_\lambda\}_{\lambda=1}^{\infty}\subset G$ then $\{t\in[t_0,t_0+\delta]\longrightarrow \langle X_\lambda(t)u_n\otimes y(f_N),u_k\otimes y(f_K)\rangle\}_{\lambda=1}^{\infty}$ is, for each $n,k,N,K\in\mathbb{N}$, a uniformly bounded and equicontinuous family of complex valued functions. By the Ascoli-Arzela theorem and the diagonal process we can select a subsequence, say $\{X_\lambda\}_{\lambda=1}^{\infty}$, again, such that for all $n,k,N,K\in\mathbb{N}$ $\{t \longrightarrow \langle X_\lambda(t)u_n\otimes y(f_N),u_k\otimes y(f_K)\rangle\}_{\lambda=1}^{\infty}$ converges uniformly on

$[t_0,t_0+\delta]$ to a continuous function. For each $t\in[t_0,t_0+\delta]$ let

$$<X(t)u_n\otimes y(f_N),u_k\otimes y(f_K)> \stackrel{def}{=} \lim_\lambda <X_\lambda(t)u_n\otimes y(f_N),u_k\otimes y(f_K)> \quad (2.5)$$

and extend to $H_0\otimes\mathrm{span}\{y(f)/f\in B\}$ by

$$<X(t)u\otimes y(f),v\otimes y(g)> = \sum_{n,k=1}^{\infty} <u,u_n><u_k,v> \lim_{N,K} <X(t)u_n\otimes y(f_N),u_k\otimes y(f_K)>$$

with the infinite sum converging by (2.4) and Bessel's inequality. Thus defined, $X=\{X(t)/t_0\le t\le t_0+\delta\}$ is a weakly continuous adapted process to which $\{X_\lambda\}_{\lambda=1}^{\infty}$ converges in the topology of S. Moreover, for each $n,k,N,K\in\mathbb{N}$, ${}_{n,N}\rho_{k,K}(X)\le{}_{n,N}\rho_{k,K}(X-X_\lambda)+{}_{n,N}\rho_{k,K}(X_\lambda)\le{}_{n,N}\rho_{k,K}(X-X_\lambda)+\delta M$ which upon letting $\lambda\longrightarrow\infty$ implies ${}_{n,N}\rho_{k,K}(X)\le\delta M$. Finally, for $t,t'\in[t_0,t_0+\delta]$, $|<(X(t)-X(t'))u_n\otimes y(f_N),u_k\otimes y(f_K)>|=\lim_\lambda|<(X_\lambda(t)-X_\lambda(t'))u_n\otimes y(f_N),u_k\otimes y(f_K)>|\le$

$\le (M-\|X_0\|)\cdot|t-t'|^{1/2}$ since $\{X_\lambda\}_{\lambda=1}^{\infty}\subset G$. Thus $X\in G$ and so G is compact.

3. THE EXISTENCE AND UNIQUENESS THEOREM

Theorem 3.1. For each $(t_0,X_0)\in[0,\infty)\times L(H_0\otimes\Gamma(L^2([0,\infty),\mathbb{C})))$ the initial value problem.

$$\begin{aligned} dX(t) &\quad A_1X(t)dA^+(t)+A_2X(t)dA(t)+A_3X(t)dt \\ X(t_0)&=X_0 , \quad t_0\le t\le t_0+\delta \end{aligned} \quad (3.1)$$

where δ,A_1,A_2,A_3 are as in section 2, has a unique solution X which belongs to the set G defined in (2.4).

Proof. For $X\in G$ and for $t\in[t_0,t_0+\delta]$ define

$$(PX)(t) = X_0+\int_{t_0}^{t} A_1X(s)dA^+(s)+A_2X(s)dA(s)+A_3X(s)ds . \quad (3.2)$$

PX is an adapted process which, by the way δ was chosen and by the first fundamental theorem of the Hudson-Parthasarathy calculus ([1], Thm. 4.1) along with the triangle and Holder's inequalities, belongs to G. Moreover, if $\{X_\lambda\}_{\lambda=1}^{\infty}\subset G$ and $X_\lambda\longrightarrow X$ in the topology of Thm. 2.1, as $\lambda\longrightarrow\infty$, then in view of (2.2) we have for each $n,k,N,K\in\mathbb{N}$

$${}_{n,N}\rho_{k,K}(PX_\lambda-PX)\le[\sqrt{2}(\|A_1\|^2+\|A_2\|^2)^{1/2}\delta^{1/2}+\|A_3\|\delta]\,{}_{n,N}\rho_{k,K}(X_\lambda-X). \quad (3.3)$$

which implies that $P:G\longrightarrow G$ is continuous. By Tichonov's fixed point theorem P has a fixed point $X\in G$ which solves (3.1). By (2.2) and Gronwall's inequality such a solution is unique.

REFERENCES

[1] HUDSON, R.L., PARTHASARATHY, K.R.: Quantum Ito's formula and stochastic evolutions, Comm. in Math. Phys. 93, 301-323 (1984).

[2] SMART, D.R.: "66 Fixed Point Theorems", Cambridge University Press (1974).

[3] VON NEUMANN, J.: Zur Algebra der Funktionaloperationen und Theorie der Normalen Operatoren, Math. Ann. 102, 370-427, (1929-1930).

Quantum Probability and Related Topics Vol. VIII (pp. 99–108)

Matrix elements stability of quantum stochastic differential equations

by

ANDREAS BOUKAS
Department of Mathematics, American College of Greece, 6 Gravias Str., Aghia Paraskevi, GR-15342, Athens, Greece

Abstract The stability of the zero solution of the quantum stochastic differential equation $dX(t)=A_1(t)X(t)dB(t)+A_2(t)X(t)dt$, in the presence of Fock space white noise B, is defined and studied with the use of the Hudson-Parthasarathy quantum stochastic calculus. An explicit formula is obtained for the solution of the associated initial value problem in the case when A_1, A_2 are compact self-adjoint adapted processes. A variation of parameters formula is obtained in the case of constant A_2.

1. Introduction

The Hudson-Parthasarathy Fock space Γ over $L^2([0,\infty),\mathbb{C})$ is defined (along with the associated quantum stochastic calculus, familiarity with which is assumed by the reader) in [1], [2] as the Hilbert space completion of (span E, <,>) where $E=\{y(f)/f\in L^2([0,\infty),\mathbb{C})\}$ is a family of exponential vectors with $\langle y(f),y(g)\rangle=\exp(\int_0^\infty f(s)\bar{g}(s)ds)$. If $a^+=\{a^+(t)/t\geq 0\}$ and $a=\{a(t)/t\geq 0\}$ are respectively the creation and annihilation processes then $B\stackrel{def}{=} i(a^+-a)$ is a quantum Brownian motion. Moreover, if $A_1=\{A_1(t)/t\geq 0\}$ and $A_2=\{A_2(t)/t\geq 0\}$ are *adapted processes* (in the sense of [2]) which are also *locally bounded* and *measurable* in the sense that if H_0 is the initial Hilbert space then $A_1(t)$, $A_2(t)\in L(H_0\otimes\Gamma)$ for each $t\geq 0$, $\sup_{0\leq t\leq T}\|A_1(t)\|_{L(H_0\otimes\Gamma)}<\infty$ and $\sup_{0\leq t\leq T}\|A_2(t)\|_{L(H_0\otimes\Gamma)}<\infty$ for each $T>0$, and the maps $t\to A_1(t)$ and $t\to A_2(t)$ are strongly measurable, where $L(H_0\otimes\Gamma)$ is the algebra of bounded linear operators: $H_0\otimes\Gamma\to H_0\otimes\Gamma$, then the initial value problem

$$\begin{aligned} dX(t) &= A_1(t)X(t)dB(t)+A_2(t)X(t)dt \\ X(t_0) &= X_0 \end{aligned} \tag{1.1}$$

associated with the quantum stochastic differential equation (SDE)

$$dX(t) = A_1(t)X(t)dB(t)+A_2(t)X(t)dt \tag{1.2}$$

can be shown, by modifying the proof in [1] for $X(0) = I$, to have a unique strongly continuous solution. That is, for each $t_0\geq 0$ and for each

$X_0 \in L(H_0 \otimes \Gamma)$ with $X_0 = \tilde{X}_0 \otimes I$, where I is the identity of $L(\Gamma)$, there exists a unique adapted process $X = \{X(t)/t \geq t_0\}$ defined on spanE, such that for each $u \in H_0$, $f \in L^2([0,\infty),\mathbb{C})$ and $t \geq t_0$

$$X(t)u\otimes y(f) = [X_0 + \int_{t_0}^{t} A_1(s)X(s)dB(s) + \int_{t_0}^{t} A_2(s)X(s)ds]u\otimes y(f) \tag{1.3}$$

and the map $t \to X(t)y(f)$ is strongly continuous on $[t_0,\infty)$. Here, and in what follows, u and v denote arbitrary elements of H_0 and y(f), y(g) denote arbitrary exponential vectors.

Since in Fock space stochastic analysis an adapted process X is usually considered through its *matrix elements* $\langle X(t)u\otimes y(f), v\otimes y(g)\rangle$, we define the stability of X in the following way adapted from the classical theory of ordinary and stochastic differential equations ([5], [7]).

Definition 1.1. *Let $\{X(t,t_0,X_0)/t \geq t_0\}$ be the solution of (1.1) and let $D \subset H_0 \times H_0 \times L^2([0,\infty),\mathbb{C}) \times L^2([0,\infty),\mathbb{C})$. The zero solution of (1.2) is:*

(i) stable on D if and only if for every $\varepsilon>0$, $t_0 \geq 0$ and $(u,v,f,g) \in D$ there exists $\delta>0$ depending on them such that for all $t \geq t_0$, $\|X_0\| < \delta$ implies $|\langle X(t,t_0,X_0)u\otimes y(f), v\otimes y(g)\rangle| < \varepsilon$.

(ii) uniformly stable on D if and only if it is stable on D and δ is independent of t_0.

(iii) globally asymptotically stable on D if and only if it is stable on D and for each (t_0,X_0) we have $\lim_{t\to\infty} \langle X(t,t_0,X_0)u\otimes y(f), v\otimes y(g)\rangle = 0$ for all $(u,v,f,g) \in D$.

The stability of (1.2) in the case of constant compact self-adjoint A_1, A_2 is considered with the help of the spectral resolution theorem for such operators. The case of constant A_2 is considered via a variation of parameters formula by treating $A_1(t)X(t)dB(t)$ as a perturbation. The case of nonconstant A_1, A_2 is handled with the use of Gronwall's inequality. The results obtained constitute quantum stochastic analogues of stability results known for linear systems of first order ODE ([5], Ch. 1.1) as well as for classical SDE ([7]).

In what follows, subscripts to norms are omitted since the underlying normed space is in each case obvious. All "time" integrals of complex-valued functions are in the sense of Lebesgue. All integrals of vector valued functions are in the sense of [1], [2]. In particular, if $f \in L^2([0,\infty),\mathbb{C})$ and $g \in L^1([0,\infty),\mathbb{C})$ then

$\|f\|_2 = [\int_0^\infty |f(t)|^2 dt]^{1/2}$ and $\|g\|_1 = \int_0^\infty |g(t)| dt$. For an operator A, the notions of positivity ($A \geq 0$) and positive definiteness ($A>0$) as well as the associated notions of negativity and negative definiteness, defined through $-A$, are as in [3]. Finally if $A = \tilde{A}\otimes I \in L(H_0\otimes\Gamma)$ then $\|A\|=\|\tilde{A}\|$ ([8], Prop. 43.1).

2. The basic Lemmas

Lemma 2.1. *(Variation of parameters formula). If $A\in L(H_0\otimes\Gamma)$ is a constant adapted process and $p=\{p(t)/t\geq t_0\}$ is an adapted process for which $\int_{t_0}^{t} p(s)dB(s)$ makes sense then the solution of*

$$dX(t) = p(t)dB(t)+A\,X(t)dt \qquad (2.1)$$
$$X(t_0) = X_0$$

is given by

$$X(t) = \exp((t-t_0)A)X_0+\int_{t_0}^{t}\exp((t-s)A)p(s)dB(s). \qquad (2.2)$$

Proof. *By (2.2) we have $X(t_0) = X_0$ and*

$$dX(t)=d(\exp((t-t_0)A))X_0+d(\exp(tA))\int_{t_0}^{t}\exp(-sA)p(s)dB(s)+$$

$$+\exp(tA)d(\int_{t_0}^{t}\exp(-sA)p(s)dB(s))+d(\exp(tA))d(\int_{t_0}^{t}\exp(-sA)p(s)dB(s))$$

$$=A\exp((t-t_0)A)X_0dt+A\exp(tA)dt\int_{t_0}^{t}\exp(-sA)p(s)dB(s)+p(t)dB(t)+$$

$$+A\,p(t)dt\,dB(t) = A\,X(t)dt+p(t)dB(t) \quad \text{since} \quad dt\cdot dB(t) \quad 0.$$

Thus the unique solution of (2.1) is given by (2.2). ■

Lemma 2.2. *(i) If $t_0,T\in[0,\infty)$ and $\{A_T(t)/t\geq t_0\}$ is an adapted process for which $M_T(t) \stackrel{\text{def}}{=} \int_{t_0}^{t}A_T(s)dB(s)$ and $\int_{t_0}^{t}\|A_T(s)u\otimes y(f)\|^2ds$ make sense for $t\in[t_0,T], u\in H_0$ and $f\in L^2([0,\infty),\mathbb{C})$ then*

$$\|M_T(t)u\otimes y(f)\|^2 \leq 2\exp(4\|f\|_2^2)\int_{t_0}^{t}\|A_T(s)u\otimes y(f)\|^2ds. \qquad (2.3)$$

(ii) Let $t_0 \geq 0$ *and let* $\{A(t) \in L(H_0 \otimes \Gamma)/t \geq 0\}$ *and* $\{X(t)/t \geq t_0\}$ *be adapted processes for which* $N(t) \overset{\text{def}}{=} \int_{t_0}^{t} A(s)X(s)ds$, $\int_0^{\infty} \|A(s)\|ds$, *and* $\int_{t_0}^{t} \|A(s)\|\|X(s)u\otimes y(f)\|^2 ds$ *make sense for* $t \geq t_0$, $u \in H_0$ *and* $f \in L^2([0,\infty),\mathbb{C})$. *If for each* $t \geq t_0$, $A(t) \geq 0$ *or* $A(t) \leq 0$ *then*

$$\|N(t)u\otimes y(f)\|^2 \leq \exp\left(\int_0^{\infty} \|A(s)\|ds\right)\int_{t_0}^{t} \|A(s)\|\|X(s)u\otimes y(f)\|^2 ds. \tag{2.4}$$

Proof. (i) By Theorem 4.2 of [2], as in the derivation of (5.2) of [2], for $t_0 \leq t \leq T, u \in H_0$, and $f \in L^2([0,\infty),\mathbb{C})$ we have

$$\frac{d}{dt}\|M_T(t)u\otimes y(f)\|^2 = 2\mathrm{Re}\langle i(\bar{f}-f)(t)M_T(t)u\otimes y(f), A_T(t)u\otimes y(f)\rangle + \|A_T(t)u\otimes y(f)\|^2 \leq$$

$$\leq 4|f(t)|^2\|M_T(t)u\otimes y(f)\|^2 + 2\|A_T(t)u\otimes y(f)\|^2$$

from which the result follows by using the integrating factor $\exp(-4\int_s^t |f(w)|^2 dw)$.

(ii) By Theorem 4.2 of [2], for each $t \geq t_0$, $u \in H_0$ and $f \in L^2([0,\infty),\mathbb{C})$ we have

$$\frac{d}{dt}\|N(t)u\otimes y(f)\|^2 \quad 2\mathrm{Re}\langle N(t)u\otimes y(f), A(t)X(t)u\otimes y(f)\rangle. \tag{2.5}$$

Without loss of generality we may assume $A(t) \geq 0$. If $A(t) \leq 0$ then the right hand side of (2.5) is written as $-2\mathrm{Re}\langle N(t)u\otimes y(f), -A(t)X(t)u\otimes y(f)\rangle$ and the rest of the proof is exactly the same. Now, (2.5) implies

$$\frac{d}{dt}\|N(t)u\otimes y(f)\|^2 \quad 2\mathrm{Re}\langle A(t)^{1/2}N(t)u\otimes y(f), A(t)^{1/2}X(t)u\otimes y(f)\rangle \leq$$

$$\leq \|A(t)\|\|N(t)u\otimes y(f)\|^2 + \|A(t)\|\|X(t)u\otimes y(f)\|^2$$

and the proof is completed by using the integrating factor $\exp(-\int_s^t \|A(w)\|dw)$. ■

Lemma 2.3.: *If* A_1, A_2 *are constant adapted processes and* $(Y_k)_{k=1}^{\infty}$ *is a family of adapted processes such that for* $t \geq t_0$

$$\begin{aligned} dY_k(t) &= A_1 Y_k(t)dB(t) + A_2 Y_k(t)dt \\ Y_k(t_0) &= L_k \end{aligned} \tag{2.6}$$

where $L_k \in L(H_0 \otimes \Gamma)$ *satisfies* $\lim_k \| L_k u \otimes y(f) \| = 0$ *for all* $u \in H_0$, $f \in L^2([0,\infty),\mathbb{C})$ *then*

$$\lim_k \langle Y_k(t)\, u \otimes y(f),\, v \otimes y(g) \rangle = 0 \tag{2.7}$$

uniformly on compact subsets of $[t_0,\infty)$, *for all* $(u,v,f,g) \in H_0 \times H_0 \times L^2([0,\infty),\mathbb{C}) \times L^2([0,\infty),\mathbb{C})$.

Proof: By Lemma 2.2, if $t \in [t_0,\rho]$, $\rho \geq t_0$, then

$$\|Y_k(t)u \otimes y(f)\|^2 \leq 3\|Y_k(t_0)u \otimes y(f)\|^2 + 6\exp(4\|f\|_2^2) \cdot \int_{t_0}^{t} \|A_1\|^2 \cdot \|Y_k(s)u \otimes y(f)\|^2 ds +$$

$$+2\rho \int_{t_0}^{t} \|A_2\|^2 \|Y_k(s)u \otimes y(f)\|^2 ds \leq 3\|Y_k(t_0)u \otimes y(f)\|^2 + c \cdot \int_{t_0}^{t} \|Y_k(s)u \otimes y(f)\|^2 ds$$

where $c = \max(6\exp(4\|f\|_2^2)\|A_1\|^2,\ 2\rho\|A_2\|^2)$.

Thus by Gronwall's inequality ([5]),

$$\|Y_k(t)\ u \otimes y(f)\|^2 \leq 3\|L_k u \otimes y(f)\|^2 \exp(c\rho) \to 0 \quad \text{as} \quad k \to \infty\ . \quad \blacksquare$$

Lemma 2.4. *If* $A_1 = \tilde{A}_1 \otimes I$, $A_2 = \tilde{A}_2 \otimes I \in L(H_0 \otimes \Gamma)$ *are constant adapted processes such that* $\tilde{A}_1, \tilde{A}_2$ *commute and are compact and self-adjoint then the matrix elements of the solution of (1.1) are given for each* $t \geq t_0 \geq 0$ *by*

$$\langle X(t,t_0,X_0)u \otimes y(f), v \otimes y(g) \rangle =$$

$$= \sum_{n=1}^{\infty} \langle \tilde{X}_0 u, e_n \rangle \langle e_n \otimes y(f), v \otimes y(g) \rangle \exp[i\lambda_n \int_{t_0}^{t} (\bar{f}-g)(s)ds + (t-t_0)\mu_n] \tag{2.8}$$

where $(e_n)_{n=1}^{\infty}$ *is an orthonormal basis for* H_0 *consisting of eigenvectors of* $\tilde{A}_1$ *and* $\tilde{A}_2$ *corresponding to eigenvalues* $(\lambda_n)_{n=1}^{\infty}$, $(\mu_n)_{n=1}^{\infty}$ *respectively.*

Remark. The existence of an individual basis of eigenvectors for each of $\tilde{A}_1$ and $\tilde{A}_2$ is a well-known fact for compact self-adjoint operators while the existence of a common basis for commuting $\tilde{A}_1$, $\tilde{A}_2$ follows from a general spectral resolution theorem of Von Neumann ([3],[4]).

Proof of lemma. For n=1,2,... define Fourier projection like operators $E_n(X_0):H_0\otimes\Gamma \to H_0\otimes\Gamma$ by

$$E_n(X_0)u\otimes\gamma \quad <e_n,\bar{X}_0 u>e_n\otimes\gamma \tag{2.9}$$

If X_n is the solution of

$$\begin{aligned} dX_n(t) &\quad A_1X_n(t)dB(t)+A_2X_n(t)dt \\ X_n(t_0) &= \ _n(X_0) \end{aligned} \tag{2.10}$$

and

$$Y_n \quad [(A_1-\lambda_nI)+(A_2-\mu_nI)]X_n,$$

$$Z_n \quad [(A_1-\lambda_nI)-(A_2-\mu_nI)]X_n,$$

$$W_n \quad [-(A_1-\lambda_nI)+(A_2-\mu_nI)]X_n$$

where I is the identity of $L(H_0\otimes\Gamma)$ then for $K_n\in\{Y_n,Z_n,W_n\}$,

$$dK_n(t) = A_1K_n(t)dB(t)+A_2K_n(t)dt$$

$$K_n(t_0) \quad 0.$$

Thus $K_n(t)=0$ for all $t\geq t_0$ and so $(Y_n+Z_n)(t)=(Y_n+W_n)(t)=0$ for all $t\geq t_0$. Hence, for each $t\geq t_0$ and $n=1,2,...$

$$A_1X_n(t) = \lambda_nX_n(t), \quad A_2X_n(t) = \mu_nX_n(t). \tag{2.11}$$

For each $u\in H_0$ and $\gamma\in\Gamma$,

$$X_0u\otimes\gamma \quad \sum_{n=1}^{\infty}E_n(X_0)u\otimes\gamma \tag{2.12}$$

By applying Lemma 2.3 to $\{Y_k = X \quad \sum_{n=1}^{k}X_n\}_{k=1}^{\infty}$ we see that

$$<X(t)u\otimes y(f),v\otimes y(g)> \quad \sum_{n=1}^{\infty}<X_n(t)\ u\otimes y(f),v\otimes y(g)> \tag{2.13}$$

uniformly on compact subsets of $[t_0,\infty)$, where $X(t)=X(t,t_0,X_0)$.
By (1.3), (2.9), (2.10), (2.11), for $t\geq t_0$ and $n\geq 1$

$<X_n(t)u\otimes y(f),v\otimes y(g)> =$

$$<\bar{X}_0u,e_n> <e_n\otimes y(f),v\otimes y(g)> \exp\left[i\lambda_n\int_{t_0}^{t}(\bar{f}-g)(s)ds+(t-t_0)\mu_n\right]. \tag{2.14}$$

from which (2.8) follows in view of (2.13).■

3. The stability results

Proposition 3.1. *Let $A_1=\bar{A}_1\otimes I$, $A_2=\bar{A}_2\otimes I \in L(H_0\otimes\Gamma)$ be constant adapted processes such that $\bar{A}_1,\bar{A}_2$ commute and are compact and self-adjoint and let $D = H_0\times H_0\times D_0$ where*

$$D_0=\{(f,g)\in L^2([0,\infty),\mathbb{C})\times L^2([0,\infty),\mathbb{C})/\ \Psi \overset{\text{def}}{=} Im(g-\bar{f})\in L^1([0,\infty),\mathbb{R})\}.$$

(i) If $\bar{A}_2\le 0$ then the zero solution of (1.2) is uniformly stable on D.

(ii) If $\bar{A}_2<0$ then the zero solution of (1.2) is globally asymptotically stable on D.

Proof. (i) With the notation of Lemma 2.4, $\bar{A}_2\le 0$ implies $\mu_n\le 0$ for all n. Thus for $t\ge t_0$, $\exp(\mu_n(t-t_0))\le 1$ and so for $(f,g)\in D_0$ using $\|A_1\| = \sup_n|\lambda_n|$ we obtain

$$|\exp(i\lambda_n\int_{t_0}^{t}(\bar{f}-g)(s)ds+(t-t_0)\mu_n| \le \exp(\|A_1\|\|\Psi_1\|). \tag{3.1}$$

By Lemma 2.4, (3.1), and Bessel's inequality we have for $(u,v,f,g)\in D$ $|\langle X(t,t_0,X_0)u\otimes y(f),v\otimes y(g)\rangle|\le\|X_0\|\|u\otimes y(f)\|\|v\otimes y(g)\|\exp(\|A_1\|\|\Psi_1\|)$ from which the result follows by taking in Definition 1.1 (ii)

$\delta=\varepsilon[\|u\otimes y(f)\|\|v\otimes y(g)\|\exp(\|A_1\|\|\Psi_1\|)]^{-1}$.

(ii) As in the proof of (i), $A_2<0$ implies $\mu_n<0$ for all n. Thus for $(f,g)\in D_0$, $t_0\ge 0$ and $n=1,2,\dots$ we have

$$|\exp(i\lambda_n\int_{t_0}^{t}(\bar{f}-g)(s)ds+(t-t_0)\mu_n)|\le\exp(\|A_1\|\|\Psi\|_1)\exp(\mu_n(t-t_0))\to 0,\ t\to\infty. \tag{3.2}$$

Letting $f_n(t)=\langle\bar{X}_0u,e_n\rangle\langle e_n\otimes y(f),v\otimes y(g)\rangle\exp(i\lambda_n\int_{t_0}^{t}(\bar{f}-g)(s)ds+(t-t_0)\mu_n)$ and

$g_n=|\langle\bar{X}_0u,e_n\rangle||\langle e_n\otimes y(f),v\otimes y(g)\rangle|\exp(\|A_1\|\|\Psi\|_1)$ we see that by (3.2) $f_n(t)\to 0$ as $t\to\infty$, $|f_n(t)|\le g_n$ for all $t\ge t_0$ and $n=1,2,\dots$, and

$\sum_{n=1}^{\infty} g_n \le \exp(\|A_1\|\|\Psi\|_1)\|X_0\|\|u\|\|y(f)\|\|v\otimes y(g)\|<\infty$. Thus by Lebesgue's bounded convergence theorem, $\lim_{t\to\infty} \sum_{n=1}^{\infty} f_n(t) = \sum_{n=1}^{\infty} \lim_{t\to\infty} f_n(t)$ 0 which, in view of (2.8), implies $<X(t,t_0,X_0)u\otimes y(f), v\otimes y(g)> \to 0$ as $t\to\infty$ for all $(u,v,f,g)\in D$ and $(t_0,X_0)\in[0,\infty)\times L(H_0\otimes\Gamma)$. Combining this with (i) we see that the zero solution of (1.2) is globally asymptotically stable on D. ■

Proposition 3.2. *Let $A_2 = \tilde{A}_2\otimes I \in L(H_0\otimes\Gamma)$ be a constant adapted process, let $\{A_1(t)/t\ge 0\}$ be a locally bounded measurable adapted process such that $\int_0^{\infty}\|A_1(t)\|^2 dt<\infty$, and let $D = H_0\times H_0\times L^2([0,\infty),\mathbb{C})\times L^2([0,\infty),\mathbb{C})$.*

(i) If $\tilde{A}_2\le 0$ then the zero solution of (1.2) is uniformly stable on D.

(ii) If there exists $\beta\in(0,\infty)$ such that $\tilde{A}_2+\beta I\le 0$, where I is the identity of $L(H_0)$, then the zero solution of (1.2) is globally asymptotically stable on D.

Proof. (i) By the Lumer-Phillips theorem for contraction semigroups ([6]), $\tilde{A}_2\le 0$ implies $\sup_{t\ge 0}\|\exp(t\tilde{A}_2)\|\le 1$.

By Lemmas 2.1, 2.2 (i), for $t\ge t_0$, $u\in H_0$ and $f\in L^2([0,\infty),\mathbb{C})$ we have

$$\|X(t,t_0,X_0)u\otimes y(f)\|^2 \le 2\|X_0\|^2\cdot\|y(f)\|^2 +$$

$$+4\exp(4\|f\|_2^2)\int_{t_0}^{t}\|A_1(s)\|^2\|X(s,t_0,X_0)u\otimes y(f)\|^2 d]s$$

and by Gronwall's inequality,

$$\|X(t,t_0,X_0)u\otimes y(f)\| \le \sqrt{2}\|X_0\|\|y(f)\|\exp(2\exp(4\|f\|_2^2)\int_0^{\infty}\|A_1(s)\|^2 ds)$$

from which the result follows by applying the Cauchy-Schwartz inequality to $<X(t,t_0,X_0)u\otimes y(f), v\otimes y(g)>$ and taking in Definition 1.1 (ii) $\delta=\varepsilon[\sqrt{2}\|u\otimes y(f)\|\|v\otimes y(g)\|\exp(2\exp(4\|f\|_2^2)\int_0^{\infty}\|A_1(s)\|^2 ds)]^{-1}$.

(ii) For every $t\ge 0$, $\|\exp(t(\tilde{A}_2+\beta I))\|\le 1$ and so $\|\exp(t\tilde{A}_2)\|\le\exp(-\beta t)$.

By Lemmas 2.1, 2.2 (i), for $t\ge t_0$, $u\in H_0$ and $f\in L^2([0,\infty),\mathbb{C})$ we have

$$\|X(t,t_0,X_0)u\otimes y(f)\|^2 \le 2\exp(-2\beta(t-t_0))\|X_0\|^2\|u\otimes y(f)\|^2 + 4\exp(4\|f\|_2^2)\cdot$$

$$\cdot\int_{t_0}^{t}\exp(-2\beta(t-s))\|A_1(s)\|^2\|X(s,t_0,X_0)u\otimes y(f)\|^2ds. \tag{3.3}$$

Multiplying both sides of (3.3) by $\exp(2\beta(t-t_0))$, applying Gronwall's inequality to the result and then multiplying by $\exp(-2\beta(t-t_0))$ we obtain

$$\|X(t,t_0,X_0)u\otimes y(f)\|^2\le 2\exp(-2\beta(t-t_0))\|X_0\|^2\|y(f)\|^2\exp(4\exp(4\|f\|_2^2)\cdot$$

$\cdot\int_0^\infty\|A_1(s)\|^2ds)\to 0$ as $t\to\infty$, which by the Cauchy-Schwartz inequality

implies $\langle X(t,t_0,X_0)u\otimes y(f),v\otimes y(g)\rangle\to 0$ as $t\to\infty$ for all $(u,v,f,g)\in D$ and $(t_0,X_0)\in[0,\infty)\times L(H_0\otimes\Gamma)$. Since by part (i) the zero solution is uniformly stable on D the proof is complete. ■

Proposition 3.3 *Suppose that $(A_1(t)/t\ge 0)$ and $(A_2(t)/t\ge 0)$ are locally bounded measurable adapted processes such that $\int_0^\infty\|A_1(t)\|^2dt<\infty$ and $\int_0^\infty\|A_2(t)\|dt<\infty$. If for each $t\ge 0$, $A_2(t)\ge 0$ or $A_2(t)\le 0$ then the zero solution of (1.2) is uniformly stable on $D=H_0\times H_0\times L^2([0,\infty),\mathbb{C})\times L^2([0,\infty),\mathbb{C})$.*

Proof. By (1.3) and Lemma 2.2, for $u\in H_0$, $f\in L^2([0,\infty),\mathbb{C})$ and $t\ge t_0\ge 0$ we have

$$\|X(t,t_0,X_0)u\otimes y(f)\|^2\le 3\|X_0\|^2\|y(f)\|^2+\int_{t_0}^{t}[6\exp(4\|f\|_2^2)\|A_1(s)\|^2+$$

$$+3\exp(\int_0^\infty\|A_2(s)\|ds)\|A_2(s)\|]\|X(s,t_0,X_0)u\otimes y(f)\|^2ds. \tag{3.4}$$

Applying Gronwall's inequality to (3.4) we obtain

$$\|X(t,t_0,X_0)u\otimes y(f)\|^2 \le 3\|X_0\|^2\|y(f)\|^2\exp\Big[6\exp(4\|f\|_2^2)\int_0^\infty\|A_1(s)\|^2ds+$$

$$+3\exp(\int_0^\infty\|A_2(s)\|ds)\int_0^\infty\|A_2(s)\|ds\Big] \tag{3.5}$$

In view of (3.5) the proof is completed by applying the Cauchy-Schwartz inequality to $\langle X(t,t_0,X_0)u\otimes y(f),v\otimes y(g)\rangle$ and by taking in Definition 1.1 (ii)

$$\delta = \varepsilon\left[\sqrt{3}\exp(3\exp(4\|f\|_2^2)\int_0^\infty \|A_1(s)\|^2 ds + \frac{3}{2}\exp\left(\int_0^\infty \|A_2(s)\|ds\right)\int_0^\infty \|A_2(s)\|ds\right)\cdot$$

$$\cdot\|u\otimes y(f)\|\|v\otimes y(g)\|\Big]^{-1}. \blacksquare$$

Acknowledgement: The author wishes to thank Professors R.L. Hudson and V.P. Belavkin for helpful discussions.

References

[1] HUDSON, R.L., PARTHASARATHY, K.R.: Quantum Ito's formula and stochastic evolutions, *Comm. Math. Physics* 93, 301-323, (1984).

[2] HUDSON, R.L.: Quantum stochastic calculus in Fock space: a review, *Fundamental aspects of quantum theory,* ed V. Gorini and A. Frigerio, 111-124, Plenum (1986).

[3] DUNFORD, N., SCHWARTZ, J.T.: "Linear Operators", Interscience Publishers 1966.

[4] VON NEUMANN, J.: Zur Algebra der Funktionaloperationen und Theorie der Normalen Operatoren, *Math. Ann.* 102, 370-427, (1929-1930).

[5] BURTON, T.A.: "Stability and Periodic Solutions of Ordinary and Functional Differential Equations", Academic Press, 1985.

[6] YOSIDA, K.: "Functional Analysis", Springer, 1980.

[7] ARNOLD, L.: "Stochastische Differentialgleichungen. Theorei und Anwendung.", R. Oldenbourgh Verlag, 1973.

[8] TREVES, F.: "Topological Vector Spaces, Distributions and Kernels", Academic Press, 1967.

Quantum Probability and Related Topics Vol. VIII (pp. 109–118)

A non-commutative Markov equivalence theorem

by

Carlo Cecchini

Dipartimento di Matematica e Informatica
dell'Università di Udine
Via Zanon 6 - 33100 Udine - Italy

1. Introduction

In [6] several notions of markovianity were introduced and studied for states on a Neumann algebra A with respect to a given localization (cfr. also [1] and [3]). By this we denote a triple (A_1, A_2, A_3) of von Neumann subalgebras of A which mutually commute, have trivial intersection and whose union generates A. The aim of this paper is to continue the study of this subject by introducing still one more notion of markovianity, giving the relations with the previous ones and proving for it the generalization of the classical Markov equivalence theorem.

2. Notations and preliminaries

Let A be a von Neumann algebra, acting standardly on a Hilbert space H, Ω be a cyclic and separating vector in H for A associated with a normal faithful state ω on A and J_Ω (or simply J when no confusion arises) the canonical isometrical involution for Ω. For $a \in A$, we set $\omega_a(\cdot) = \omega(a^+ \cdot a)$. When using this notation we shall implicitely assume $\omega(|a|^2) = 1$, so ω_a is a normal state on A.

Whenever φ is a normal state on A such $\varphi \leq \alpha\omega$ for some $\alpha \in R^+$ we denote by $[\varphi/\omega]$ the extension to the point i/2 of the Connes' cocycle in A for φ and ω.

If A is a von Neumann subalgebra of A, $\omega_o = \omega|_A$ and $E^o{}_\Omega$ (or simply E_o) is the projection onto the closure $H^o{}_\Omega$ (or simply H_o) of the linear space $\{a_o\,\Omega : a_o \in A_o\}$, then $E_o\, A_o$ acts standardly on H_o and we denote by $J^o{}_\Omega$ (or J_o) the isometrical involution on H_o associated to Ω and A.

In [2] it was shown that for each $a \in A$ there is a (unique) operator $\varepsilon(a) \in A$, called the ω-conditional expectation of a in A_o, such that

$$\varepsilon(a)\,\Omega = J_o\, E_o\, J\, a\, \Omega$$

For the properties of the positive linear contraction $\varepsilon\ .\ A \rightarrow A_o$ we refer to [2], [7] and [8], and only recall that for A abelian it reduces to the usual conditional expectation.

We shall set

$R(\varepsilon) = \{a \in A: \varepsilon(a^+ a) = \varepsilon(a)^+\, \varepsilon(a)\}$,

$M(\varepsilon) = R(\varepsilon) \cap R(\varepsilon)^+$

$F(\varepsilon) = \{a \in A: \varepsilon(a) = a\}$.

We recall that $M(\varepsilon)$ and $F(\varepsilon)$ are von Neumann subalgebras of A, and we have $M(\varepsilon) \supseteq A_o \supseteq F(\varepsilon)$. Let now A be a von Neumann algebra generated by two von Neumann subalgebras A_1 and A_2. In [4] we defined the (A_2, A_1) stochastic coupling associated to a normal state ω on A as the completely antipositive weak operator continuous contraction $\lambda^{2,1}: A_2 \to A_1$ defined by the equality

$$J_1 \lambda^{2,1} (a_2)^+ \Omega = E_1 a_2 \Omega.$$

In the above formula Ω is a vector associated to ω, E_1 the projection on H_1, the closure of $\{a_1 \Omega: a_1 \in A_1\}$ and J_1 an Ω preserving isometrical involution for A_1 on H_1. We recall that

$$\lambda^{2,1} (|a_2|^2) = \|[(\omega_{a2})_1/\omega_1]\| \quad a_2 \in A_2.$$

In this paper we shall consider (as in [10], 38) a linearly ordered index set T (typically, a set of reals). We shall say, for $U, V \subset T$, that $U < V$ if for each $r \in U$, $s \in V$ we have $r < s$.

We shall consider a von Neumann algebra A (which will be assumed to act standardly on a Hilbert space H) containing for each $U \subset T$ a von Neumann subalgebra A_U.

We shall assume that the mapping $U \to A_U$ satisfies the following properties:

1. If $U \cap V = 0$ then $A_U \cap A_V = CI$
2. If $U \subset V$ then $A_U \subset A_V$
3. Let M be an index set and $\cup_{\alpha \in M} U_\alpha = U$. Then A_u is generated by $\cup_{\alpha \in M} M_{U_\alpha}$.

Let U, V, W be disjoint subsets of T such that $U < V < W$. Then we say that the triple (A_U, A_V, A_W) defines a <u>localization</u> for $M_{U \cup V \cup W}$. It is useful to interpret U as "past", V as "present" and W as "future".

In the following we shall label all objects introduced in this section by the indices rather than by the corresponding von Neumann algebras.

For example, if ω is a state on A we shall write ω_V for $\omega|_{AV}$ $(V \subset T)$ instead of ω_{AV}; $\varepsilon^{u,v}$ will denote the ω (ω_U) preserving conditional expectation from A_U to A_V $(U \subseteq V \subseteq T)$, $\lambda^{A,B}$ shall denote the (A_A, A_B) transition coupling for $\omega_{A \cup B}$, etc. Whenever we use the states ω_a $(a \in A)$ we shall drop the ω in the related objects (e.q. ε_a instead of $\varepsilon_{\omega a}$, $\lambda^{A,B}$ instead of $\lambda^{A,B}{}_{\omega a}$, etc.).

3. Some notions of markovianity

In this section we shall take $T = \{1, 2, 3\}$ with the usual order and omit in our notations the brackets for the singleton subsets of T. We shall assume all the states we consider to be normal and, in order to simplify technically our treatment, ω_i to be faithful on A_i. The last one is, however, only a technical hypothesis. We start by recalling two definitions from [6].

3.1 Definition A state ω on A is called strongly markovian with respect to the localization (A_1, A_2, A_3) if $\lambda^{3,\{1,2\}} = \lambda^{3,2}$.

3.2 Definition A state ω on A is called symmetric markovian with respect to the localization (A_1, A_2, A_3) if $[(\omega_{a_1a_3})_2/\omega_2] = [(\omega_{a_1})_2/\omega_2]\,[(\omega_{a_3})_2/\omega_2]$

Equivalent definitions for strong markovianity (symmetric markovianity) are given in [6], th. 3.2 (th. 3.4).

3.3 Theorem The following statements are equivalent, for $a_i \in A_i$ $(i = 1,3)$:

a) $\lambda^{3,\{1,2\}}(a_3) \in R(\varepsilon^{\{1,2\},2})$

a') $\lambda^{1,\{2,3\}}(a_1) \in R(\varepsilon^{\{2,3\},3})$

b) $\lambda^{3,\{1,2\}}(a_3) \in M(\varepsilon^{\{1,2\},2})$

b') $\lambda^{1,\{2,3\}}(a_1) \in M(\varepsilon^{\{2,3\},2})$

c) $E_{\{1,2\}}E_3 = E_2E_3$

c') $E_{\{2,3\}}E_1 = E_2E_1$

d) $\lambda^{\{1,3\},2}(a_1a_3) = \lambda^{1,2}(a_1)\,\lambda^{3,2}(a_3)$

d') $\lambda^{\{1,3\},2}(a_1a_3) = \lambda^{3,2}(a_3)\,\lambda^{1,2}(a_1)$.

In c), c') above E_U $(U \subset \{1,2,3\})$ is the orthogonal projection from H to H_U.

Proof The primed statements are obtained from the corresponding non primed statements by interchange of the indices 1 and 3; it suffices therefore to prove the equivalence of a), b), c), d), d').

a) $\Rightarrow$ b) As λ is a positive mapping, its range is selfadjoint, and the von Neumann subalgebra $M(\varepsilon^{\{1,2\},2}) = R(\varepsilon^{\{1,2\},2}) \cap R(\varepsilon^{\{1,2\},2})^*$ is the biggest selfadjoint subset of $R(\varepsilon^{\{1,2\},2})$.

b) $\Rightarrow$ c) We have (cfr. [2], th. 4.3):
$E_2\, a_3\, \Omega = E_2\, E_{\{1,2\}}\, a_3\, \Omega = E_2\, J_{\{1,2\}}\, \lambda^{3,\{1,2\}}\, (a^+_3)\, \Omega = E_{\{1,2\}}\, a_3\, \Omega,$
and then our claim follows by the density of $\{a_3\, \Omega: a_3 \in A_3\}$ in H_3.

c) $\Rightarrow$ a) We have:
$E_2\, J_{\{1,2\}}\, \lambda^{3,\{1,2\}}\, (a_3)\, \Omega = E_2\, E_{\{1,2\}}\, a^+_3\, \Omega = E_2\, a^+_3\, \Omega = E_{\{1,2\}}\, a^+_3\, \Omega =$
$= J^{\{1,2\}}\, \lambda^{3,\{1,2\}}\, (a_2)\, \Omega,$
and, again by [2], 4.3, we get our claim.

c) $\Leftrightarrow$ d) We have:
$\lambda^{\{1,3\},2}\, (a_1\, a_3)\, \Omega = J_2\, E_2\, (a_1\, a_3)^+\, \Omega = J_2\, E_2\, E_{\{1,2\}}\, a^+_1\, a^+_3\, \Omega =$
$= J_2\, E_2\, a^+_1\, E_{\{1,2\}}\, a^+_3\, \Omega$
and
$\lambda^{1,2}\, (a_1)\, \lambda^{3,2}\, (a_3)\, \Omega = J_2\, E_2\, a^+_1\, E_2\, a^+_2\, \Omega;$
so c) is immediately equivalent to d).

d) $\Leftrightarrow$ d') As $\lambda^{\{1,3\},2}$ is a positive mapping, we have:
$\lambda^{3,2}\, (a_3)\, \lambda^{1,2}\, (a_1) = [\lambda^{1,2}\, (a^+_1)\, \lambda^{3,2}\, (a^+_3)]^+ = [\lambda^{\{1,3\},2}\, (a^+_1\, a^+_3)]^+ =$
$= \lambda^{\{1,3\},2}\, (a_1\, a_3).$
The converse follows by interchanging the indices.

3.4 Definition We shall call markovian with respect to the localization (A_1, A_2, A_3) any state on A satisfying the conditions in th. 3.3.

3.5 Proposition

a) markovianity with respect to the localization (A_1, A_2, A_3) implies markovianity with respect to the localization (A_1, A_2, A_3).
b) strong markovianity implies markovianity.
c) if $F\, (\varepsilon^{\{1,2\},2}) = R\, (\varepsilon^{\{1,2\},2})$ then markovianity implies strong markovianity.
d) symmetric markovianity implies markovianity.

Proof

a) is an immediate consequence of the equivalence of d) and d') in th. 3.3.
b) and c) follow from th. 3.2 (cond. e).

d) we have by th. 3.4 in [6]:
$\lambda^{\{1,3\},2}(a_1 a_3) = \|[(\omega_{a_1} a_3)_2 / \omega_2]\|^2 = \|[(\omega_{a_1})_2/\omega_2]\,[(\omega_{a_3})_2/\omega_2]\|^2 =$
$= \|[(\omega_{a_1})_2/\omega_2]\|^2\,\|[(\omega_{a_3})_2/\omega_2]\|^2 = \lambda^{1,2}(a_1)\,\lambda^{3,2}(a_3).$

3.6 Theorem The strong markovianity of a state ω on A with respect to the localization (A_1, A_2, A_3) implies its symmetric markovianity with respect to the same localization.

Proof: By [6], 3.2 $\varepsilon^{\{2,3\},2}{}_{a_1}$ does not depend on $a_1 \in A_1$; this implies $\lambda^{3,2}{}_{a_1} = \lambda^{3,2}$, and one of the requirements in [6], 3.4 is therefore satisfied. Using this and 3.3 d) and d^1), since ω is (A_1, A_2, A_3) markovian by 3.3, we get:

$< J_2\,[(\omega_{a_1})_2/\omega_2]^+\,\lambda^{3,2}(a_3)\,[(\omega_{a_1})_2/\omega_2]\,\Omega,\ a_2\,\Omega > =$
$= < J_2\,\lambda^{3,2}(a_3)\,[(\omega_{a_1})_2/\omega_2]\,\Omega,\ a_2\,[(\omega_{a_1})_2/\omega_2]\,\Omega > =$
$= \omega\,(|a_1|^2\,a_2\,a_3) = < J_2\,\lambda^{\{1,3\},2}(|a_1|^2\,a_3)\,\Omega,\ a_2\,\Omega > =$
$= < J_2\,\lambda^{3,2}(a_3)\,\lambda^{1,2}(|a_1|^2)\,\Omega,\ a_2\,\Omega >,$
which implies
$\lambda^{3,2}(a_3)\,\|[(\omega_{a_1})_2/\omega_2]\|^2\,\Omega =$
$= \lambda^{3,2}(a_3)\,\lambda^{1,2}(|a_1|^2)\,\Omega =$
$= [(\omega_{a_1})_2/\omega_2]^+\,\lambda^{3,2}(a_3)\,[(\omega_{a_1})_2/\omega_2]\,\Omega.$
If supp $a_1 = I$, then $[(\omega_{a_1})_2/\omega_2]\,\Omega$ is a separating vector for A_2, and $[(\omega_{a_1})_2/\omega_2]^+$ (and therefore $[(\omega_{a_1})_2/\omega_2]$) commutes with $\lambda^{3,2}(a_3)$ for all $a_3 \in A_3$.

For the general $a_1 \in A_1$ we notice that the sequence
$a_n = (a_1 + 1/n\ I)/\|a_1 + 1/n\ I\|$
tends in norm to a_1; this implies $(\omega_{a_n})_2 \to (\omega_{a_1})_2$ in norm, and $[(\omega_{a_{1n}})_2/\omega_2] \to [(\omega_{a_1})_2/\omega_2]$ in the strong operator topology.
Now the commutation of $\lambda^{3,2}(a_3)$ with $[(\omega_{a_1})_2/\omega_2]$ for all $a_1 \in A_1$, $a_3 \in A_3$ follows by continuity. This means that for all $a_3 \in A_3$ $\lambda^{3,2}(a_3)$ commutes with the von Neumann algebra $A^1{}_2$ generated by the set $\{[(\omega_{a_1})_2/\omega_2] : a_1 \in A_1\}$. From [5], it follows that there is an ω_2-preserving norm one projection from A_2 to $A^1{}_2$.
Let us consider now the smallest von Neumann algebra $A^3{}_2$ which contains $\lambda^{3,2}(a_3)$ for all $a_3 \in A_3$ and onto which there is an ω_2-preserving norm one projection from A_2.

A^3_2 is generated by the operators $\{\sigma^t_2(\lambda^{3,2}(a_3)), a_3 \in A_3, t \in R\}$, if σ^t_2 is the automorphism group for ω_2 on A_2. We have, for $a_3 \in A_3$, $a^1_2 \in A^1_2$, $t \in R$:
$\sigma^t_2(\lambda^{3,2}(a_3))\, a^1_2 = \sigma^t_2(\lambda^{3,2}(a_2)\, \sigma^{-t}_2(a^1_2)) =$
$= \sigma^t_2(\sigma^{-t}_2(a^1_2)\, \lambda^{3,2}(a_2)) = a^1_2(\sigma^t_2(\lambda^{3,2}(a_2))$,
since A^1_2 is invariant under σ^t_2. This proves that A^1_2 and A^3_2 commute. On the other hand, since A^3_2 is ω_2 expected in A_2, it contains $[(\omega_{a3})_2/\omega_2]$ for each $a_3 \in A_3 3$; so for each $a_1 \in A_1$, $a_3 \in A_3$ $[(\omega_{a3})_2/\omega_2]$ and $[(\omega_{a1})_2/\omega_2]$, and our proof is complete.

3.8 Corollary Let A_2 be a finite von Neumann algebra. Then strong markovianity, symmetric markovianity and markovianity for ω with respect to the localization (A_1, A_2, A_3) are equivalent.

Proof We need only to prove that markovianity implies strong markovianity. As A_2 is finite, $M(\varepsilon^{\{1,2\},2}) = F(\varepsilon^{\{1,2\},2})$ (cfr [9]), and then our result follows by [6].

3.9 Examples

a) if A is abelian, then, as already remarked in [6], the three equivalent notions of markovianity are equivalent to classical markovianity

b) let A_i be matrix algebras $(i = 1,2,3)$, $A_1 = A_1 \otimes I \otimes I$, $A_2 = I \otimes A_2 \otimes I$, $A_3 = A_2 \otimes I \otimes I$, and ω a state on M. Let $[\omega]$ ($[\omega_{\{1,2\}}]$) be the density of ω ($\omega_{\{1,2\}}$) with respect to the trace tr on A ($A_{\{1,2\}}$). After some lengthy but straightforward computations our condition for markovianity becomes:
$\varepsilon^{\{1,2,3\},\{1,2\}}_{tr}([\omega]\,[\omega_{\{1,2\}}]^{-1/2}\, a_3\, [\omega_{\{1,2\}}]^{-1/2}) \in A_2$ for all $a_3 \in A_3$.

We plan to deal in another paper with the structure of markovian chains.

4. A Markov equivalence theorem

4.1 Lemma The ω is markovian on A with respect to the localization (A_1, A_2, A_3) if and only if

$$(*)\ E_{\{1,2\}}\, E_{\{2,3\}} = E_2\, E_{\{2,3\}}.$$

Proof Let (*) be satisfied. Then

$E_{\{1,2\}} E_3 = E_{\{1,2\}} E_{\{2,3\}} E_3 = E_2 E_{\{2,3\}} E_3 = E_2 E_3$, which implies by 3.3, c) markovianity.

Let now $a_i \in A_i$, $i = 2,3$, ω markovian. Then $E_{\{1,2\}} a_2 a_3 \Omega = a_2 E_{\{1,2\}} a_3 \Omega = a_2 E_2 a_3 \Omega = E_2 a_2 a_3 \Omega$, and our claim follows from the density of the linear combinations of vectors of the form $a_2 a_3 \Omega$ in $H_{\{2,3\}}$.

4.2 Lemma Let $I = \{1,2,3,4\}$ and ω a state on A such that

a) $\omega_{\{1,2,3\}}$ is markovian on $A_{\{1,2,3\}}$ with respect to the localization (A_1, A_2, A_3).

b) ω is markovian on A with respect to the localization $(A_{\{1,2\}}, A_3, a_4)$.
Then ω is markovian on A with respect to both the localizations $(A_1, A_{\{2,3\}}, A_4)$ and $(A_1, A_2, A_{\{3,4\}})$.

Proof We have, using

$E_{\{1,2,3\}} E_4 = E_3 E_4 = E_{\{2,3\}} E_3 E_4 = E_{\{2,3\}} E_{\{1,2,3\}} E_4 = E_{\{2,3\}} E_4$.

We prove our second claim by noticing that:

$E_{\{1,2\}} E_{\{3,4\}} = E_{\{1,2\}} E_{\{1,2,3\}} E_{\{3,4\}} = E_{\{1,2\}} E_{\{2,3\}} E_{\{3,4\}} = E_2 E_{\{2,3\}} E_{\{3,4\}} =$
$= E_2 E_{\{3,4\}}$.

The second equality above follows from lemma 4.1 and markovianity with respect to $(A_1, A_{\{2,3\}}, A_4)$, and the third from lemma 4.1 again and markovianity of $\omega_{\{1,2,3\}}$ with respect to (A_1, A_2, A_3).

4.3 Theorem (Markov equivalence theorem, cfr [10], 3.8.1.A)

The following Markov properties are equivalent for a state ω on a von Neumann algebra A:

a) markovianity on A_Z with respect to the localization (A_U, A_V, A_W) for all $U, V, W \subset T$ with $U < V < W$, $U \cup V \cup W = Z$

b) markovianity on A_Z with respect to the localization (A_U, A_V, A_W), where
$$V = \{s\},\ U = \{r : r < s,\ r \in T\},$$
$$W = \{t : t > s,\ t \in T\} \text{ for all } s \in T$$

c) markovianity on A_Z with respect to the localization (A_U, A_V, A_W), where $V = \{s\}$, $W = \{t\}$, U is any finite set in T such that $U < s$ and $Z = U \cup V \cup W$, for all $s, t \in T$ with $s < t$.

Proof Clearly a) $\Rightarrow$ b) $\Rightarrow$ c).

To prove c) $\Rightarrow$ a) let us consider first the situation in which U is finite and V contains only one element. Then by lemma 4.2, we obtain markovianity with respect to the localization (A_U, A_V, A_{W_n}), where W_n is any finite subset of W. As A_W is generated by the union of all the subalgebras M_{W_n}, we get our claim.

In our next step we assume only $V = \{t\}$, $t \in T$. We note that

$E_{U \cup \{t\}} = \sup \{E_{U_n \cup \{t\}} : U_n \subset U, U_n \text{ finite}\}$.

So, for all $a_W \in A_W$:

$\|E_{U \cup \{t\}} a_W \Omega\| = \sup \{\|E_{U_n \cup \{t\}} a_W \Omega\| : U_n \subset U, U_n \text{ finite}\}$.

As by our first step $E_{U_n \cup \{t\}} a_W \Omega = E_t a_w \Omega$ for all $U_n \subset U$, U_n finite, the above supremum is simply $\|E_t a_W \Omega\|$.

Since $E_t \leq E_{U \cup \{t\}}$, this implies $E_{U \cup \{t\}} a_W \Omega = E_t a_W \Omega$, and our claim is proved.

To complete our proof, let us consider first the case in which V has a last element t. We consider a localization ($A_{U'}$, $A_{V'}$, A_W), where

$U' = U \cup V/\{t\}$, $V' = \{t\}$. Our state is markovian with respect to this localization by our previous step. On the other hand, for $a_W \in A_W$:

$E_{U \cup V} a_W \Omega = E_{U' \cup V'} a_W \Omega = E_{V'} a_W \Omega = E_V E_{V'} a_W \Omega = E_V E_{U \cup V} a_W \Omega =$

$= E_V a_W \Omega$ and our claim follows.

Finally, if V has no last element, let $V = V_1 \cup V_2$, with $V_1 < V_2$ and V_1 endowed with a last element. Take a localization (A_U, $A_{V'}$, $A_{W'}$), with $V' = V_1$, $W' = V_2 \cup W$. As seen in our previous step, $E_{U \cup V'} a_{W'} \Omega = E_{V'} a_{W'} \Omega$ for all $a_{W'} \in A_{W'}$. By th. 3.3 we can interchange U and V' and get $E_{V \cup W} a_U \Omega =$ $= E_{V'} a_U \Omega$ for all $a_U \in A_U$. By application of E_V on both sides, we get $E_V a_U \Omega = E_V E_{V \cup W} a_U \Omega = E_V E_{V'} a_U \Omega = E_V a_U \Omega = E_{V \cup W} a_U \Omega$.

The above equality gives the markovianity of ω with respect to the localization (A_W, A_V, A_U) and hence for the localization (A_U, A_V, A_W).

Aknowledgement

This paper was completed during a visit at the University of Heidelberg, with the financial support of SFB123. The author would like to thank Prof. W. von Waldenfels for his most kind hospitality.

Bibliography

1. L. Accardi, Noncommutative Markov claims. International School of Mathematical Physics, Camerino (1974), 268-295.

2. L. Accardi C. Cecchini, Conditional expectations in von Neumann algebras and a theorem of Takasaki. J. Funct. Anal. 45 (1982), 245-273.

3. L. Accardi - A. Frigerio, Markovian cocycles. Proc. Royal Irish Acad. 83 A (1983), 251-263.

4. C. Cecchini, Stochastic couplings for von Neumann algebras. Quantum probability and applications IV. Springer Verlag Lecture Notes 1396 (1989), 128-142.

5. C. Cecchini, Fixed points subalgebras and invariant states for ω conditional expectations. To be published in the Lith. J. of Math.

6. C. Cecchini, Markovianity for states on von Neumann algebras. Preprint.

7. C. Cecchini - D. Petz, State extensions and a Radon Nikodyn theorem for conditional expectations on von Neumann algebras. Pac. J. Math. 138 (1989), 9-23.

8. C. Cecchini D. Petz, Classes of conditional expectations over von Neumann algebras. J. Funct. Anal. 92 (1990), 8-29.

9. R.V. Kadison, Remarks on the type of von Neumann algebras of local observables in quantum field theory. J. Math. Phys. 4 (1963), 1511-1516.

10. M. Loeve, Probability theory. Van Nostrand (1963).

Quantum Probability and Related Topics Vol. VIII (pp. 119–122)

A Note on the Cohomology of Quantum Flows

P.K.Das
Indian Statistical Institute
Calcutta

March,1992

Let $\mathcal{A}$ be a *-algebra and let $\mathcal{C}$ be a second *-algebra which is also a 2-sided $\mathcal{A}$-*-module so that there are actions $\mathcal{A} \times \mathcal{C} \rightarrow \mathcal{C}$, $\mathcal{C} \times \mathcal{A} \rightarrow \mathcal{C}$ satisfying

$$(ac)^* = c^*a^*$$

in addition to the usual linearity,associativity and compatibility relations of a two-sided action.

By defition,a 0-chain is an element ω of $\mathcal{C}$ satisfying

$$\omega^* + \omega + \omega^*\omega = \omega + \omega^* + \omega\omega^* = 0. \tag{1}$$

The 0-chains form a group under the composition

$$\omega_1 * \omega_2 = \omega_1 + \omega_2 + \omega_1\omega_2.$$

A 1-cocycle is a linear *-map $\lambda : \mathcal{A} \rightarrow \mathcal{C}$ satisfying the identity

$$\lambda(xy) = \lambda(x)y + x\lambda(y) + \lambda(x)\lambda(y). \tag{2}$$

It may be verified that given a 0-chain ω, the map $\delta\omega : \mathcal{A} \rightarrow \mathcal{C}$ given by

$$(\delta\omega)(x) = \omega^*x + x\omega + \omega^*x\omega$$

is a 1-cocycle.Denote the set of 0-chains by $C^o(\mathcal{A}, \mathcal{C})$ and of 1-cocycles by $Z^1(\mathcal{A}, \mathcal{C})$, and let $B^1(\mathcal{A}, \mathcal{C})$ be the range of the map δ.

$Z^1(\mathcal{A}, \mathcal{C})$ is not a group,but the group $C^o(\mathcal{A}, \mathcal{C})$ acts on $Z^1(\mathcal{A}, \mathcal{C})$ through $(\lambda, \omega) \rightarrow \lambda^\omega$, where

$$\lambda^\omega(x) = \lambda(x) + \omega^*(x + \lambda(x)) + (x + \lambda(x))\omega + \omega^*(x + \lambda(x))\omega. \tag{3}$$

Now, let ϑ be the *-algebra generated by differentials of integrator processes satisfying a closed Ito multiplication table,for example, that given by (2.3)[Das and Sinha] with the involution

$$(d\Lambda_\nu^\mu)^* = d\Lambda_\mu^\nu.$$

From now on we let $\mathcal{A}$ be a unital *-subalgebra of $\mathcal{B}(\mathcal{H}_0)$ which may be called a *system algebra* and let $\mathcal{C}$ be the tensor product *-algebra $\mathcal{C} = \mathcal{A} \otimes \vartheta$ on which $\mathcal{A}$ acts through the multiplication in $\mathcal{A}$ that is, $\forall a \in \mathcal{A}, b \in \mathcal{A}, d\Lambda \in \vartheta$

$$a(b \otimes d\Lambda) = ab \otimes d\Lambda, (b \otimes d\Lambda)a = ba \otimes d\Lambda.$$

A family of linear bounded maps $\lambda_\nu^\mu : \mathcal{A} \to \mathcal{A}$ is called *structure map* if

$$\begin{array}{llll} (i) & \lambda_\nu^\mu(I) & = & 0 \\ (ii) & \lambda_\nu^\mu(x)^* & = & \lambda_\mu^\nu(x^*) \\ (iii) & \lambda_\nu^\mu(xy) & = & x\lambda_\nu^\mu(y) + \lambda_\nu^\mu(x)y + \lambda_n^\mu(x)\lambda_\nu^n(y), \end{array} \tag{4}$$

where $x, y \in \mathcal{A}$,and

(iv) if there exists for each $\nu \geq 0$ constants $\alpha_\nu > 0$,a countable index set I_ν and a family $\{D_\nu^\mu\}_{\mu\in I_\nu}$ of bounded operators in $\mathcal{H}_0$ such that for $x \in \mathcal{A}$, $u \in \mathcal{H}_0$

$$\begin{array}{rcl} \sum_{\mu\geq 0} \|\lambda_\nu^\mu(x)u\|^2 & \leq & \sum_{\mu\in I_\nu} \|xD_\nu^\mu u\|^2 \quad \textit{with} \\ \sum_{\mu\in I_\nu} \|D_\nu^\mu u\|^2 & \leq & \alpha_\nu^2\|u\|^2. \end{array} \tag{5}$$

We define a *quantum stochastic flow* on $\mathcal{A}$ as a family $\{j_t\}_{t\geq 0}$ of contractive *-homomorphism from $\mathcal{A}$ into $\mathcal{B}(\tilde{\mathcal{H}})$ satisfying for each $x \in \mathcal{A}$:

(i) $j_0(x) = x$

(ii) $j_t(I) = I$ and $j_t(x)$ is an adapted process,

(iii) there exist structure maps λ_ν^μ obeying (4) such that $j_t(x)$ satisfy the q.s.d.e.

$$dj_t(x) = j_t(\lambda_\nu^\mu(x))d\Lambda_\mu^\nu(t). \tag{6}$$

A theorem proved by Mohari and Sinha [Th.3.5,Mohari and Sinha] ensures the existence of a quantum stochastic flow j on $\mathcal{A}$ satisfying (6) if the structure maps λ_ν^μ satisfy (4) and (5).

Now by writting each element λ in the form

$$\lambda(x) = \sum_{\mu,\nu=0}^{\infty} \lambda_\nu^\mu(x) \otimes d\Lambda_\mu^\nu$$

where each λ_ν^μ maps $\mathcal{A}$ to $\mathcal{A}$, we see that the set $Z^1(\mathcal{A}, \mathcal{C})$ of 1-cocyles is co-extensive with the set of solutions to the structure relations (4)and(6). Thus quantum stochastic flows are generated by elements of $Z^1(\mathcal{A}, \mathcal{C})$.

To interpret the set of 0-chains, we write an element $\omega \in \mathcal{C}$ in the form

$$\omega = \sum_{\mu,\nu=0}^{\infty} \omega_\nu^\mu \otimes d\Lambda_\mu^\nu, \omega_\nu^\mu \in \mathcal{A};$$

then it may be verified that the defining relation (1) is equivalent to the following relations among the ω_ν^μ :

$$(\omega_\mu^\nu)^* + \omega_\nu^\mu + \sum_{k=1}^{\infty}(\omega_k^\nu)^*\omega_\mu^k = \omega_\mu^\nu + (\omega_\nu^\mu)^* + \sum_{k=1}^{\infty}\omega_k^\nu(\omega_k^\mu)^* = 0. \qquad (7)$$

The structure maps corresponding to $\delta\omega$ are given by

$$\lambda_\nu^\mu(x) = (\omega_\mu^\nu)^* x + x\omega_\nu^\mu + \sum_{k=1}^{\infty}(\omega_\mu^k)^* x\omega_\nu^k. \qquad (8)$$

provided each ω_ν^μ is bounded,the conditions (7) are necessary and sufficient for the existence of a unique unitary-operator valued process u satisfying the stochastic differential equation

$$du(t) = (\omega_\nu^\mu \otimes 1)u(t)d\Lambda_\mu^\nu.$$

The inner quantum stochastic flow corresponding to the bounded structure maps (8) is then given by

$$j_t(x) = u(t)^* x \otimes 1u(t).$$

Thus quantum stochastic flows are generated by elements of $Z^1(\mathcal{A}, \mathcal{C})$ and inner flows by elements of $B^1(\mathcal{A}, \mathcal{C})$.

The group action (3) can be interpreted as follows : the flow generated by λ^ω is an inner perturbation of that generated by λ in the following sense.Denoting the corresponding flows by j^ω and j respectively we have

$$j^\omega(x) = u^{(j)*} j(x) u^{(j)}$$

where $u^{(j)}$ is the solution of

$$du^{(j)}(x) = j_t(\omega_\mu^\nu)u(x)d\Lambda_\nu^\mu.$$

In the case of bounded ω_μ^ν and λ_μ^ν the existence,uniqueness and unitarity of a solution of this quantum stochastic differential equation was proved by Das and Sinha recently.

REFERENCES

1. R.L.Hudson *Quantum Stochastic Flows* .Prob. Theo. and Math. Stat. proceedings; Vilnius 1989,vol 1,1990.pp512-525.

2. P.K.Das and K.B.Sinha *Quantum flows with infinite degrees of freedom and their perturbations* . To appear in Q.P(7):Proceeding of Q.P(Delhi)(1990);World Scientific, 1992.

3. A.Mohari and K.B.Sinha *Quantum Stochastic Flows With Infinite Degrees of Freedom and Countable State Markov Processes.* Sankhya : The Indian Journal of Statistics 1990,vol 52 A,Pt.1,pp. 43-57.

Quantum Probability and Related Topics Vol. VIII (pp. 123–142)

CHEBOTAREV'S SUFFICIENT CONDITIONS FOR CONSERVATIVITY OF QUANTUM DYNAMICAL SEMIGROUPS

Franco Fagnola
Università di Trento, Dipartimento di Matematica
I 38050 Povo (TN) Italia

Abstract. We expose Chebotarev's sufficient conditions for the conservativity of a quantum dynamical semigroup with some extensions and simplifications needed in the applications to quantum stochastic calculus.

1. Introduction

If $\mathcal{B}(h)$ is the Banach space of bounded operators on a Hilbert space h a quantum dynamical semigroup (q.d.s.) $\mathcal{T} = (\mathcal{T}_t)_{t\geq 0}$ on $\mathcal{B}(h)$ is a semigroup of completely positive linear maps on $\mathcal{B}(h)$ with some continuity properties (see Definition 2.1) satisfying the condition $\mathcal{T}_t(I) \leq I$ for all $t \geq 0$. Here I denotes the identity operator. The q.d.s. $\mathcal{T}$ is conservative (or Markovian, or identity preserving) if $\mathcal{T}_t(I) = I$. When $\mathcal{T}$ is conservative and uniformily continuous (see Definition 2.2) the canonical form of its infinitesimal generator $\mathcal{L}$

$$\mathcal{L}(X) = i\,[X, H] - \frac{1}{2}XC + \sum_{j=1}^{\infty} L_j^* X L_j - \frac{1}{2}CX, \tag{1.1}$$

with $C = \sum_{j=1}^{\infty} L_j^* L_j$ and H selfadjoint operator, is well known (see Theorem 2.3). As shown by Frigerio [9] and Davies [5], the infinitesimal generator has the same form also in many cases in which $\mathcal{T}$ is conservative and ultraweakly continuous. Given such an operator on $\mathcal{B}(h)$ one can construct a q.d.s. with this infinitesimal generator (see [2], [10], [5]). However the q.d.s. may not be conservative (see the example 3.3 in [4]). In [2] Chebotarev gave a necessary and sufficient condition for conservativity of a q.d.s. which is a quantum analogue of Feller's condition of non-explosion of a classical process (see also [10], [11]). This condition is practically impossible to check in many interesting examples. Extending some sufficient conditions for non-explosion of Markov jump processes [1], Chebotarev gave also sufficient conditions for the conservativity of a q.d.s.

with infinitesimal generator of the form (1.1) and $L_l \neq 0$ only for a finite number of indices l. These are essentially the lower relative boundedness with respect to $C^{1/2}$ of the commutator of H and C and the lower boundedness of the operator $C^2 - \sum_{l=1}^{\infty} L_l^* C L_l$. These conditions are very easy to check when the operators H, L_l are differential operators. However some operator domain problems and other technical problems related to the use of quadratic forms is left open in Chebotarev's paper [2].

In this note we extend Chebotarev's sufficient conditions to the case of infinite nonzero L_l under the assumption (see the condition C) that the quadratic form

$$(v,u) \mapsto \sum_{l=1}^{\infty} \langle L_l v, L_l u\rangle$$

is closed. We weaken also the assumptions on the set of vectors on which one should check some quadratic forms inequalities (see Theorem 3.2). Moreover we simplify some steps of the proof.

These extensions and simplifications are necessary to apply this conservativity result to quantum stochastic calculus [3], [6], [7], [8], [10]. Some applications of the main theorem of the present note have been given in [3]. The application to the realisation of classical diffusion processes and Azéma-Emery martingales as quantum flows in the Boson Fock space will be studied in the forthcoming paper [7].

2. Quantum dynamical semigroups

Let h be a complex separable Hilbert space endowed with the scalar product $\langle\cdot,\cdot\rangle$ and norm $\|\cdot\|$. Consider a dense linear submanifold $\mathcal{D}$ of h. Let $\mathcal{B}(h)$ denote the Banach space of bounded operators on h. The uniform norm in $\mathcal{B}(h)$ will be denoted also by $\|\cdot\|_\infty$. Let I be the identity operator on h.

Definition 2.1. *A quantum dynamical semigroup on $\mathcal{B}(h)$ is a family $\mathcal{T} = (\mathcal{T}_t)_{t\geq 0}$ of operators in $\mathcal{B}(h)$ with the following properties:*

i) $\mathcal{T}_0(X) = X$, *for all* $X \in \mathcal{B}(h)$,

ii) $\mathcal{T}_{t+s}(X) = \mathcal{T}_t(\mathcal{T}_s(X))$, *for all* $s, t \geq 0$ *and all* $X \in \mathcal{B}(h)$,

iii) $\mathcal{T}_t(I) \leq I$, *for all* $t \geq 0$,

iv) (complete positivity) for all $t \geq 0$, *for all integer* n *and all finite sequences* $(X_j)_{j=1}^n$, $(Y_l)_{l=1}^n$ *of elements of* $\mathcal{B}(h)$ *we have*

$$\sum_{j,l=1}^{n} Y_l^* \mathcal{T}_t(X_l^* X_j) Y_j \geq 0, \tag{2.1}$$

v) *for every sequence $(X_n)_{n\geq 0}$ of elements of $\mathcal{B}(h)$ converging weakly to an element X of $\mathcal{B}(h)$ the sequence $(\mathcal{T}_t(X_n))_{n\geq 0}$ converges weakly to $\mathcal{T}_t(X)$ for all $t \geq 0$,*

vi) *(ultraweak continuity) for all trace class operator ρ on h and all $X \in \mathcal{B}(h)$ we have*

$$\lim_{t \to 0+} \mathrm{Tr}(\rho \mathcal{T}_t(X)) = \mathrm{Tr}(\rho X).$$

Let us remark that, as a consequence of properties iii), iv) for all $X \in \mathcal{B}(h)$ we have the inequalities between bounded operator on $h \oplus h$

$$0 \leq \begin{pmatrix} \mathcal{T}_t(I) & \mathcal{T}_t(X) \\ \mathcal{T}_t(X^*) & \mathcal{T}_t(X^*X) \end{pmatrix} \leq \begin{pmatrix} I & \mathcal{T}_t(X) \\ \mathcal{T}_t(X^*) & \mathcal{T}_t(X^*X) \end{pmatrix}$$

Hence we obtain the identity $\mathcal{T}_t(X^*) = \mathcal{T}_t(X)^*$ and inequality

$$\mathcal{T}_t(X)^*\mathcal{T}_t(X) \leq \mathcal{T}_t(X^*X).$$

Since $\mathcal{T}_t$ is a positive operator because of property iii) we have the inequalities

$$\mathcal{T}_t(X^*X) \leq \|X\|_\infty^2 \, \mathcal{T}_t(I) \leq \|X\|_\infty^2 .$$

Therefore, for all $X \in \mathcal{B}(h)$, we have

$$\|\mathcal{T}_t(X)\|_\infty \leq \|X\|_\infty . \tag{2.2}$$

Let us recall also that the ultraweak topology and the weak topology induce the same topology on the spheres of $\mathcal{B}(h)$.

Definition 2.2. *A quantum dynamical semigroup $\mathcal{T}$ is said to be* uniformly continuous *if*

$$\lim_{t \to 0} \sup_{\|X\|_\infty \leq 1} \|\mathcal{T}_t(X) - X\|_\infty = 0.$$

We refer to [12] for the proof of the following characterisation of the infinitesimal generator of a uniformly continuous q.d.s..

Theorem 2.3. *(Gorini, Kossakowski, Sudarshan, Lindblad) A bounded operator $\mathcal{L}$ on $\mathcal{B}(h)$ is the infinitesimal generator of a uniformly continuous quantum dynamical semigroup such that $\mathcal{T}_t(I) = I$ for all $t \geq 0$ if and only if there exists a bounded selfadjoint operator H and a sequence $(L_l)_{l=1}^\infty$ of bounded operators in h such that the series $\sum_{l=1}^\infty L_l^* L_l$ is strongly convergent and*

$$\mathcal{L}(X) = i[H, X] + \frac{1}{2}\sum_{l=1}^{\infty}(-L_l^* L_l X + 2L_l^* X L_l - X L_l^* L_l)$$

for all $X \in \mathcal{B}(h)$.

Letting

$$G = -\frac{1}{2}\sum_{l=1}^{\infty} L_l^* L_l - iH$$

the infinitesimal generator is given by

$$\mathcal{L}(X) = XG + \sum_{l=1}^{\infty} L_l^* X L_l + G^* X.$$

The condition $\mathcal{T}_t(I) = I$ for all $t \geq 0$ implies then

$$G + G^* + \sum_{l=1}^{\infty} L_l^* L_l = 0.$$

In the applications of quantum stochastic calculus one a has q.d.s. $\mathcal{T}$ with unbounded generator of a similar form. The operators G, L_l often satisfy the following assumptions

A - The operator G is the infinitesimal generator of a strongly continuous contraction semigroup $(P(t))_{t\geq 0}$ in h and $\mathcal{D}$ is a core for G. The domain of the operators $(L_l)_{l=1}^{\infty}$ contains $\mathcal{D}$ and for all $u, v \in \mathcal{D}$ we have

$$\langle v, Gu\rangle + \langle Gv, u\rangle + \sum_{l=1}^{\infty} \langle L_l v, L_l u\rangle = 0. \tag{2.3}$$

B - The operator G^2 is densely defined and its domain contains $\mathcal{D}$.

We consider a q.d.s. $\mathcal{T}$ with infinitesimal generator $\mathcal{L}$ given by

$$\langle v, \mathcal{L}(X)u\rangle = \langle v, XGu\rangle + \langle Gv, Xu\rangle + \sum_{l=1}^{\infty} \langle L_l v, XL_l u\rangle = 0 \tag{2.4}$$

satisfying the equation

$$\begin{aligned}\langle v, \mathcal{T}_t(X)u\rangle &= \langle v, Xu\rangle + \\ &+ \int_0^t \left(\langle v, \mathcal{T}_s(X)Gu\rangle + \langle Gv, \mathcal{T}_s(X)u\rangle + \sum_{l=1}^{\infty} \langle L_l v, \mathcal{T}_s(X)L_l u\rangle \right) ds\end{aligned} \tag{2.5}$$

for all $X \in \mathcal{B}(h)$ and all $u, v \in \mathcal{D}$. For all $t \geq 0$ denote $\mathcal{T}_t(I)$ by $T(t)$. The following proposition can be proved as in [6].

Proposition 2.4. *Let us suppose that conditions (A), (B) hold and let $(T(t))_{t\geq 0}$ be a strongly continuous family of positive contractions on h. The following conditions are equivalent:*

i) for all $v, u \in \mathcal{D}$ we have

$$\langle v, T(t)u\rangle = \langle v, u\rangle + \int_0^t \left(\langle Gv, T(s)u\rangle + \langle v, T(s)Gu\rangle + \sum_{l=1}^{\infty} \langle L_l v, T(s)L_l u\rangle\right) ds,$$

ii) for all $v, u \in \mathcal{D}$ we have

$$\langle v, T(t)u\rangle = \langle P(t)v, P(t)u\rangle + \sum_{l=1}^{\infty} \int_0^t \langle L_l P(t-s)v, T(s)L_l P(t-s)u\rangle \, ds.$$

Definition 2.5. *Suppose that the conditions (A), (B) hold. The semigroup $\mathcal{T}$ is called* conservative *(or* Markovian*) if the only solution of the equation (2.6) is given by $T(t) = I$ for all $t \geq 0$.*

Let us consider the operator $\mathcal{Q}$ on $\mathcal{B}(h)$ defined by

$$\langle v, \mathcal{Q}(X)u\rangle = \int_0^{\infty} \exp(-s) \sum_{l=1}^{\infty} \langle L_l P(s)v, X L_l P(s)u\rangle \, ds \tag{2.6}$$

Using (2.3) and integrating by parts we can estimate this by $\|X\|^2$ times

$$\begin{aligned}\int_0^{\infty} \exp(-s) \sum_{l=1}^{\infty} \|L_l P(s)u\|^2 \, ds &= -\int_0^{\infty} \exp(-s) \frac{d}{ds} \|P(s)u\|^2 \, ds \\ &= \|u\|^2 - \int_0^{\infty} \exp(-s) \|P(s)u\|^2 \, ds\end{aligned} \tag{2.7}$$

Hence $\mathcal{Q}(X)$ belongs to $\mathcal{B}(h)$. Moreover the map $\mathcal{Q} : \mathcal{B}(h) \mapsto \mathcal{B}(h)$ is normal and monotone.

The following theorem has been proved in [2]. (See also [6], [10]; the parameter λ used there can be chosen to be 1.)

Theorem 2.6. *(Chebotarev) Suppose that the conditions (A) and (B) hold. Then the following conditions are equivalent:*

i) the semigroup $\mathcal{T}$ is conservative,

ii) the decreasing sequence $(\mathcal{Q}^n(I))_{n\geq 1}$ of positive operators in $\mathcal{B}(h)$ converges strongly to zero,

iii) the equation $\mathcal{Q}(X) = X$ has no positive solutions in $\mathcal{B}(h)$,

iv) the equation $\mathcal{L}(X) = X$ has no positive solutions in $\mathcal{B}(h)$.

The above conditions are difficult to check in many cases. Therefore it is interesting to give simpler easily verifiable conditions.

3. Sufficient conservativity conditions

In this section we will give a simplified version of Chebotarev's sufficient conditions for conservativity. Let us first fix some technical preliminaries:

Proposition 3.1. *For all $l \geq 1$, the operator L_l has a unique extension as an operator with domain coinciding with $D(G)$ satisfying, for all $v, u \in D(G)$*

$$\langle v, Gu\rangle + \langle Gv, u\rangle + \sum_{l=1}^{\infty} \langle L_l v, L_l u\rangle = 0, \tag{3.1}$$

moreover, for all $u \in D(G)$, the following map is continuous

$$t \mapsto L_l P(t) u. \tag{3.2}$$

Proof. Since $\mathcal{D}$ is a core for G, for all $u \in D(G)$ there exists a sequence $(u_n)_{n\geq 1}$ of elements of $\mathcal{D}$ such that

$$\lim_{n\to\infty} u_n = u, \qquad \lim_{n\to\infty} Gu_n = Gu.$$

Using (2.3) it is easy to see that the sequence $(L_l u_n)_{n\geq 1}$ is Cauchy in h for all $l \geq 1$. Therefore L_l has an extension to the domain of G that we will denote always by L_l. The formula (3.1) can be easily checked. The uniqueness of the extension can be proved using again (3.1). Moreover, for all $s, t \geq 0$ and all $u \in D(G)$, the identity

$$\sum_{l=1}^{\infty} \|L_l P(t)u - L_l P(s)u\|^2 = -2\Re \langle (P(t) - P(s))u, G(P(t) - P(s))u\rangle$$

shows the continuity of the map (3.2), for all $l \geq 1$. □

We introduce now a transformation of the operators G, $(L_l)_{l\geq 1}$ which leaves the infinitesimal generator invariant. This allows to skip some technical problems without loss of generality. Let us consider the operators with domain $D(G)$

$$\tilde{G} = G - \frac{1}{2}I$$

$$\tilde{L}_l = \begin{cases} I & \text{if } l = 1 \\ L_{l-1} & \text{if } l > 1 \end{cases}$$

Differentiating with respect to t it is easy to see that a strongly continuous family of positive contractions on h satisfyies the condition i) of Proposition 2.4 if and only if it satisfies the equation

$$\langle v, T(t)u\rangle = \langle v, u\rangle + \int_0^t \left(\left\langle \tilde{G}v, T(s)u\right\rangle + \left\langle v, T(s)\tilde{G}u\right\rangle + \sum_{l=1}^{\infty} \left\langle \tilde{L}_l v, T(s)\tilde{L}_l u\right\rangle \right) ds.$$

Therefore from now on we will denote $\tilde{G}$ and $\tilde{L}_l$ by G and L_l respectively and we will suppose that $L_1 = I$. Under this assumption it follows then from (3.1) that for all $t \geq 0$, $u \in D(G)$ we have

$$\frac{d}{dt} \|P(t)u\|^2 = 2\Re e \langle P(t)u, GP(t)u\rangle \leq - \|P(t)u\|^2$$

hence we have the inequality

$$\|P(t)\| \leq \exp(-t/2). \tag{3.3}$$

We shall consider operators G, L_l satisfying also the following assumption:

C There exists a selfadjoint operator C with domain containing $\mathcal{D}$ such that $\mathcal{D}$ is a core for C and, for all $v, u \in \mathcal{D}$, $t \geq 0$ we have

$$\sum_{l=1}^{\infty} \langle L_l v, L_l u\rangle = \langle v, Cu\rangle = -\langle v, Gu\rangle - \langle Gv, u\rangle, \tag{3.4}$$

$$P(t)u \in D(C) \tag{3.5}$$

Clearly, since we supposed $L_1 = I$, for all $u \in D(C)$ we have the inequality

$$\langle u, Cu\rangle \geq \|u\|^2. \tag{3.6}$$

Therefore C has a positive bounded inverse C^{-1}; the inequality (3.6) implies then

$$0 \leq C^{-1} \leq I \tag{3.7}$$

Moreover, for all $\varepsilon > 0$, we can consider bounded operators

$$C_\varepsilon = C(I + \varepsilon C)^{-1} = \left(C^{-1} + \varepsilon I\right)^{-1}. \tag{3.8}$$

For all $u \in D(C)$ we have

$$\lim_{\varepsilon \to 0+} C_\varepsilon u = Cu. \tag{3.9}$$

The result we shall prove is contained in the following which is essentially due to A. Chebotarev.

Theorem 3.2. *Suppose that the assumptions (A), (B), (C) are satisfied and there exists a positive constant b such that:*
i) for all $u \in \mathcal{D}$ we have

$$2\Re e \left\langle Gu, C^{-1}u \right\rangle \geq -b\|u\|^2, \tag{3.11}$$

ii) for all $u \in \mathcal{D}$ we have

$$2\Re e \langle Cu, Gu \rangle \leq -\|Cu\|^2 + b\langle u, Cu \rangle, \tag{3.12}$$

iii) for all $u \in \mathcal{D}$ and all $\varepsilon > 0$ we have

$$\langle Cu, Cu \rangle - \sum_{l=1}^{\infty} \langle L_l u, C_\varepsilon L_l u \rangle \geq -b\langle u, Cu \rangle. \tag{3.13}$$

Then a quantum dynamical semigroup satisfying the equation (2.5) is conservative.

The above inequalities are more easily verifiable under some additional domain assumptions on the operators G, $(L_l)_{l=1}^{\infty}$, C.

Corollary 3.3. *Suppose that the assumptions (A), (B), (C) are satisfied and there exists a positive constant b such that the following conditions hold:*
a) the linear manifold $C(\mathcal{D})$ coincides with $\mathcal{D}$,
b) for all $l \geq 1$, $L_l(\mathcal{D}) \subseteq D(C)$,
c) for all $u \in \mathcal{D}$ we have

$$2\Re e \langle Cu, Gu \rangle \leq -\|Cu\|^2 + b\langle u, Cu \rangle, \tag{3.14}$$

d) for all $u \in \mathcal{D}$ we have

$$\langle Cu, Cu \rangle - \sum_{l=1}^{\infty} \langle L_l u, CL_l u \rangle \geq -b\langle u, Cu \rangle. \tag{3.15}$$

Then a quantum dynamical semigroup satisfying the equation (2.5) is conservative.

Proof. The conditions b), d) and the inequality

$$\langle L_l u, CL_l u \rangle \geq \langle L_l u, C_\varepsilon L_l u \rangle$$

for all $\varepsilon > 0$, $l \geq 1$, $u \in \mathcal{D}$ imply that the condition iii) of Theorem 3.2. Since $C(\mathcal{D}) = \mathcal{D}$, for all $u \in \mathcal{D}$ the vector $C^{-1}u$ belongs to $\mathcal{D}$ and the condition c) gives

$$2\Re e \left\langle u, GC^{-1}u \right\rangle \leq -\|u\|^2 + b\left\langle u, C^{-1}u \right\rangle. \tag{3.16}$$

Using the identity (3.4) we have

$$\langle u, GC^{-1}u\rangle = -\|u\|^2 - \langle Gu, C^{-1}u\rangle .$$

Thus, from (3.16), we obtain the inequality

$$2\Re\langle u, GC^{-1}u\rangle \geq -\|u\|^2 - b\langle u, C^{-1}u\rangle \geq -(b+1)\|u\|^2$$

Therefore the conditions of Theorem 3.3 are satisfied with the positive constant $b+1$. □

Remark. Let H be the operator with domain $\mathcal{D}$ defined by

$$iH = G + \frac{1}{2}C.$$

Because of (3.4) H is symmetric. Then (3.14) is equivalent to the following inequality

$$i(\langle Hu, Cu\rangle - \langle Cu, Hu\rangle) \geq -b\langle u, Cu\rangle .$$

Thus, roughly speeking, (3.14) is an assumption on the commutator of H and C.

In a similar way, if $L_l = 0$ for all $l \geq 2$, then, under some more technical conditions,

$$Cu = u + L_1^* L_1 u$$

for all $u \in \mathcal{D}$. It is easy to see that (3.15) is an assumption on the commutator of L_1^* and L_1.

4. Chebotarev's inequalities for completely positive maps

In this section $\mathcal{Q}$ will denote an arbitrary completely positive map $\mathcal{Q} : \mathcal{B}(h) \mapsto \mathcal{B}(h)$ (not necessarily the map $\mathcal{Q}$ defined by (2.6)).

Lemma 4.1. *Let A, B, C be elements of $\mathcal{B}(h)$ such that B and C have bounded inverses. Suppose that the bounded operator on $h \oplus h$*

$$\begin{pmatrix} A & B \\ B^* & C \end{pmatrix}$$

is positive. Then the following inequality holds

$$BC^{-1}B^* \leq A. \tag{4.1}$$

Proof. The following identities

$$B^*(B^{-1})^* = (B^{-1}B)^* = I = (B(B^{-1}))^* = (B^{-1})^*B^*$$

imply that B^* has the bounded inverse $(B^{-1})^*$. Moreover the positivity assumption implies that A and C are selfadjoint. The inequality (4.1) follows then from the positivity of the operator on $h \oplus h$

$$\begin{pmatrix} B^{-1}A(B^{-1})^* - C^{-1} & 0 \\ 0 & C \end{pmatrix} = \begin{pmatrix} -B^{-1} & C^{-1} \\ 0 & I \end{pmatrix} \begin{pmatrix} A & B \\ B^* & C \end{pmatrix} \begin{pmatrix} -(B^*)^{-1} & 0 \\ C^{-1} & I \end{pmatrix}. \ \square$$

Lemma 4.2. *Let $\mathcal{Q}$ be a completely positive map on $\mathcal{B}(h)$ and let X be a positive invertible element of $\mathcal{B}(h)$ with bounded inverse X^{-1} Suppose that $\mathcal{Q}(X^{-1})$ and $\mathcal{Q}(I)$ have also bounded inverses. We have then the inequality*

$$\mathcal{Q}(I)\mathcal{Q}(X^{-1})^{-1}\mathcal{Q}(I) \leq \mathcal{Q}(X)$$

Proof. Since $\mathcal{Q}$ is completely positive, the operator on $h \oplus h$

$$\begin{pmatrix} \mathcal{Q}(X) & \mathcal{Q}(I) \\ \mathcal{Q}(I) & \mathcal{Q}(X^{-1}) \end{pmatrix} = \begin{pmatrix} \mathcal{Q}(X^{1/2}X^{1/2}) & \mathcal{Q}(X^{1/2}X^{-1/2}) \\ \mathcal{Q}(X^{-1/2}X^{1/2}) & \mathcal{Q}(X^{-1/2}X^{-1/2}) \end{pmatrix}$$

is positive. Therefore it suffices to apply Lemma 4.1 with $A = \mathcal{Q}(X)$, $B = \mathcal{Q}(I)$, $C = \mathcal{Q}(X^{-1})$. $\square$

Lemma 4.3. *Let $\mathcal{Q}$ be a completely positive map on $\mathcal{B}(h)$ and let X be a positive element of $\mathcal{B}(h)$. Suppose that the operators X, $\mathcal{Q}(I)$, $\mathcal{Q}(X)$ have bounded inverses. Then the following inequality holds*

$$\mathcal{Q}\left((I+X^{-1})^{-1}\right) \leq \left(\mathcal{Q}(I)^{-1}+\mathcal{Q}(X)^{-1}\right)^{-1}. \tag{4.2}$$

Proof. Denote by B the positive square root of $\mathcal{Q}(I)^{1/2}$ We have then

$$\begin{aligned}
\mathcal{Q}\left((I+X^{-1})^{-1}\right) &= \mathcal{Q}(I-(I+X)^{-1}) \\
&= B^2 - \mathcal{Q}((I+X)^{-1}) \\
&= B\left(I - B^{-1}\mathcal{Q}((I+X)^{-1})B^{-1}\right)B
\end{aligned}$$

Using Lemma 4.2 we obtain the inequality

$$B^2\left(\mathcal{Q}(I+X)\right)^{-1}B^2 \leq \mathcal{Q}\left((I+X)^{-1}\right).$$

Therefore we have

$$\begin{aligned}
\mathcal{Q}\left((I+X^{-1})^{-1}\right) &\leq B\left(I - B\left(\mathcal{Q}(I)+\mathcal{Q}(X)\right)^{-1}B\right)B \\
&= B\left(I - \left(I + B^{-1}\mathcal{Q}(X)B^{-1}\right)^{-1}\right)B \\
&= B\left(I + B\mathcal{Q}(X)^{-1}B\right)^{-1}B \\
&= \left(\mathcal{Q}(I)^{-1} + \mathcal{Q}(X)^{-1}\right)^{-1}
\end{aligned}$$

This proves the inequality (4.2). □

Theorem 4.4. *(Chebotarev) Let $\mathcal{Q} : \mathcal{B}(h) \mapsto \mathcal{B}(h)$ be a completely positive map and, for all $n \geq 1$, let $(X_k)_{k=1}^n$ be a family of positive bounded operators. Suppose that the operators $X_1, \ldots, X_n$, $\mathcal{Q}(I), \mathcal{Q}(X_1), \ldots, \mathcal{Q}(X_n)$ have bounded inverses. Then the following inequality holds*

$$\mathcal{Q}\left((I+X_1^{-1}+\ldots+X_n^{-1})^{-1}\right) \leq \left(\mathcal{Q}(I)^{-1}+\mathcal{Q}(X_1)^{-1}+\ldots+\mathcal{Q}(X_n)^{-1}\right)^{-1} \tag{4.3}$$

Proof. The inequality (4.3) follows from Lemma 4.3 when $n = 1$. Suppose it has been established for an integer $n \geq 1$. Consider the bounded operator Y defined by

$$Y^{-1} = I + X_1^{-1} + \ldots + X_n^{-1}.$$

For all $X \in \mathcal{B}(h)$ let

$$\widetilde{X} = Y^{-1/2}XY^{-1/2}, \qquad \widetilde{\mathcal{Q}}(X) = \mathcal{Q}(Y^{1/2}XY^{1/2}).$$

Clearly $\tilde{\mathcal{Q}}$ is a completely positive map on $\mathcal{B}(h)$, hence, applying Lemma 4.3 to the map $\tilde{\mathcal{Q}}$ and (4.3) for the integer n n and the map $\tilde{\mathcal{Q}}$ we get the inequalities

$$\begin{aligned}\mathcal{Q}\left((I+X_1^{-1}+\ldots+X_{n+1}^{-1})^{-1}\right) &= \tilde{\mathcal{Q}}\left(\left(I+\tilde{X}_{n+1}\right)^{-1}\right)\\ &\le \left(\tilde{\mathcal{Q}}(I)^{-1}+\tilde{\mathcal{Q}}(\tilde{X}_{n+1})^{-1}\right)^{-1}\\ &= \left(\mathcal{Q}(Y)^{-1}+\mathcal{Q}(X_{n+1})^{-1}\right)^{-1}\\ &\le \left(I+\mathcal{Q}(X_1)^{-1}+\ldots+\mathcal{Q}(X_{n+1})^{-1}\right)^{-1}\end{aligned}$$

This completes the induction argument. □

5. Proof of Theorem 3.2

Let $\mathcal{Q}$ be the completely positive linear map on $\mathcal{B}(h)$ defined by (2.6) and let Q denote $\mathcal{Q}(I)$. Let us first prove some properties of Q.

Lemma 5.1. *The operator Q has a positive bounded inverse satisfying the following inequalities*

$$\frac{1}{2}I \le Q \le I, \qquad I \le Q^{-1} \le 2I. \tag{5.1}$$

Moreover $Qu = u$ if and only if $u = 0$.

Proof. From the inequalities (2.7) and (3.3) we obtain

$$\frac{1}{2}\|u\|^2 = \|u\|^2\left(1-\int_0^\infty \exp(-2s)ds\right) \le \langle u, Qu\rangle \le \|u\|^2$$

Therefore Q has a bounded inverse Q^{-1} and (5.1) holds. Moreover it follows from (2.7) that $Qu = u$ if and only if $P(t)u = 0$ for all $t \ge 0$, i.e. $u = 0$. □

Lemma 5.2. *Let R be the positive selfadjoint operator on h defined by*

$$Ru = Q(I-Q)^{-1}u$$

for all $u \in D(R)$. For all $\varepsilon > 0$ the bounded operators R_ε, C_ε defined by

$$R_\varepsilon = R(I+\varepsilon R)^{-1}, \qquad C_\varepsilon = C(I+\varepsilon C)^{-1}$$

satisfy the inequalities

$$(2+\varepsilon)^{-1}I \le R_\varepsilon \le \varepsilon^{-1}I, \qquad (1+\varepsilon)^{-1}I \le C_\varepsilon \le \varepsilon^{-1}I. \tag{5.2}$$

For all $\varepsilon > 0$ and all $n \geq 1$ the operators $\mathcal{Q}^n(R_\varepsilon)$, $\mathcal{Q}^n(C_\varepsilon)$ have bounded inverses and the following inequalities hold

$$(2+\varepsilon)^{-n} I \leq \mathcal{Q}^n(R_\varepsilon) \leq \varepsilon^{-n} I, \qquad (1+\varepsilon)^{-n} I \leq \mathcal{Q}^n(C_\varepsilon) \leq \varepsilon^{-n} I \tag{5.3}$$

Proof. The operator R can be defined by functional calculus using the spectral family of Q. Then (5.1) yields the inequalities

$$R_\varepsilon = \left(\varepsilon I + R^{-1}\right)^{-1} = \left(\varepsilon I + Q^{-1} - I\right)^{-1} \geq \left(\varepsilon I + Q^{-1}\right)^{-1} \geq (2+\varepsilon)^{-1}$$

Moreover, because of (5.1), R has a bounded inverse. Hence, for all $\varepsilon > 0$, the operators R_ε, C_ε are bounded from above by $\varepsilon^{-1} I$. The operator C_ε is bounded from below by $(1+\varepsilon)^{-1} I$ because of (3.6). This proves (5.2). Now (5.3) follows easily by induction. □

Lemma 5.3. *For all positive bounded operator X and all $\varepsilon \in (0, 1/2)$ we have*

$$\left(Q^{-1} + X\right)^{-1} \leq \varepsilon I + \left(I + R_\varepsilon^{-1} + X\right)^{-1}$$

Proof. Let us first compute

$$\begin{aligned}\left(Q^{-1} + X\right)^{-1} &= \left(I + R_\varepsilon^{-1} + X - \varepsilon I\right)^{-1} \\ &= \left(I + R_\varepsilon^{-1} + X\right)^{-1} \left(I - \varepsilon \left(I + R_\varepsilon^{-1} + X\right)^{-1}\right)^{-1}\end{aligned}$$

Using the inequality (5.2) we find the lower bound

$$I + R_\varepsilon^{-1} + X \geq (3+\varepsilon)(2+\varepsilon)^{-1}$$

For all $\varepsilon \in (0, 1/2)$ and all $t \geq (3+\varepsilon)(2+\varepsilon)^{-1}$ we have

$$\left(1 - \varepsilon t^{-1}\right)^{-1} \leq 1 + \varepsilon t \tag{5.4}$$

In fact the right-hand side is an increasing function, the left-hand side is a decreasing function of t and (5.4) holds for $t = (3+\varepsilon)(2+\varepsilon)^{-1}$ because of the elementary inequalities

$$\begin{aligned}(3+\varepsilon)(2+\varepsilon) &\leq (2\varepsilon+1)(6-3\varepsilon) \\ &= (4+2\varepsilon)(3-3\varepsilon/2) \\ &\leq (2+4\varepsilon+\varepsilon^2)(3-\varepsilon-\varepsilon^2)\end{aligned}$$

which hold for all $\varepsilon \in (0, 1/2)$. Therefore, using (5.4), we obtain the inequality

$$\begin{aligned}\left(Q^{-1} + X\right)^{-1} &\leq \left(I + R_\varepsilon^{-1} + X\right)^{-1} \left(I + \varepsilon \left(I + R_\varepsilon^{-1} + X\right)\right) \\ &= \varepsilon I + \left(I + R_\varepsilon^{-1} + X\right)^{-1}\end{aligned}$$

which completes the proof. □

We shall prove now the inequality needed to prove Chebotarev's sufficient conditions.

Theorem 5.4. *For each integer $n \geq 0$ and each finite sequence $(\varepsilon_k)_{k=0}^n$ of elements of $(0, 1/2)$ such that $\sum_{k=0}^n \varepsilon_k \leq 1/2$ we have*

$$\mathcal{Q}^{n+1}(I) \leq I \sum_{k=0}^{n} \varepsilon_k + \left(I + \sum_{k=0}^{n} \mathcal{Q}^k \left(R_{\varepsilon_k}\right)^{-1} \right)^{-1} \tag{5.5}$$

Proof. The inequality (5.5) holds for $n = 0$ because of (5.1). Suppose that it has been established for an integer $n-1$. Using (5.1) and Chebotarev's inequality (4.3) we have

$$\begin{aligned}
\mathcal{Q}^{n+1}(I) &= \mathcal{Q}(\mathcal{Q}^n(I)) \\
&\leq I \sum_{k=0}^{n-1} \varepsilon_{k+1} + \mathcal{Q}\left(\left(I + \sum_{k=0}^{n-1} \mathcal{Q}^k \left(R_{\varepsilon_{k+1}}\right)^{-1} \right)^{-1} \right) \\
&\leq I \sum_{k=1}^{n} \varepsilon_k + \left(Q^{-1} + \sum_{k=1}^{n} \mathcal{Q}^k \left(R_{\varepsilon_k}\right)^{-1} \right)^{-1}
\end{aligned}$$

Applying Lemma 5.3 with $\varepsilon = \varepsilon_0$ we get the inequality

$$\mathcal{Q}^{n+1}(I) \leq I \sum_{k=0}^{n} \varepsilon_k + \left(I + R_{\varepsilon_0}^{-1} + \sum_{k=1}^{n} \mathcal{Q}^k \left(R_{\varepsilon_k}\right)^{-1} \right)^{-1}$$

This completes the induction argument. □

Proposition 5.5. *Suppose that the conditions (A), (B) and (C) hold and there exists a positive constant b such that*

$$2\Re e \left\langle Gu, C^{-1}u \right\rangle \geq -b \|u\|^2 \tag{5.6}$$

for all $u \in \mathcal{D}$. For all $\varepsilon > 0$ we have then

$$(I - Q) \geq (b+1)^{-1} C^{-1}, \tag{5.7}$$

$$R_\varepsilon \leq (b+1) C_\varepsilon. \tag{5.8}$$

Proof. Since $\mathcal{D}$ is a core for G, (5.6) holds for all $u \in D(G)$. Moreover, using the inequality $C^{-1/2} \leq 1$, we have

$$\langle u, (I - Q)u \rangle = \int_0^\infty \exp(-s) \|P(s)u\|^2 \, ds \geq \int_0^\infty \exp(-s) \left\| C^{-1/2} P(s) u \right\|^2 ds$$

Integrating by parts and using the inequality (5.6) we have

$$
\begin{aligned}
\int_0^\infty \exp(-s)\left\|C^{-1/2}P(s)u\right\|^2 ds &= \left\langle u, C^{-1}u\right\rangle \\
&\quad + \int_0^\infty \exp(-s)2\Re\left\langle GP(s)u, C^{-1}P(s)u\right\rangle ds \\
&\geq \left\langle u, C^{-1}u\right\rangle - b\int_0^\infty \exp(-s)\left\|P(s)u\right\|^2 ds \\
&= \left\langle u, C^{-1}u\right\rangle - b\langle u, (I-Q)u\rangle
\end{aligned}
$$

Hence we obtain the inequality

$$
(b+1)\langle u, (I-Q)u\rangle \geq \left\langle u, C^{-1}u\right\rangle
$$

and (5.7) follows. Moreover, using (5.1), for all $\varepsilon > 0$, we get the inequality

$$
R_\varepsilon^{-1} = \varepsilon + Q^{-1}(I-Q) \geq \varepsilon + (I-Q) \geq \varepsilon + (b+1)^{-1}C^{-1} \geq (b+1)^{-1}(\varepsilon + C^{-1})
$$

This implies (5.8). □

Proposition 5.6. *Suppose that the conditions (A), (B), (C) hold. Then:*

a) *for all $l \geq 1$, the operator L_l has an extension with domain containing $D(C)$,*

b) *for all sequence $(u_n)_{n\geq 1}$ of elements of $\mathcal{D}$ that converges to a vector u such that the sequence $(Cu_n)_{n\geq 1}$ converges to Cu we have*

$$
\lim_{n\to\infty} L_l u_n = L_l u,
$$

c) *the domain $D(G)$ of G is contained in the domain $D(C^{1/2})$ of $C^{1/2}$,*

d) *for all $v \in D(G), u \in D(G)\cap D(C)$ we have*

$$
\sum_{l=1}^{\infty} \langle L_l v, L_l u\rangle = \langle v, Cu\rangle = -\langle v, Gu\rangle - \langle Gv, u\rangle . \tag{5.9}
$$

Proof. Since $\mathcal{D}$ is a core for C, for all $u \in D(C)$ there exists a sequence $(u_n)_{n\geq 1}$ of elements of $\mathcal{D}$ such that

$$
\lim_{n\to\infty} u_n = u, \qquad \lim_{n\to\infty} Cu_n = Cu.
$$

Therefore, by the identity (3.4), for all integers n, m, we have

$$
\sum_{l=1}^{\infty} \left\|L_l(u_n - u_m)\right\|^2 = \left\|C^{1/2}(u_n - u_m)\right\|^2 = -2\Re\left\langle u_n - u_m, G(u_n - u_m)\right\rangle .
$$

As in Proposition 3.1 this shows that the operators L_l have as an extension to $D(C)$. Using the above identity one can prove also b). Consider now a sequence $(u_n)_{n\geq 1}$ of elements of $\mathcal{D}$ such that

$$\lim_{n\to\infty} u_n = u, \qquad \lim_{n\to\infty} Gu_n = Gu.$$

For all integers n, m, the identity (3.4) yields

$$\left\|C^{1/2}(u_n - u_m)\right\|^2 = -2\Re\left\langle u_n - u_m, G(u_n - u_m)\right\rangle. \tag{5.10}$$

Therefore the sequence $(C^{1/2}u_n)_{n\geq 1}$ is Cauchy. Since the operator $C^{1/2}$ is selfadjoint, u belongs to the domain $D(C^{1/2})$ of $C^{1/2}$ and the sequence $(C^{1/2}u_n)_{n\geq 1}$ converges to $C^{1/2}u$. This proves the statement c). Moreover, for all $v, u \in D(G)$, we have

$$\left\langle C^{1/2}v, C^{1/2}u\right\rangle = -\langle v, Gu\rangle - \langle Gv, u\rangle.$$

This implies that (5.9) holds for all $v \in D(G), u \in D(G)\cap D(C)$. □

Proposition 5.7. *Suppose that the conditions (A), (B), (C) are satisfied and that the inequality (3.12) holds for all $u \in \mathcal{D}$. Then (3.12) holds for all $u \in D(G^2)\cap D(C)$.*

Proof. Let H be the operator with domain coinciding with $\mathcal{D}$ defined by

$$iH = G + \frac{1}{2}C.$$

Since C is selfadjoint, (3.4) implies that H is symmetric. For all $u \in \mathcal{D}$ the inequality (3.12) yields

$$2\Re\langle Cu, iHu\rangle \leq b\langle u, Cu\rangle.$$

Therefore, for all $u \in \mathcal{D}$, we have

$$\|Gu\|^2 = \frac{1}{4}\|Cu\|^2 - \Re\langle Cu, iHu\rangle + \|Hu\|^2 \geq \frac{1}{4}\|Cu\|^2 - \frac{b}{2}\langle u, Cu\rangle$$

which implies

$$\|Cu\|^2 \leq 4\|Gu\|^2 + 2b\left\|C^{1/2}u\right\|^2. \tag{5.11}$$

Clearly, form (3.12), for all $\varepsilon > 0$ and all $u \in \mathcal{D}$ we have

$$2\Re\langle Cu, Gu\rangle \leq -\|C_\varepsilon u\|^2 + b\left\langle C^{1/2}u, C^{1/2}u\right\rangle. \tag{5.12}$$

Let u be an element of $D(G^2) \cap D(C)$ and let $(u_n)_{n \geq 1}$ be a sequence of elements of $\mathcal{D}$ such that

$$\lim_{n \to \infty} u_n = u, \qquad \lim_{n \to \infty} Gu_n = Gu.$$

Then, by (5.10) and (5.11), we have

$$\lim_{n \to \infty} C^{1/2} u_n = C^{1/2} u, \qquad \sup_{n \geq 1} \|Cu_n\| < \infty. \tag{5.13}$$

Because of (5.9), for all $v \in D(G)$, we have

$$\lim_{n \to \infty} \langle Cu_n, v \rangle = \langle Cu, v \rangle .$$

Therefore, since $u \in D(G^2)$, using (5.13) we obtain

$$\lim_{n \to \infty} \langle Cu_n, Gu_n \rangle = \lim_{n \to \infty} \langle Cu_n, Gu_n - Gu \rangle + \lim_{n \to \infty} \langle Cu_n, Gu \rangle = \langle Cu, Gu \rangle .$$

Writing the inequality (5.12) for the vector u_n and letting n tend to infinity, using (5.13), we obtain the inequality

$$2\Re e \langle Cu. Gu \rangle \leq - \|C_\varepsilon u\|^2 + b \left\langle C^{1/2} u, C^{1/2} u \right\rangle .$$

Letting ε tend to 0 we obtain (3.12) for all vectors $u \in D(G^2) \cap D(C)$. □

Proposition 5.8. *Suppose that the conditions (A), (B), (C) are satisfied and there exists a positive constant b such that the assumption ii) of Theorem 3.2 holds. Then, for all $u \in \mathcal{D}$, we have*

$$\int_0^\infty \exp(-t) \|CP(t)\|^2 \, dt \leq \langle u, (C + bI)u \rangle . \tag{5.14}$$

Proof. Let us note first that, for all $u \in \mathcal{D}$, we have

$$\frac{d^2}{dt^2} \|P(t)u\|^2 = -2\Re e \langle CP(t)u, GP(t)u \rangle .$$

Moreover the function $CP(\cdot)u$ is measurable because it can be approximated by continuous functions $(C_\varepsilon P(\cdot)u)_{\varepsilon > 0}$. Since $u \in \mathcal{D} \subseteq D(G^2)$, due to Proposition 5.7, for all $t \geq 0$, the inequality (3.12) holds for the vector $P(t)u \in D(G^2) \cap D(C)$, hence

$$\int_0^\infty \exp(-t) \|CP(t)u\|^2 \, dt \leq \int_0^\infty \exp(-t) \frac{d^2}{dt^2} \|P(t)u\|^2 \, dt$$
$$+ b \int_0^\infty \exp(-t) \langle P(t)u, CP(t)u \rangle \, dt.$$

Integrating by parts we can easily compute the first integral

$$\int_0^\infty \exp(-t) \frac{d^2}{dt^2} \|P(t)u\|^2 \, dt = \langle u, Cu \rangle - \langle u, Qu \rangle$$

The second one can be computed using (2.6) and (3.4). Thus we obtain (5.14).□

Proposition 5.9. *Suppose that the conditions (A), (B), (C) and the assumptions i), ii) of Theorem 3.2 hold. Then, for every $\varepsilon > 0$, for every integer $n \geq 1$ and every $u \in \mathcal{D}$, we have*

$$\langle u, \mathcal{Q}^n(C_\varepsilon)u\rangle \leq \langle u, (C + nbI)\, u\rangle \tag{5.15}$$

$$\langle u, \mathcal{Q}^n(R_\varepsilon)u\rangle \leq (b+1)\,\langle u, (C + nbI)\, u\rangle \tag{5.16}$$

Proof. Let us first notice the following property of the regularisation of positive operators used in Lemma 5.2. For all $\varepsilon, \eta > 0$ and all positive bounded operator X with bounded inverse we have

$$(X_\varepsilon)_\eta = \left(\eta I + X_\varepsilon^{-1}\right)^{-1} = \left((\varepsilon + \eta)I + X^{-1}\right)^{-1} = X_{\varepsilon+\eta}.$$

Therefore, by Chebotarev's inequality (4.3) and the inequalities (5.1), we have

$$\begin{aligned} \mathcal{Q}(X_{\varepsilon+\eta}) &= \mathcal{Q}\left(\left(\varepsilon I + X_\eta^{-1}\right)^{-1}\right) \\ &\leq \left(\varepsilon Q^{-1} + \mathcal{Q}(X_\eta)^{-1}\right)^{-1} \\ &\leq \left(\varepsilon I + \mathcal{Q}(X_\eta)^{-1}\right)^{-1} = \left(\mathcal{Q}(X_\eta)\right)_\varepsilon \end{aligned} \tag{5.17}$$

We prove now (5.15) by induction. By the same argument of Proposition 5.6 one can see that the condition iii) of Theorem 3.2 holds for all vector $u \in D(C)$. Using (5.9) and (3.13), for all $u \in \mathcal{D}$ we easily obtain the inequalities

$$\begin{aligned} \langle u, \mathcal{Q}(C_\varepsilon)u\rangle &\leq \int_0^\infty \exp(-t) \sum_{l=1}^\infty \langle L_l P(t)u, C_\varepsilon L_l P(t)u\rangle\, dt \\ &\leq \int_0^\infty \exp(-t) \left(\|CP(t)u\|^2 + b\,\langle P(t)u, CP(t)u\rangle\right) dt \\ &= \langle u, (C + bI)u\rangle \end{aligned} \tag{5.18}$$

Suppose now that (5.15) has been established for a certain integer $n \geq 1$ and let us prove it for the integer $n+1$. Using (5.17) and (5.18) we obtain

$$\begin{aligned} \left\langle u, \mathcal{Q}^{n+1}(C_{2\varepsilon})u\right\rangle &= \langle u, \mathcal{Q}^n\left(\mathcal{Q}(C_{2\varepsilon})\right)u\rangle \\ &\leq \langle u, \mathcal{Q}^n\left(\left(\mathcal{Q}(C_\varepsilon)\right)_\varepsilon\right)u\rangle \\ &\leq \langle u, \left(\mathcal{Q}^n(C_\varepsilon) + bI\right)u\rangle \\ &\leq \langle u, \left(C + (n+1)bI\right)u\rangle \end{aligned}$$

This completes the induction argument. Now (5.16) follows from (5.15) and the inequality (5.8). □

Proof. (of Theorem 3.2) For all $\varepsilon > 0$ and all integer n consider a finite sequence $(\varepsilon_k)_{k=0}^n$ such that $\varepsilon_0 + \ldots + \varepsilon_n = \varepsilon$. The inequality (5.5) and Chebotarev's inequality (4.3) applied to the completely positive map

$$X \mapsto \langle u, Xu \rangle$$

with $u \in \mathcal{D}$ yield the inequality for all nonzero u

$$\langle u, \mathcal{Q}^{n+1}(I)u \rangle \leq \varepsilon \|u\|^2 + \left(\|u\|^{-2} + \sum_{k=0}^{n} \langle u, \mathcal{Q}^k(R_{\varepsilon_k})u \rangle^{-1} \right)^{-1}$$

Using (5.16) for all $u \in \mathcal{D}$ we have then

$$\langle u, \mathcal{Q}^{n+1}(I)u \rangle \leq \varepsilon \|u\|^2 + \left(\|u\|^{-2} + (b+1)^{-1} \sum_{k=0}^{n} \langle u, (C + kbI)u \rangle^{-1} \right)^{-1}.$$

The divergence of the series

$$\sum_{k=0}^{\infty} \langle u, (C + kbI)u \rangle^{-1}$$

implies then

$$\lim_{n \to \infty} \langle u, \mathcal{Q}^{n+1}(I)u \rangle \leq \varepsilon \|u\|^2$$

Therefore the condition ii) of Theorem 2.6 is satisfied and the q.d.s. satisfying the equation (2.5) is conservative. □

References

[1] Chebotarev, A.M.: Sufficient conditions for the regularity of jump Markov processes. *Theory Probab. Appl.* **33**, No.1 (1990).

[2] Chebotarev, A.M.: *The theory of the conservative dynamical semigroups and its applications.* Preprint MIEM n.1. March 1990.

[3] Chebotarev, A.M., Fagnola, F., Frigerio, A.: Towards a stochastic Stone's theorem. To appear in: *Stochastic Partial Differential Equations and Applications III*, Pitman Research Notes in Mathematics.

[4] Davies, E.B.: Quantum dynamical semigroups and the neutron diffusion equation. *Rep. Math. Phys.* **11** (1977), 169–188

[5] Davies, E.B.: Generators of Dynamical Semigroups. *J. Funct. Anal.* **34** (1979), 421–432.

[6] Fagnola, F.: Unitarity of solutions of quantum stochastic differential equations and conservativity of the associated semigroups. To appear in: *Quantum Probability and Related Topics VII.*

[7] Fagnola, F.: Realisation of diffusion processes and Azema-Emery martingales as quantum flows in Fock space. In preparation.

[8] Fagnola, F.: Characterisation of isometric and unitary weakly differentiable cocycles in Fock space. Preprint UTM n.358 October 1991.

[9] Frigerio, A.: Quantum dynamical semigroups and approach to equilibrium. Preprint, 1977.

[10] Mohari, A.: Quantum stochastic calculus with infinite degrees of freedom and its applications. Ph.D. Thesis I.S.I., New Delhi 1991.

[11] Mohari, A., Parthasarathy, K.R.: A quantum probabilistic analogue of Feller's condition for the existence of unitary Markovian cocycles in Fock space. Preprint, I.S.I., Delhi Centre. (1991).

[12] Parthasarathy, K.R.: *An Introduction to Quantum Stochastic Calculus.* I.S.I., New Delhi 1990.

Quantum Probability and Related Topics Vol. VIII (pp. 143-164)

CHARACTERIZATION OF ISOMETRIC AND UNITARY WEAKLY DIFFERENTIABLE COCYCLES IN FOCK SPACE

Franco Fagnola
Dipartimento di Matematica, Università di Trento
I - 38050 POVO (TN) ITALY

Abstract. We prove a characterization theorem for weakly differentiable (contractive, isometric and unitary) Markovian cocycles in the Boson Fock space analogue to Stone's theorem on strongly continuous groups of unitary operators.

1. Introduction

A Markovian cocycle is a family $(V(t))_{t\geq 0}$ of contractive operators with a time shift covariance and localisation properties. This notion was introduced by Accardi in [1] to construct Feynman-Kac perturbations of quantum Markovian semigroups and exploited by Accardi, Frigerio and Lewis in [2] to construct quantum stochastic processes.

Unitary Markovian cocycles in the Boson Fock space with norm continuous reduced semigroup can be obtained solving a quantum stochastic differential equation of the form

$$dV(t) = \sum_{i,j} V(t)L^i_j d\Lambda^j_i, \qquad V(0) = I \tag{1.1}$$

where Λ^j_i are the basic integrators of Hudson-Parthasarathy Boson Fock space quantum stochastic calculus and the operators L^i_j are bounded operators playing the role of the infinitesimal generator. Conversely it has been shown by Hudson and Lindsay [12] that Markovian cocycles with norm continuous reduced semigroup are all solutions of equations of this form.

Clearly norm continuity is not satisfied by the most important semigroups. Thus it is interesting to weaken this assumption to establish a quantum stochastic analogue of Stone's theorem on one-parameter group of unitary operators. Journé [13] investigated the strongly continuous case and showed, under a regularity condition, that a cocycle can be reconstructed from the infinitesimal generator by a recursive procedure. However the cocycle obtained by this procedure may not satisfy any q.s.d.e. when the operator L^i_j have $\{ 0 \}$ as common domain. This cannot happen when the cocycle V satisfies a weak differentiability condition introduced by Accardi, Journé and Lindsay in [3] (see Definition 2.3 here). In this case the cocycle V satisfies a q.s.d.e. of the form (1.1) and, moreover, the operators L^i_j have the property (A) (see Definition 2.6, Theorem 2.4). If V is isometric it has the properties (B) (see Section 5), and (C) (the conservativity of the associated quantum dynamical semigroup, see Section 5). If V is unitary and its dual cocycle $\widetilde{V}$ (see [13]) is also weakly differentiable then $\widetilde{V}$ has the properties (A), (B), (C).

In this paper we establish a quantum stochastic Stone's theorem for weakly differentiable cocycles proving the converse result: given the operators L^i_j,satisfying the property (A), there exists a unique solution of (1.1) which is a weakly differentiable cocycle. Since every contractive solution of (1.1), when unique, is a weakly differentiable cocycle this generalizes also all previous existence results on q.s.d.e. of the form (1.1) with unbounded coefficients (see [4],[7],[8],[11],[15],[16],[17]) which have been obtained recently and applied to the construction of stochastic processes in the Boson Fock space. Moreover we show that conditions (B), (C) on the operators L^i_j (resp. (B) and (C) on the operators $\widetilde{L}^i_j$) are necessary and sufficient in order V to be isometric (resp. unitary). Thus isometry (and by dualisation unitarity) of a contractive cocycle satisfying (1.1) can be obtained from the characterization theorem of Chebotarev [4] showing that conservativity of a quantum dynamical semigroup is equivalent to a quantum analogue of Feller's classical condition or from quite general and easily verifiable sufficient conditions given also in [4] (see also Mohari [15] for a more detailed study of quantum dynamical semigroups arising from classical countable state Markov chains).

The proof is done approximating the operators L^i_j by bounded operators using the resolvent family of the reduced semigroup (and multiplying the coefficients of creation and annihilation operators by a real number $\eta \in (0,1)$ as in Da Prato [6], in the case of classical s.d.e., and Frigerio [11]). We construct a solution of (1.1) as a weak limit of solutions of the approximating equations using the equicontinuity method suggested by Frigerio as described by the author in [8]. This solution is unique by a result of Mohari-Parthasarathy [16]. Then, using some results on the minimal solution of the classical equation for a quantum dynamical semigroup (Langevin equation), we show that the unique solution of (1.1) gives a dilation of the classical minimal solution and give necessary and sufficient conditions for isometry.

Conservativity of the associated quantum dynamical semigroups for unitarity of the solution of (1.1), whose importance was pointed out in [4],[5] and [9], is necessary and sufficient for unitarity of the solution also for q.s.d.e. of the form (1.1) driven by "free" integrators as shown by Mancino [14] using a time reversal property for cocycles in the full Fock space proved in [10]. "Free" integrators, however, have a different Ito table.

The new notation of Belavkin-Parthasarathy (see [19],[16]) for Boson Fock space quantum stochastic calculus we use here allows to find easily a necessary and sufficient conditions for contractivity of solutions to (1.1) and to deal with infinite integrator processes.

The characterisation theorem of weakly differentiable unitary cocycles obtained in the present paper can be easily applied to the realisation of a large class of classical Markov processes as inner quantum flows in Fock space. This will be done in a forthcoming paper.

2. Notations and preliminaries

We recall here some notations and results in Boson Fock space quantum stochastic calculus as exposed in [16], [18], [19]. Let h, k be two complex separable Hilbert spaces and let $\{ e_l \mid l \in S \}$ be an orthonormal basis in k. Let $\mathcal{M}_S$ denote the dense linear submanifold of $L^2(\mathbf{R}_+) \otimes k$ of elements satisfying $\langle e_l, u(t)\rangle_k = 0$ for all $t \geq 0$ for all but a finite number of indices k and let $\mathcal{M}_S^0$ be the submanifold of continuous functions in $\mathcal{M}_S$. For all Hilbert space $\mathcal{K}$ let $\Gamma(\mathcal{K})$ denote the Boson Fock space over $\mathcal{K}$. Let $\bar{S} = S \cup \{-\infty, +\infty\}$ and consider the Hilbert spaces

$$\begin{gathered} \hat{k} = \mathbf{C}e_{-\infty} \oplus k \oplus \mathbf{C}e_{+\infty}, \\ \mathcal{H} = h \otimes \Gamma\left(L^2(\mathbf{R}_+) \otimes k\right), \qquad \widehat{\mathcal{H}} = h \otimes \hat{k} \otimes \Gamma\left(L^2(\mathbf{R}_+) \otimes k\right). \end{gathered} \tag{2.1}$$

In the following all the operators defined on a tensor product factor of $\widehat{\mathcal{H}}$ will be identified with their canonical extension to the whole space. Let F be the unique unitary operator in $\hat{k}$ defined by

$$Fe_l = e_l \quad \text{if} \quad l \in S, \qquad Fe_{+\infty} = e_{-\infty}, \qquad Fe_{-\infty} = e_{+\infty}.$$

and let $E_{-\infty}$, E, $E_{+\infty}$ be the orthogonal projections of $\hat{k}$ onto $\mathbf{C}e_{-\infty}$, k, $\mathbf{C}e_{+\infty}$ respectively. We denote by $\mathbb{B}$ the *Belavkin algebra* associated with (h, k)

$$\mathbb{B} = \left\{ L \in \mathcal{B}(h \otimes \hat{k}) \mid Lue_{-\infty} = L^* ue_{+\infty} = 0 \text{ for all } u \in h \right\}$$

The map $L \mapsto L^b$ where $L^b = FL^*F$ is an involution of $\mathbb{B}$.

Let us introduce the integrator processes in $\widehat{\mathcal{H}}$, $\{\Lambda_m^l \mid l, m \in \bar{S}\}$ defined by

$$\begin{aligned} \Lambda_m^{+\infty}(t) &= A^+(\chi_{(0,t)} \otimes |e_m\rangle), && \text{if} \quad m \in S,\ l = +\infty \\ \Lambda_m^l(t) &= \Lambda(\chi_{(0,t)} \otimes |e_m\rangle\langle e_l|), && \text{if} \quad m, l \in S \\ \Lambda_{-\infty}^l(t) &= A(\chi_{(0,t)} \otimes \langle e_l|), && \text{if} \quad l \in S,\ m = -\infty \\ \Lambda_{-\infty}^{+\infty}(t) &= tI && \text{if} \quad l = +\infty,\ m = -\infty \\ \Lambda_m^l(t) &= 0 && \text{otherwise} \end{aligned}$$

where A^+, A, Λ denote creation, annihilation and gauge operators in $\Gamma(L^2(\mathbf{R}_+) \otimes k)$. Let D be a dense linear submanifold of h and let D_S, $\mathcal{D}_S$ and $\widehat{\mathcal{D}}_S$ be the dense linear submanifolds of $h \otimes k$, $\mathcal{H}$ and $\widehat{\mathcal{H}}$ generated by $\{ ue_l \mid u \in D, l \in \bar{S} \}$, $\{ ue(f) \mid u \in D, f \in \mathcal{M}_S \}$, $\{ ue_l e(f) \mid u \in D, l \in \bar{S}, f \in \mathcal{M}_S \}$ respectively. We recall the definition of $(D_S, \mathcal{M}_S)$- *adapted process* as a family $\{ L(t) \mid t \geq 0 \}$ of operators in $\widehat{\mathcal{H}}$ satisfying the following conditions:

the domain of $L(t)$ contains $\widehat{\mathcal{D}}_S$ for all $t \geq 0$,

the map $t \mapsto L(t)\xi e(f)$ is measurable for each $\xi \in D_S$, $f \in \mathcal{M}_S$,

for each $\xi \in D_S$, $f \in \mathcal{M}_S$, $t \geq 0$ we have the tensor product factorization

$$L(t)\xi e(f) = \left\{ L(t)\xi e(\chi_{(0,t)} f) \right\} \otimes e(\chi_{(t,+\infty)} f).$$

A $(D_S, \mathcal{M}_S)$-adapted process L is called *regular* if the following condition holds the map $t \mapsto L(t)\xi e(f)$ is continuous for all $\xi \in D_S$, $f \in \mathcal{M}_S$.

In a similar way one can define $(D, \mathcal{M}_S)$-adapted regular processes.

When $D = h$ a $(D, \mathcal{M}_S)$-adapted process $X = (X(t))_{t \geq 0}$ in $\mathcal{H}$ is called *unitary, isometric, coisometric, contractive* if, for all $t \geq 0$, the operators $X(t)$ are unitary, isometric, coisometric, contractive.

Let $\mathcal{I}(D_S, \mathcal{M}_S)$ be the vector space of $(D_S, \mathcal{M}_S)$-adapted regular processes L in $\widehat{\mathcal{H}}$ satisfying

$$\langle ve_{-\infty} e(g), L(t) ue_l e(f) \rangle = 0, \qquad L(t) ue_{-\infty} e(f) = 0$$

for all $u, v \in D$, $f, g \in \mathcal{M}_S$, $l \in \bar{S}$. With every $L \in \mathcal{I}(D_S, \mathcal{M}_S)$ it is possible to associate the family $\{ L^l_m \mid l, m \in \bar{S} \}$ of $(D, \mathcal{M}_S)$-adapted regular processes in $\widehat{\mathcal{H}}$ defined by

$$\left\langle ve(g), L^l_m(t) ue(f) \right\rangle = \langle ve_l e(g), L(t) ue_m e(f) \rangle .$$

Moreover one can define the stochastic integral

$$\Lambda_L(t) = \int_0^t \sum_{l,m \in \hat{S}} L^l_m(t) d\Lambda^m_l(t)$$

on the domain $\mathcal{D}_S$ as a $(D, \mathcal{M}_S)$-regular adapted process. The following fundamental formulas of Boson Fock space quantum stochastic calculus and results on q.s.d.e. can be proved as in [16], [18].

Proposition 2.1. *Let $L, M \in \mathcal{I}(D_S, \mathcal{M}_S)$ and let X_0, Y_0 be operators on h with domain containing D. Consider the processes*

$$X(t) = X_0 + \Lambda_L(t), \qquad Y(t) = Y_0 + \Lambda_M(t).$$

For each $v, u \in D$, $f, g \in \mathcal{M}_S$ the following formulas hold:

$$\begin{aligned} \langle ve(g), X(t)ue(f) \rangle &= \langle ve(g), X_0 ue(f) \rangle \\ &+ \int_0^t \langle v(e_{-\infty} + g(s)) e(g), L(s) u(f(s) + e_{+\infty}) e(f) \rangle \, ds \end{aligned} \tag{2.2}$$

$$\langle Y(t)ve(g), X(t)ue(f)\rangle = \langle Y_0 ve(g), X_0 ue(f)\rangle$$
$$+ \int_0^t \Big\{ \langle (Y(s) + FM(s)F)v(e_{-\infty} + g(s))e(g), (X(s) + L(s))u(f(s) + e_{+\infty})e(f)\rangle$$
$$- \langle Y(s)v(e_{-\infty} + g(s))e(g), X(s)u(f(s) + e_{+\infty})e(f)\rangle \Big\} ds \qquad (2.3)$$

Proposition 2.2. *For each $L \in \mathbb{B}$ there exists a unique $(h, \mathcal{M}_S)$-regular adapted process X satisfying the q.s.d.e.*

$$dX(t) = X(t)d\Lambda_L(t), \qquad X(0) = I.$$

The process X is unitary if and only if

$$L + L^b + L^b L = 0, \qquad L + L^b + LL^b = 0.$$

For each $t \geq 0$ let θ_t be the right shift and let ρ_t be the time reversal on the interval $[0, t]$ defined on $L^2(\mathbf{R}_+) \otimes k$ by

$$(\theta_t f)(x) = \begin{cases} f(x - t) & \text{if } x > t \\ 0 & \text{if } x \leq t \end{cases}, \qquad (\rho_t f)(x) = \begin{cases} f(t - x) & \text{if } x \leq t \\ f(x) & \text{if } x > t \end{cases}.$$

Let $\mathcal{S}_t$, $\mathcal{R}_t$ be the operators in $\Gamma\left(L^2(\mathbf{R}_+) \otimes k\right)$ defined respectively by the second quantizations of θ_t, ρ_t

$$\mathcal{S}_t e(f) = e(\theta_t f), \qquad \mathcal{R}_t e(f) = e(\rho_t f)$$

for all $f \in L^2(\mathbf{R}_+) \otimes k$. The operators $\mathcal{S}_t$, θ_t are isometric and the operators $\mathcal{R}_t$, ρ_t are unitary. Remark that, for all $s, t \geq 0$ we have

$$\mathcal{S}_s \mathcal{S}_t = \mathcal{S}_{s+t}, \qquad \mathcal{R}_s \mathcal{R}_s = I$$

For all $s \geq 0$ and all bounded operator $X \in \mathcal{B}(\mathcal{H})$ the operator $\mathcal{S}_s X \mathcal{S}_s^*$ maps $h \otimes \Gamma(L^2(s, \infty) \otimes k)$ into itself. The canonical extension to $\mathcal{H}$ will be denoted by $\overline{\mathcal{S}_s X \mathcal{S}_s^*}$.

Definition 2.3. *An $(h, \mathcal{M}_S)$-regular adapted contractive process V in $\mathcal{H}$ is called a cocycle if, for all $s, t \geq 0$, we have*

$$V(t + s) = V(s)\overline{\mathcal{S}_s V(t) \mathcal{S}_s^*}.$$

A cocycle V is $(D_S, \mathcal{M}_S^0)$-weakly differentiable if for all $u \in D$ the function

$$t \mapsto \langle ve(g), V(t)ue(f)\rangle$$

belongs to $C^1(\mathbf{R}_+;\mathbf{C})$ *for all* $v \in h$ *and all functions* $g, f \in \mathcal{M}_S^0$.

Let us remark that $(D_S, \mathcal{M}_S^0)$-weak differentiability for a contraction cocycle V implies weak and hence strong continuity for both V and V^*. Thus we can associate with V the strongly continuous contraction semigroup P (called also the reduced semigroup) on h defined by

$$\langle v, P(t)u\rangle = \langle ve(0), V(t)ue(0)\rangle . \qquad (2.4)$$

By a well-known fact in semigroup theory the infinitesimal generator G of P can be defined as the set of $u \in h$ such that the limit

$$\lim_{t\to 0+} t^{-1}\left(P(t)u - u\right)$$

exists in the strong or equivalently weak topology of h. Hence $(D_S, \mathcal{M}_S^0)$-weak differentiability of V implies that D is contained in the domain of G.

Theorem 2.4. *Let* V *be a* $(D_S, \mathcal{M}_S^0)$*-weakly differentiable cocycle. Let* P *and* G *be respectively the strongly continuous contraction semigroup on* h *associated with* V *by (2.4) and its infinitesimal generator. Let* $D_{S,G}$ *be the domain*

$$D_{S,G} = \{\, we_{-\infty} + uf + ve_{+\infty} \mid w \in h, u \in D, v \in D(G)\,\}$$

There exists an operator L *on* $h \otimes \hat{k}$ *with domain containing* D_S *such that the cocycle* V *satisfies the q.s.d.e. in* $\mathcal{D}_S$

$$dV(t) = V(t)d\Lambda_L(t), \qquad V(0) = I. \qquad (2.5)$$

The operator L *has the following properties:*

i) L *can be extended to* $D_{S,G}$ *and, for all* $u \in D$, $v \in D(G)$, $w \in h$, *we have*

$$Lwe_{-\infty} = 0, \qquad E_{+\infty}L(ve_{+\infty} + uf) = 0 \qquad (2.6)$$

$$\langle w, Gv\rangle = \langle we_{-\infty}, Lve_{+\infty}\rangle \qquad (2.7)$$

ii) for all $x \in D_{S,G}$ *we have*

$$\langle Fx, Lx\rangle + \langle Lx, Fx\rangle + \langle FLFx, Lx\rangle \le 0. \qquad (2.8)$$

Proof. The first statement can be proved by the same technique of Theorem p. 62 [3]. The operator L with domain D_S is defined by (2.6) and, for all $w \in h$, $u \in D$, $g, f \in k$,

$$\langle we_{-\infty}, Lue_{+\infty}\rangle = \lim_{t\to 0} t^{-1}\langle we(0), (V(t) - I)ue(0)\rangle = \lim_{t\to 0} t^{-1}\langle w, (P(t) - I)u\rangle$$

$$\langle wg, Lue_{+\infty}\rangle = \lim_{t\to 0} t^{-1}\left\langle w(e(g\chi_{(0,t)}) - e(0)), (V(t) - I)ue(0)\right\rangle$$

$$\langle we_{-\infty}, Luf\rangle = \lim_{t\to 0} t^{-1}\left\langle we(0), (V(t) - I)u(e(f\chi_{(0,t)}) - e(0))\right\rangle$$

$$\langle wg, Luf\rangle = \lim_{t\to 0} t^{-1}\left\langle w(e(g\chi_{(0,t)}) - e(0)), (V(t) - I)u(e(f\chi_{(0,t)}) - e(0))\right\rangle$$

Thus to prove i) it is sufficient to show that the right-hand side limit in the second formula exists for all $u \in D(G)$. This can be done by the same argument of the proof of Lemma 2.1 in [13]. To prove ii) consider $v \in D(G)$, $u \in D$, $f \in \mathcal{M}_S^0$ and compute the norm

$$t \mapsto \|V(t)(ve(0) + ue(f))\|^2$$

using (2.3) which makes sense also for $v \in D(G)$ because of part i). Since V is a cocycle, the above function is decreasing. Therefore its derivative at $t = 0$ is negative and (2.8) follows. □

Moreover, from [16] Propositions 4.3, 4.4, we have the following

Proposition 2.5. *The following facts hold:*

i) if D is a core for G, then V is the unique solution of (2.5),

ii) if V is isometric then, for all $x, y \in D_{S,G}$ we have

$$\langle Fy, Lx\rangle + \langle Ly, Fx\rangle + \langle FLFy, Lx\rangle = 0. \tag{2.9}$$

Definition 2.6. *Let P be a strongly continuous contraction semigroup on h and let G be the infinitesimal generator. Suppose that D is contained in the domain of G and is a core for G. An operator L on $h \otimes k$ with domain containing D_S satisfies the condition (A) if it has the properties i), ii) of Theorem 2.4.*

Clearly operators L satisfying the condition (A) are the natural candidates as infinitesimal generators of $(D_S, \mathcal{M}_S^0)$-weakly differentiable cocycles.

Proposition 2.7. *(Journé, Prop.1 p.294) Let $V = (V(t))_{t\geq 0}$ be a cocycle. Then the process $\widetilde{V} = (\widetilde{V}(t))_{t\geq 0}$ defined by*

$$\widetilde{V}(t) = \mathcal{R}_t V^*(t)\mathcal{R}_t \tag{2.10}$$

is a cocycle.

The process $\widetilde{V}$ is called the *dual cocycle* associated with V. The following result follows from Theorem 2.4 and Proposition 2.7.

Proposition 2.8. *If a $(D_S, \mathcal{M}_S^0)$-weakly differentiable cocycle V with associated operator L as in Theorem 2.4 satisfies the q.s.d.e. (2.5) in $\mathcal{D}_S$ and the dual cocycle $\widetilde{V}$ is $(D_S, \mathcal{M}_S^0)$-weakly differentiable then $\widetilde{V}$ satisfies the q.s.d.e.*

$$d\widetilde{V}(t) = \widetilde{V}(t)d\Lambda_{\widetilde{L}}(t), \qquad \widetilde{V}(0) = I \tag{2.11}$$

in $\mathcal{D}_S$ and its associated operator $\widetilde{L}$ satisfies the condition

$$\left\langle y, \widetilde{L}x\right\rangle = \langle FLFy, x\rangle.$$

for all $x, y \in D_S$. In particular, if L is bounded, then $\widetilde{L} = L^b$

Remark. It should be noted that the $(D_S, \mathcal{M}_S^0)$-weakly differentiabilty of V and $\widetilde{V}$ implies that the domains of G and G^* have D as a common dense intersection.

3. Characterization of weakly differentiable contractive cocycles

In this section we give a general existence theorem for contractive solutions of q.s.d.e. with unbounded coefficient. This result will be used to construct contractive cocycles with a given infinitesimal generator.

Proposition 3.1. *Let $L \in \mathbb{B}$ and let V be the unique regular $(h, \mathcal{M}_S)$-adapted process satisfying the q.s.d.e.*

$$\begin{cases} dV(t) = V(t)d\Lambda_L(t) \\ V(0) = I \end{cases} \tag{3.1}$$

The following conditions are equivalent:

i) the process V is contractive,

ii) the following inequality holds

$$LF + FL^* + LFL^* \leq 0 \tag{3.2}$$

iii) the following inequality holds

$$FL + L^*F + L^*FL \leq 0 \tag{3.3}$$

Proof. The existence of a unique regular $(h, \mathcal{M}_S)$-adapted process satisfying (3.1) follows from Proposition 2.2. For all finite sets $(u_j)_{j=1}^J \subset h$, $(f_j)_{j=1}^J \subset \mathcal{M}_S$ denote by ξ the vector in $\mathcal{H}$ given by $\sum_{j=1}^J u_j e(f_j)$. The adjoint process U satisfies the q.s.d.e. $dU(t) = d\Lambda_{L^b}(t)U(t)$ therefore by (2.3) we have

$$\|U(t)\xi\|^2 = \|\xi\|^2 + \int_0^t \langle U(r)\sum_j u_j(e_{-\infty} + f_j(r))e(f_j), \\ (LF + FL^* + LFL^*)U(r)\sum_j u_j(e_{-\infty} + f_j(r))e(f_j)\rangle\, dr$$

Therefore condition ii) implies that, for all $t \geq 0$, the operator $U(t)$ and $V(t)$ are contractions. Conversely, if condition i) holds by the cocycle property of V, we can show that the function

$$t \mapsto \|U(t)\xi\|^2$$

is decreasing. Therefore, for all $(f_j)_{j=1}^J$ such that t is a continuity point for every function f_j, we have

$$\left\langle \sum_j u_j(e_{-\infty} + f_j(t))e(f_j), (FL + L^*F + L^*FL)\sum_j u_j(e_{-\infty} + f_j(t))e(f_j)\right\rangle$$
$$= \lim_{h\to 0}\frac{1}{h}\left(\|U(t+h)\xi\|^2 - \|U(t)\xi\|^2\right) \leq 0$$

This shows that i) implies ii).

The equivalence of i) and iii) can be shown in the same way considering the dual cocycle $\widetilde{V}$. In fact the contractivity of $\widetilde{V}$ is equivalent to that of V. □

In order to deal with q.s.d.e. with unbounded operators L on $h\otimes\hat{k}$ let us first prove the following technical lemma inspired by the regularity conditions of [6] Th. 4.4 for coefficients of classical s.d.e. and by [11] Th. 4. The parameter η will be useful to study isometry of cocycles in Section 5.

Lemma 3.2. *Let $L\in\mathbb{B}$ satisfying the inequalities (3.2),(3.3). For all $\eta\in[0,1]$ consider the operators on $h\otimes\hat{k}$*

$$I^{(\eta)} = \eta E_{-\infty} + E + \eta E_{+\infty}, \qquad L^{(\eta)} = I^{(\eta)}LI^{(\eta)} + (1-\eta^2)E_{-\infty}LE_{+\infty}$$

The operator $L^{(\eta)}$ is an element of $\mathbb{B}$ and satisfies the inequalities (3.2), (3.3).

Proof. Clearly $L^{(\eta)}$ is an element of $\mathbb{B}$. The operators $I^{(\eta)}$ and F commute. Moreover we have the following identities

$$E_{-\infty}LE_{+\infty}FI^{(\eta)}L^*I^{(\eta)} = 0, \qquad I^{(\eta)}LI^{(\eta)}FE_{+\infty}L^*E_{-\infty} = 0,$$
$$LI^{(\eta)}FI^{(\eta)}L^* = LEI^{(\eta)}FI^{(\eta)}EL^* = LEL^* = LFL^* = 0.$$

Hence we obtain the inequality

$$\begin{aligned} L^{(\eta)}F + FL^{(\eta)*} + L^{(\eta)}FL^{(\eta)*} &= \\ &= I^{(\eta)}(LF + FL^* + LFL^*)I^{(\eta)} + (1-\eta^2)E_{-\infty}(LF + FL^*)E_{-\infty} \\ &\leq -(1-\eta^2)E_{-\infty}(LFL^*)E_{-\infty} = -(1-\eta^2)E_{-\infty}LEL^*E_{-\infty} \leq 0 \end{aligned}$$

This completes the proof. □

Let L be an operator on $h\otimes\hat{k}$ satisfying the condition (A) and let G be the associated operator defined by (2.6). For all strictly positive integer n, let $R(n;G)$ denote the bounded operator $(nI-G)^{-1}$ ([20] Ch.IX, Th.1 p.240). It is to be noted that the adjoint operator G^* is the infinitesimal generator of the dual contraction semigroup and we have $R(n;G^*) = R(n;G)^*$. Moreover we have the well-known relations for all $w\in h$, $v\in D(G)$ ([20] Ch.IX, Th. p.246)

$$\lim_{n\to\infty} nR(n;G)w = w, \qquad \lim_{n\to\infty} nGR(n;G)v = Gv.$$

Proposition 3.3. *Let L be an operator on $h \otimes \hat{k}$ satisfying the condition (A). For each strictly positive integer n let I_n, L_n be the operators on $h \otimes \hat{k}$ with domain $\mathcal{D}_S$ defined by*

$$I_n = nR(n;G)E_{-\infty} + E + nR(n;G)E_{+\infty}, \qquad L_n = I_n^* L I_n. \tag{3.4}$$

Then the operator L_n has a bounded extension which is an element of $\mathbb{B}$ satisfying the inequalities (3.2) (3.3) and its uniform norm is estimated by

$$\|L_n\| \leq 2\left(n + 4\sqrt{n} + 1\right). \tag{3.5}$$

Moreover, for all $\xi \in D_S$, we have

$$\lim_{n\to\infty} L_n\xi = L\xi \tag{3.6}$$

Proof. The operator L_n satisfies the inequality (2.8). In fact it is sufficient to write (2.8) for vectors of the form $I_n x$ with $x \in D_S$ (since $I_n x \in D_{S,G}$ by well-known propeties of the operators $R(n;G)$) and use the commutation of the operator F with the operators I_n and I_n^*. To prove that L_n is bounded, because of (2.6), we need only to estimate the norm of $L_n\xi$ when ξ is a vector in $h \otimes \hat{k}$ of the form $\sum_{j=1}^J u_j(f_j + e_{+\infty})$ with $u_j \in D$, $f_j \in \mathcal{M}_S$ for all j. In this case we have

$$\|L_n\xi\| \leq \|EL_nE\xi\| + \|E_{-\infty}L_nE\xi\| + \|EL_nE_{+\infty}\xi\| + \|E_{-\infty}L_nE_{+\infty}\xi\| \tag{3.7}$$

The norm of $E_{-\infty}L_nE_{+\infty}$ can be estimated using (2.7). We have in fact, for each $v, u \in D$,

$$\langle ve_{-\infty}, E_{-\infty}L_nE_{+\infty}ue_{+\infty}\rangle = n^2\langle R(n;G)v, GR(n;G)u\rangle$$

Using the identity $GR(n;G) = nR(n;G) - I$ and the contractivity of $nR(n;G)$ ([20] Ch.IX Th.1, Cor.2 p.240–241) we can estimate the norm of $GR(n;G)$ by 2. Therefore we obtain the inequality

$$|\langle ve_{-\infty}, E_{-\infty}L_nE_{+\infty}ue_{+\infty}\rangle| \leq 2n\,\|v\|\cdot\|u\|$$

which implies that

$$\|E_{-\infty}L_nE_{+\infty}\| \leq 2n. \tag{3.8}$$

Taking a vector $x \in D_S$ in the range of E from the inequality (2.8) for L_n we obtain immediately the estimates

$$\|E + EL_nE\| \leq 1, \qquad \|EL_nE\| \leq 2. \tag{3.9}$$

Consider now a vector $x = ue_{+\infty}$ with $u \in D$. The inequality (2.8) for L_n yields

$$2n^2 \Re e \langle R(n;G)u, GR(n;G)u\rangle + \|EL_nE_{+\infty}ue_{+\infty}\|^2 \leq 0$$

Hence, using again the equality $GR(n;G) = nR(n;G) - I$, and the contractivity of $nR(n;G)$ we get the inequality

$$\|EL_nE_{+\infty}ue_{+\infty}\|^2 \leq 4n\,\|ue_{+\infty}\|^2$$

which implies

$$\|EL_nE_{+\infty}\| \leq 2\sqrt{n}. \tag{3.10}$$

To estimate the norm of $E_{-\infty}L_nE$ consider the inequality (2.8) for L_n with vectors x of the form $x_0 + \lambda ve_{+\infty}$ with x_0 finite linear combination of vectors in D_S in the range of E, $v \in D$ and $\lambda \in \mathbf{R}$. Denote

$$\begin{aligned}
a &= -2\Re e \langle ve_{-\infty}, E_{-\infty}L_nE_{+\infty}ve_{+\infty}\rangle - \|EL_nE_{+\infty}ve_{+\infty}\|^2 \\
c &= -2\Re e \langle x_0, EL_nEx_0\rangle - \|EL_nEx_0\|^2 = -\|(E+EL_nE)x_0\|^2 + \|x_0\|^2 \\
b &= \Re e \{\langle ve_{-\infty}, E_{-\infty}L_nEx_0\rangle + \langle (E+EL_nE)x_0, EL_nE_{+\infty}ve_{+\infty}\rangle\}
\end{aligned}$$

From the inequality (2.8) for L_n it follows that a and c are nonnegative. Moreover the estimates (3.8), (3.9) yield

$$a \leq 4n\|v\|^2, \qquad c \leq \|x_0\|^2 \tag{3.11}$$

Formula (2.8) for L_n with $x = x_0 + \lambda ve_{+\infty}$ gives us the inequality

$$a\lambda^2 - 2b\lambda + c \geq 0$$

for all $\lambda \in \mathbf{R}$. This implies that $|b| \leq 2\sqrt{ac}$ hence we get

$$|\Re e \langle ve_{-\infty}, E_{-\infty}L_nEx_0\rangle| \leq 2\sqrt{ac} + |\langle (E+EL_nE)x_0, EL_nE_{+\infty}ve_{+\infty}\rangle|$$

Therefore multiplying v by a complex number of modulus 1 such that the scalar product in the right-hand side becomes real, from (3.9), (3.10), (3.11) we obtain the inequality

$$|\langle ve_{-\infty}, E_{-\infty}L_nEx_0\rangle| \leq 6\sqrt{n}\,\|v\| \cdot \|x_0\|$$

for all $v \in D$ which implies

$$\|E_{-\infty}L_nE\| \leq 6\sqrt{n}$$

Using (3.9), (3.10), (3.11), and the above inequality from (3.7) we obtain (3.5). This shows that L_n is bounded. Then $L_n \in \mathbb{B}$ because of (2.6). Now (3.6) can be proved by the same decomposition of $L_n\xi$ used to obtain (3.7) using the properties of the operators $(R(n;G^*))_{n>0}$, the inequality

$$\begin{aligned}\left\|(EL_nE_{+\infty} - EL_nE_{+\infty})ue_{+\infty}\right\|^2 &= \|EL(nR(n;G)u - u)e_{+\infty}\| \\ &\leq -2\Re\langle nR(n;G)u - u, G(nR(n;G)u - u)\rangle\end{aligned}$$

and the closedness of G. □

The operators $(L_n)_{n>0}$ will be used in the following proposition to find a sequence of cocycles approximating the cocycle generated by L by the equicontinuity method of [8].

Proposition 3.4. *Let L be an operator on $h\hat{\otimes}\hat{k}$ with domain containing D_S such that there exists a sequence $(L_n)_{n>0}$ of elements of $\mathbb{B}$ satisfying the inequalities (3.2), (3.3) and the condition (3.6). For all $n \in \mathbf{N}$, $\eta \in [0,1]$ let*

$$L_n^{(\eta)} = I^{(\eta)} L_n I^{(\eta)}, \qquad L^{(\eta)} = I^{(\eta)} L I^{(\eta)} \tag{3.12}$$

and let $V_n^{(\eta)}$ be the unique regular $(h, \mathcal{M}_S)$-adapted contractive process satisfying the q.s.d.e.

$$dV_n^{(\eta)}(t) = V_n^{(\eta)}(t)d\Lambda_{L_n^{(\eta)}}(t), \qquad V_n^{(\eta)}(0) = I$$

in h. The following facts hold:

i) for all $u \in D$, $f \in \mathcal{M}_S$, there exists a positive constant $c(u,f)$ such that, for all $n \in \mathbf{N}$ and all $\eta \in [0,1]$ we have

$$\left\|\left(V_n^{(\eta)}(t) - V_n^{(\eta)}(s)\right) ue(f)\right\|^2 \leq c(u,f)\int_s^t \left(1 + \|f(r)\|^2\right) dr \tag{3.13}$$

ii) for each $\eta \in [0,1]$ there exists a subsequence $(V_{n_k}^{(\eta)})_{k\geq 0}$ and an $(h, \mathcal{M}_S)$-adapted contractive process $V^{(\eta)}$ such that

$$w - \lim_{k\to+\infty} V_{n_k}^{(\eta)}(t) = V^{(\eta)}(t)$$

for all $t \geq 0$. Moreover, for each $u \in D$, $f \in \mathcal{M}_S$ we have the inequality

$$\left\|\left(V^{(\eta)}(t) - V^{(\eta)}(s)\right) ue(f)\right\|^2 \leq c(u,f)\int_s^t \left(1 + \|f(r)\|^2\right) dr \tag{3.14}$$

iii) for each $\eta \in [0,1]$ the $(h, \mathcal{M}_S)$-adapted process $V^{(\eta)}$ is regular and satisfies the q.s.d.e. in $\mathcal{D}_S$

$$dV^{(\eta)}(t) = V^{(\eta)}(t)d\Lambda_{L^{(\eta)}}(t), \qquad V^{(\eta)}(0) = I. \tag{3.15}$$

Proof. For all $\eta \in [0,1]$, $n \in \mathbf{N}$ and each $u \in D$, $f \in \mathcal{M}_S$ by the Ito formula (2.3) , the Schwartz inequality and the contractivity of $V_n^{(\eta)}$ we have

$$\left\| \left(V_n^{(\eta)}(t) - V_n^{(\eta)}(s) \right) ue(f) \right\|^2 \leq$$
$$\leq \exp(\|f\|^2) \int_s^t \left(\|u(f(r) + e_{+\infty})\|^2 + 2 \left\| L_n^{(\eta)} u(f(r) + e_{+\infty}) \right\|^2 \right) dr$$

A straightforward computation yields

$$\left\| L_n^{(\eta)} u(f(r) + e_{+\infty}) \right\|^2 = \|EL_n u(f(r) + \eta e_{+\infty})\|^2 + \|E_{-\infty} L_n u(\eta f(r) + e_{+\infty})\|^2$$
$$\leq 2 \left(\|L_n u e_{+\infty}\|^2 + \sum_{j \in S} |f_j(r)|^2 \|L_n u e_j\|^2 \right)$$

Therefore, denoting by $c'(u,f)$ the constant

$$\max_{n \in \mathbf{N}} \left\{ \max_{j \in \{ k \in S | f_k \not\equiv 0 \}} \left\{ \|L_n u e_j\|^2 \right\}, \|L_n u e_{+\infty}\|^2, \|u\|^2, 1 \right\}$$

we have

$$\left\| L_n^{(\eta)} u(f(r) + e_{+\infty}) \right\|^2 \leq 2c'(u,f) \left(1 + \|f(r)\|^2 \right)$$

and (3.13) follows taking $c(u,f) = 5 \exp(\|f\|^2) c'(u,f)$.

The inequality (3.13) implies that for each $y \in \mathcal{H}$, $u \in D$, $f \in \mathcal{M}_S$, the sequence of functions on $\mathbf{R}_+$

$$t \mapsto \left\langle y, V_n^{(\eta)}(t) ue(f) \right\rangle$$

is equicontinuous and equibounded. Hence by the Ascoli-Arzelà theorem ([20] Th. p. 85) there exists a subsequence uniformly convergent on every bounded interval of $\mathbf{R}_+$. The diagonalisation argument and the separability of $\mathcal{H}$ allow to find a subsequence $(V_{n_k}^{(\eta)})_{k \geq 0}$ weakly convergent to an $(h, \mathcal{M}_S)$-adapted contractive process $V^{(\eta)}$. The inequality (3.14) follows from (3.13) and the lower semicontinuity of the norm with respect to the weak convergence. This implies, in particular, that $V^{(\eta)}$ is strongly continuous. Moreover, for all $n \in \mathbf{N}$, by (2.2) we have

$$\left\langle ve(g), V_{n_k}^{(\eta)}(t) ue(f) \right\rangle = \langle ve(g), ue(f) \rangle$$
$$+ \int_0^t \left\langle v(e_{-\infty} + g(r)) e(g), V_{n_k}^{(\eta)}(r) L_{n_k}^{(\eta)} u(f(r) + e_{+\infty}) e(f) \right\rangle dr$$

Letting k tend to infinity and using the weak convergence of contractions $V_{n_k}^{(\eta)}(s)$ and the strong convergence of $L_{n_k}^{(\eta)} u(f(r) + e_{+\infty})$ to $L^{(\eta)} u(f(r) + e_{+\infty})$ we obtain iii). □

Proposition 3.5. *For each $\eta \in [0,1]$, the regular $(h, \mathcal{M}_S)$-adapted contractive process $V^{(\eta)}$ in Proposition 3.4 is the unique solution of the q.s.d.e. (3.15). The process $V^{(\eta)}$ is a $(D_S, \mathcal{M}_S)$-weakly differentiable cocycle. Moreover, for all $t \in \mathbf{R}_+$, we have*

$$w - \lim_{\eta \to 1^-} V^{(\eta)}(t) = V^{(1)}(t)$$

Proof. Uniqueness can be proved as in [16] Proposition 4.3. The cocycle property for $V^{(\eta)}$ follows from uniqueness. Weak differentiability follows from (2.2). We can show as in the proof of Proposition (3.4) ii) that, for all sequence $(\eta_n)_{n\geq 0}$ converging to 1, there exists a subsequence $(\eta_{n_l})_{l\geq 0}$ such that the limit

$$w - \lim_{l \to \infty} V^{(\eta_{n_l})}(t)$$

exists for all $t \geq 0$ and is a regular $(h, \mathcal{M}_S)$-adapted contractive process satisfying (3.15) with $\eta = 1$. Therefore it coincides with $V^{(1)}(t)$ by the first statement. □

Thus we have proved the following

Theorem 3.6. *Let L be an operator on $h \otimes \hat{k}$ satisfying the condition (A). Then L is the infinitesimal generator of the unique $(D_S, \mathcal{M}_S^0)$-weakly differentiable cocycle satisfying the q.s.d.e.*

$$dV(t) = V(t) d\Lambda_L(t), \qquad V(0) = I. \tag{3.16}$$

4. Some results on the classical differential equation for a completely positive semigroup

In this section we shall suppose that the operator L satisfies the condition (A). The following propositions give some results on the classical Feller-Kolmogorov equation for a completely positive semigroup we will need in the following. We are interested in the existence and properties of families $T^{(\eta)} = (T^{(\eta)}(t))_{t\geq 0}$ of strongly continuous contractions on h satisfying one of the following equations:

$$\begin{aligned}\left\langle v, T^{(\eta)}(t)u \right\rangle = \langle v, u\rangle + \int_0^t \Big\{ &\left\langle v, T^{(\eta)}(s)Gu \right\rangle \\ &+ \eta^2 \left\langle ELve_{+\infty}, T^{(\eta)}(s)ELue_{+\infty} \right\rangle + \left\langle Gv, T^{(\eta)}(s)u \right\rangle \Big\} ds\end{aligned} \tag{4.2}$$

$$\begin{aligned}\left\langle v, T^{(\eta)}(t)u \right\rangle = &\langle P(t)v, P(t)u \rangle \\ &+ \eta^2 \int_0^t \left\langle ELP(t-s)ve_{+\infty}, T^{(\eta)}(s)ELP(t-s)ue_{+\infty} \right\rangle ds\end{aligned} \tag{4.3}$$

for all $u, v \in D$, $\eta \in [0,1]$. Let us recall that for contractions on h strong and weak continuity are equivalent. The following proposition shows that equation (4.3) is a weak form of equation (4.2).

Proposition 4.1. *Let L satisfy the condition (A). Then, if $T^{(\eta)}$ satisfies (4.2), it also satisfies (4.3).*

Proof. Since D is a core for G and (2.8) holds, if $T^{(\eta)}$ satisfies (4.2) for each $v, u \in D$ then, it also satisfies (4.2) for each v, u in the domain of G. Moreover, for all such u, v, the function

$$t \mapsto \left\langle v, T^{(\eta)}(t)u \right\rangle$$

is differentiable. Therefore, for every $s \in [0, t)$, and all $u, v \in D$ we have

$$\begin{aligned}\frac{d}{ds}\Big\langle P(t-s)v, &T^{(\eta)}(s)P(t-s)u\Big\rangle = \\ &= \eta^2 \left\langle ELP(t-s)ve_{+\infty}, T^{(\eta)}(s)ELP(t-s)ue_{+\infty}\right\rangle\end{aligned}$$

Hence we obtain (4.3) integrating on $[0, t]$. □

Proposition 4.2. *Let L satisfy the condition (A). Then, for all $\eta \in [0, 1]$, there exists a strongly continuous family $T^{(\eta)}$ of positive contractions on h satisfying the equation (4.3). Moreover, if $\eta_1 \leq \eta_2 \leq 1$, then, for all $t \geq 0$, we have*

$$T^{(\eta_1)}(t) \leq T^{(\eta_2)}(t). \tag{4.4}$$

Proof. We follow the iterative method for the construction of the minimal solution of a Feller-Kolmogorov equation exposed in [4]. Let us define a sequence $T_n^{(\eta)}$ by

$$\begin{aligned} T_0^{(\eta)}(t) &= P(t)^* P(t) \\ \left\langle u, T_{n+1}^{(\eta)}(t)u \right\rangle &= \langle P(t)u, P(t)u \rangle \\ &+ \eta^2 \int_0^t \left\langle ELP(t-s)ue_{+\infty}, T_n^{(\eta)}(s)ELP(t-s)ue_{+\infty}\right\rangle ds \end{aligned} \tag{4.5}$$

where $u, v \in D$. We will prove by induction that, for all $n \in \mathbf{N}$, $T_n^{(\eta)}$ is a strongly continuous family of positive contractions on h. Moreover, for each $\eta \in [0, 1]$, $t \geq 0$, the sequence of contractions $(T_n^{(\eta)}(t))_{n\geq 0}$ is increasing. This fact clearly holds for $n = 0$. Suppose it holds also for an integer n. Because of condition (A) and the strong continuity of $T_n^{(\eta)}$ the argument of the right-hand side integral in (4.5) is a continuous function. Therefore the integral makes sense. Moreover, using the inequality $0 \leq T_n^{(\eta)} \leq I$ and (2.8) we can majorize it by

$$\begin{aligned} \eta^2 \int_0^t \|LP(t-s)ue_{+\infty}\|^2 \, ds &\leq -2\eta^2 \Re e \int_0^t \langle P(s)v, P(s)Gu \rangle \, ds \\ &= -\eta^2 \int_0^t \frac{d}{ds} \|P(s)u\|^2 \, ds \\ &= \eta^2 \left(\|u\|^2 - \|P(t)u\|^2 \right) \end{aligned}$$

Hence the left-hand side of the above inequality can be majorized by

$$\eta^2 \|u\|^2 - (1-\eta^2) \|P(t)u\|^2 \le \|u\|^2$$

This implies by polarization that the right-hand side of (4.5) defines a contraction $T_{n+1}^{(\eta)}(t)$ on h. Clearly the family $T_{n+1}^{(\eta)}$ is strongly continuous. Moreover, since $T_n^{(\eta)}(t) - T_{n-1}^{(\eta)}(t) \ge 0$ for every $t \ge 0$, we have

$$\begin{aligned}&\left\langle u, \left(T_{n+1}^{(\eta)}(t) - T_n^{(\eta)}(t)\right) u\right\rangle = \\ &\quad = \eta^2 \int_0^t \left\langle ELP(t-s)ue_{+\infty}, \left(T_n^{(\eta)}(s) - T_{n-1}^{(\eta)}(s)\right) ELP(t-s)ue_{+\infty}\right\rangle \ge 0\end{aligned}$$

Finally, for each $\eta \in [0,1]$, $t \ge 0$ and each $u \in D$, let

$$T^{(\eta)}(t) = \sup_n T_n^{(\eta)}(t).$$

Since the sequence $T_n^{(\eta)}(t)$ converges strongly to $T^{(\eta)}(t)$ it follows that $T^{(\eta)}$ satisfies (4.3). If $0 \le \eta_1 \le \eta_2 \le 1$ then we can show by induction that

$$T_n^{(\eta_1)}(t) \le T_n^{(\eta_2)}(t)$$

for all $n \in \mathbf{N}$ and all $t \in \mathbf{R}_+$. Hence we obtain (4.4). □

Proposition 4.3. *Let L satisfy the condition (A) and (B_0). The following facts hold:*

i) Let $\eta = 1$. If T is a strongly continuous family of positive contractions solving (4.3), for all $t \ge 0$ we have

$$T^{(1)}(t) \le T(t) \le I \tag{4.6}$$

ii) For each $t \ge 0$ we have

$$s - \lim_{\eta \to 1^-} T^{(\eta)}(t) = T^{(1)}(t).$$

iii) For all $\eta \in [0,1)$ $T^{(\eta)}$ is the unique solution of (4.3).

Proof. For every $\eta \in [0,1]$ let $S^{(\eta)}$ be another strongly continuous family of positive contractions solving (4.3) and let $T_n^{(\eta)}$ be as in the proof of Proposition 4.2. It easily seen by induction that

$$T_n^{(\eta)}(t) \le S^{(\eta)}(t) \le I$$

for all $n \in \mathbf{N}$ and all $t \ge 0$. In particular, for $\eta = 1$, i) follows.

Because of (4.4) the strong limit $\lim_{\eta\to 1^-} T^{(\eta)}(t)$ exists, is less than $T^{(1)}(t)$, and defines a solution of (4.3) with $\eta = 1$. Therefore ii) follows from i).

If $\eta \in [0,1)$ putting $R^{(\eta)}(t) = S^{(\eta)}(t) - T^{(\eta)}(t)$ we obtain a strongly continuous family of positive contractions on h satisfying

$$\left\langle u, R^{(\eta)}(t)u \right\rangle = \eta^2 \int_0^t \left\langle ELP(t-s)ue_{+\infty}, R^{(\eta)}(s)ELP(t-s)ue_{+\infty} \right\rangle ds \quad (4.7)$$

Then from (4.4), (4.6) we obtain

$$\left\langle u, R^{(\eta)}(t)u \right\rangle \leq \eta^2 \left(\|u\|^2 - \|P(t)u\|^2 \right) \leq \eta^2 \|u\|^2$$

which implies that the positive operator $R^{(\eta)}(t)$ is bounded from above by $\eta^2 I$. Using this fact to majorize the right-hand side of (4.7) by the same argument we can prove by induction on n the inequality

$$0 \leq R^{(\eta)}(t) \leq \eta^{2n} I$$

for all $n \in \mathbf{N}$. This proves that $R^{(\eta)}(t) = 0$. □

The following condition is necessary in order $T(t) = I$ for all $t \geq 0$ to be a solution of (4.2) and (4.3)

$\mathbf{B}_0$ - for all $v, u \in D$, we have

$$2\Re\langle v, Gu\rangle + \langle ELve_{+\infty}, ELue_{+\infty}\rangle = 0$$

Corollary 4.4. *Let L satisfy the conditions (A) and (B_0) and let $\eta = 1$. Then $T(t) = I$ for all $t \geq 0$ is the unique solution of (4.3) if and only if $T^{(1)}(t) = I$ for all $t \geq 0$. In this case the equation (4.2) has the unique positive solution $T(t) = I$ for all $t \geq 0$. Moreover the equation*

$$\left\langle v, T(t)u \right\rangle = \int_0^t \left\{ \langle v, T(s)Gu\rangle + \langle ELve_{+\infty}, T(s)ELue_{+\infty}\rangle + \langle Gv, T(s)u\rangle \right\} ds \quad (4.8)$$

has the unique positive solution $T(t) = 0$ for all $t \geq 0$.

Proof. In fact, if T satisfies (4.8), then $(I + T(t))_{t\geq 0}$ satisfies (4.2). □

The same arguments of [4] Th. 3.4 (see also [9]) allow to prove the following theorem giving a necessary and sufficient condition for uniqueness of the solution of (4.3). This condition is the quantum analogue of Feller condition used in [16].

Theorem 4.5. *(Chebotarev) Let L satisfy the conditions (A) and (B_0). For each $\lambda > 0$ let $\mathcal{B}_\lambda^+$ denote the set of positive operators $X \in \mathcal{B}(h)$ such that*

$$\langle v, XGu\rangle + \langle ELve_{+\infty}, XELue_{+\infty}\rangle + \langle Gv, Xu\rangle = \lambda \langle v, Xu\rangle$$

for all $v, u \in D$. The following conditions are equivalent:

i) the equation (4.3) has the only solution $T(t) = I$ for all $t \geq 0$,

ii) for each $\lambda > 0$ we have $\mathcal{B}_\lambda^+ = \{0\}$.

5. Characterisation of isometry of cocycles satisfying a q.s.d.e.

As a first step we will show that, under our assumptions, the vacuum expectation of a contractive cocycle satisfying a q.s.d.e. coincides with the miminal solution of the associated Feller-Kolmogorov (or Lindblad) equation. In the following $V^{(1)}$ will be denoted simply by V

Proposition 5.1. *Let L satisfy the conditions (A), (B_0) and let $T^{(1)}$ be the minimal solution of (4.3), with $\eta = 1$ constructed in Proposition 4.2. For each $u \in D$ we have then*

$$\langle V(t)ue(0), V(t)ue(0)\rangle = \left\langle u, T^{(1)}(t)u\right\rangle.$$

Proof. Since $V^{(\eta)}$ satisfies the q.s.d.e. (3.15) in $\mathcal{D}_S$, using (2.3) it easy to compute, for every $\eta \in [0,1]$, $u \in D$, $t \geq 0$, $s \in [0,t)$,

$$\frac{d}{ds}\left\langle V^{(\eta)}(s)P(t-s)ue(0), V^{(\eta)}(s)P(t-s)ue(0)\right\rangle$$

and show that the bilinear form on h given by $(v,u) \mapsto \left\langle V^{(\eta)}(t)v, V^{(\eta)}(t)u\right\rangle$ defines a family of positive contractions on h satisfying the equation (4.3). Hence, from Proposition 4.3 iii) , for each $\eta \in [0,1)$ and $t \geq 0$ we have:

$$\left\langle V^{(\eta)}(t)ue(0), V^{(\eta)}(t)ue(0)\right\rangle = \left\langle u, T^{(\eta)}(t)u\right\rangle.$$

Because of (4.4), (4.6) and the lower semicontinuity of the norm with respect to weak convergence we have the inequalities

$$\left\|V^{(\eta)}(t)ue(0)\right\|^2 \leq \|V(t)ue(0)\|^2 \leq \liminf_{\eta\to 1^-}\left\|V^{(\eta)}(t)ue(0)\right\|^2$$

for all $\eta \in [0,1]$, $u \in h$, $t \geq 0$. Therefore, using Proposition 4.3 ii), we get

$$\begin{aligned}\left\langle u, T^{(1)}(t)u\right\rangle &= \lim_{\eta\to 1^-}\left\langle u, T^{(\eta)}(t)u\right\rangle \\ &= \lim_{\eta\to 1^-}\left\|V^{(\eta)}(t)ue(0)\right\|^2 = \langle V(t)ue(0), V(t)ue(0)\rangle\end{aligned}$$

This completes the proof. □

To study the isometry of the cocycle V we need the following assumption on L strenghtening (B_0) which has been proved to be necessary in [16] Theorem 4.4.

B - for each $x, y \in D_S$ we have

$$\langle y, Lx\rangle + \langle FLFy, x\rangle + \langle FLFy, Lx\rangle = 0. \tag{5.1}$$

We will consider also the following condition

C - for all $t \geq 0$ we have $T^{(1)}(t) = I$.

Proposition 5.2. *Let L satisfy the conditions (A), (B). Let V be the unique regular $(D_S, \mathcal{M}_S^0)$-weakly differentiable cocycle satisfying the q.s.d.e. (3.16). The following conditions are equivalent:*

i) for each $v, u \in D$ and each $t \geq 0$ we have $\langle V(t)ve(0), V(t)ue(0)\rangle = \langle v, u\rangle$,

ii) the process V is isometric.

Proof. Clearly we need only to prove that the first condition implies the second one. By Proposition 5.1, condition i) implies that $T^{(1)}(t) = I$ for all $t \geq 0$. Hence, by Corollary 4.4, the equation (4.3) has only the solution $T(t) = I$ for all $t \geq 0$ and the equation (4.8) has only the trivial positive solution zero.

Using these facts, following the proof of Th. 2.4 in [9], we will prove by induction that for each $N \in \mathbf{N}$, $n, m \in \mathbf{N}$ with $n + m = N$, $v, u \in D$, $g, f \in \mathcal{M}_S$ we have

$$\left\langle V(t)vg^{\otimes m}, V(t)uf^{\otimes n}\right\rangle = \delta_{m,n}\langle v, u\rangle \langle g, f\rangle^n$$

This clearly holds for $N = 0$ because of condition i). Suppose it holds also for all integers not greater than a fixed $N > 0$ and take n, m such that $n + m = N + 1$. Then by (5.1), writing the Ito formula (2.3) of Boson Fock space quantum stochastic calculus using l-fold tensor product vectors, we obtain the equation

$$\begin{aligned}\left\langle V(t)vg^{\otimes m}, V(t)uf^{\otimes n}\right\rangle &= \delta_{m,n}\langle v, u\rangle\langle g, f\rangle^n + \int_0^t \Big\{ \left\langle V(s)vg^{\otimes m}, V(s)Guf^{\otimes n}\right\rangle \\ &+ \left\langle V(s)ELve_{+\infty}g^{\otimes m}, V(s)ELue_{+\infty}f^{\otimes n}\right\rangle + \left\langle V(s)Gvg^{\otimes m}, V(s)uf^{\otimes n}\right\rangle \Big\}\, ds\end{aligned}$$

Replacing f by $f/\langle g, f\rangle$ if $\langle g, f\rangle \neq 0$, we obtain the strongly continuous family of positive contractions on h defined by

$$\langle v, T(t)u\rangle = \left\langle V(t)vg^{\otimes m}, V(t)uf^{\otimes n}\right\rangle$$

which satisfies the equation

$$\begin{aligned}\Big\langle v, T(t)u\Big\rangle &= \delta_{m,n}\langle v, u\rangle \\ &+ \int_0^t \Big\{ \langle v, T(s)Gu\rangle + \langle ELve_{+\infty}, T(s)ELue_{+\infty}\rangle + \langle Gv, T(s)u\rangle \Big\}\, ds\end{aligned}$$

Therefore, by Corollary 4.4, we have $T(t) = \delta_{m,n}I$. This completes the proof. □

The above results are summarized in the following

Theorem 5.3. *Let L satisfy the conditions (A) and (B). There exists a unique regular $(D_S, \mathcal{M}_S^0)$-weakly differentiable cocycle V satisfying the q.s.d.e.*

$$dV(t) = V(t)d\Lambda_L(t), \qquad V(0) = I.$$

The following conditions are equivalent:
i) the process V is isometric,
ii) for each $v, u \in D$ and each $t \geq 0$ we have

$$\langle V(t)ve(0), V(t)ue(0)\rangle = \langle v, u\rangle ,$$

iii) condition (C) holds,
iv) for each $\lambda > 0$ we have $\mathcal{B}_\lambda^+ = \{0\}$.

The characterisation of unitary cocycles which are $(D_S, \mathcal{M}_S^0)$ -weakly differentiable toghether with their dual cocycle can be obtained from a dualisation argument with the help of the following proposition. We shall refer to condition $(\widetilde{\mathrm{A}})$ as to condition (A) with L, P, G replaced by $\widetilde{L}$, P^*, G^* respectively.

Proposition 5.4. *Let L satisfy the conditions (A) and let $\widetilde{L}$ satisfy condition $(\widetilde{\mathrm{A}})$. Let V and Z be the unique $(D_S, \mathcal{M}_S^0)$-weakly differentiable cocycles satisfying the q.s.d.e.*

$$dV(t) = V(t)d\Lambda_L(t), \qquad V(0) = I, \qquad \text{in } \mathcal{D}_S$$
$$dZ(t) = Z(t)d\Lambda_{\widetilde{L}}(t), \qquad Z(0) = I \qquad \text{in } \mathcal{D}_S$$

Suppose that:
i) the domain of the operator L^* contains D_S,
ii) we have $\widetilde{L} \subseteq (FL^*F)$,
iii) for all $u \in D$ and all $n \geq 1$, $R(n;G)u$ belongs to the domain of G^* and the sequence $(G^*R(n;G)u)_{n\geq 1}$ converges.
Then Z coincides with the dual cocycle $\widetilde{V}$ of V

Proof. Remark that, because of conditions (A), $(\widetilde{\mathrm{A}})$, and assumptions i), ii) the operator $\widetilde{G}$ coincides with the adjoint G^* of G. Consider the operators $(L_n)_{n>0}$ defined in Proposition 3.3 and the contractive cocycles $\widetilde{V}_n$ which are the dual cocycles of V_n. These satisfy the q.s.d.e.

$$d\widetilde{V}_n(t) = \widetilde{V}_n(t)d\Lambda_{L_n^b}(t), \qquad \widetilde{V}_n(0) = I.$$

We want to show that, under the conditions i), ii) the sequence $\left(L_n^b\right)_{n>0}$ converges strongly to $\widetilde{L}$ on D_S. In fact, for all $\xi \in D_S$ we have

$$\begin{aligned} L_n^b\xi &= EL_n^bE\xi + E_{-\infty}L_n^bE\xi + EL_n^bE_{+\infty}\xi + E_{-\infty}L_n^bE_{+\infty}\xi \\ &= (ELE)^*\xi + nR(n;G^*)FE_{+\infty}L^*E\xi + (E_{-\infty}L_nE)^*\,F\xi + E_{-\infty}L_n^bE_{+\infty} \end{aligned}$$

Clearly the first two terms converge strongly when n tends to infinity. The vector $E_{+\infty}\xi$ can be written in the form $ue_{+\infty}$ with $u \in D$, for all $n, m \in \mathbf{N}$, hence we have

$$\left\|\left(E_{-\infty}L_n^b E_{+\infty} - E_{-\infty}L_m^b E_{+\infty}\right)\xi\right\| = \|(G_n)^* u - (G_m)^* u\|.$$

Since $(G_n)^* u = n^2 R(n; G^*) G^* R(n; G) u$, the limit of the fourth term exists by condition *iii)*. The third term can be written in the form

$$EL_n^b E_{+\infty} = (E_{-\infty}L_n E)^* F = -(E + ELE)^* EL_n E_{+\infty} F$$

where the second identity follows from the equality (5.1) for the operator L_n. This shows that the sequence $\left(L_n^b\right)_{n>0}$ is strongly convergent. To show that the limit coincides with $\widetilde{L}$ on D_S consider $y \in D_S$, $x \in D(\widetilde{L})$, we have then, because of condition ii)

$$\lim_{n\to\infty} \langle y, L_n^b x\rangle = \lim_{n\to\infty} \langle L_n F y, F x\rangle = \langle LFy, Fx\rangle = \langle y, FL^* F x\rangle = \left\langle y, \widetilde{L} x\right\rangle$$

By Propositions 3.4, 3.5 the sequence $\widetilde{V}_n$ converges weakly to Z this implies that Z satisfies the condition

$$Z(t) = \mathcal{R}_t V^*(t) \mathcal{R}_t$$

for all $t \geq 0$ and completes the proof. □

References

[1] Accardi, L.: On the quantum Feynman-Kac formula. *Rend. Sem. Matematico e Fisico di Milano*. Vol. XLVIII (1978).

[2] Accardi, L., Frigerio A., Lewis J.T.: Quantum stochastic processes. *Publ. R.I.M.S. Kyoto Univ.*, **18**, 97–133 (1982).

[3] Accardi, L., Journé, J.L., Lindsay, J.M.: On multidimensional Markovian cocycles. In: Accardi, L., von Waldenfels, W., (eds.) Quantum Probability and Applications IV. Proceedings Rome 1987 (Lect. Notes Math. Vol. 1396, pp. 59-67) Berlin Heidelberg New York, Springer, 1989.

[4] Chebotarev, A.M.: The theory of the conservative dynamical semigroups and its applications. Preprint MIEM n.1. March 1990.

[5] Chebotarev, A.M., Fagnola, F., Frigerio A.: Towards a stochastic Stone's theorem. Trento, Italia, Preprint (1990).

[6] Da Prato, G.: Some results on linear stochastic evolution equations in Hilbert space by the semigroup method. *Stoch. Anal. Appl.*, **1** 57–88 (1983).

[7] Fagnola, F.: On quantum stochastic differential equations with unbounded coefficients. *Probab. Th. Rel. Fields*, **86**, 501–516 (1990).

[8] Fagnola, F.: Pure birth and pure death processes as quantum flows in Fock space. *Sankhyā*, **53**, Series **A** (1991).

[9] Fagnola, F.: Unitarity of solutions to quantum stochastic differential equations and conservativity of the associated quantum dynamical semigroup. Quantum Probability and Related Topics VII. (1991).

[10] Fagnola, F., Mancino, M.: Free noise dilation of semigroups of countable state Markov processes. Quantum Probability and Related Topics VII. (1991).

[11] Frigerio, A.: Positive contraction semigroups on $\mathcal{B}(\mathcal{H})$ and quantum stochastic differential equations. In: Clément, P., Invernizzi, S., Mitidieri, E., Vrabie, I. (eds.) Semigroup Theory and Applications. Proceedings, Trieste 1987. pp. 175–188. New York and Basel: Marcel Dekker, Inc. 1989.

[12] Hudson, R.L., Lindsay, J.M.: On characterizing quantum stochastic evolutions. *Math. Proc. Camb. Phil. Soc.*, **102**, 363–369 (1987).

[13] Journé, J.-L.: Structure des cocycles markoviens sur l'espace de Fock. *Probab. Th. Rel. Fields* **75**, 291–316 (1987).

[14] Mancino M.: Quantum stochastic differential equations driven by free noises and dilation of Markovian semigroups with unbounded generators. To appear in: *Rend. Sem. Mat. Univ. Padova.*

[15] Mohari, A.: Quantum stochastic differential equations with unbounded coefficients and dilation of Feller's minimal solution. Preprint, I.S.I., Delhi Centre. To appear in *Sankya.*

[16] Mohari, A., Parthasarathy, K.R.: A quantum probabilistic analogue of Feller's condition for the existence of unitary Markovian cocyclyes in Fock space. Preprint, I.S.I., Delhi Centre (1991).

[17] Mohari, A., Sinha, K.B.: Quantum stochastic flows with infinite degrees of freedom and countable state Markov processes. *Sankhyā*, **52**, Series **A**, 43–57 (1990).

[18] Parthasarathy, K.R.: *An Introduction to Quantum Stochastic Calculus.* New Delhi 1990, Birkhauser-Verlag, Basel (1992).

[19] Parthasarathy, K.R.: Realisation of a class of Markov processes through unitary evolutions in Fock space. Preprint, I.S.I., Delhi Centre (1990).

[20] Yosida, K.: *Functional Analysis.* Fifth Edition. Springer-Verlag. Berlin, Heidelberg, New York 1978.

Quantum Probability and Related Topics Vol. VIII (pp. 165–188)

ON QUANTUM RESONANCE

CLAUDIO FERNÁNDEZ AND ROLANDO REBOLLEDO

ABSTRACT. In a preceding research [4], [5], we characterized resonance for pure states by means of Shannon entropy of their spectral measures. In the present article we extent this investigation to general states ρ as considered in Quantum Probability [9].

Together with resonance phenomena we investigate the complementary behaviour, that is, the characterization of *non* resonant states. Namely, we establish that resonant states are not smooth in the sense of Kato [6] and derive main properties of smooth states.

1. INTRODUCTION

Resonance phenomena appear in many fields in Physics. They are usually considered as unstable limit behaviour of a family of perturbations of a given system. Many attemps to give a precise formulation of Quantum Resonance have been done. One of the most common approach is via poles of a certain analytic continuation of the Hamiltonian resolvent (see [7]). Since there is no unique way to construct such an analytic continuation the latter approach has an intrinsic weakness. In this paper we use a phenomenological viewpoint (see the literature in Scattering Theory, for example [8]) and formulate the phenomenon in term of physical quantities that can be defined in a precise mathematical way. Two main physical facts have to be considered in this direction: when the system attains a resonant state it seems to stay there for a very long time (say, the sojourn time tends to infinity); furthermore,

Key words and phrases. Entropy functionals, quantum probability, resonance, smooth operators.
This research has been supported by DIUC Research Program "Analytic contributions to Mathematical Physics" and FONDECYT grants nr. 807–91 and 815–91.

a high concentration of the energy is observed. In our model this two facts are a consequence of the definition we provide for a resonant perturbation.

In Quantum Mechanics a system is described by states and their evolution represented by a group of operators [1]. Our problem is then connected to the Theory of Perturbations of operators and also with Probability Theory. So that a natural framework to analyze this kind of phenomenon is Quantum Probability. That is the formalism used throughout this paper.

For the sake of completeness we state below main notions and properties from Quantum Probability that will be used. For this we follow [9].

1.1. Hilbert space notations. $\mathcal{H}$ will denote a complex Hilbert space. The customary symbol $\langle .,. \rangle$ will be used for the scalar product on $\mathcal{H}$. In what follows we use T^* to mean the adjoint of an operator T.

We will use some customary notations for some remarkable families of operators: $\mathcal{B}(\mathcal{H})$ is the algebra of all linear bounded operators on $\mathcal{H}$; $\mathcal{P}(\mathcal{H})$ is the space of orthogonal projectors; $\mathcal{U}(\mathcal{H})$, the group of all unitary operators on $\mathcal{H}$; $\mathcal{I}_1(\mathcal{H})$ and $\mathcal{I}_\infty(\mathcal{H})$ denote the trace class (or nuclear) and compact operators on $\mathcal{H}$ respectively.

1.2. Observables. Given a measurable space $(\Omega, \mathcal{F})$, an Ω-*observable* is a mapping $\xi : \mathcal{F} \to \mathcal{P}(\mathcal{H})$ that satisfies $\xi(\Omega) = I$, where I denotes the identity in $\mathcal{P}(\mathcal{H})$ and for any countable collection of pairwise disjoint elements (F_i) in $\mathcal{F}$, it holds

$$\xi(\bigcup_i F_i) = \sum_i \xi(F_i) \tag{1.1}$$

In other words, we make no difference between *observables* and *spectral (vector) measures* associated to selfadjoint operators. Any selfadjoint operator X on $\mathcal{H}$ is identified with the observable $\xi : \mathcal{B}(\mathbb{R}) \to \mathcal{P}(\mathcal{H})$ defined by its spectral measure ξ which satisfies the representation formula

$$X = \int x\xi(dx) \tag{1.2}$$

This is simply Von Neumann's spectral theorem.

The space of all selfadjoint operators (or observables) is denoted $\mathcal{O}(\mathcal{H})$.

1.3. States. The trace of any nuclear operator T is given by

$$trT = \sum_{n=0}^{\infty} \langle e_n, Te_n \rangle, \tag{1.3}$$

where, (e_n) is any orthonormal basis of the Hilbert space $\mathcal{H}$.

A positive operator $\rho \in \mathcal{I}_1(\mathcal{H})$ of unit trace is called a *state*.Positivity means here that $\langle u, \rho u \rangle \geq 0$ for all $u \in \mathcal{H}$.

We let denote $\mathcal{S}(\mathcal{H})$ the convex set of all states on $\mathcal{H}$. Any extreme point of $\mathcal{S}(\mathcal{H})$ is called a *pure state*. A state ρ is pure if and only if $\rho = |\varphi\rangle\langle\varphi|$ for some unit vector φ in $\mathcal{H}$. The pure state $|\varphi\rangle\langle\varphi|$ will be written as φ to simplify notations.

For any observable ξ its *distribution* in the state ρ is the probability measure μ_ρ^ξ over $(\Omega, \mathcal{F})$ defined by the expression

$$\mu_\rho^\xi(E) = tr\rho\xi(E) \tag{1.4}$$

for each $E \in \mathcal{F}$, where "tr" means the trace. For a pure state $\rho = |\varphi\rangle\langle\varphi|$, expression (1.4) reduces to

$$\mu_\rho^\xi(E) = \langle \varphi, \xi(E)\varphi \rangle \tag{1.5}$$

The *expectation* of a selfadjoint operator $X \in \mathcal{I}_1(\mathcal{H})$ or, similarly, *the expectation of the trace class observable* ξ, is given by $tr\rho X$

Given any σ-finite measure λ on $(\Omega, \mathcal{F})$, a state ρ and an observable ξ, the measure μ_ρ^ξ can be decomposed in three parts by Lebesgue's Theorem, namely, a measure absolutely continuous with respect to λ, another singular continuous and a third one atomic with respect to λ. Fixing an observable ξ, this provides a decomposition of $\mathcal{S}(\mathcal{H})$ the convex set of all states on $\mathcal{H}$: we have three convex subsets $\mathcal{S}_{ac}(\xi)$ (respectively $\mathcal{S}_{sc}(\xi)$, respectively $\mathcal{S}_{pp}(\xi)$) that correspond to all states ρ such that μ_ξ^ρ is absolutely continuous with respect to λ (resp. singular continuous, resp. atomic). Extremal points of each convex subset above can be identified with an element in $\mathcal{H}$ (pure states) thus leading to a decomposition of the latter in three closed subspaces $\mathcal{H}_{ac}, \mathcal{H}_{sc}, \mathcal{H}_{pp}$.

1.4. The Dynamics. In what follows the dynamics is described by a group $(U(t); t \in \mathbb{R})$ of unitary transformations or, equivalently, by the Hamiltonian H defined over a domain $D(H)$ in $\mathcal{H}$ that generates the group. So that $U(t) = \exp(-itH)$. The spectral measure of H is our observable ξ. The space Ω is taken to be the real line endowed with Lebesgue measure.

We refer to $(H, \rho, t) \mapsto \theta_t(H, \rho) = \exp(-itH)\rho\exp(itH)$ from $\mathcal{O}(\mathcal{H}) \times \mathcal{S}(\mathcal{H}) \times \mathbb{R}$ into $\mathcal{S}(\mathcal{H})$ as the *Quantum Flow*. We write $\theta = (\theta_t;\ t \in \mathbb{R})$.

The decomposition of the set of states introduced in 1.3 is then referred to H. So that we say the Hamiltonian is *absolutely continuous* if $\mathcal{S}(\mathcal{H}) = \mathcal{S}_{ac}(\xi)$, where ξ is the spectral measure of H.

2. Weak topology on the set of states

We now turn to weak topologies. This material has been presented in [11].

It is well known that $\mathcal{B}(\mathcal{H})$ (respectively $\mathcal{I}_1(\mathcal{H})$) is a Banach space with the customary norm $\|\cdot\|$ for bounded operators (resp. $\|T\|_1 = tr|T|$, where $|T| = (T^*T)^{\frac{1}{2}}$). The set $\mathcal{I}_\infty(\mathcal{H})$ is a closed subspace of $\mathcal{B}(\mathcal{H})$ for the strong topology.

2.1. Definition. The *weak topology* over the set $\mathcal{I}_1(\mathcal{H})$ is the finest among all topologies for which all mappings $T \mapsto trTX$ are continuous, for all $X \in \mathcal{I}_\infty(\mathcal{H})$. Therefore, a sequence (T_n) converges weakly to T in $\mathcal{I}_1(\mathcal{H})$ if and only if,

$$\lim_n trT_nX = trTX, \tag{2.1}$$

for all $X \in \mathcal{I}_\infty(\mathcal{H})$. We write $T_n \xrightarrow{w} T$.

We have the following main result:

Theorem 1. *The set C of all positive nuclear operators ρ such that $tr\rho \leq 1$ is a weakly compact convex set.*

Proof. The convexity property of C is inmediate; let us prove it is weakly compact.

Since the Hilbert space is separable, the same property holds for $\mathcal{I}_\infty(\mathcal{H})$ in the strong topology of bounded operators. Thus, assume (X_k) to be a dense sequence in $\mathcal{I}_\infty(\mathcal{H})$.

Consider (ρ_n) to be any sequence of states. We will prove that it admits a weakly convergent subsequence whose limit belongs to $\mathcal{S}(\mathcal{H})$.

Remark that for all $n, k \in \mathbb{N}$,

$$|tr\rho_n X_k| \leq \|X_k\|. \tag{2.2}$$

Therefore, there exists a subsequence (n_m^0) of indexes such that $tr\rho_{n_m^0} X_0$ converges to a limit $\lambda(X_0)$ as $n_m^0 \uparrow \infty$. Let N_0 denote the set of all indexes n_m^0. Since,

$$|tr\rho_{n_m^0} X_1| \leq \|X_1\|, \tag{2.3}$$

we obtain the existence of a sequence n_m^1 of indexes in N_0 such that $tr\rho_{n_m^1} X_1$ converges to $\lambda(X_1)$ as $n_m^1 \uparrow \infty$ over N_0. We denote N_1 the subset of all $n_m^1 \in N_0$. By induction we construct a sequence $(\rho_{n_m^k})$ such that:

$$tr\ \rho_{n_m^k} X_l \underset{m}{\rightarrow} \lambda(X_l), \tag{2.4}$$

for all $l \leq k$, all $k \in \mathbb{N}$; and,

$$|\lambda(X_l) - \lambda(X_k)| \leq \|X_l - X_k\|. \tag{2.5}$$

From (2.4) and (2.5) it follows both the linearity of λ and its continuity on the set (X_k). Since this sequence is dense in $\mathcal{I}_\infty(\mathcal{H})$, we can extend the functional λ to the whole space $\mathcal{I}_\infty(\mathcal{H})$ in a continuous linear functional. By Schatten's Theorem [9] (4.14), it follows that there exists a unique nuclear operator ρ such that

$$tr\rho X = \lambda(X), \tag{2.6}$$

for all $X \in \mathcal{I}_\infty(\mathcal{H})$. From (2.5) it also follows that $|tr\rho X| \leq \|X\|$.

To finish the proof, let us show first that ρ before is the weak limit of a convergent subsequence of (ρ_n). From this we will in particular derive that ρ is positive and $tr\rho \leq 1$.

Given any $X \in \mathcal{I}_\infty(\mathcal{H})$, we have:

$$\begin{aligned} |tr\rho X - tr\rho_{n^k_m} X| &\leq |tr\rho(X - X_l)| \\ &+ |tr\rho X_l - tr\rho_{n^k_m} X_l| \\ &+ |tr\rho_{n^k_m}(X - X_l)|. \end{aligned}$$

Now if $\epsilon > 0$, we choose first l such that $\|X - X_l\| < \frac{\epsilon}{3}$ and then take $k \geq l$ and n^k_k such that $|tr\rho X - tr\rho_{n^k_k}| < \frac{\epsilon}{3}$. This proves that $\rho_{n^k_k} \xrightarrow{w} \rho$.

Finally, given any $u \in \mathcal{H}$, we have $\langle u, \rho u\rangle = tr\rho|u\rangle\langle u|$ and since $|u\rangle\langle u| \in \mathcal{I}_\infty(\mathcal{H})$, we obtain that

$$\langle u, \rho u\rangle = \lim_k \langle u, \rho_{n(k)} u\rangle, \tag{2.7}$$

therefore, $\langle u, \rho u\rangle \geq 0$.

Furthermore, notice that $tr\rho = \sum_{i=0}^{\infty} tr\rho|e_i\rangle\langle e_i|$, where (e_i) is any orthonormal basis. Then, (2.7) and Fatou's Lemma imply

$$\begin{aligned} tr\rho &= \sum_i \liminf_k tr\rho_{n(k)}|e_i\rangle\langle e_i| \\ &\leq \liminf_k \sum_i tr\rho_{n(k)}|e_i\rangle\langle e_i| \\ &\leq 1. \end{aligned}$$

Therefore $\rho \in \mathcal{C}$ and the proof is complete. □

In connection with the analysis of weak topology for states, we will introduce a concept that is important in its own because it deals with Information Theory.

2.2. Definition. Given a state ρ, and a finite dimensional subspace $\mathcal{H}'$ of $\mathcal{H}$ we let $tr_{\mathcal{H}'}\rho$ denote the trace $tr\rho P$ where P is the projection over $\mathcal{H}'$.

Now we call the *Geometrical Information Coefficient of order n* ($n \geq 1$), the expression

$$\Gamma(\rho, n) = \sup\{tr_{\mathcal{H}'}\rho : dim\mathcal{H}' = n\}. \tag{2.8}$$

$\Gamma(\rho, n)$ is bounded by 1, and gives an idea about how "spread" the state ρ is. In [13], Thirring uses this concept to introduce an ordering in the set of states, the ordering of "chaoticity". We prefer to call it "complexity": we say that a state ρ is *more complex* than another state ρ' if

$$\Gamma(\rho, n) \leq \Gamma(\rho', n), \tag{2.9}$$

for all $n \geq 1$. We use the symbol $\rho' \prec \rho$ to express this partial order.

2.3. Remark. For any $n \geq 1$, the mapping $\rho \mapsto \Gamma(\rho, n)$ is lower semicontinuous in the weak topology, since it is the supremum of weak continuous functions. Notice $\Gamma(\cdot, n)$ can be extended to $\mathcal{C}$ for all integer n, in particular this procedure leads to a maximal element for the order $\prec$: $0 \in \mathcal{C}$ will be more complex than any other element in $\mathcal{C}$.

Moreover, $\Gamma(\cdot, n)$ is convex for all $n \geq 1$. Given any convex combination $\rho = a\rho_1 + (1-a)\rho_2, (0 \leq a \leq 1)$ of elements in $\mathcal{C}$, it follows that

$$\Gamma(\rho, n) \leq a\Gamma(\rho_1, n) + (1-a)\Gamma(\rho_2, n). \tag{2.10}$$

Given any state ρ call

$$\mathcal{C}(\rho) = \{\rho' \in \mathcal{C} : \rho \prec \rho'\}. \tag{2.11}$$

This is a convex set and it is weakly closed because of the lower semicontinuity of Γ. Since $C(\rho)$ is included in $\mathcal{C}$, we conclude it is a compact convex set from Theorem 1 even if limit points could have a trace less than 1. Notice that ρ is an extremal point of $C(\rho)$ and any other extremal point will be of the form $V\rho V^*$, where V is an isometry . Indeed, given any other extremal point ρ', we have that $\Gamma(\rho', n) = \Gamma(\rho, n)$ for all n. By ordering eigenvalues λ_i^ρ of ρ by their magnitudes, we obtain $\Gamma(\rho, n) = \lambda_1^\rho + \ldots + \lambda_n^\rho$, and similarly for ρ'. Then from the equivalence of geometrical information it follows $\lambda_i^{\rho'} = \lambda_i^\rho$ for all i. This means there exists an isometry V such that $\rho' = V\rho V^*$.

These are main properties of $C(\rho)$ we will use in the sequel. However we finish this section by a criterion for weak compactness of a *set of states* that is interesting in its own.

2.4. Definition. A family $\mathcal{K}$ of states is *tight* if for any ϵ there exists a finite-dimensional subspace $\mathcal{H}_\epsilon$ of $\mathcal{H}$ such that

$$\sup_{\rho \in \mathcal{K}} tr \rho P_\epsilon \geq 1 - \epsilon, \tag{2.12}$$

where P_ϵ is the projection over $\mathcal{H}_\epsilon$.

Notice that $\mathcal{K} = \{\rho\}$ satisfies (2.12). Take any $\epsilon > 0$. Given an orthonormal basis (e_n), choose $N(\epsilon) \geq 1$ such that

$$\sum_{n > N(\epsilon)} \langle e_n, \rho e_n \rangle < \epsilon, \tag{2.13}$$

then the space $\mathcal{H}_\epsilon$ spanned by the finite set $(e_1, \ldots, e_{N(\epsilon)})$ makes the job for (2.12).

Theorem 2. *A set $\mathcal{K}$ of states is relatively weakly compact if and only if it is tight.*

Proof. Assume first that $\mathcal{K}$ is relatively weakly compact. Then for any projection P over a finite dimensional space the continuous mapping $\rho \mapsto tr\ \rho P$ is uniformly bounded on $\mathcal{K}$, and (2.12) follows.

To prove the converse we use Theorem 1. $\mathcal{K}$ is included in C and it is weakly relative compact. We only need to prove that limit points are states. Take a weakly convergent subsequence $\rho^n \in \mathcal{K}$ and denote ρ its limit. Given any $\epsilon > 0$, the definition of weak convergence and (2.12) imply that

$$tr\ \rho P_\epsilon \geq 1 - \epsilon. \tag{2.14}$$

Since ϵ is arbitrary it follows that $tr\ \rho = 1$. □

2.5. Definition. We say that a family $\mathcal{K}$ of states is *uniformly complex or chaotic* if for all $\epsilon > 0$ there exists $n(\epsilon) \in \mathbb{N}$ such that

$$\Gamma(\rho, n(\epsilon)) \geq 1 - \epsilon, \tag{2.15}$$

for all $\rho \in \mathcal{K}$.

We derive the following corollaries

Corollary 1. *A family $\mathcal{K}$ of states is weakly relative compact if and only if it is uniformly complex.*

This is straightforward from the definition and Theorem 2.

Corollary 2. *Assume $\mathcal{K}$ to be a family of states that is less complex (or chaotic) than a given state ρ. Then $\mathcal{K}$ is relatively weakly compact.*

Proof. It suffices to remark that

$$\Gamma(\rho, n) \leq \Gamma(\rho', n), \tag{2.16}$$

for all $n \in \mathbb{N}$, all $\rho' \in \mathcal{K}$. Then apply the above corollary. □

To use weak topologies in the analysis of asymptotic behaviour of both states and quantum dynamics, we need to work with couples (H, ρ) where $H \in \mathcal{O}(\mathcal{H})$ is a Hamiltonian and $\rho \in \mathcal{S}(\mathcal{H})$.

2.6. Definition. On the set $\mathcal{O}(\mathcal{H}) \times \mathcal{S}(\mathcal{H})$ we introduce the weakest topology for which the Quantum Flow θ_t is weakly continuous for all $t \in \mathbb{R}$. For simplicity, we also call it the *weak topology on* $\mathcal{O}(\mathcal{H}) \times \mathcal{S}(\mathcal{H})$ (abreviated w) .

Thus $(H^n, \rho^n) \xrightarrow{w} (H^\infty, \rho^\infty)$ if and only if for all $t \in \mathbb{R}$, we have

$$tr\ \exp(-itH^n)\rho^n \exp(itH^n)X \to tr\ \exp(-itH^\infty)\rho^\infty \exp(itH^\infty)X, \quad (2.17)$$

for all $X \in \mathcal{I}_\infty(\mathcal{H})$.

3. Sojourn Times

Our aim now is to analyze the time spent by a system in a given state. To do this we introduce the following definition:

3.1. Definition. The *sojourn time* of a positive observable X on a given state ρ is defined by

$$\tau(\rho, X) = \int_{-\infty}^{\infty} tr\ \rho X_t dt = \int_{-\infty}^{\infty} tr\ \rho_t X dt, \quad (3.1)$$

where $\rho_t = U(t)\rho U(-t)$, $X_t = U(t)XU(-t)$, $t \in \mathbb{R}$.

Recall that $tr\ \rho X_t$ represents the mean or expectation of X_t in the state ρ. So that $\tau(\rho, X)$ measures the mean value of the total time spent by X on the state ρ.

The functional $\tau(\rho, X)$ takes values in $[0, \infty]$. Moreover, $(\rho, \rho') \mapsto \tau(\rho, \rho')$ is a quadratic form on $\mathcal{S}(\mathcal{H})$ and $\rho \mapsto \sqrt{\tau(\rho, \rho)}$ is convex. Denote $\tau(\rho) = \tau(\rho, \rho)$ the sojourn time of ρ in the state ρ.

Theorem 3. *The mapping* $\rho \mapsto \tau(\rho)$ *is weakly upper semicontinuous.*

If the state ρ' *is more complex than* ρ, *then* $\tau(\rho') \leq \tau(\rho)$. *Moreover,*

$$\tau(\rho) = \sup_{\rho' \in C(\rho)} \tau(\rho'). \quad (3.2)$$

Proof. To prove the upper semicontinuity,take ρ^n that converge weakly to ρ. Notice that for another fixed state ρ',

$$\tau(\rho, \rho') \leq \liminf \tau(\rho^n, \rho'), \tag{3.3}$$

by Fatou's Lemma. So that $\tau(\cdot, \rho')$ is upper semicontinuous for all state ρ' and the property follows for $\tau(\cdot)$.

To finish the proof, it suffices to remark that $\sqrt{\tau(\cdot)}$ is weakly upper semicontinuous and convex, then it attains maximum values on extremal points of the compact convex set $C(\rho)$. As we discussed in Remark 2.3, ρ is extremal and all other extremal points are isometrically equivalent to ρ so they have the same sojourn time. □

4. Spectral Entropy

We fix a Hamiltonian H with spectral measure ξ and assume $\rho \in \mathcal{S}_{ac}(\xi)$. Therefore μ_ρ^ξ introduced in (1.4) is absolutely continuous with respect to Lebesgue measure. Denote $p_\rho(x)$ its Radon–Nikodym derivative. (To avoid complications we identify the *class* of functions p_ρ with any representative in it). We will introduce a notion of spectral entropy for this state.

4.1. Definitions. The *spectral entropy* of ρ is defined if $p_\rho(x) \log p_\rho(x)$ is integrable by the expression

$$\mathbb{S}(p_\rho) = - \int_{-\infty}^{\infty} p_\rho(x) \log p_\rho(x) dx \tag{4.1}$$

where we adopt the convention $0 \log 0 = 0$.

For a pure state $\rho = |\varphi\rangle\langle\varphi| \in \mathcal{S}_{ac}$ we simply write p_φ the density $p_{|\varphi\rangle\langle\varphi|}$. Similarly we write $\tau(\varphi)$ the sojourn time $\tau(|\varphi\rangle\langle\varphi|)$. It reduces now to the expression

$$\tau(\varphi) = \int_{-\infty}^{\infty} |\langle\varphi, U(t)\varphi\rangle|^2 dt. \tag{4.2}$$

Notice we have used the classical definition of Shannon entropy for a probability density function on the real line. It includes a dependence upon the group $(U(t))$ (or equivalently, on the Hamiltonian) describing the dynamics of the system. Moreover, the existence of the density p_ρ and the integrability condition with respect to Lebesgue measure depend on the evolution of the state ρ under $U(t)$. Furthermore, it is interesting to compare the entropy functional for two particular cases. Take first $p(x) = 1/n$ if $x \in [0, n]$ and $p(x) = 0$ otherwise. In this case $\mathbb{S}(p) = \log n$ and tends to ∞ as n grows. This is a "flat" density. If on the contrary we consider $q(x) = n$ if $x \in [0, 1/n]$ and $q(x) = 0$ otherwise, then $\mathbb{S}(q) = -\log n$ which tends to $-\infty$ as n grows. This density q approaches a "delta" function, that is it concentrates the probability in one point.

It is natural to wonder if $\tau(\rho) < \infty$ and if $\mathbb{S}(p_\rho)$ exists for a wide class of states ρ. The next proposition answers this question.

Proposition 1. *The set $D_c(H) = \{\rho \in \mathcal{S}_{ac} : p_\rho$ has compact support$\}$ is dense in $\mathcal{S}_{ac}$.*

Furthermore $D_c(H)$ is included in $D(H)$, $\mathbb{S}(p_\rho)$ is well defined and $\tau(\rho) < \infty$, for all $\rho \in D_c(H)$.

Proof. First of all we notice that $\rho \mapsto p_\rho$ is linear from $\mathcal{S}_{ac}$ into $L^1(\mathbb{R})$. Therefore $D_c(H)$ is a convex set and its extremal points are pure states φ for which p_φ has compact support.

Remark that given any $\rho \in \mathcal{S}_{ac}$, then $\rho_n = \xi([-n,n])\rho\xi([-n,n])/tr\ [\xi([-n,n])\rho]$ is a sequence in $D_c(H)$ that approaches ρ in norm as $n \to \infty$. This proves the density of $D_c(H)$.

Moreover, $\mathbb{S}(p_\rho)$ is obviously well defined if p_ρ has compact support.

Notice that given a pure state φ in $D_c(H)$ we can apply Plancherel's Theorem:

equation (4.2) is equivalent to

$$\tau(\varphi) = 2\pi \int_{-\infty}^{\infty} |\hat{p}_\varphi(t)|^2 = \int_{-\infty}^{\infty} p_\varphi(x)^2 dx < \infty, \tag{4.3}$$

where $\hat{p}_\varphi$ is the Fourier Transform of p_φ.

Now, given any state $\rho \in D_c(H)$ it can be written as a convex combination of pure states in $D_c(H)$:

$$\rho = \sum_i c_i |\varphi_i\rangle\langle\varphi_i|, \tag{4.4}$$

where $c_i \geq 0$ for all i and $\sum_i c_i = 1$.

Since $\sqrt{\tau(\cdot)}$ is convex we then obtain from (4.3)

$$\tau(\rho) < \infty, \tag{4.5}$$

and the proof is complete □

Call $D_s(H)$ the set of all $\rho \in \mathcal{S}_{ac}$ for which $p_\varphi \log p_\varphi$ is integrable. This set is dense since $D_c(H) \subset D_s(H)$.

Proposition 2. *Under the above hypotheses the following inequalities hold for any initial pure state* $\varphi \in D_s(H)$*:*

$$1 \leq \tau(\varphi) \exp \mathbb{S}(p_\varphi) \leq \tau(\varphi)(\|(H - \lambda I)\varphi\| \sqrt{2\pi e}), \tag{4.6}$$

for all real number λ.

Proof. First of all consider two probability measures with densities p, q with respect to Lebesgue measure, and with finite entropy. By the elementary inequality $\log x \leq x - 1$, we have

$$\log \frac{q}{p} \leq \frac{q}{p} - 1$$

or, equivalently,

$$1 + \log \frac{1}{p} \leq \frac{q}{p} + \log \frac{1}{q}.$$

Therefore,

$$\mathbb{S}(p) \leq \int_{-\infty}^{\infty} p(x) \log \frac{1}{q(x)} dx \tag{4.7}$$

By choosing q in (4.7) as the density of a Gaussian distribution of mean m and variance σ^2, that is

$$q(x) = \frac{1}{\sigma\sqrt{2\pi}} \exp(-\frac{(x-m)^2}{2\sigma^2})$$

we obtain

$$\mathbb{S}(p) \leq \frac{1}{2}\log(2\pi\sigma^2) + \frac{1}{2} = \log\sigma + \frac{1}{2}(\log(2\pi) + 1) = \log(\sigma\sqrt{2\pi e}) \tag{4.8}$$

Thus we derive the conclusion that any density p with a first moment m and finite variance σ^2, has its entropy bounded as in (4.8). The bound in the right-hand side of inequality (4.8) is the value of the entropy for a Gaussian density.

In our case we first apply the above inequality to the density p_φ with zero mean. In that case, σ is given by

$$\sigma^2 = \int_{-\infty}^{\infty} x^2 p_\varphi(x) dx = \|H\varphi\|^2, \tag{4.9}$$

Therefore,

$$\mathbb{S}(p_\varphi) \leq \log(\|H\varphi\|\sqrt{2\pi e}) \tag{4.10}$$

As the reader can easily verify, inequality (4.8) is invariant by a change on the mean m. The same holds in (4.10) if H is replaced by the operator $H - mI$, where m is the mean of H in state $|\varphi\rangle\langle\varphi|$ which is given by $m = \langle\varphi, H\varphi\rangle$. The variance is then $\|(H - mI)\varphi\|^2$. We then obtain

$$\mathbb{S}(p_\varphi) \leq \log(\|(H - mI)\varphi\|\sqrt{2\pi e}). \tag{4.11}$$

But it is well known that the mean m gives the minimum value to the deviation $\|(H - \lambda I)\varphi\|^2$ when λ runs over $\mathbb{R}$. In particular, inequality (4.11) holds also if we replace m by any $\lambda \in \mathbb{R}$.

Now consider the sojourn time associated to the state φ. Since μ_φ is a probability measure on the real line, we may apply Jensen's inequality to obtain:

$$\begin{aligned}\log \tau(\varphi) &= \log \int_{-\infty}^{\infty} p_\varphi(x)\mu_\varphi(dx)\\ &\geq \int_{-\infty}^{\infty} \log p_\varphi(x)\mu_\varphi(dx)\end{aligned}$$

Then,

$$\log \tau(\varphi) \geq -\mathbb{S}(p_\varphi) \tag{4.12}$$

From (4.11) and (4.12) we finally derive the announced inequality:

$$1 \leq \tau(\varphi)\exp \mathbb{S}(p_\varphi) \leq \tau(\varphi)\|(H-\lambda I)\varphi\|\sqrt{2\pi e}.$$

□

We remark that if p_φ is a Gaussian density with mean $m = \langle\varphi, H\varphi\rangle = \lambda$ and variance $\sigma^2 = \|(H-\lambda)\varphi\|^2$, then the equality is attained in the second inequality in (4.6).

Proposition before leads to our main "uncertainty principle"

Theorem 4. *For any state $\rho \in D_s(H)$ and any real number λ it holds*

$$1 \leq \tau(\rho)\exp(\mathbb{S}(\rho)) \leq \tau(\rho)\sqrt{2\pi e\; tr\; \rho(H-\lambda I)^2}. \tag{4.13}$$

Proof. Assume $\rho = \sum_i c_i|\varphi_i\rangle\langle\varphi_i|$ a convex combination of pure states in $D_s(H)$.

Since the function $x \mapsto -x\log x$ is concave and $\rho \mapsto p_\rho$ is linear, it follows that $\rho \mapsto \mathbb{S}(p_\rho)$ is concave. Therefore

$$\sum_i c_i\mathbb{S}(p_{\varphi_i}) \leq \mathbb{S}(p_\rho), \tag{4.14}$$

From (4.6) it follows that

$$-\sum_i c_i \log\tau(\varphi_i) \leq \sum_i c_i\mathbb{S}(p_{\varphi_i}). \tag{4.15}$$

Now since $-\frac{1}{2}\log\tau(\cdot)$ is a convex function, we get

$$-\log\tau(\rho) \leq -\sum_i c_i \log\tau(\varphi_i), \tag{4.16}$$

after a multiplication by 2. Hence, from (4.14), (4.15), (4.16) it follows

$$1 \leq \tau(\rho)\exp(\mathbb{S}(p_\varphi)). \tag{4.17}$$

To get the second part of inequality (4.13) we remark that if $\rho \in D_c(H)$ then p_ρ is a density that has moments of any order. The mean m is given by $m = tr\ \rho H$ and the variance is $\sigma^2 = tr\ \rho(H - mI)^2$. By the same argument used to derive (4.11) we get

$$\mathbb{S}(p_\rho) \leq \log(\sqrt{2\pi e\ tr\ \rho(H - mI)^2}), \tag{4.18}$$

for all $\rho \in D_c(H)$, but indeed (4.18) is trivially satisfied if $tr\ \rho(H - mI)^2$ is infinite, thus it also holds in the more general case $\rho \in D_s(H)$. Also m can be replaced in (4.18) by any $\lambda \in \mathbb{R}$ since $tr\ \rho(H - mI)^2 \leq tr\ \rho(H - \lambda I)^2$.

From (4.16) and (4.18) we obtain (4.13) and the proof is complete. □

4.2. Energy width. Consider the family $(q_\epsilon; \epsilon > 0)$ of Cauchy densities on the real line, defined by

$$q_\epsilon(x) = \frac{1}{\pi}\frac{\epsilon}{\epsilon^2 + x^2} \tag{4.19}$$

$x \in \mathbb{R}$ and $\epsilon > 0$.

Assume $\rho \in \mathcal{S}_{ac}$. Denote $m = tr\ \rho H$ We define the *energy width* $\Delta E(\rho)$ of ρ as the infimum over the set of all $\epsilon > 0$ such that

$$q_\epsilon \star p_\rho(m) = \int_{-\infty}^{\infty} q_\epsilon(x - m)p_\rho(x)dx \geq \frac{1}{2\pi\epsilon} \tag{4.20}$$

The next proposition gives the relationship between the energy width and the entropy of the state ρ.

Proposition 3. *Under the above hypotheses, the following inequality holds:*

$$\exp(\mathbb{S}(p_\rho)) \le \pi[\Delta E(\rho) + tr\ (\rho(H - mI)^2)(\Delta E(\rho))^{-1}] \tag{4.21}$$

Proof. We use (4.7) applied to densities p_ρ and $q_\epsilon(\cdot - m)$ to derive

$$\mathbb{S}(p_\rho) \le \int_{-\infty}^{\infty} p_\rho(x) \log \frac{1}{q_\epsilon(x-m)} dx, \tag{4.22}$$

where $m = tr\ \rho H$.

Now, by Jensen's inequality (used with the concave function log) we obtain

$$\int_{-\infty}^{\infty} p_\rho(x) \log \frac{1}{q_\epsilon(x-m)} dx \le \log \int_{-\infty}^{\infty} p_\rho(x) \log \frac{1}{q_\epsilon(x-m)} dx. \tag{4.23}$$

Remark that

$$\int_{-\infty}^{\infty} p_\rho(x)(x-m)^2 dx = tr\ \rho(H - mI)^2,$$

so that (4.21) follows if we put $\epsilon = \Delta(\rho)$ and we take exponentials in both sides of (4.22) and (4.23). □

5. Resonance

We are now in position to characterize resonance as a limit behaviour of the Quantum Flow.

5.1. Definitions. A *perturbation of the flow* in $(H^\infty, \rho^\infty) \in \mathcal{O}(\mathcal{H}) \times \mathcal{S}(\mathcal{H})$ is a sequence $((H^n, \rho^n);\ n \in \mathbb{N})$ such that $(H^n, \rho^n) \xrightarrow{w} (H^\infty, \rho^\infty)$.

The perturbation (H^n, ρ^n) is *resonant* if $\rho^n \in D_s(H^n)$ and

$$\limsup_n \mathbb{S}(p^n_{\rho^n}) = -\infty, \tag{5.1}$$

where $p^n_{\rho^n}$ is the spectral density of ρ^n obtained with the spectral measure of the Hamiltonian H^n.

5.2. Remark. Given a perturbation (H^n, ρ^n) of (H^∞, ρ^∞) it follows

$$\tau_\infty(\rho^\infty) \leq \liminf_n \tau_n(\rho^n), \tag{5.2}$$

by a simple application of Fatou's Lemma.

Theorem 5. *If* $((H^n, \rho^n);\ n \in \mathbb{N})$ *is a resonant perturbation of the flow, then* $\liminf_n \tau_n(\rho^n) = \infty$.

Proof. This theorem follows straightforward from (4.13) and resonance's definition. □

The theorem before says that a resonant perturbation "sojourns a long time" in a state with higly concentrated energy. Thus, our definition of resonance agrees with the phenomenological analysis in scattering theory (see [8]). As we refered in the introduction, this intrinsic characterization avoids ambiguities connected to definitions of resonance via poles commonly used in physics literature.

The relationship between this approach and that of Lavine is illustrated by an application of (4.21) as follows. In general $\Delta E(\rho) \leq \|(H - mI)\varphi\|$ ([8], Theorem 2.3), for a pure state $\rho = |\varphi\rangle\langle\varphi|$. This inequality can be generalized easily to our context as $\Delta E(\rho) \leq \sqrt{tr\ \rho(H - mI)^2}$. Consider a perturbation (H^n, ρ^n) of (H^∞, ρ^∞). If $\Delta E_n(\rho^n)$ is close to $\sqrt{tr\ \rho^n(H^n - m_n I)^2}$, then (4.21) says that the exponential of the spectral entropy is dominated by $\Delta E_n(\rho^n)$. In this case if the energy width goes to zero, the state is resonant. And (4.13) gives a finer estimate of the convergence to $+\infty$ of sojourn times.

In other cases it is not possible to compare explicitly Lavine's criterion on resonance and ours.

Pure resonant states very often appear as "asymptotic eigenvectors" of a given sequence of Hamiltonians (H^n). We say that φ is *an asymptotic eigenvector* for the sequence (H^n) if there exists $\lambda \in \mathbb{C}$, called *asymptotic eigenvalue* such that

$$\limsup_n \|(H^n - \lambda I)\varphi\| = 0. \tag{5.3}$$

From (4.6) it follows

Corollary 3. *For any ρ which is a convex combinaton of pure states $|\varphi_i\rangle\langle\varphi_i|$ where φ_i are asymptotic eigenvectors of $(H^n)_n$, the perturbation (H^n, ρ) of (H^∞, ρ) is resonant.*

In the next section we introduce an important case of *non* resonant perturbations.

6. Smooth Operators

We follow [6] to introduce the following concept connected with a Hamiltonian H.

6.1. Definition. A closed operator A on the Hilbert space $\mathcal{H}$ is *H-smooth* if for any $\varphi \in \mathcal{H}$ it holds that $\exp(-itH)\varphi \in D(A)$ and

$$\int_{-\infty}^{\infty} \|A\exp(-itH)\varphi\|^2 dt \le c\|\varphi\|^2, \tag{6.1}$$

where $c > 0$ is a constant. Then the smallest constant c for which (6.1) holds is called the H-norm of A and is denoted $\|A\|_H$.

We say a perturbation $((H^n, \rho^n) : n \in \mathbb{N})$ of (H^∞, ρ^∞) is *smooth* if ρ^∞ is H^∞-smooth.

6.2. Remark. To check H-smoothness of a closed operator A it suffices to prove that condition (6.1) holds for all φ in a dense subset.

Furthermore in [6] (or [12]) it is proven that A is H-smooth if and only if satisfies:

$$\sup_{a<b} \frac{\|A\xi([a,b])\|^2}{|b-a|} < \infty. \tag{6.2}$$

Proposition 4. *The set $\Sigma(H)$ of all H-smooth states is convex; extremal points of $\Sigma(H)$ are H-smooth pure states and $D_c(H) \subset \Sigma(H)$.*

Moreover, $\tau(\varphi) < \infty$ for all pure state φ in $\Sigma(H)$.

If the Hamiltonian is absolutely continuous, then $\Sigma(H)$ is dense in the set of all states in the sense of the strong topology of operators.

Proof. The set of H-smooth operators is a vector space and $\Sigma(H)$ is the intersection of this space with the convex set $\mathcal{S}(\mathcal{H})$. This remark proves the first two statements.

To prove $D_c(H) \subset \Sigma(H)$ it suffices to prove that pure states in $D_c(H)$ are smooth and for a general state the result will follow by convexity. Notice the numerator in (6.2) is equal to $\mu_\rho^\xi([a,b])$ for a pure state $A = \rho = |\varphi\rangle\langle\varphi|$.

Under the hypothesis we have $\mu_\rho^\xi([a,b]) \leq Const.|b-a|$, for all $-\infty < a < b < \infty$. Therefore condition (6.2) holds and the pure state ρ is H-smooth. Moreover, for a pure state $\rho = |\varphi\rangle\langle\varphi|$ the expression (6.1) evaluated over the vector φ reduces to

$$\tau(\varphi) = \int_{-\infty}^{\infty} |\langle\varphi, \exp(-itH)\varphi\rangle|^2 dt \leq c < \infty. \tag{6.3}$$

The last statement follows from the density of $D_c(H)$ in $\mathcal{S}_{ac}$. □

We then have the following main result on smooth perturbations:

Theorem 6. *Suppose (H^n, ρ^n) is a perturbation of (H^∞, ρ^∞) in $\mathcal{O}(\mathcal{H}) \times \mathcal{S}(\mathcal{H})$ The perturbation is smooth if*

$$\sup_n \sup_{a<b} \frac{\mu_{\xi^n}^{\rho^n}([a,b])}{|b-a|} < \infty. \tag{6.4}$$

Proof. To simplify notations, we write μ^n (respectively μ^∞) measure $\mu_{\xi^n}^{\rho^n}$ (resp. $\mu_{\xi^\infty}^{\rho^\infty}$.

By weak convergence we have

$$\mu^n(B) = tr\ \rho^n\xi^n(B) \to tr\ \rho^\infty\xi^\infty(B) = \mu^\infty(B), \tag{6.5}$$

for all borelian set B in the real line.

Therefore, μ^∞ satisfies

$$\sup_{a<b} \frac{\mu^\infty([a,b])}{|b-a|} < \infty. \tag{6.6}$$

Use then convexity of the set of all states: ρ^∞ is written as a convex combination $\sum_i c_i|\varphi_i\rangle\langle\varphi_i|$ of pure states. The same holds for measure μ^∞: it is a convex combination $\sum_i c_i\mu_i^\infty$ of measures $\mu_i^\infty(\cdot) = \langle\varphi_i, \xi^\infty(\cdot)\varphi_i\rangle$. For each measure μ_i^∞ (6.6) holds. That is, each pure state φ_i satisfies (6.2) and it is H^∞–smooth. Therefore, ρ^∞ is also H^∞–smooth. □

Corollary 4. *Given a perturbation (H^n, ρ^n) of (H^∞, ρ^∞) assume the following two hypotheses hold:*

(1) *$\rho^n \in \mathcal{S}_{ac}(\xi^n)$ for all $n \in \mathbb{N}$ and*

(2) *The sequence of spectral densities $(p^n_{\rho^n})$ is Lebesgue–uniformly integrable.*

Then the perturbation is smooth.

Proof. Both hypotheses imply in particular that $(p^n_{\rho^n})$ is uniformly bounded in $L^1(\mathbb{R})$ and (6.4) holds. □

6.3. Remark. Smoothness avoids resonances in the case of pure states. Indeed, Proposition 4 and Remark 5.2 imply that a smooth perturbation $((H^n, \rho^n); n \in \mathbb{N})$ of (H^∞, ρ^∞) is not resonant if $\rho^n = |\varphi^n\rangle\langle\varphi^n|$, $(n \in \mathbb{N} \cup \{\infty\})$.

In the next section we study a family of applications connected with barrier problems or tunnelling effect.

7. Application to tunnelling resonance

In what follows, we shall deal with the tunnel effect, that is, the penetration of a quantum particle through a classicaly forbidden region. We consider this model as an example of the characterization of the resonance phenomena given in the previous sections.

It is well known that a particle incident on a potential barrier from one side will tunnel through the other side and, as remarked in [14], [8], resonance will occur if the particle encounters a second barrier and if it has a suitable energy. Moreover, the set of these resonant energies is discrete and it is contained in the absolutely continuous part of the spectrum.

7.1. Formulation of the model. Let a, b and h be real numbers such that $h > 0$ and $0 < a < b$. For each positive integer n, let $V^n(x)$ be a smooth function of compact support satisfying $V^n(x) = 0$, for $|x| \leq a$ and $V^n(x) = h$, for $b \leq |x| \leq b + n$. Thus, $V^n(x)$ is a potential with two symmetric barriers of heigth h and width at least n. As n converges to infinity, $V^n(x)$ approaches to the potential $V^\infty(x)$ which has two barriers of the same height but with an infinite length.

For each n (including $n = \infty$), we denote H^n the selfadjoint realization of $-\frac{d^2}{dx^2} + V^n(x)$ on the Hilbert space $\mathcal{H} = L^2(\mathbb{R})$. For n finite, the operator H^n is unitarily equivalent to the Laplacian $-\frac{d^2}{dx^2}$ in $\mathcal{H}$ and therefore it is absolutely continuous and its spectrum consists of the interval $[0, \infty[$. On the other hand, the spectrum of H^∞ in $[0, h]$ is discrete, while its absolutely continuous part is $[h, \infty[$.

Let φ be an eigenvector of H^∞, with $\|\varphi\| = 1$. Take $\rho = |\varphi\rangle\langle\varphi|$, so that (H^n, ρ) is a perturbation of (H^∞, ρ) by the construction of potentials made before. We claim that this perturbation is resonant in the sense of definition 5.1.

From the eigenvalue equation, it follows that for $|x| \geq b$, $\varphi(x) = c\exp(-\sqrt{h - \lambda}|x|)$, where λ is the eigenvalues associated to φ, and C is a suitable constant. Hence, φ is concentrated in bounded sets and for large h it is mainly localized between the barriers. Therefore a quantum particle with initial state φ should spent a very long time between the barriers.

Moreover, for any positive integer n, we have that

$$
\begin{aligned}
(H^n - \lambda)\varphi &= -\varphi'' + V^n\varphi - \lambda\varphi \\
&= (V^n - V)\varphi \, .
\end{aligned}
$$

It follows that

$$\begin{aligned}\|(H^n - \lambda)\varphi\|^2 &= 2c^2 \int_{b+n}^{\infty} (V^n - V)^2 (\varphi\lambda)^2 dx \\ &\leq 2c^2 h^2 \int_{b+n}^{\infty} e^{-2\sqrt{h-\lambda}x} dx \\ &= \frac{c^2 h^2}{\sqrt{h-\lambda}} e^{-2\sqrt{h-\lambda}(b+n)}.\end{aligned}$$

By (4.6), this gives an explicit upper bound on the entropy of the density p^n_φ associated to the pure state φ, with respect to the Hamiltonian H^n

Theorem 7. *The entropy of p^n_φ verifies*

$$\mathbb{S}(p^n_\varphi) \leq \log\left[\frac{kh}{(h-\lambda)^{1/4}} e^{-\sqrt{h-\lambda}(b+n)}\right] \tag{7.1}$$

In particular, the perturbation $(H^n, |\varphi\rangle\langle\varphi|)$ is resonant.

Proof. It follows immediately from above calculations and (4.6). □

REFERENCES

1. Cohen–Tanoudji, C. – Diu, B. – Laloe, F. (1977) *Quantum Mechanics*, John Wiley and Sons.
2. Dellacherie,C.–Meyer,P.A. (1983) *Probabilités et Potentiel. Théorie Discrète du Potentiel Chap. IX–XI.* Hermann, Paris. 229p.
3. Fernández, C.–Lavine, R. (1990) Lower Bounds for Resonance Widths in Potential and Obstacle Scattering. Comm. in Math. Phys. **128**, 263–284.
4. Fernández, C.–Rebolledo, R. (1991) Entropy functional and resonant pure states. Informe Técnico P.U.C.–F.M. 91/15. (To appear in *Notas de la Sociedad de Matemática de Chile*).
5. Fernández, C.–Rebolledo, R. (1992) Sur l'entropie spectrale et la résonance quantique. *C.R.Acad.Sci.Paris (to appear)*.
6. Kato, T. (1966) *Perturbation Theory for Linear Operators.* Springer–Verlag, N.Y.
7. Kukulin, V.I.–Krasnopol'sky, V.M.–Horacek, J. (1989) *Theory of Resonances.* Kluwer Ac.Publishers, 355p.
8. Lavine, R. (1978) Spectral density and Sojourn times, in *Atomic Scattering Theory* ed. J.Nutall, Univ. Western Ontario.
9. Parthasarathy, K.R. (1988) *I.S.I. Lectures on Quantum Stochastic Calculus.* Indian Statistical Institute, Delhi Centre (mimeographed notes), 331 p.
10. Peierls, R. (1976) *Surprises in Theoretical Physics.* Princeton University Press.
11. Rebolledo, R. (1992) Complexity and Quantum Dynamics. Inf.Técnico P.U.C./F.M.–92/1. (Lectures delivered at CIMPA School on Dynamical Systems and Frustrated Systems).
12. Reed, M.–Simon, B. (1978) *Analysis of Operators.* Vol. IV, Acad.Press, N.Y., 396 p.

13. Thirring, W. (1983) *A Course in Mathematical Physics 4. Quantum Mechanics of Large Systems*.Springer-Verlag, New-York, Wien, 290 p.
14. Zakhariev, B.N.-Suzko, A.A. (1990) *Direct and Inverse Problems. Potentials in Quantum Scattering*, Springer-Verlag, Berlin, 223 p.

FACULTAD DE MATEMÁTICAS, PONTIFICIA UNIVERSIDAD CATÓLICA DE CHILE, CASILLA 306,CORREO 22,SANTIAGO

Quantum Probability and Related Topics Vol. VIII (pp. 189–210)

QUANTUM STOCHASTIC INTEGRALS ON A GENERAL FOCK SPACE

Wolfgang Freudenberg (Jena)

Abstract

We introduce quantum stochastic integrals with respect to creation and annihilation operators on a general symmetric Fock space. Conditions for existence and boundedness of these integrals are given. It is shown that for local operators from a quasilocal algebra the quantum stochastic integrals exist. In the special case of non-anticipating kernel processes on an interval $[0, T]$ these operators coincide with the integrals introduced by Maassen.

1 Introduction

The aim of this paper is to discuss quantum stochastic integrals with respect to creation and annihilation operators on a general (symmetric) Fock space $\mathcal{M} = L_2(M, F)$ where M is the set of all finite counting measures on a Polish space G which is assumed to be equipped with a locally finite diffuse measure ν. The σ-finite measure F restricted to configurations in a bounded area $\Lambda \in \mathcal{G}$ is just the (non-normalized) Poisson measure with intensity measure ν. We consider the densely defined operators $\mathcal{D}^c$ from $\mathcal{M}$ into $\mathcal{M} \otimes \mathcal{M}$ resp. $\mathcal{S}^c$ from $\mathcal{M} \otimes \mathcal{M}$ into $\mathcal{M}$ given on their domains by

$$\begin{aligned} \mathcal{D}^c\Psi(\varphi_1, \varphi_2) &= \Psi(\varphi_1 + \varphi_2) && (\varphi_1, \varphi_2 \in M) \\ \mathcal{S}^c U(\varphi) &= \sum_{\hat{\varphi} \subseteq \varphi} U(\hat{\varphi}, \varphi - \hat{\varphi}) && (\varphi \in M). \end{aligned}$$

In previous papers (cf. [1, 2, 3, 5]) we considered operators of the type $\mathcal{S}^c(A \otimes B)\mathcal{D}^c$ and applied them to the description and construction of locally normal states of boson systems. In the case that B is an integral operator and $A = \mathbf{1}$ is the identity the kernel of B is called by Maassen an integral kernel of $\mathcal{S}^c(\mathbf{1} \otimes B)\mathcal{D}^c$. For non-anticipating families $(b_t)_{t \in [0, T]}$ of such integral kernels on the positive half-line in [11] there are defined quantum stochastic integrals with respect to creation and annihilation operators. Motivated by these definitions we introduce for a general family $(X_t)_{t\in G}$ of operators on $\mathcal{M}$ and sets $\Lambda \in \mathcal{G}$ integrals $\int_\Lambda X\,dA$ and $\int_\Lambda dA^* X$ by

$$\int_\Lambda X\,dA\,\Psi(\varphi) = \int_\Lambda (X_t\mathcal{D}^c\Psi(\delta_t,\cdot))(\varphi)\nu(dt) \tag{1}$$

and

$$\int_\Lambda dA^*\,X\Psi(\varphi) = \int_\Lambda X_t\Psi(\varphi-\delta_t)\varphi(dt). \tag{2}$$

The aim of this paper is to discuss existence and possible domains of definition for operators of the type (1) and (2). For instance if the operators X_t are bounded and $\sup_{t\in\Lambda}\|X_t\| < \infty$ then $\int_\Lambda X\,dA$ and $\int_\Lambda dA^* X$ are densely defined linear operators. Observe that we don't need any smoothness conditions with respect to the family $(X_t)_{t\in G}$ except certain assumption of measurability.

Further, we look for conditions ensuring that the integrals of operators of the form $\mathcal{S}^c(A_t\otimes B_t)\mathcal{D}^c$ exist and are again of the type $\mathcal{S}^c(A\otimes B)\mathcal{D}^c$. We will see that non-anticipating processes $X_t = \mathcal{S}^c(\mathbf{1}\otimes B_t)\mathcal{D}^c$ have quantum stochastic integrals of the same form as in [11] without the need of requiring the existence of kernels. We introduce families of so-called weakly non–anticipating operators for which still the quantum stochastic integrals exist (Proposition 4.4). These operators are combinations of local measurements with position measurements outside and seem to represent a reasonable class of integrands for quantum stochastic integrals. For any bounded area $\Lambda \in \mathcal{G}$ the operators from the local algebra $\mathcal{A}_\Lambda = \mathcal{L}(\mathcal{M}_\Lambda)\otimes\mathbf{1}_{\Lambda^c}$ where $\mathcal{M}_\Lambda$ is the Fock space over Λ can be written in the form $X = \mathcal{S}^c(\mathbf{1}_{\Lambda^c}\otimes B)\mathcal{D}^c$ where $B \in \mathcal{L}(\mathcal{M}_\Lambda)$ and $\mathbf{1}_{\Lambda^c}$ is the identity operator on configurations outside Λ. We will prove that for all $\Lambda \in \mathcal{B}$ and families $(X_t)_{t\in G}$ with the property that for all $t \in \Lambda$ the operator X_t belongs to the local algebra $\mathcal{A}_\Lambda$ and $\sup_{t\in\Lambda}\|X_t\| < \infty$ the quantum stochastic integrals $\int_\Lambda X\,dA$ and $\int_\Lambda dA^* X$ exist (Theorem 3.7). So the results of the present paper may give a contribution to the study of quantum stochastic integrals for operators on quasilocal algebras. An extensive study of such integrals finally could lead to a quantum stochastic calculus for infinite boson systems. That's why we also tried never to leave the general Fock space over G also when we touched problems of adapted processes. This caused to introduce a somewhat artificial notion of non-anticipating processes.

In the present paper we consider only the operators $\int_\Lambda X\,dA$ and $\int_\Lambda dA^* X$.

But one should also define quantum stochastic integrals $\int_\Lambda dA\, X$, $\int_\Lambda X\, dA^*$ and the integrals $\int_\Lambda X\, d\mathcal{N}$ and $\int_\Lambda d\mathcal{N}\, X$ where $\mathcal{N}$ is the number operator. However, especially the integrals $\int_\Lambda dA\, X$ and $\int_\Lambda X\, dA^*$ seem to behave quite complicated.

The assumption that the measure ν on the phase space G has to be non-atomic is not necessary for most things in this paper (except the results and notions in section 4). So one could also treat lattice systems. However, in this case combinatorical problems lead to an unpleasant eyesore in the proofs. That's why we restricted ourselves here to the non-atomic case.

As mentioned above in the case of non-anticipating kernel processes on the line the integrals (1) and (2) are the same as in [11]. Also in [7] one can find these integrals for special integrands. After having finished this paper we noticed a (still unpublished) paper by Lindsay [9] where quantum stochastic integrals are introduced in the same way as above. The intersection of the results in both papers is non-void but their union is much bigger than the contents of the single papers. So the papers may complement each other.

I would like to thank K.-H. Fichtner for fruitful discussions and suggestions on the subject of this paper.

2 Basic Notions and Notations

2.1 The Symmetric Fock space

Let (G,ρ) be a complete separable metric space. By $\mathcal{G}$ we denote the σ-algebra of Borel subsets of G and by $\mathcal{B}$ the Borel sets bounded with respect to the metric ρ. Further, let ν be an arbitrary locally finite diffuse measure on $[G,\mathcal{G}]$, i.e. $\nu(B) < \infty$ for $B \in \mathcal{B}$ and $\nu(\{x\}) = 0$ for $x \in G$. G will represent the phase space of the considered quantum systems. In applications G usually will be an Euclidean space $\mathbf{R}^d, d \geq 1$ and ν the corresponding Lebesgue measure.

Since we will deal only with the symmetric Fock space and to avoid using operators of symmetrizations we identify point configurations $(x_1, \cdots, x_n)$ from G with the counting measures $\varphi = \sum_{j=1}^n \delta_{x_j}$ where δ_x is the Dirac measure in x. Denote by M_Λ the set of all finite configurations in Λ, i.e. the set of counting measures of the form $\varphi = \sum_{j=1}^n \delta_{x_j}$ with $x_j \in \Lambda$. Since G was assumed to be a Polish space we have for all $\Lambda \in \mathcal{G}$

$$M_\Lambda = \{\varphi : \varphi \text{ is an integer-valued finite measure on } [G,\mathcal{G}] \text{ such that } \varphi(\Lambda^c) = 0\}.$$

We equip M_Λ with the σ-algebra $\mathcal{F}_\Lambda$ generated by all sets of the type $\{\varphi \in M_\Lambda : \varphi(B) = n\}$ with $B \in \mathcal{G}$ and $n \in \mathrm{N} := \{0, 1, \cdots\}$. For each $\Lambda \in \mathcal{G}$ we define a σ-finite measure F_Λ on $[M_\Lambda, \mathcal{F}_\Lambda]$ by

$$F_\Lambda = \chi_Y(\theta) + \sum_{n\geq 1} \frac{1}{n!} \int_{\Lambda^n} \chi_Y\Big(\sum_{j=1}^{n} \delta_{x_j}\Big)\nu(d[x_1, \cdots, x_n]) \quad (Y \in \mathcal{F}_\Lambda) \qquad (3)$$

where χ_Y denotes the indicator function of the set Y and θ the empty configuration in M_G, i.e. $\theta \in M_G, \theta(G) = 0$. Let us remark that for $\Lambda \in \mathcal{B}$ the measure F_Λ is a finite one, $F_\Lambda(M_\Lambda) = \exp\{\nu(\Lambda)\}$ and $F_\Lambda(\cdot)\cdot\exp\{-\nu(\Lambda)\}$ is the Poisson point process with intensity measure ν.

Definition 2.1 *Let $\Lambda \in \mathcal{G}$. The Hilbert space $\mathcal{M}_\Lambda = L_2(M_\Lambda, F_\Lambda)$ is called the (symmetric) Fock space over Λ.*

In the case $\Lambda = G$ we omit the index Λ in all of the notations above. Sometimes we also consider F_Λ to be a measure on $[M, \mathcal{F}]$ concentrated on M_Λ. A similar definition of the symmetric Fock space (in the case that G is an interval on the real axis and ν the Lebesgue measure) one can find for instance in [6] and [11]. One easily checks (cf. [1, Remark 2.5] that $\mathcal{M}_\Lambda$ is isomorphic to the usual definition of the Fock space in physics books.

2.2 Compound Malliavin Derivatives and Skorohod Integrals

Definition 2.2 *Let $\mathcal{D}^c : \mathcal{M} \longrightarrow \mathcal{M}^2$ be given by*

$$\mathcal{D}^c\Psi(\varphi_1, \varphi_2) = \Psi(\varphi_1 + \varphi_2) \qquad (\varphi_1, \varphi_2 \in M, \Psi \in Dom(\mathcal{D}^c))$$

where

$$Dom(\mathcal{D}^c) := \{\Psi \in \mathcal{M} : \int |\Psi(\varphi)|^2 2^{\varphi(G)} F(d\varphi) < \infty\}.$$

The operator $\mathcal{D}^c$ will be called compound Malliavin derivative.

Definition 2.3 *Let $\mathcal{S}^c : \mathcal{M}^2 \longrightarrow \mathcal{M}$ be given by*

$$\mathcal{S}^c U(\varphi) := \sum_{\hat{\varphi} \subseteq \varphi} U(\hat{\varphi}, \varphi - \hat{\varphi}) \qquad (\varphi \in M, U \in Dom(\mathcal{S}^c))$$

where

$$Dom(\mathcal{S}^c) := \{U \in \mathcal{M}^2 : \int\int |U(\varphi_1, \varphi_2)|^2 2^{\varphi_1(G)+\varphi_2(G)} F(d\varphi_1)F(d\varphi_2) < \infty\}.$$

$\mathcal{S}^c$ wil be called compound Skorohod integral.

In the above definition $\hat{\varphi} \subseteq \varphi$ means that $\hat{\varphi}$ is a subconfiguration of φ, i.e. $\varphi - \hat{\varphi} \in M$. In [3] it is shown that the operators $\mathcal{D}^c$ and $\mathcal{S}^c$ defined by 2.2 and 2.3 are densely defined operators from $\mathcal{M}$ into $\mathcal{M}^2$ resp. from $\mathcal{M}^2$ into $\mathcal{M}$. We mention briefly some properties of these operators. Proofs and more details one can find in [3, section 3].

Lemma 2.4 *$\mathcal{D}^c$ and $\mathcal{S}^c$ are mutually adjoint, i.e. for all $\Psi \in Dom(\mathcal{D}^c)$ and $U \in Dom(\mathcal{S}^c)$*

$$\langle \mathcal{D}^c \Psi, U \rangle_{\mathcal{M}^2} = \langle \Psi, \mathcal{S}^c U \rangle_{\mathcal{M}}.$$

For $f \in L_2(G, \nu)$ the function $\Psi_f : M \to \mathbf{C}$ defined by

$$\Psi_f(\varphi) := \prod_{x \in \varphi} f(x)$$

is called *coherent function of f*. Hereby $x \in \varphi$ means $\varphi(\{x\}) \geq 1$, the product is taken according to the multiplicity of the mass-points of φ and $\prod_{x \in \theta} f(x) = 1$. Observe that linear combinations of coherent functions are dense in $\mathcal{M}$ and that all coherent functions belong to $Dom(\mathcal{D}^c)$. Indeed, one immediately checks that for $f \in L_2(G, \nu)$

$$\int |\Psi_f(\varphi)|^2 2^{\varphi(G)} F(d\varphi) = \exp\{2\|f\|^2\} < \infty.$$

For coherent functions one gets the simple well-known relation

$$\mathcal{D}^c \Psi_f = \Psi_f \otimes \Psi_f \quad \text{and} \quad \mathcal{S}^c \Psi_f \otimes \Psi_g = \Psi_{f+g} \quad (f, g \in L_2(G, \nu)).$$

Definition 2.5 *Let $\mathcal{D} : \mathcal{M} \longrightarrow L_2(G, \nu) \otimes \mathcal{M}$ be given by*

$$\mathcal{D}\Psi(x, \varphi) = \Psi(\delta_x + \varphi) \qquad (\varphi \in M, x \in G, \Psi \in Dom(\mathcal{D}))$$

where

$$Dom(\mathcal{D}) := \{\Psi \in \mathcal{M} : \int |\Psi(\varphi)|^2 \varphi(G) F(d\varphi) < \infty\}.$$

The operator $\mathcal{D}$ is called Malliavin derivative.

Definition 2.6 *Let $\mathcal{S} : L_2(G, \nu) \otimes \mathcal{M} \longrightarrow \mathcal{M}$ be given by*

$$\mathcal{S}U(\varphi) := \int U(x, \varphi - \delta_x) \varphi(dx) \qquad (\varphi \in M, U \in Dom(\mathcal{S}))$$

where

$$Dom(\mathcal{S}) := \{U \in L_2(G, \nu) \otimes \mathcal{M} : \int\int |U(x, \varphi)|^2 \varphi(G) \nu(dx) F(d\varphi) < \infty\}.$$

$\mathcal{S}$ is called Skorohod integral.

Using the isomorphism between the Fock space and Wiener space one can prove that $\mathcal{D}$ and $\mathcal{S}$ are isomorphic to the "usual" Malliavin derivative and Skorohod integral. Since we are interested in this paper only in operators on the Fock space we will not touch these problems here and refer e.g. to [8, 12, 13, 14, 15, 16, 17, 4].
Observe that $\mathcal{D}$ and $\mathcal{S}$ are "restrictions" of the operators $\mathcal{D}^c$ and $\mathcal{S}^c$, i.e. we have that $\mathcal{D}$ restricted to $Dom(\mathcal{D}^c)$ is equal $\mathbf{1}_{L_2(G,\nu)\otimes\mathcal{M}}\mathcal{D}^c$ and $\mathcal{S}$ can be identified with $\mathcal{S}^c\mathbf{1}_{L_2(G,\nu)\otimes\mathcal{M}}$. Hereby $\mathbf{1}_{(\cdot)}$ again denotes the projection.

2.3 Operators of the Type $\mathcal{S}^c(A_1 \otimes A_2)\mathcal{D}^c$

For a set Y of configurations denote by $\mathcal{O}_Y$ the operator of multiplication by the indicator function of Y, i.e.

$$\mathcal{O}_Y\Psi(\varphi) := \chi_Y(\varphi)\Psi(\varphi) \qquad (\Psi \in \mathcal{M}, \varphi \in M).$$

In [1, 2, 3, 5] we considered in detail operators of the type $\mathcal{S}^c(A_1 \otimes A_2)\mathcal{D}^c$ and we discussed applications of these operators to the characterization of normal and locally normal states of boson systems (they are denoted in the above mentioned papers by $S(Y, A)$).

Let us consider the special case that $B \in \mathcal{L}(\mathcal{M})$ and $Y = M$, i.e. $\mathcal{O}_Y = \mathbf{1}$. Provided $\mathcal{S}^c(\mathbf{1} \otimes B)\mathcal{D}^c$ is a densely defined (usually unbounded) operator on $\mathcal{M}$ we get for each Ψ from the domain of this operator and all $\varphi \in M$

$$(\mathcal{S}^c(\mathbf{1} \otimes B)\mathcal{D}^c\Psi)(\varphi) = \sum_{\hat{\varphi}\subseteq\varphi} (B(\mathcal{D}^c\Psi)(\varphi - \hat{\varphi}, \cdot))(\hat{\varphi}).$$

If in addition B is an integral operator with kernel b we may continue

$$\begin{aligned}(\mathcal{S}^c(\mathbf{1} \otimes B)\mathcal{D}^c\Psi)(\varphi) &= \sum_{\hat{\varphi}\subseteq\varphi} \int F(d\tilde{\varphi}) b(\hat{\varphi}, \tilde{\varphi})\mathcal{D}^c\Psi(\varphi - \hat{\varphi}, \tilde{\varphi}) \\ &= \int F(d\tilde{\varphi}) \sum_{\hat{\varphi}\subseteq\varphi} b(\hat{\varphi}, \tilde{\varphi})\Psi(\varphi - \hat{\varphi} + \tilde{\varphi}). \qquad (4)\end{aligned}$$

Operators of this type represent the basic tool of Maassen's kernel calculus on the Fock space over $\mathbf{R}$. In [11] the function b is called the *integral kernel* of the operator $\mathcal{S}^c(\mathbf{1} \otimes B)\mathcal{D}^c$ and in [11, 10] there are discussed quantum stochastic integrals with respect to families of such integral kernels and applications to quantum stochastic differential equations and the characterization of quantum noises.

Starting with a bounded operator $B \in \mathcal{L}(\mathcal{M})$ and $Y \in \mathcal{F}$ the operator $\mathcal{S}^c(\mathcal{O}_Y \otimes B)\mathcal{D}^c$ usually will be unbounded. There even exist bounded operators B and sets Y such that $\mathcal{S}^c(\mathcal{O}_Y \otimes B)\mathcal{D}^c$ does not lead to a linear operator from $\mathcal{M}$ to $\mathcal{M}$ for any dense domain in $\mathcal{M}$. Since this class of operators play a crucial rôle in this paper we will give now some conditions ensuring $\mathcal{S}^c(\mathcal{O}_Y \otimes B)\mathcal{D}^c$ to be bounded resp. densely defined.

Proposition 2.7

(i) Let $\Lambda, \Lambda' \in \mathcal{G}, B \in \mathcal{L}(\mathcal{M}_\Lambda), Y \in \mathcal{F}_{\Lambda'}, \Lambda \cap \Lambda' = \emptyset$. *Then*

$$X := \mathcal{S}^c(\mathcal{O}_Y \otimes B)\mathcal{D}^c \in \mathcal{L}(\mathcal{M}_{\Lambda \cup \Lambda'}) \subseteq \mathcal{L}(\mathcal{M}),$$

and one has $\|X\| \leq \|B\|$.

(ii) For all $Y_1, Y_2 \in \mathcal{F}$ *the operator* $\mathcal{S}^c(\mathcal{O}_{Y_1} \otimes \mathcal{O}_{Y_2})\mathcal{D}^c$ *is a densely defined operator.*

The proof of Proposition 2.7 (except the assertion $\|X\| \leq \|B\|$) one can find in [2, section 3]. We will show this inequality at the end of this section.

Let us remark that each local operator in a quasilocal algebra is an operator of this type. To make this more precise, consider the quasilocal algebra $\mathcal{A} = \overline{\bigcup_{\Lambda \in \mathcal{B}} \mathcal{A}_\Lambda}$ where $\mathcal{A}_\Lambda = \mathcal{L}(\mathcal{M}_\Lambda) \otimes \mathcal{O}_{M_{\Lambda^c}}$. Locally normal states ω over the quasilocal algebra are determined by the expectations $\omega(B)$ for all local operators B from some $\mathcal{A}_\Lambda$. But these operators can be written in the form $\mathcal{S}^c(\mathcal{O}_{M_{\Lambda^c}} \otimes \hat{B})\mathcal{D}^c$ with $\hat{B} \in \mathcal{L}(\mathcal{M}_\Lambda)$. Denote by M^m the set of configurations having at most m points, i.e. $M^m := \{\varphi \in M : \varphi(G) \leq m\}$. Analogously, we set $\mathcal{M}^m = \{\Psi \in \mathcal{M} : \Psi = \mathcal{O}_{M^m}\Psi\}$. We denote by $\mathcal{A}^f$ the operators from $\mathcal{L}(\mathcal{M})$ which are concentrated on some $\mathcal{M}^m$, i.e.

$$\mathcal{A}^f := \{B \in \mathcal{L}(\mathcal{M}) : \text{There exists an } m \in \mathbf{N} \text{ such that } B = \mathcal{O}_{M^m} B \mathcal{O}_{M^m}\}.$$

Proposition 2.8 *Let* $B \in \mathcal{A}^f$. *Then for each* $Y \in \mathcal{F}$ *we have that* $\mathcal{S}^c(\mathcal{O} \otimes B)\mathcal{D}^c$ *is a densely defined operator on* $\mathcal{M}$. *The set* $\mathcal{M}^f := \bigcup_{m \in \mathbf{N}} \mathcal{M}^m$ *is contained in the maximal domain of definition of this operator.*

Observe that $\mathcal{M}^f$ is already dense in $\mathcal{M}$. Since in the proof we need some lemmas given below we postpone the proof of Proposition 2.8 to the end of this section. We still express the number operator, creation and annihilation operators in the form $\mathcal{S}^c(\mathcal{O} \otimes B)\mathcal{D}^c$. Setting $M_n := \{\varphi \in M : \varphi(G) = n\}$, $M_1 = G$ we get

Proposition 2.9 [2, 3]

(i) Let $\mathcal{N} : \mathcal{M} \longrightarrow \mathcal{M}$ be defined by $\mathcal{S}^c(\mathbf{1} \otimes \mathcal{O}_{M_1})\mathcal{D}^c$ with

$$Dom(\mathcal{N}) := \{\Psi \in \mathcal{M} : \int \varphi^2(G)|\Psi(\varphi)|^2 F(d\varphi) < \infty\}.$$

$\mathcal{N}$ is the number operator on $\mathcal{M}$, i.e.

$$\mathcal{N}\Psi(\varphi) = \varphi(G)\Psi(\varphi).$$

(ii) Let $f \in L_2(G,\nu)$ and $A_f^* \in \mathcal{L}(\mathcal{M})$ be the operator

$$A_f^*\Psi(\varphi) := \chi_{M_1}(\varphi) f(\varphi)\Psi(\varphi).$$

Then $a_f^* : \mathcal{M} \longrightarrow \mathcal{M}$ given by

$$a_f^* := \mathcal{S}^c(\mathbf{1} \otimes A_f^*)\mathcal{D}^c$$

with $Dom(a_f^*) = Dom(\mathcal{D}) = Dom(\mathcal{N}^{1/2})$ is the creation operator corresponding to f, i.e.

$$a_f^*\Psi(\varphi) = \int f(x)\Psi(\varphi - \delta_x)\varphi(dx).$$

(iii) Let $f \in L_2(G,\nu)$ and $A_f \in \mathcal{L}(\mathcal{M})$ be the operator

$$A_f\Psi(\varphi) := \chi_{\{\theta\}}(\varphi) \int \overline{f(x)}\Psi(\delta_x)\nu(dx).$$

Then $a_f : \mathcal{M} \longrightarrow \mathcal{M}$ given by

$$a_f := \mathcal{S}^c(\mathbf{1} \otimes A_f)\mathcal{D}^c$$

with $Dom(a_f) = Dom(\mathcal{D}) = Dom(\mathcal{N}^{1/2})$ is the annihilation operator corresponding to f, i.e.

$$a_f\Psi(\varphi) = \int \overline{f(x)}\Psi(\varphi + \delta_x)\nu(dx).$$

From Proposition 2.8 we already know that a_f and a_f^* are densely defined because A_f and A_f^* belong to $\mathcal{A}^f$. For generalizations of these operators cf. [2, 3]. Just observe that the above description leads immediately to the well-known relations $\mathcal{N} = \mathcal{S}\,\mathcal{D}$, $a_f^*\Psi = \mathcal{S}(\Psi \otimes f)$ and $a_f\Psi = \langle f, \mathcal{D}\Psi(\cdot,\varphi)\rangle_{L_2(G,\nu)}$.

2.4 Some Useful Lemmas

At the end of this introductory section we collect some lemmas on integration in the Fock space which will be used very often in the sequel. For $\Lambda \in \mathcal{G}$ and $\varphi \in M$ denote by φ_Λ the restriction of φ from M to M_Λ, i.e. $\varphi_\Lambda = \varphi(\cdot \cap \Lambda)$.

Lemma 2.10 *Let $\Lambda, \Lambda' \in \mathcal{G}$, $\Lambda \cap \Lambda' = \emptyset$ and let $h : M \times M \longrightarrow \mathbf{C}$ be a measurable function. Then*

$$\int\int h(\varphi_1, \varphi_2) F_\Lambda(\varphi_1) F_{\Lambda'}(d\varphi_2) = \int h(\varphi_\Lambda, \varphi_{\Lambda'}) F_{\Lambda \cup \Lambda'}(d\varphi)$$

(provided the integrals exist).

Lemma 2.11 *Let $\Lambda \in \mathcal{G}$ and let $h : M \times M \longrightarrow \mathbf{C}$ be a measurable function. Then*

$$\int\int h(\varphi_1, \varphi_2) F_\Lambda(\varphi_1) F_\Lambda(d\varphi_2) = \int \sum_{\hat{\varphi} \subseteq \varphi} h(\hat{\varphi}, \varphi - \hat{\varphi}) F_\Lambda(d\varphi)$$

(provided the integrals exist).

Immediately from Lemma 2.11 one obtains

Lemma 2.12 *Let $\Lambda \in \mathcal{G}$ and let $h : G \times M \longrightarrow \mathbf{C}$ be a measurable function. Then*

$$\int_M \int_\Lambda h(x, \varphi) \nu(dx) F_\Lambda(d\varphi) = \int_M \int_\Lambda h(x, \varphi - \delta_x) \varphi(dx) F_\Lambda(d\varphi)$$

(provided the integrals exist).

Lemma 2.13 *Let $\Lambda \in \mathcal{G}$ and let $h : M \longrightarrow \mathbf{C}$ be a measurable function. Then*

$$\int\int h(\varphi_1 + \varphi_2) F_\Lambda(d\varphi_1) F_\Lambda(d\varphi_2) = \int 2^{\varphi(\Lambda)} h(\varphi) F_\Lambda(d\varphi)$$

(provided the integrals exist).

The proof of these lemmas can be found in [1] and [2, sections 11 and 12]. Lemma 2.11 is also contained in [10].

Proof of Proposition 2.7.

As mentioned above because of the results in [2, section 3] we only have to prove the relation $\|X\| \leq \|B\|$. Observe that the assumptions $B = \mathcal{O}_{M_\Lambda} B \mathcal{O}_{M_\Lambda}$ and $Y \subseteq \mathcal{F}_{\Lambda^c}$ imply that for all $\Psi \in \mathcal{M}$ and $\varphi \in M$

$$X\Psi(\varphi) = \chi_Y(\varphi_{\Lambda^c})\big(B\mathcal{D}^c\Psi(\varphi_{\Lambda^c},\cdot)\big)(\varphi_\Lambda)$$

Thus we get using Lemma 2.10

$$\begin{aligned}
\|X\|^2 &= \int_M \chi_Y(\varphi_{\Lambda^c})|\big(B\mathcal{D}^c\Psi(\varphi_{\Lambda^c},\cdot)\big)(\varphi_\Lambda)|^2 F(d\varphi) \\
&= \int_{M_{\Lambda^c}} \chi_Y(\varphi_1) \int_{M_\Lambda} |\big(B\mathcal{D}^c\Psi(\varphi_1),\cdot)\big)(\varphi_2)|^2 F(d\varphi_2)F(d\varphi_1) \\
&\leq \|B\|^2 \int_{M_{\Lambda^c}} \int_{M_\Lambda} \chi_Y(\varphi_1)|\Psi(\varphi_1+\varphi_2)|^2 F(d\varphi_2)F(d\varphi_1) \\
&\leq \|B\|^2 \int_{M_{\Lambda^c}} \int_{M_\Lambda} |\Psi(\varphi_1+\varphi_2)|^2 F(d\varphi_2)F(d\varphi_1) \\
&\leq \|B\|^2 \int_M |\Psi(\varphi)|^2 F(d\varphi) \;=\; \|B\|^2\|\Psi|^2.
\end{aligned}$$

Consequently, we get $\|X\| \leq \|B|$. □

Proof of Proposition 2.8.

Let Ψ be from $\mathcal{M}^f$, $B \in \mathcal{A}^f$. There exists an $m \in \mathbf{N}$ such that $\Psi \in \mathcal{M}^m$ and $\mathcal{O}_{M^m} B \mathcal{O}_{M^m} = B$. Using several times Schwarz' inequality, Lemma 2.11 and Lemma 2.13 we get

$$\begin{aligned}
&\|\mathcal{S}^c(\mathcal{O}_Y \otimes B)\mathcal{D}^c\psi\|^2 \\
&= \int \sum_{\varphi_1 \subseteq \varphi} \chi_Y(\varphi-\varphi_1)\overline{(B\mathcal{D}^c\Psi(\varphi-\varphi_1,\cdot))(\varphi_1)} \\
&\times \sum_{\varphi_2 \subseteq \varphi} \chi_Y(\varphi-\varphi_2)(B\mathcal{D}^c\Psi(\varphi-\varphi_2,\cdot))(\varphi_2)F(d\varphi) \\
&= \int \sum_{\varphi_1 \subseteq \varphi} \sum_{\varphi_2 \subseteq \varphi-\varphi_1} \sum_{\varphi_3 \subseteq \varphi-\varphi_1-\varphi_2} \chi_Y(\varphi-\varphi_1-\varphi_2)\chi_Y(\varphi-\varphi_1-\varphi_3) \\
&\times \overline{(B\mathcal{D}^c\Psi(\varphi-\varphi_1-\varphi_2,\cdot))(\varphi_1+\varphi_2)} \\
&\times (B\mathcal{D}^c\Psi(\varphi-\varphi_1-\varphi_3,\cdot))(\varphi_1+\varphi_3)F(d\varphi)
\end{aligned}$$

$$
\begin{aligned}
&= \iiiint \chi_Y(\varphi+\varphi_3)\chi_Y(\varphi+\varphi_2)\overline{(B\mathcal{D}^c\Psi(\varphi+\varphi_3,\cdot))(\varphi_1+\varphi_2)} \\
&\times \ (B\mathcal{D}^c\Psi(\varphi+\varphi_2,\cdot))(\varphi_1+\varphi_3)F^4(d[\varphi,\varphi_1,\varphi_2,\varphi_3]) \\
&\leq \iiiint \chi_Y(\varphi+\varphi_3)|(B\mathcal{D}^c\Psi(\varphi+\varphi_3,\cdot))(\varphi_1+\varphi_2)|^2F^4(d[\varphi,\varphi_1,\varphi_2,\varphi_3]) \\
&\leq \iiiint |(B\mathcal{D}^c\Psi(\varphi+\varphi_3,\cdot))(\varphi_1+\varphi_2)|^2F^4(d[\varphi,\varphi_1,\varphi_2,\varphi_3]) \\
&= \iint |(B\mathcal{D}^c\Psi(\varphi_1,\cdot))(\varphi_2)|^2 2^{\varphi_1(G)+\varphi_2(G)}F^2(d[\varphi_1,\varphi_2]) \\
&\leq 2^{2m}\iint |(B\mathcal{D}^c\Psi(\varphi_1,\cdot))(\varphi_2)|^2F^2(d[\varphi_1,\varphi_2]) \\
&\leq 2^{2m}\|B\|^2\iint |\Psi(\varphi_1+\varphi_2)|^2F^2(d[\varphi_1,\varphi_2]) \\
&= 2^{2m}\|B\|^2\int |\Psi(\varphi)|^2 2^{\varphi(G)}F(d\varphi) \ \leq \ 2^{3m}\|B\|^2\|\Psi\|^2 \ < \ \infty.
\end{aligned}
$$

Hereby we used the fact that for $\Psi \in \mathcal{M}^m$ we have $\mathcal{D}^c\Psi(\varphi,\cdot) \equiv 0$ if $\varphi \notin M^m$. □

3 Stochastic Integrals on the Fock Space

3.1 Definitions

For a family $(X_t)_{t\in G}$ of operators on $\mathcal{M}$ and $\Lambda \in \mathcal{G}$ we want to define stochastic integrals $\int_\Lambda X\,dA$ and $\int_\Lambda dA^*X$ that can be interpreted as integrals with respect to annihilation and creation operators. First we will give quite a formal definition assuming the existence of such integrals and by specialization to concrete cases we will motivate the definition.

Definition 3.1 *Let $(X_t)_{t\in G}$ be a family of linear densely defined operators on $\mathcal{M}$. $(X_t)_{t\in G}$ is called measurable if*

(i) $D := \bigcap_{t\in G} Dom(X_t)$ is dense in $\mathcal{M}$ and

(ii) for all $\Psi \in D$and $\Phi \in \mathcal{M}$ the mapping $t \longmapsto \langle\Phi, X_t\Psi\rangle$ is a measurable mapping from G into $\mathbf{C}$.

Remark. It is easy to check that condition *(ii)* ensures that there exist measurable versions of the integrands in the integrals defined below. Especially, for all $\Psi \in D$ and $\varphi \in M$ the mappings $t \longmapsto (X_t\mathcal{D}\Psi(t,\cdot))(\varphi)$ and

$t \longmapsto X_t\Psi(\varphi - \delta_t)$ may be assumed to be measurable mappings from G to $\mathbf{C}$.

Definition 3.2 *Let $X = (X_t)_{t\in G}$ be a measurable family of operators defined on a dense subset D of $\mathcal{M}$ such that for a set $\Lambda \in \mathcal{G}$ and for all $\Psi \in D$*

$$\int_\Lambda (X_t\mathcal{D}\Psi(t,\cdot))(\varphi)\nu(dt) \qquad\qquad (\varphi \in M)$$

defines an element of $\mathcal{M}$. The operator $\int_\Lambda X dA$ from $\mathcal{M}$ into $\mathcal{M}$ defined on D by

$$\Big(\int_\Lambda X dA\, \Psi\Big)(\varphi) := \int_\Lambda (X_t\mathcal{D}\Psi(t,\cdot))(\varphi)\nu(dt) \tag{5}$$

is called quantum stochastic integral of X with respect to annihilation.

Definition 3.3 *Let $X = (X_t)_{t\in G}$ be a measurable family of operators defined on a dense subset $\tilde{D}$ of $\mathcal{M}$ such that for a set $\Lambda \in \mathcal{G}$ and for all $\Psi \in \tilde{D}$*

$$\int_\Lambda X_t\Psi(\varphi - \delta_t)\varphi(dt) \qquad\qquad (\varphi \in M)$$

defines an element of $\mathcal{M}$. The operator $\int_\Lambda dA^* X$ from $\mathcal{M}$ into $\mathcal{M}$ defined on $\tilde{D}$ by

$$\Big(\int_\Lambda dA^* X\, \Psi\Big)(\varphi) := \int_\Lambda X_t\Psi(\varphi - \delta_t)\varphi(dt) \tag{6}$$

is called quantum stochastic integral of X with respect to creation.

The integrals in Definitions 3.2 and 3.3 are correctly defined. This can be seen by considering first step functions.

Proposition 3.4 *Let $\Lambda \in \mathcal{G}$, X a measurable family of operators on $\mathcal{M}$ and suppose $\int_\Lambda X dA$ and $\int_\Lambda dA^* X$ are operators defined on dense domains D rep. $\tilde{D}$. Then $\int_\Lambda X dA$ and $\int_\Lambda dA^* X$ are mutually disjoint, i.e. for all $\Psi \in D$ and $\Phi \in \tilde{D}$ we have*

$$\langle \int_\Lambda X dA\, \Psi, \Phi\rangle = \langle \Psi, \int_\Lambda dA^* X\, \Phi\rangle.$$

Proof.
Using Lemma 2.12 we get

$$\begin{aligned}
\langle \int_\Lambda X dA\, \Psi, \Phi \rangle &= \int_M \overline{(\int_\Lambda X dA\, \Psi)(\varphi)}\Phi(\varphi)\, F(d\varphi) \\
&= \int_M \int_\Lambda \overline{(X_t \mathcal{D}\Psi(t,\cdot))(\varphi)}\Phi(\varphi)\, \nu(dt)\, F(d\varphi) \\
&= \int_\Lambda \int_M \overline{(X_t \mathcal{D}\Psi(t,\cdot))(\varphi)}\Phi(\varphi)\, F(d\varphi)\, \nu(dt) \\
&= \int_\Lambda \langle X_t \mathcal{D}\Psi(t,\cdot), \Phi \rangle \nu(dt) \;=\; \int_\Lambda \langle \mathcal{D}\Psi(t,\cdot), X_t^* \Phi \rangle \nu(dt) \\
&= \int_\Lambda \int_M \overline{\Psi(\varphi + \delta_t)} X_t^* \Phi(\varphi) F(d\varphi) \nu(dt) \\
&= \int_M \int_\Lambda \overline{\Psi(\varphi)} X_t^* \Phi(\varphi - \delta_t) \varphi(dt) F(d\varphi) \\
&= \int_M \overline{\Psi(\varphi)} \int_\Lambda X_t^* \Phi(\varphi - \delta_t) \varphi(dt) F(d\varphi) \;=\; \langle \Psi, \int_\Lambda dA^* X, \Phi \rangle.
\end{aligned}$$

□

3.2 Domains of Definition

Suppose that for an $\Psi \in \mathcal{M}$ and $\Lambda \in \mathcal{G}$ we have that $\int_\Lambda X dA\, \Psi \in \mathcal{M}$. Then using several times Schwarz' inequality an easy calculation shows that

$$\| \int_\Lambda X dA\, \Psi \|^2 \leq \nu(\Lambda) \int_M \int_\Lambda |X_t \Psi(\varphi + \delta_t)|^2 \nu(dt)\, F(d\varphi). \tag{7}$$

Analogously, if for $\Psi \in \mathcal{M}$ and $\Lambda \in \mathcal{G}$ $\int_\Lambda dA^* X\, \Psi$ is square integrable one gets using Lemma 2.12

$$\| \int_\Lambda dA^* X\, \Psi \|^2 \leq \int_M \int_\Lambda (\varphi(\Lambda) + 1) |X_t \Psi(\varphi)|^2 \nu(dt)\, F(d\varphi). \tag{8}$$

Consequently, $\int_\Lambda X dA$ is a well-defined operator if $\Lambda \in \mathcal{B}$ and

$$Dom(\int_\Lambda X dA) \subseteq \{\Psi \in \mathcal{M} : \int_M \int_\Lambda |X_t \Psi(\varphi + \delta_t)|^2 \nu(dt)\, F(d\varphi) < \infty\}$$

is dense in $\mathcal{M}$. Analogously, one defines a maximal domain of definition for the integral $\int_\Lambda dA^* X$ by the right-hand side of (8).
If the operators X_t are bounded one can give more explicit domains of definition.

Proposition 3.5 *Let $X = (X_t)_{t\in G}$ be a measurable family of bounded operators on $\mathcal{M}$. Then for all $\Lambda \in \mathcal{G}$ such that $\nu(\Lambda) < \infty$ and $\sup_{t\in\Lambda} \|X_t\| < \infty$ the operator $\int_\Lambda X dA$ is a densely defined linear operator whose maximal domain of definition contains the set*

$$E := \{\Psi \in \mathcal{M} : \int_M \varphi(\Lambda)|\Psi(\varphi)|^2 F(d\varphi) < \infty\}. \tag{9}$$

Observe that E is just the maximal domain of definition of the operator $\mathcal{D}_\Lambda$ being the restriction of $\mathcal{D}$ to Λ, i.e. $\mathcal{D}_\Lambda\Psi(x,\varphi) = \chi_\Lambda(x)\Psi(\varphi+\delta_x)$. So one has the relation $Dom(\mathcal{D}) \subseteq Dom(\mathcal{D}_\Lambda) = E$. If $\mathcal{N}_\Lambda$ denotes the number operator on $\mathcal{M}_\Lambda$ we also have $E = Dom(\mathcal{N}_\Lambda^{1/2})$.

Proof.
Let $\Psi \in E$. From the estimation (7) one gets using Lemma 2.12

$$\begin{aligned}
\|\int_\Lambda X dA\, \Psi\|^2 &\le \nu(\Lambda)\int_M\int_\Lambda |X_t\Psi(\varphi+\delta_t)|^2 \nu(dt)\, F(d\varphi) \\
&= \nu(\Lambda)\int_\Lambda \|X_t\mathcal{D}\Psi(t,\cdot)\|^2\nu(dt) \le \nu(\Lambda)\int_\Lambda \|X_t\|^2\|\mathcal{D}\Psi(t,\cdot)\|^2\nu(dt) \\
&\le \nu(\Lambda)\sup_{t\in\Lambda}\|X_t\|^2 \int_\Lambda\int_M |\Psi(\varphi+\delta_t)|^2 F(d\varphi)\nu(dt) \\
&\le \nu(\Lambda)\sup_{t\in\Lambda}\|X_t\|^2 \int_M \varphi(\Lambda)|\Psi(\varphi)|^2 F(d\varphi)
\end{aligned}$$

□

Proposition 3.6 *Let $X = (X_t)_{t\in G}$ be a measurable family of bounded operators on $\mathcal{M}$. Then for all $\Lambda \in \mathcal{G}$ such that $\nu(\Lambda) < \infty$ and $\sup_{t\in\Lambda} \|X_t\| < \infty$ the operator $\int_\Lambda dA^*X$ is a densely defined linear operator the maximal domain of which contains the set E.*

Thus we have that in the case $\nu(\Lambda) < \infty$ and $\sup_{t\in\Lambda}\|X_t\| < \infty$ the operators $\int_\Lambda dA^*X$ and $\int_\Lambda XdA$ could be defined on the same dense domain E given by (9).

Proof.
For $\Psi \in E$ we get using the estimation (8)

$$\begin{aligned}
\|\int_\Lambda dA^*X\, \Psi\|^2 &\le \int_M\int_\Lambda (\varphi(\Lambda)+1)|X_t\Psi(\varphi)|^2\nu(dt)\, F(d\varphi) \\
&\le \|\Psi\|^2\int_\Lambda \|X_t\|^2\nu(dt) + \int_\Lambda\int_M \varphi(\Lambda)|X_t\Psi(\varphi)|^2\, F(d\varphi)\nu(dt) \qquad (10)
\end{aligned}$$

The second term on the right-hand side of (10) can be estimated by

$$\begin{aligned}
&\leq \int_\Lambda \int_M \int_\Lambda |X_t\Psi(\varphi)|^2 \,\varphi(ds)F(d\varphi)\nu(dt) \\
&= \int_\Lambda \int_M \int_\Lambda |X_t\Psi(\varphi+\delta_s)|^2 \,\nu(ds)F(d\varphi)\nu(dt) \\
&\leq \int_\Lambda \int_\Lambda \|X_t\mathcal{D}\Psi(s,\cdot)\|^2 \,\nu(ds)\nu(dt) \leq \nu(\Lambda)\sup_{t\in\Lambda}\|X_t\|^2 \int_\Lambda \|\mathcal{D}\Psi(s,\cdot)\|^2\nu(ds) \\
&\leq \nu(\Lambda)\sup_{t\in\Lambda}\|X_t\|^2 \int_M \varphi(\Lambda)|\Psi(\varphi)|^2 F(d\varphi) \qquad (11)
\end{aligned}$$

Combining (10) and (11) we finally get the estimation

$$\|\int_\Lambda dA^*X\ \Psi\|^2 \leq \nu(\Lambda)\sup_{t\in\Lambda}\|X_t\|^2\Big(\|\Psi\|^2 + \int_M \varphi(\Lambda)|\Psi(\varphi)|^2F(d\varphi)\Big).$$

□

Now, consider the local algebras $\mathcal{A}_\Lambda = \mathcal{L}(\mathcal{M}_\Lambda)\otimes \mathbf{1}_{\Lambda^c}$ introduced in section 1. Since each operator from $\mathcal{A}_\Lambda$ is of the form $\mathcal{S}^c(\mathbf{1}_{\Lambda^c}\otimes B)\mathcal{D}^c$ we have that $\mathcal{A}_\Lambda \subseteq \tilde{\mathcal{A}}_\Lambda$ where

$$\tilde{\mathcal{A}}_\Lambda := \{X \in \mathcal{L}(\mathcal{M}) : X = \mathcal{S}^c(\mathcal{O}_Y \otimes B)\mathcal{D}^c \text{ with } Y \in \mathcal{F}_{\Lambda^c}, B \in \mathcal{L}(\mathcal{M}_\Lambda)\}.$$

We have the following proposition:

Theorem 3.7 *Let $\Lambda \in \mathcal{B}$ and let $(X_t)_{t\in G}$ be a measurable family such that $X_t = \mathcal{S}^c(\mathcal{O}_{Y_t}\otimes B_t)\mathcal{D}^c \in \tilde{\mathcal{A}}_\Lambda$ for all $t \in \Lambda$ and $\sup_{t\in\Lambda}\|B_t\| < \infty$. Then the operators $\int_\Lambda dA^*X$ and $\int_\Lambda X\,dA$ are densely defined linear operators the maximal domain of which contain the set E.*

The proof of Theorem 3.7 is an immediate consequence of Propositions 2.7, 3.5 and 3.6.

Observe that all of the above integrals could also be defined on the same (dense) domain $Dom(\mathcal{D})$ of the Malliavin derivative because $Dom(\mathcal{D}) \subseteq E$.

4 Integrals for Non-Anticipating Operators

In the sequel let x_0 be an arbitrarily chosen fixed point from G. By L_x we denote the ball with center x_0 and radius $\rho(x,x_0)$, i.e. $L_x := \{t \in G : \rho(t,x_0) < \rho(x,x_0)\}$. Since the measure ν was assumed to be diffuse we have that F – *a.a.* configurations $\varphi \in M$ consist only of points with different distances from x_0, i.e.

$$F(\{\varphi \in M : \textit{There exist } t \in \varphi, s \in \varphi \textit{ such that } \rho(t,x_0) = \rho(s,x_0)\}) = 0.$$

Hereby $t \in \varphi$ again means that $\varphi(\{x\}) \geq 1$. By $\max_{L_x} \varphi$ or briefly $\max_x \varphi$ we denote the mass-point of φ_{L_x} with maximal distance from x_0, i.e. for all $\varphi \in M$ and $x \in G$ we have

$$(y = \max_x \varphi) \textit{ if and only if } (y \in \varphi_{L_x} \textit{ and } \rho(y,x_0) = \max_{t \in \varphi_{L_x}} \rho(t,x_0)).$$

Now we want to examine integrals for families of operators as considered first time by Maassen in [11].

Definition 4.1 *Let $X = (X_t)_{t \in G}$ be a measurable family of operators on $\mathcal{M}$. X is called non-anticipating with respect to x_0 if there exists a measurable family of operators $(B_t)_{t \in G}$ with $B_t \in \mathcal{L}(\mathcal{M}_{L_t})$ such that $X_t = \mathcal{S}^c(\mathbf{1} \otimes B_t)\mathcal{D}^c$.*

Remark. If $G = [0,\infty)$, $x_0 = 0$, $L_t = [0,t)$ and for each $t \in G$ the operator B_t has a kernel b_t this definition coincides with Maassen's definition of *non-anticipating kernel processes.*
Observe that even for bounded operators B_t the operator X_t may be unbounded or not exist at all as a linear densely defined operator. So one should require at least that all B_t are bounded.

Proposition 4.2 *Let $X = (X_t)_{t \in G}$ be non-anticipating with respect to x_0, and let x be from G.*

(i) For all $\Psi \in \mathcal{M}$ such that $\int_{L_x} dA^* X \Psi$ given by (3.3) defines an element of $\mathcal{M}$ we have

$$\int_{L_x} dA^* X \Psi = \mathcal{S}^c(\mathbf{1} \otimes B^x)\mathcal{D}^c \Psi \qquad (12)$$

with

$$\begin{aligned} B^x \Psi(\varphi) &= \int_{L_x} dA^* B \Psi(\varphi) \\ &= \chi_{M_{L_x} \setminus \{\theta\}}(\varphi) B_{\max_x \varphi} \Psi(\varphi - \delta_{\max_x \varphi}) \quad (\varphi \in M) \end{aligned} \qquad (13)$$

(ii) If $\sup_{t \in L_x} \|B_t\| < \infty$ then B^x is bounded and belongs to $\mathcal{L}(\mathcal{M}_{L_x})$.

(iii) If $B^x \in \mathcal{A}^f$ then $\int_{L_x} dA^* X$ is a densely defined linear operator.

(iv) If for all $t \in \Lambda$ the operator B_t has a kernel b_t then B^x has a kernel b^x fulfilling

$$b^x(\varphi_1, \varphi_2) = b_{\max_x \varphi_1}(\varphi_1 - \delta_{\max_x \varphi_1}, \varphi_2)\chi_{M_{L_x}\setminus\{\theta\}}(\varphi_1) \quad (\varphi_j \in M) \tag{14}$$

Proof.
For all $\Psi \in \mathcal{M}$ such that $\int_{L_x} dA^* X\Psi$ is an element of $\mathcal{M}$ we have

$$\begin{aligned}
\int_{L_x} dA^* X\Psi(\varphi) &= \int_{L_x} X_t\Psi(\varphi - \delta_t)\varphi(dt) \\
&= \int_{L_x} \sum_{\hat{\varphi} \subseteq \varphi - \delta_t} \big(B_t\mathcal{D}^c\Psi(\varphi - \delta_t - \hat{\varphi}, \cdot)\big)(\hat{\varphi})\varphi(dt) \\
&= \int_{L_x} \sum_{\hat{\varphi} \subseteq \varphi - \delta_t} \mathcal{O}_{M_{L_t}}(\hat{\varphi})\big(B_t\mathcal{D}^c\Psi(\varphi - \delta_t - \hat{\varphi}, \cdot)\big)(\hat{\varphi})\varphi(dt) \\
&= \sum_{\hat{\varphi} \subseteq \varphi} \chi_{M_{L_x}\setminus\{\theta\}}(\hat{\varphi})B_{\max_x \hat{\varphi}}\mathcal{D}^c\Psi(\varphi - \hat{\varphi}, \cdot)(\hat{\varphi} - \delta_{\max_x \hat{\varphi}}) \\
&= \mathcal{S}^c(1 \otimes B^x)\mathcal{D}^c\Psi(\varphi).
\end{aligned}$$

This proves (i). The corresponding representation (14) of the kernel b^x in the case (iv) is immediately checked.
Now assume $\sup_{t\in L_x} \|B_t\| < \infty$. We will prove that in this case the operator B^x is bounded. Using Lemma 2.12 we get for all $\Psi \in \mathcal{M}$

$$\begin{aligned}
&\|B^x\Psi\|^2 \\
&= \int_{M_{L_x}\setminus\{\theta\}} |B_{\max_x \varphi}\Psi(\varphi - \delta_{\max_x \varphi})|^2 F(d\varphi) \\
&= \int_{M_{L_x}} \int_{L_x} |B_t\Psi(\varphi - \delta_t)|^2 \varphi(dt)F(d\varphi) \\
&= \int_{M_{L_x}} \int_{L_x} |B_t\Psi(\varphi)|^2 \nu(dt)F(d\varphi) \\
&\le \int_{L_x} \|B_t\|^2\nu(dt) \int_{M_{L_x}} |\Psi(\varphi)|^2 F(d\varphi) \\
&\le \nu(L_x) \sup_{t\in L_x} \|B_t\|^2\|\Psi\|^2.
\end{aligned}$$

The assertion (iii) is an immediate consequence of Proposition 2.8. □

Proposition 4.3 *Let $X = (X_t)_{t \in G}$ be non-anticipating with respect to x_0, and let x be from G.*

(i) For all $\Psi \in \mathcal{M}$ such that $\int_{L_x} X\, dA\, \Psi$ given by (3.2) defines an element of $\mathcal{M}$ we have

$$\int_{L_x} X\, dA\, \Psi = \mathcal{S}^c(\mathbf{1} \otimes \tilde{B}^x)\mathcal{D}^c\, \Psi \tag{15}$$

with

$$\tilde{B}^x \Psi = \int_{L_x} B\, dA\, \Psi \tag{16}$$

(ii) If $\sup_{t \in L_x} \|B_t\| < \infty$ then $\tilde{B}^x$ is densely defined and its maximal domain contains $Dom(\mathcal{D}_\Lambda)$.

(iii) If $\tilde{B}^x \in \mathcal{A}^f$ then $\int_{L_x} X\, dA$ is a densely defined linear operator.

(iv) If for all $t \in \Lambda$ the operator B_t has a kernel b_t then $\tilde{B}^x$ has a kernel $\tilde{b}^x$ fulfilling

$$\tilde{b}^x(\varphi_1, \varphi_2) = b_{\max_x \varphi_2}(\varphi_1, \varphi_2 - \delta_{\max_x \varphi_2})\chi_{M_{L_x} \setminus \{\theta\}}(\varphi_2) \quad (\varphi_j \in M) \tag{17}$$

Proof.

First observe that for $\Psi \in \mathcal{M}$ and $\varphi_1, \varphi_2 \in M$ we have the relation

$$\Big(\mathcal{D}^c\big(\mathcal{D}\Psi(t,\cdot)\big)\Big)(\varphi_1, \varphi_2) = \Big(\mathcal{D}\big(\mathcal{D}^c\Psi(\varphi_1,\cdot)\big)\Big)(t, \varphi_2) \tag{18}$$

(As an operator on $\mathcal{M}$ $\mathcal{D}^c \circ \mathcal{D}$ obviously is not defined for all Ψ but e.g. on the dense domain $\mathcal{M}^f$). For all Ψ such that $\int_{L_x} X\, dA \Psi \in \mathcal{M}$ we get using equation (18)

$$\begin{aligned}
\Big(\int_{L_x} X\, dA\Psi\Big)(\varphi) &= \int_{L_x} (X_t \mathcal{D}\Psi(t,\cdot))(\varphi)\nu(dt) \\
&= \int_{L_x} \sum_{\hat{\varphi} \subseteq \varphi} \Big(B_t \mathcal{D}^c\big(\mathcal{D}\Psi(t,\cdot)\big)(\varphi - \hat{\varphi}, \cdot)\Big)(\hat{\varphi})\nu(dt) \\
&= \int_{L_x} \sum_{\hat{\varphi} \subseteq \varphi} \Big(B_t \mathcal{D}\big(\mathcal{D}^c\Psi(\varphi - \hat{\varphi}, \cdot)\big)(t,\cdot)\Big)(\hat{\varphi})\nu(dt) \\
&= \sum_{\hat{\varphi} \subseteq \varphi} \int_{L_x} \Big(B_t \mathcal{D}\big(\mathcal{D}^c\Psi(\varphi - \hat{\varphi}, \cdot)\big)(t,\cdot)\Big)(\hat{\varphi})\nu(dt) \\
&= \mathcal{S}^c\Big(\mathbf{1} \otimes \int_{L_x} B\, dA\Big)\mathcal{D}^c\Psi(\varphi)
\end{aligned}$$

The assertion (ii) is a consequence of Proposition 3.5, and (iii) follows from Proposition 2.8. We still have to prove (iv). Assume all operators B_t are integral operators with kernel b_t. Then we get

$$\begin{aligned}
&\int_{L_x} B\,dA\Psi(\varphi_1) \\
&= \int_{L_x} (B_t\mathcal{D}\Psi(t,\cdot))(\varphi_1)\nu(dt) \\
&= \int_{L_x}\int_M b_t(\varphi_1,\varphi_2)\Psi(\varphi_2+\delta_t)\chi_{M_{L_t}}(\varphi_2)F(d\varphi_2)\nu(dt) \\
&= \int_M\int_{L_x} b_t(\varphi_1,\varphi_2-\delta_t)\Psi(\varphi_2)\chi_{M_{L_t}}(\varphi_2-\delta_t)\varphi_2(dt)F(d\varphi_2) \\
&= \int_M\int_{L_x} \chi_{M_{L_t}\setminus\{\theta\}}b_t(\varphi_1,\varphi_2-\delta_t)\Psi(\varphi_2)(\varphi_2)(dt)F(d\varphi_2) \\
&= \int_M \chi_{M_{L_x}\setminus\{\theta\}}b_{\max_x\varphi_2}(\varphi_1,\varphi_2-\delta_{\max_x\varphi_2})\Psi(\varphi_2)F(d\varphi_2)
\end{aligned}$$

This proves the statement concerning the kernel of $\tilde{B}^x$. □

Remark. Quite formally (without taking care about domains of definitions) one could illustrate the relation between both integrals in the following way:
If $X_t = \mathcal{S}^c(\mathbf{1}\otimes B_t)\mathcal{D}^c$ then $X_t^* = \mathcal{S}^c(\mathbf{1}\otimes B_t^*)\mathcal{D}^c$. Consequently, we have $\int_{L_x} dA^*X^* = \mathcal{S}^c(\mathbf{1}\otimes(B^*)^x)\mathcal{D}^c$ what implies because of Proposition 3.4 $\int_{L_x} X\,dA = \mathcal{S}^c\Big(\mathbf{1}\otimes\big((B^*)^x\big)^*\Big)\mathcal{D}^c$. Combined with the relation

$$\langle\Phi,(B^*)^x\Psi\rangle = \langle\int_{L_x} BdA\Phi,\Psi\rangle$$

this gives $\tilde{B}^x = ((B^*)^x)^*$.
From Proposition 4.2 we know that

$$(B^*)^x\Psi(\varphi) = \chi_{M_{L_x}\setminus\{\theta\}}(\varphi)B^*_{\max_x\varphi}\Psi(\varphi-\delta_{\max_x\varphi}) \quad (\varphi\in M).$$

Observe that $\tilde{B}^x = ((B^*)^x)^*$ has not such a nice representation. For integral operators B_t the kernels b^x and $\tilde{b}^x$ are the same as in [11].

Even for operators of the type $X_t = \mathcal{S}^c(\mathbf{1}\otimes B_t)\mathcal{D}^c$ which are non-anticipating the quantum stochastic integrals may not exist resp. these integrals may have very strange domains of definition. On the other hand we saw in section 3 that for local operators from the quasilocal algebra $\mathcal{A}$ natural and weak conditions imply the existence of the stochastic integrals as densely defined operators. The operators B_t above are also local — they all belong to $\mathcal{L}(\mathcal{M}_{L_x})$. So replacing $\mathbf{1}$ by $\mathbf{1}_{(\mathbf{L_x})^c}$ or $\mathcal{O}_{Y_t}$ with $Y_t \in \mathcal{F}_{(L_x)^c}$ the quantum stochastic integrals will be well-behaved densely defined operators (Theorem 3.7). But one could consider also operators of the form $X_t = \mathcal{S}^c(\mathcal{O}_{Y_t} \otimes B_t)\mathcal{D}^c$ with $Y_t \in \mathcal{F}_{(L_t)^c}$ and $B_t \in \mathcal{L}(\mathcal{M}_{L_t})$. This type of operators representing a combination of non-anticipating local measurements B_t with position measurements $\mathcal{O}_{Y_t}$ outside L_t seems to be the appropriate class of operators for quantum stochastic calculus. Let us call families of such operators *weakly non-anticipating*. There is some hope that for weakly non-anticipating integrands one can derive handable Itô formulae. In any case finally we will see that the integrals for such families of operators exist.

Proposition 4.4 *Let* $(X_t)_{t\in G}$ *be a measurable weakly non-anticipating family of operators on* $\mathcal{M}$ *and assume that for an* $x \in G$ *we have* $\sup_{t\in L_x} \|B_t\| < \infty$. *Then* $\int_{L_x} X\,dA$ *and* $\int_{L_x} dA^* X$ *are densely defined operators on* $\mathcal{M}$ *whose maximal domain of definition contains* $Dom(\mathcal{D}_\Lambda)$.

Proof.

From Proposition 2.7 we know that $X_t \in \mathcal{L}(\mathcal{M})$ and $\|X_t\| \leq \|B_t\|$. Consequently, we have $\sup_{t\in L_x} \|X_t\| < \infty$. The assertion follows now directly from Propositions 3.5 and 3.6. □

References

[1] K.-H. Fichtner and W. Freudenberg. Point processes and the position distribution of infinite boson systems. *J. Stat. Phys.*, **47**, 959—978, 1987.

[2] K.-H. Fichtner and W. Freudenberg. Characterization of states of infinite boson systems I. On the construction of states of boson systems. *Commun. Math. Phys.*, **137**, 315—357, 1991.

[3] K.-H. Fichtner and W. Freudenberg. Remarks on stochastic calculus on the Fock space. In L. Accardi, editor, *Quantum Probability and Related Topics*, pages 305 - 323, Singapore, New Jersey, London, Hong Kong, 1991. World Scientific Publishing Co.

[4] K.-H. Fichtner and G. Winkler. Generalized Brownian motion, point processes and stochastic calculus for random fields. *Preprint*, 1990.

[5] W. Freudenberg. Characterization of states of infinite boson systems II. On the existence of the conditional reduced density matrix. *Commun. Math. Phys.*, **137**, 461—472, 1991.

[6] A. Guichardet. *Symmetric Hilbert Spaces and Related Topics*, volume 231 of *Lecture Notes in Mathematics*. Springer-Verlag, Berlin, Heidelberg, New York, 1972.

[7] R.L. Hudson and K.R. Parthasarathy. Quantum Itô's formula and stochastic evolution. *Commun. Math. Phys.*, **93**, 301 - 323, 1984.

[8] K. Itô. Multiple Wiener integrals. *J. Math. Soc. Japan*, **3**, 157 - 169, 1951.

[9] J.M. Lindsay. On generalised quantum stochastic integrals. *Preprint*, 1991.

[10] M. Lindsay and H. Maassen. An integral kernal approach to noise. In L. Accardi and W. von Waldenfels, editors, *Quantum Probability and Applications III*, volume 1303 of *Lecture Notes in Mathematics*, pages 192 -208, Heidelberg, 1988. Springer-Verlag.

[11] H. Maassen. Quantum Markov processes on Fock space described by integral kernels. In L. Accardi and W. von Waldenfels, editors, *Quantum Probability and Applications II*, volume 1136 of *Lecture Notes in Mathematics*, pages 361 - 374, Berlin, Heidelberg, New York, 1985. Springer-Verlag.

[12] D. Nualart and E. Pardoux. Stochastic calculus with anticipating integrands. *Prob. Theory Rel. Fields*, **78**, 535 -581, 1988.

[13] D. Nualart and M. Zakai. Generalized stochastic integrals and the Malliavin calculus. *Prob. Theory Rel. Fields*, **73**, 255 - 280, 1986.

[14] A.V. Skorohod. On a generalization of a stochastic integral. *Theory Probab. Appl.*, **20**, 219 - 233, 1975.

[15] D.W. Stroock. The Malliavin calculus. A functional analytic approach. *J. Funct. Anal.*, **44**, 212 - 257, 1980.

[16] M. Zakai. The Malliavin calculus. *Acta Applicandae Mat.*, **3**, 175 – 207, 1985.

[17] M. Zakai. Malliavin derivatives and derivatives of functionals of the Wiener process with respect to a scale parameter. *Ann. Probab.*, **13**, 609 – 615, 1985.

Wolfgang Freudenberg
Friedrich–Schiller–Universität
Mathematische Fakultät
UHH, 17. OG
D-O–6900 Jena
Germany

Quantum Probability and Related Topics Vol. VIII (pp. 211–222)

THE STATISTICAL INVARIANTS IN QUANTUM PROBABILITY AND DE FINETTI'S COHERENCE PRINCIPLE

Angelo Gilio, Universita' "La Sapienza", Roma.

ABSTRACT. The concept of *statistical invariant*, introduced in the recent years in the field of *Quantum Probability*, is examined from the viewpoint of de Finetti's *coherence principle*. It is argued that, in the Kolmogorovian case, the statistical invariants are nothing but particular coherence conditions for conditional probability assessments: the subjective theory provides the most suitable framework to study them with full generality. Then, using some results on conditional probabilities, a few examples of statistical invariants are examined.

1. INTRODUCTION

In the last ten years, with the aim of explaining some classical paradoxes, a new approach about the foundations of quantum theory, the Quantum Probability, has been developed (see Accardi, 1981, 1984, 1985, 1986, 1990. See also Accardi and Fedullo, 1982). Some general points in Quantum Probability are the following.

a) The checking of the predictions of the theories is based on some conditional probabilities derived from empirical data;

b) If conditional probabilities are estimated by frequencies in mutually incompatible experiments, then the application of classical probability rules to the statistical data is not warranted. Then the existence (or non existence) of a Kolmogorovian probabilistic model for these data cannot be postulated, but must proved evaluating the so-called *Statistical Invariants*;

c) By adopting the Quantum Probability approach, no new unplausible physical properties need to be introduced to explain the difficulties arising in the interpretation of the experimental data.

The fact that one cannot postulate the existence of a Kolmogorovian probabilistic model, when the data come from different experiments, is also claimed in a paper of van den Berg, Hoekzema and Radder (1990). See also Scozzafava (1992) for a critical analysis of the (misleading) tendency to give frequencies the same "status" of probabilities.

The aim of this paper is to show that the notion of statistical invariant is naturally related to the concept of coherence introduced by de Finetti in his subjective theory of probability (see de Finetti, 1974). As it is well known, using the coherence principle as the unique axiom, de Finetti's approach allows to introduce a probability P on *arbitrary* families of (conditional or unconditional) events, even if some (or possibly all) *conditioning* events have zero probability (see Scozzafava (1990) for a comparison of de Finetti's and Kolmogorovian approaches to the notion of conditional probability). Each time we estimate from experimental data some conditional probabilities, we need to check consistency of the probabilistic assessment. This process produces some numerical condition which must be satisfied. This condition is precisely what in the Quantum Probability language is called a *statistical invariant* and the subjective framework is the most general one to fully exploit this concept. In order to deepen this aspect, in Sect. 2 we give a concise review of de Finetti's coherence principle and of the betting and penalty criteria. Then, in Sect. 3, the relationship between the two concepts of coherence and statistical invariant is examined and, finally, in Sect. 4 some examples are considered.

2. COHERENCE PRINCIPLE, BETTING AND PENALTY CRITERIA

The coherence principle of de Finetti can be based on one of the two following methods: the betting scheme or the penalty criterion. The application of these methods to the case of conditional events has been studied by many authors: for the betting scheme see Holzer, (1984), (1985), Regazzini (1985), (1987); for the penalty criterion see Gilio, (1990), (1992). A concise review of both methods follows.

(2.a). **Penalty criterion.** Denote by $\mathcal{K}$ an arbitrary class of conditional events and by P a real function defined on $\mathcal{K}$. Then, given a family $\mathcal{F} = (E_1|H_1, E_2|H_2, \dots, E_n|H_n) \subseteq \mathcal{K}$, consider a probability assessment $\mathcal{P} = (p_1, p_2, \dots, p_n)$, with $p_i = P(E_i|H_i)$, $i = 1,\dots,n$. As it is well known, on the basis of de Finetti's penalty criterion, to the point $\mathcal{P}$ it is associated a *loss*

$$\mathfrak{L} \quad \sum_{i=1}^{n} H_i \left(E_i - p_i\right)^2 .$$

We denote the events and their indicators by the same symbol.

We can interpret geometrically the values of $\mathfrak{L}$ using the concept of *generalized atom* introduced in Gilio (1990). Denoting by $C_1, C_2, \dots, C_s$ those atoms generated by the family $(E_1, \dots, E_n, H_1, \dots, H_n)$ such that $C_h \subseteq H_1 \vee H_2 \vee \dots \vee H_n$, $h = 1,2,\dots,s$, the *generalized atoms* $Q_h = (\alpha_{h1}, \alpha_{h2}, \dots, \alpha_{hn})$ are obtained in $\mathbb{R}^n$ from the C_h's putting

$$\alpha_{hi} \quad \begin{cases} 1, & \text{if } C_h \subseteq E_i \wedge H_i \\ 0, & \text{if } C_h \subseteq E_i^c \wedge H_i , \\ p_i, & \text{if } C_h \subseteq H_i^c \end{cases} \qquad \begin{array}{l} i=1,2,\dots,n \\ h=1,2,\dots,s \end{array} . \tag{1}$$

Then to each constituent C_h there corresponds through (1) a value L_h of $\mathfrak{L}$ given by

$$L_h = \sum_{i=1}^{n} \left(\alpha_{hi} - p_i\right)^2 = \overline{\mathcal{P}Q_h}^{\,2} . \tag{2}$$

In Gilio (1990) the following definition (C_o*-coherence*) and results are given .

(2.1). Definition. The real function $P: \mathcal{K} \to \mathbb{R}$ is a coherent conditional probability on $\mathcal{K}$ if, for each n and for each finite family $\mathcal{F} \subseteq \mathcal{K}$, putting $\mathcal{P} = (p_1, p_2, \ldots, p_n)$, with $p_i = P(E_i|H_i)$, there does not exist a point $\mathcal{P}^* \neq \mathcal{P}$ such that $L_h^* \leq L_h$ for each h and with $L_h^* < L_h$ for at least one index h ; i. e., such that $\mathfrak{L}^* \leq \mathfrak{L}$, $\mathfrak{L}^* \neq \mathfrak{L}$.

Denoting, respectively, by $\mathfrak{I}$ the convex hull of points $Q_h \in \mathbb{R}^n$, relative to the family $\mathcal{F}$ and point $\mathcal{P}$, and, for each $J \subset \{1,2,\ldots,n\}$, by $\mathfrak{I}_J$ the convex hull of generalized atoms relative to the family $\mathcal{F}_J = \{E_j|H_j : j \in J\}$ and point $\mathcal{P}_J = (p_j : j \in J)$, the following results can be proved.

(2.2). Theorem. If $\mathcal{P}$ is coherent then, for each $J \subset \{1, 2, \ldots, n\}$, $\mathcal{P}_J$ is coherent.

(2.3). Theorem. If $\mathcal{P}$ is coherent, then $\mathcal{P} \in \mathfrak{I}$.

Note that $\mathcal{P} \in \mathfrak{I}$ amounts to say that there exist s non-negative real numbers $\lambda_1, \lambda_2, \ldots, \lambda_s$, with $\sum_{h=1}^{s} \lambda_h = 1$, such that $\mathcal{P} = \sum_{h=1}^{s} \lambda_h Q_h$.

(2.4). Theorem. If there exist s *positive* numbers $\lambda_1, \lambda_2, \ldots, \lambda_s$, with $\sum_{h=1}^{s} \lambda_h = 1$, such that $\mathcal{P} = \sum_{h=1}^{s} \lambda_h Q_h$, then $\mathcal{P}$ is coherent.

(2.5). Theorem. The probability assessment $\mathcal{P}$ is coherent *if and only if* $\mathcal{P}_J \in \mathfrak{I}_J$ for every $J \subseteq \{1,2,\ldots,n\}$.

(2.b). **Betting scheme.** For each integer n and for each family $\mathcal{F} = \{E_1|H_1, E_2|H_2, \ldots, E_n|H_n\} \subseteq \mathcal{K}$, given n arbitrary real numbers $s_1, s_2, \ldots, s_n$, we consider the random quantity G defined as

$$G = \sum_{i=1}^{n} s_i H_i (E_i - p_i),$$

which can be interpreted as the random gain corresponding to a combination of n bets on $E_1|H_1, \ldots, E_n|H_n$, with stakes $s_1, \ldots, s_n$.

Putting $H_o = H_1 \vee H_2 \vee \ldots \vee H_n$ and denoting by $G|H_o$ the *restriction* of G to H_o, we have the following Definition (see Holzer, 1984, 1985, Regazzini, 1985, 1987).

(2.6) Definition. The real function $P: \mathcal{K} \to \mathbb{R}$ is a $d\mathcal{F}$-coherent conditional probability on $\mathcal{K}$ if, for each n, for every $s_1, \ldots, s_n$ and for each finite family $\mathcal{F} \subseteq \mathcal{K}$, it is

$$\min G|H_o \cdot \max G|H_o \leq 0 . \tag{3}$$

Denoting by $C_1, C_2, \dots, C_s$ the *atoms* generated by the events $E_1, \dots, E_n, H_1, \dots, H_n$ and contained in H_o, the corresponding values $g_1, g_2, \dots, g_s$ of G represent the possible values of $G|H_o$.

Then, condition (3) is equivalent to the following one.

$$\operatorname*{Min}_h g_h \cdot \operatorname*{Max}_h g_h \leq 0 \ .$$

Notice that, using the generalized atoms, from (1) it follows

$$g_h = \sum_{i=1}^{n} s_i(\alpha_{hi} - p_i) \ , \qquad h = 1,2,\dots,s.$$

It is *equivalent* to use the penalty criterion or the betting scheme (two distinct proofs of this fact are given in Gilio, 1989a, 1990). As it is well known, if the real function P defined on $\mathcal{K}$ is coherent, then P satisfies all axiomatic properties of a conditional probability, while the converse is in general not true (see Gilio, Spezzaferri, 1992, for an example). Moreover the following Proposition can be proved (see Holzer, 1985, or Regazzini, 1985).

(2.7). Proposition. Given a family $\mathcal{F} = \{E_1|H_1, E_2|H_2, \dots, E_n|H_n\}$ and a probability assessment $\mathcal{P} = (p_1, p_2, \dots, p_n)$ on $\mathcal{F}$, consider an arbitrary class of conditional events $\mathcal{K} \supseteq \mathcal{F}$. Then, there exists a (possibly not unique) coherent extension P of $\mathcal{P}$ to $\mathcal{K}$ if and only if $\mathcal{P}$ is coherent.

In particular it can be $\mathcal{K} = \mathcal{S} \times \mathcal{H}$, where $\mathcal{S}$ is the algebra generated by the events $E_1, \dots, E_n, H_1, \dots, H_n$ and $\mathcal{H} \subseteq \mathcal{K}$, with $\emptyset \notin \mathcal{H}$, in which case P is a (coherent) conditional probability on $\mathcal{S} \times \mathcal{H}$.

We recall that, given a conditional probability P on $\mathcal{S} \times \mathcal{H}$, some sufficient conditions for coherence of P are the following (see, respectively, Lehman, 1955, Regazzini, 1983, Holzer, 1984, Gilio, 1989b), from the "strongest" to the "weakest":

(a) $\mathcal{H} = \mathcal{S}^{\circ} = \mathcal{S} \setminus \{\emptyset\}$, in which case P is called *complete*;

(b) $\mathcal{H} \cup \{\emptyset\}$ is a subalgebra of $\mathcal{S}$;

(c) $\mathcal{H}$ is an additive class;

(d) $\mathcal{H}$ is a P-quasi additive class.

Finally, a necessary and sufficient condition for coherence of P is the following one (see Rigo, 1988).

$$\prod_{i=1}^{n} P(E_i|H_i) = \prod_{i=1}^{n} P(E_i|H_{i+1}) \ , \tag{4}$$

where $E_i \in \mathcal{S}$, $H_i \in \mathcal{H}$, $E_i \subseteq H_i H_{i+1}$, $i=1,2,\dots,n$, $(H_{n+1} = H_1)$.

Condition (4) has been introduced in a paper of Császár (1955).

3. STATISTICAL INVARIANTS AS COHERENCE CONDITIONS

In Accardi (1986), p. 11, the notion of statistical invariant is intuitively defined in the following way:

"By analogy with the geometrical invariants, we call a "statistical invariant" - for a given set of statistical data, with respect to a given probabilistic model- a mathematical constraint on the given set of data, whose realization is a necessary and sufficient condition for this set of data to admit a description within the given probabilistic model."

Notice that in the above definition the actual problem is to examine the possibility of describing within the given probabilistic model some conditional probabilities, which has been estimated on the basis of some experimental data. But this problem can be looked on as that of studying the existence of a probabilistic model P compatible with a given probability assessment $\mathcal{P}$ $(p_1, p_2, \ldots, p_n)$, relative to a family of conditional events $\mathcal{F}$ $(E_1|H_1,\ldots,E_n|H_n)$. And, as shown by Proposition (2.7), the probability assessment $\mathcal{P}$ can be described within a (possibly not unique) probabilistic model P if and only if $\mathcal{P}$ is coherent.

Therefore the mathematical constraints corresponding to a given statistical invariant are nothing but some particular coherence conditions on the set of statistical data at hand.

From this point of view, the most general statistical invariant for a set of statistical data is represented by the *necessary and sufficient* condition given in Theorem (2.5), while Theorems (2.3), (2.4) provide respectively a *necessary* condition and a *sufficient* one. Also notice that Theorem (2.4) can be looked on as an algorithm providing a statistical invariant, corresponding to a sufficient mathematical condition on the given set of data.

Theorem (2.2) points out the obvious circumstance that if a set of statistical data can be described within a probabilistic model, the same happens for each subset of the given set.

We observe that the subjective theory provides the most general framework to study the notion of statistical invariant: in fact, within this approach, we can *directly* assign (and check coherence of) a set of conditional probabilities, even if some (or possibly all) *conditioning* events have zero probability.

In other words, to investigate the existence of a probabilistic model describing a given set of statistical data, the subjective methodology does not necessarily require the use of the *compound probability theorem*. In Accardi (1985), p. 13, it is claimed that

"The theory of statistical invariants provides for the first time a model independent solution of the problem: the existence or not of a Kolmogorovian model can be now decided (at least in principle) uniquely in terms of the statistical data."

It seems natural to conclude that:

(i) the study of statistical invariants can be naturally embedded in de Finetti's subjective theory;

(ii) the (more general) problem to investigate is: the existence or not of a "de Finettian" model describing the given set of statistical data.

In the next Section we apply the previous considerations to the analysis of some statistical invariants.

4. EXAMPLES OF STATISTICAL INVARIANTS

(4.1). Feynman's *"two-slit experiment"*. In Accardi (1985), p. 9, the existence of a Kolmogorovian model for the probabilities $P(A|C)$, $P(A|BC)$, $P(A|B^cC)$ is studied, obtaining the following statistical invariant:

$$0 < \frac{P(A|C) - P(A|B^cC)}{P(A|BC) - P(A|B^cC)} < 1 . \tag{5}$$

Apparently, (5) cannot be used if $P(A|BC) = P(A|B^cC)$. Moreover, what about the case in which the ratio is equal to 0 or 1 ?

Using Theorem (2.5), with $\mathcal{F} = (A|C,\ A|BC,\ A|B^cC)$, $\mathcal{P} = (p,q,r)$, $p = P(A|C)$, $q = P(A|BC)$, $r = P(A|B^cC)$, $0 \le p,\ q,\ r \le 1$, in the subjective framework condition (5) can be generalized to one of the following equivalent forms.

$$\mathrm{Min}\ (q,r) \le p \le \mathrm{Max}\ (q,r) \ , \tag{6}$$

$$|p - q| \le |q - r| \ , \tag{6'}$$

$$|p - r| \le |q - r| \ . \tag{6''}$$

In fact, for $|J| = 1$, the condition $\mathcal{P}_J \in \mathfrak{I}_J$ is trivially satisfied since $0 \le p,q,r \le 1$. For $|J| = 2$, we have the three cases:

a. $\mathcal{F}_J = (A|C,\ A|BC)$; b. $\mathcal{F}_J = (A|C,\ A|B^cC)$; c. $\mathcal{F}_J = (A|BC,\ A|B^cC)$,

and the generalized atoms are respectively:

a. $(1,1)$, $(1,q)$, $(0,0)$, $(0,q)$;

b. $(1,1)$, $(1,r)$, $(0,0)$, $(0,r)$;

c. $(1,r)$, $(q,1)$, $(0,r)$, $(q,0)$.

Then, we can easily verify that in all cases it is $\mathcal{P}_J \in \mathfrak{I}_J$, for all values $p,\ q,\ r \in [0,1]$. Finally, for $|J| = 3$ we have $\mathcal{F}_J = \mathcal{F} = (A|C,\ A|BC,\ A|B^cC)$ and the generalized atoms are the points

$$(1,1,r) \ , \quad (1,q,1) \ , \quad (0,0,r) \ , \quad (0,q,0) \ .$$

Then, the condition $\mathcal{P} \in \mathfrak{I}$ amounts to the existence of a quantity $x \in [0,\ 1]$, such that $p = qx + r(1-x)$, i.e. to the three conditions (6), (6'), (6'').

(4.2). As another example to show that de Finetti's approach is more general than that of Kolmogorov, consider the family $(A|H,\ B|H,\ AB|H,\ H|\Omega)$. Then, putting $\mathcal{P} = (p_1,\ p_2,\ p_3,\ p_4)$, with

$$p_1 = P(A|H) \ , \quad p_2 = P(B|H) \ , \quad p_3 = P(AB|H) \ , \quad p_4 = P(H) \ , \tag{7}$$

let us search a statistical invariant for data (7).

If $P(H)$ 0 the compound probability theorem cannot be used. Yet, in the subjective framework the following result can be proved.

Proposition. Provided $0 \le p_i \le 1$, $i = 1,2,3,4$, there exists a probabilistic model compatible with the set of statistical data (7) if and only if the following condition is satisfied:

$$\text{Max } (0,\ p_1+p_2-1) \ \le\ p_3 \le\ \text{Min } (p_1,\ p_2)\ . \tag{8}$$

Proof. Assume that $\mathcal{P}$ is coherent. Then $\mathcal{P}_3$ (p_1,p_2,p_3) is coherent too. The generalized atoms relative to $\mathcal{F}_3 = (A|H,\ B|H,\ AB|H)$ and $\mathcal{P}_3$ are the points

$$Q_1 - (1,\ 1,\ 1)\ ,\quad Q_2 = (1,\ 0,\ 0)\ ,\quad Q_3 = (0,\ 1,\ 0)\ ,\quad Q_4\quad (0,\ 0,\ 0)\ .$$

Applying Theorem (2.3) to $\mathcal{P}_3$ it follows (8).

Conversely, assume $p_1,p_2,p_3,p_4 > 0$ and Max $(0,\ p_1+p_2-1) < p_3 <$ Min $(p_1,\ p_2)$. The generalized atoms relative to $\mathcal{F}$ and $\mathcal{P}$ are the points

$$Q_1 = (1,\ 1,\ 1,\ 1)\ ,\quad Q_2 = (1,\ 0,\ 0,\ 1)\ ,\quad Q_3\quad (0,\ 1,\ 0,\ 1)\ ,$$
$$Q_4\quad (0,\ 0,\ 0,\ 1)\ ,\quad Q_5 = (p_1,\ p_2,\ p_3,\ 0)\ .$$

Then, considering the (positive) coefficients $\lambda_1,\lambda_2,\lambda_3,\lambda_4,\lambda_5$, with $\lambda_1+\lambda_2+\lambda_3+\lambda_4+\lambda_5 = 1$ and λ_1 p_3p_4, $\lambda_2 = (p_1-p_3)p_4$, λ_3 $(p_2-p_3)p_4$, $\lambda_4 = p_4(1-p_1-p_2+p_3)$, it can be verified that $\mathcal{P}$ $\sum_{h=1}^{5}\lambda_h Q_h$. Hence, from Theorem (2.4) it follows that $\mathcal{P}$ is coherent.

The particular assessments $p_3 =$ Max $(0,\ p_1+p_2-1)$ or $p_3 =$ Min $(p_1,\ p_2)$ or p_h 0 for some subscript h can be obtained as limiting cases of the previous one. Moreover, these limiting assessments are coherent since, in the finitely additive framework, coherence satisfy the closure property with respect to the limit.

For example, letting p_4 tend to 0 , which implies λ_5 tend to 1 , we obtain the limiting (admissible) assessment $\mathcal{P} = (p_1,\ p_2,\ p_3,\ 0)$. Therefore, we can conclude that, for all $p_4 \in [0,1]$, the statistical invariant for data (7) is given by (8).

(4.3). Given two observables X, Y taking only two (arbitrary) values $(x_1,\ x_2)$, $(y_1,\ y_2)$, put $(X=x_1) = A$, $(Y=y_1) = B$. Moreover, assume that the set of statistical data for the family $\mathcal{F} = (A|B,\ B|A,\ A|B^c,\ B|A^c)$ is given by $\mathcal{P}$ $(p_1,\ p_2,\ p_3,\ p_4)$, where

$$p_1 = P(A|B)\ ,\quad p_2\quad P(A|B^c)\ ,\quad p_3 = P(B|A)\ ,\quad p_4 = P(B|A^c)\ . \tag{9}$$

Notice that to (9) there correspond the transition probability matrices

$$P(X/Y) = \begin{pmatrix} p_1 & 1-p_1 \\ p_2 & 1-p_2 \end{pmatrix},\qquad P(Y/X) = \begin{pmatrix} p_3 & 1-p_3 \\ p_4 & 1-p_4 \end{pmatrix}.$$

We can prove the following result.

Proposition. Provided $0 \le p_i \le 1$, $i = 1,2,3,4$, there exists a probabilistic model compatible with the set of statistical data (9) if and only if the following condition is satisfied:

$$p_2 p_3 (1-p_1)(1-p_4) = p_1 p_4 (1-p_2)(1-p_3) . \tag{10}$$

Proof. The generalized atoms relative to $\mathcal{F}$ and $\mathcal{P}$ are the points

$$Q_1 = (1, p_2, 1, p_4) , \; Q_2 = (p_1, 1, 0, p_4) , \; Q_3 = (0, p_2, p_3, 1) , \; Q_4 = (p_1, 0, p_3, 0) .$$

Using Theorem (2.4), each choice of four positive numbers $\lambda_1, \lambda_2, \lambda_3, \lambda_4$, with $\lambda_1 + \lambda_2 + \lambda_3 + \lambda_4 = 1$, generates a set

$$p_1 = \lambda_1/(\lambda_1+\lambda_3) , \; p_2 = \lambda_2/(\lambda_2+\lambda_4) , \; p_3 = \lambda_1/(\lambda_1+\lambda_2) , \; p_4 = \lambda_3/(\lambda_3+\lambda_4) ,$$

which satisfies (10) and, being coherent, admits a probabilistic model.

It can be verified that the range of values for p_1, p_2, p_3 is given by the closed interval $[0,1]$ (notice that the extremes of the interval are obtained as limiting points).

For each given set of statistical data $p_1, p_2, p_3, p_4 > 0$, putting $\lambda_3 = k\lambda_1$, $\lambda_4 = \delta\lambda_2$, with $k = (1-p_1)/p_1$, $\delta = (1-p_2)/p_2$, for all positive λ_1 , λ_2 such that $(1+k)\lambda_1 + (1+\delta)\lambda_2 = 1$, it can be verified that:

(i) λ_1 , λ_2 generate the values p_1 , p_2 ;

(ii) it is $\lambda_1/(\lambda_1+\lambda_2) \in (0,1)$, with $\lambda_1/(\lambda_1+\lambda_2) = p_3$ for $\lambda_2/\lambda_1 = (1-p_3)/p_3$ and with $\lim_{\lambda_1 \to 0} p_3 = 0$, $\lim_{\lambda_2 \to 0} p_3 = 1$.

Finally, from $\lambda_3/(\lambda_3 + \lambda_4) = k\lambda_1/(k\lambda_1 + \delta\lambda_2)$, since

$$k = (1-p_1)/p_1 , \qquad \delta = (1-p_2)/p_2 , \qquad \lambda_2/\lambda_1 = (1-p_3)/p_3 ,$$

it follows $\lambda_3/(\lambda_3 + \lambda_4) = p_4$ if and only if the statistical invariant (10) is satisfied.

Notice that condition (10) can also be obtained as a particular case from (4) , putting $H_1 = A^c$, $H_2 = B^c$, $H_3 = A$, $H_4 = B$, $(H_5 = H_1)$, $E_1 = A^cB^c$, $E_2 = AB^c$, $E_3 = AB$, $E_4 = A^cB$.

Remark. The condition (10) is trivially satisfied when $P(X/Y) = P(Y/X)$. In particular, for $p_1 = 1-p_2 = p_3 = 1-p_4 = p$ the standard case of symmetric transition probabilities is obtained.

(4.4). Given two observables X, Y taking (arbitrary) value $(x_1, x_2, \ldots, x_t)$, $(y_1, y_2, \ldots, y_m)$, consider the stochastic matrices $P = (p_{ij})$, $Q = (q_{ji})$, where

$$P(Y=y_j | X=x_i) = p_{ij} , \quad P(X=x_i | Y=y_j) = q_{ji} , \quad i=1,\ldots,t; \; j=1,\ldots,m.$$

In Gilio and Spezzaferri (1992) it is shown that the assessment (P, Q) is coherent if and only for each $n \le \min(t, m)$ the following condition, related to (4), is satisfied.

$$p_{i_1 j_1} q_{j_1 i_2} \cdots p_{i_h j_h} q_{j_h i_{h+1}} \cdots p_{i_n j_n} q_{j_n i_1} = \tag{11}$$
$$p_{i_1 j_n} q_{j_n i_n} \cdots p_{i_{h+1} j_h} q_{j_h i_h} \cdots p_{i_2 j_1} q_{j_1 i_1}$$

where $i_k \in \{1,\dots,t\}$, $j_k \in \{1,\dots,m\}$, $k=1,\dots,n$, and $i_k \neq i_h$, $j_k \neq j_h$ for $k \neq h$.

Notice that (11) is satisfied if $t = m$ and $p_{ij} = q_{ji}$ (symmetric case). Hence we obtain the following result.

Proposition. Given two observables X , Y taking (arbitrary) values $\{x_1, x_2, \dots, x_m\}$, $\{y_1, y_2, \dots, y_m\}$, if the set of statistical data for X , Y is represented by two stochastic matrices $P = (p_{ij})$, $Q = (q_{ji})$, with $p_{ij} = q_{ji}$ for each pair (i,j), then P and Q admit a probabilistic model.

(4.5). This example (see Accardi and Fedullo (1982) , Theorem 2, p. 165) refers to three observables X , Y , Z , assuming two (arbitrary) values $\{x_1, x_2\}$, $\{y_1, y_2\}$, $\{z_1, z_2\}$. We put

$$A = (X = x_1) , \qquad B = (Y = y_1) , \qquad C = (Z = z_1) .$$

Assuming simmetry and denoting the transition probability matrices by

$$P(X/Y) = \begin{pmatrix} p & 1-p \\ 1-p & p \end{pmatrix} , \quad P(Y/Z) = \begin{pmatrix} q & 1-q \\ 1-q & q \end{pmatrix} , \quad P(Z/X) = \begin{pmatrix} r & 1-r \\ 1-r & r \end{pmatrix} ,$$

with $0 < p,q,r < 1$, in the quoted paper the following statistical invariant is obtained:

$$|p + q - 1| \leq r \leq 1 - |p - q| . \tag{12}$$

We will derive the same result examining the example in the framework of coherence. Using for convenience tabular notations, we have the following events and respective probabilities:

$$\mathcal{F} = \left\{ \begin{matrix} A|B & B|A & A|B^c & B|A^c \\ B|C & C|B & B|C^c & C|B^c \\ A|C & C|A & A|C^c & C|A^c \end{matrix} \right\} , \qquad \mathcal{P} = \begin{bmatrix} p & p & 1-p & 1-p \\ q & q & 1-q & 1-q \\ r & r & 1-r & 1-r \end{bmatrix} .$$

The generalized atoms relative to $\mathcal{F}$ and $\mathcal{P}$ are the following points:

$$Q_1 = \begin{bmatrix} 1 & 1 & 1-p & 1-p \\ 1 & 1 & 1-q & 1-q \\ 1 & 1 & 1-r & 1-r \end{bmatrix} , \ Q_2 = \begin{bmatrix} 1 & 1 & 1-p & 1-p \\ q & 0 & 1 & 1-q \\ r & 0 & 1 & 1-r \end{bmatrix} , \ Q_3 = \begin{bmatrix} p & 0 & 1 & 1-p \\ 0 & q & 1-q & 1 \\ 1 & 1 & 1-r & 1-r \end{bmatrix} ,$$

$$Q_4 = \begin{bmatrix} p & 0 & 1 & 1-p \\ q & q & 0 & 0 \\ r & 0 & 1 & 1-r \end{bmatrix} , \ Q_5 = \begin{bmatrix} 0 & p & 1-p & 1 \\ 1 & 1 & 1-q & 1-q \\ 0 & r & 1-r & 1 \end{bmatrix} , \ Q_6 = \begin{bmatrix} 0 & p & 1-p & 1 \\ q & 0 & 1 & 1-q \\ r & r & 0 & 0 \end{bmatrix} ,$$

$$Q_7 = \begin{bmatrix} p & p & 0 & 0 \\ 0 & q & 1-q & 1 \\ 0 & r & 1-r & 1 \end{bmatrix} , \quad Q_8 = \begin{bmatrix} p & p & 0 & 0 \\ q & q & 0 & 0 \\ r & r & 0 & 0 \end{bmatrix} .$$

From Theorem (2.3) we have that, in order $\mathcal{P}$ to be coherent, it is necessary that $\mathcal{P} \in \mathfrak{I}$, that is: $\mathcal{P} = \sum_{h=1}^{8} \lambda_h Q_h$, with $\sum_{h=1}^{8} \lambda_h = 1$, $\lambda_1, \ldots, \lambda_8$ non-negative. Therefore, the following conditions must be satisfied:

$$\lambda_1+\lambda_2 = p\cdot(\lambda_1+\lambda_2+\lambda_3+\lambda_4) = p\cdot(\lambda_1+\lambda_2+\lambda_5+\lambda_6)\ ,$$

$$\lambda_7+\lambda_8 = p\cdot(\lambda_3+\lambda_4+\lambda_7+\lambda_8) = p\cdot(\lambda_5+\lambda_6+\lambda_7+\lambda_8)\ ,$$

$$\lambda_1+\lambda_5 = q\cdot(\lambda_1+\lambda_3+\lambda_5+\lambda_7) = q\cdot(\lambda_1+\lambda_2+\lambda_5+\lambda_6)\ ,$$

$$\lambda_4+\lambda_8 = q\cdot(\lambda_2+\lambda_4+\lambda_6+\lambda_8) = q\cdot(\lambda_3+\lambda_4+\lambda_7+\lambda_8)\ ,$$

$$\lambda_1+\lambda_3 = r\cdot(\lambda_1+\lambda_3+\lambda_5+\lambda_7) = r\cdot(\lambda_1+\lambda_2+\lambda_3+\lambda_4)\ ,$$

$$\lambda_6+\lambda_8 = r\cdot(\lambda_2+\lambda_4+\lambda_6+\lambda_8) = r\cdot(\lambda_5+\lambda_6+\lambda_7+\lambda_8)\ .$$

Then, recalling that $0 < p,q,r < 1$, it follows

(i) $\lambda_1+\lambda_2 = \lambda_7+\lambda_8\ ,\quad \lambda_3+\lambda_4 = \lambda_5+\lambda_6\ ,\quad p = 2(\lambda_1+\lambda_2)\ ;$

(ii) $\lambda_1+\lambda_5 = \lambda_4+\lambda_8\ ,\quad \lambda_2+\lambda_6 = \lambda_3+\lambda_7\ ,\quad q = 2(\lambda_1+\lambda_5)\ ;$

(iii) $\lambda_1+\lambda_3 = \lambda_6+\lambda_8\ ,\quad \lambda_2+\lambda_4 = \lambda_5+\lambda_7\ ,\quad r = 2(\lambda_1+\lambda_3)\ .$

Using the above relations we obtain the system:

$$\begin{cases} 2\lambda_1+2\lambda_2 = p\ , \quad 2\lambda_1+2\lambda_5 = q\ , \quad 2\lambda_1+2\lambda_3 = r\ , \\ \lambda_3+\lambda_4 = \lambda_5+\lambda_6\ , \quad \lambda_1+\lambda_5 = \lambda_4+\lambda_8\ , \quad \lambda_2+\lambda_6 = \lambda_3+\lambda_7\ , \\ \lambda_1+\lambda_2+\lambda_3+\lambda_4+\lambda_5+\lambda_6+\lambda_7+\lambda_8 = 1\ , \quad \lambda_h \geq 0\ . \end{cases} \tag{13}$$

Solving (13) with respect to p, q, r, λ_5 we obtain

$$\lambda_1 = -\lambda_5+q/2\ ,\quad \lambda_2 = \lambda_5+(p-q)/2\ ,\quad \lambda_3 = \lambda_5+(r-q)/2\ ,\quad \lambda_4 = -\lambda_5+(1-p-r+q)/2\ ,$$

$$\lambda_6 = -\lambda_5+(1-p)/2\ ,\qquad \lambda_7 = -\lambda_5+(1-r)/2\ ,\qquad \lambda_8 = \lambda_5+(p+r-1)/2\ ,$$

and from the non-negativity of the λ_h's it follows the system

$$\begin{cases} (1-p-r)/2 \leq \lambda_5 \leq q/2\ , \qquad 0 \leq \lambda_5 \leq (1-p-r+q)/2\ , \\ (q-r)/2 \leq \lambda_5 \leq (1-p)/2\ , \qquad (q-p)/2 \leq \lambda_5 \leq (1-r)/2\ , \end{cases}$$

which is equivalent to the statistical invariant (12).

Notice that if we include in the family $\mathcal{F}$ the events

$$ABC\ ,\ ABC^c\ ,\ AB^cC\ ,\ A^cBC\ ,\ AB^cC^c\ ,\ A^cBC^c\ ,\ A^cB^cC\ ,\ A^cB^cC^c,$$

it can be verified that

$$P(ABC) = \lambda_1\ ,\quad P(ABC^c) = \lambda_2\ ,\quad \ldots\ ,\quad P(A^cB^cC^c) = \lambda_8\ ,$$

from which it follows, for example:

$$P(AB^c) = \lambda_3+\lambda_4\ ,\qquad P(A^cB) = \lambda_5+\lambda_6 = P(AB^c)\ ,\quad \ldots\ ,$$

and, in particular: $P(A) = P(B) = P(C) = \lambda_1+\lambda_2+\lambda_3+\lambda_4 = 1/2$.

Then, using the additive property, we can extend the probability assessment $\mathcal{P}$ as a conditional probability P defined on $\mathcal{G} \times \mathcal{H}$, where $\mathcal{G}$ is the algebra generated by the events A, B, C and $\mathcal{H} = \{A, B, C, A^c, B^c, C^c, \Omega\}$ is P-quasi additive (see Gilio, 1989b). Therefore P is coherent and the assessment $\mathcal{P}$, as a restriction of P, is coherent too.

Notice that, putting

$$C_1 = (1,0,0)\ , \quad C_2 \quad (0,1,0)\ , \quad C_3 = (0,0,1)\ , \quad C_4 = (1,1,1)\ ,$$

the set of points (p,q,r) of the unitary cube satisfying (12) is the tetrahedron with vertices the points C_h's and with faces the four triangles

$$T_1 : C_1C_2C_3\ ; \qquad T_2 : C_1C_2C_4\ ; \qquad T_3 : C_1C_3C_4\ ; \qquad T_4 : C_2C_3C_4\ .$$

Then, to the points (p,q,r) satisfying (12), such that $0 < p,q,r < 1$, we can add the limiting (coherent) points belonging to the segments:

$$C_1C_2\ , \quad C_1C_3\ , \quad C_2C_3\ , \quad C_1C_4\ , \quad C_2C_4\ , \quad C_3C_4\ ,$$

characterized, respectively, by the conditions:

$$r \quad 0\ ,\ p+q = 1\ ; \qquad q = 0\ ,\ p+r = 1\ ; \qquad p = 0\ ,\ q+r \quad 1\ ;$$

$$p \quad 1\ ,\ q = r\ ; \qquad q \quad 1\ ,\ p = r\ ; \qquad r = 1\ ,\ p = q\ .$$

5. CONCLUSIONS

In this paper the notion of statistical invariant, introduced in Quantum Probability, has been examined and its relation to the coherence principle of subjective probability has been outlined. It has been shown that a statistical invariant is a condition of coherence that, when satisfied, allows to describe statistical data within a probabilistic model. As such, the notion of statistical invariant can be developed with full generality in the framework of subjective theory of probability.

REFERENCES

Accardi, L. (1981). *Probabilita' e teoria quantistica*, Physis, Anno XXIII - Fasc. 4, 485-524.

Accardi, L. and Fedullo, A. (1982). *On the statistical meaning of complex numbers in Quantum Theory*, Lettere al Nuovo Cimento 34, 161-172.

Accardi, L. (1984). *The probabilistic roots of the quantum mechanical paradoxes*, Diner et al. (eds) Dordrecht, Kluwer Acad. Publ.

Accardi, L. (1985). *Non-Kolmogorovian probabilistic models and Quantum Theory*, Quad. n. 8, Dip. di Matematica, II Univ. degli Studi di Roma.

Accardi, L. (1986). *Foundations of Quantum Mechanics: A Quantum Probabilistic Approach*, Quad. n. 17, Dip. di Matematica, II Univ. degli Studi di Roma.

Accardi, L. (1990). *Quantum Probability and the Foundations of Quantum Theory*, Centro Matematico V. Volterra, Quad. n. 58, Univ. degli Studi di Roma II.

van den Berg, H., Hoekzema, D., Radder H. (1990). *Accardi on Quantum Theory and the "fifth axiom" of probability*, Philosophy of Science 57, 149-157.

Császár, Á. (1955). *Sur la structure des espaces de probabilité conditionnelle*, Acta Mathematica Academiae Scientiarum Hungaricae 6, 337-361.

De Finetti, B. (1974). *Theory of Probability*, John Wiley, New York.

Gilio, A. (1989a). *Probabilita' condizionate C_o-coerenti*, Rendiconti di Matematica, Serie VII, Vol. 9, Roma, 277-295.

Gilio, A. (1989b). *Classi quasi additive di eventi e coerenza di probabilita' condizionate*, Rendiconti dell'Istituto di Matematica dell'Univ. di Trieste, Vol. XXI, Fasc. I, 22-38.

Gilio, A. (1990). *Criterio di penalizzazione e condizioni di coerenza nella valutazione soggettiva della probabilita'*. Boll. Un. Mat. Ital., Vol. 4-B, n. 3, Serie 7ª, 645-660.

Gilio, A. (1992). *C_o-coherence and extensions of conditional probabilities*, Bayesian Statistics 4, eds. J. M. Bernardo, J. O. Berger, A. P. Dawid and A. F. M. Smith, Oxford Univ. Press, 1-8.

Gilio, A. and Spezzaferri, F. (1992). *Knowledge integration for conditional probability assessments*, Proc. of The 8th Conf. on Uncertainty in Artificial Intelligence, Stanford, California, July 17-19, 1992. In press.

Holzer, S. (1984). *Sulla nozione di coerenza per le probabilita' subordinate*. Rendiconti dell'Istituto di Matematica dell'Universita' di Trieste XVI, 46-62.

Holzer, S. (1985). *On coherence and conditional prevision*, Boll. Un. Mat. Ital., Vol. IV, Serie VI, 441-460.

Lehman, R. S. (1955). *On confirmation and rational betting*, The Journal of Symbolic Logic 20, 251-262.

Regazzini, E. (1983). *Sulle probabilita' coerenti nel senso di de Finetti*. Editrice CLUEB, Bologna.

Regazzini, E. (1985). *Finitely additive conditional probabilities*, Rend. Sem. Mat. Fis. Milano, 55, 69-89.

Regazzini, E. (1987). *De Finetti's coherence and statistical inference*, Annals of Statistics 15, 845-864.

Rigo, P. (1988). *Un teorema di estensione per probabilita' condizionate finitamente additive*, Atti della XXXIV Riunione Scientifica della S. I. S., Siena, 2-1, 27-34.

Scozzafava, R. (1990). *Probabilita' condizionate: de Finetti o Kolmogorov?* Scritti in omaggio a L. Daboni, Ed. LINT, Trieste, 223-237.

Scozzafava, R. (1992). *Observed frequencies are not probabilities*, Atti della XXXVI Riunione Scientifica della S. I. S., Pescara, Vol. 2, 45-50.

Quantum Probability and Related Topics Vol. VIII (pp. 223–236)

Some Remarks on Trace of Operator Logarithm and Relative Entropy

Fumio Hiai

Department of Mathematics, Ibaraki University
Mito, Ibaraki 310, Japan

Abstract. In the setup of semifinite von Neumann algebras, we first present an additivity formula for the trace of operator logarithm, and second prove an inequality for the relative entropy under the hyperfiniteness assumption.

Introduction

The multiplicativity of matrix determinant implies the following trace formula for matrix logarithm:

$$Tr \log XY = Tr \log X^{1/2}YX^{1/2} = Tr(\log X + \log Y) \tag{0.1}$$

for positive (definite) matrices X and Y When $\mathcal{M}$ is a finite von Neumann algebra with a faithful normal finite trace τ, the (positive-valued) determinant $\Delta(x)$ of an invertible operator x in $\mathcal{M}$ was introduced in [7] by

$$\Delta(x) = \exp \tau(\log |x|).$$

Then it was proved among others in [7] that $\Delta(xy) = \Delta(x)\Delta(y)$ for any invertible $x, y \in \mathcal{M}$. This formula implies that if x, y are positive invertible in $\mathcal{M}$, then

$$\Delta(x^{1/2}yx^{1/2}) = \Delta(y^{1/2}x^{1/2})^2 = \Delta(x)\Delta(y),$$

that is,

$$\tau(\log x^{1/2}yx^{1/2}) = \tau(\log x + \log y). \tag{0.2}$$

Let us note by the way that some expositions on determinant theories for infinite dimensional operators are found in [10, 20].

In the first half (§1) of this paper we establish (0.1) and (0.2) in the setting of semifinite von Neumann algebras. More precisely, let $\mathcal{M}$ be a semifinite von Neumann algebra with a faithful normal semifinite trace τ, and $L^1(\mathcal{M};\tau)$ be the noncommutative L^1-space associated with $(\mathcal{M};\tau)$. We prove that if x and y are positive selfadjoint operators such that $\log x, \log y \in L^1(\mathcal{M};\tau)$, then $\log x^{1/2}yx^{1/2} \in L^1(\mathcal{M};\tau)$ and equality (0.2) holds.

On the other hand, when X and Y are positive matrices, we proved in [15] the following logarithmic trace inequalities:

$$\frac{1}{p} Tr\, X \log Y^{p/2}X^pY^{p/2} \le Tr\, X(\log X + \log Y) \le \frac{1}{p} Tr\, X \log X^{p/2}Y^pX^{p/2} \tag{0.3}$$

for all $p > 0$. Umegaki's relative entropy [25] for nonnegative matrices X and Y is defined by

$$S(X,Y) = Tr\,X(\log X - \log Y).$$

Inequalities (0.3) give lower and upper bounds of $S(X,Y)$ with Y being replaced by Y^{-1} In particular when $p = 1$, the second inequality in (0.3) gives

$$S(X,Y) \le Tr\,X \log X^{1/2}Y^{-1}X^{1/2} \tag{0.4}$$

for every nonnegative matrices X and Y (the right-hand side is ∞ if the support of X is not dominated by that of Y). This was first proved in [14]. The right-hand side of (0.4) was introduced in [3] as a variant of the relative entropy (in a more general setting); so we write it as $S_{\mathrm{BS}}(X,Y)$.

In the latter half (§2) of this paper we prove $S(x,y) \le S_{\mathrm{BS}}(x,y)$ for every nonnegative $x, y \in L^1(\mathcal{M};\tau)$ when a semifinite von Neumann algebra $\mathcal{M}$ is hyperfinite (i.e. AFD). But we have to be careful in defining $S_{\mathrm{BS}}(x,y)$. When x is dominated by y (i.e. $x \le \lambda y$ for some $\lambda \ge 0$), the operator $x^{1/2}y^{-1}x^{1/2} \in \mathcal{M}$ is given by

$$x^{1/2}y^{-1}x^{1/2} = \text{s-}\lim_{\varepsilon\downarrow 0} x^{1/2}(y+\varepsilon)^{-1}x^{1/2},$$

and so $S_{\mathrm{BS}}(x,y)$ is defined by

$$S_{\mathrm{BS}}(x,y) = \tau(x^{1/2}(\log x^{1/2}y^{-1}x^{1/2})x^{1/2}).$$

For general nonnegative $x, y \in L^1(\mathcal{M};\tau)$ we can define

$$S_{\mathrm{BS}}(x,y) = \lim_{\varepsilon\downarrow 0} \tau(x^{1/2}(\log x^{1/2}(y+\varepsilon x)^{-1}x^{1/2})x^{1/2}).$$

Furthermore we make clear the relation between the quantity $S_{\mathrm{BS}}(x,y)$ and the relative operator entropy introduced in [9].

Although we confine ourselves to the semifinite case in this paper, the inequality for the relative entropy in §2 can be extended to more general cases. The relative entropy $S(\omega,\varphi)$ for positive normal functionals ω and φ of a general von Neumann algebra was introduced in [1, 2], and in [23] for positive functionals of an abstract $*$-algebra. Also the quantity $S_{\mathrm{BS}}(\omega,\varphi)$ was defined in [3] for positive functionals ω and φ of a C^*-algebra. Then we can prove $S(\omega,\varphi) \le S_{\mathrm{BS}}(\omega,\varphi)$ when ω and φ are positive normal functionals of an AFD von Neumann algebra or when they are positive functionals of a nuclear (particularly AF) C^*-algebra. This extension as well as related results will be presented elsewhere.

1. Formula for trace of operator logarithm

Let $\mathcal{M}$ be a semifinite von Neumann algebra on a Hilbert space $\mathcal{H}$ and τ be a faithful normal semifinite trace on $\mathcal{M}$. A densely defined closed opreator x affiliated with $\mathcal{M}$ is said to be τ-measurable if there exists, for any $\varepsilon > 0$, a projection e in $\mathcal{M}$ such that

$e\mathcal{H} \subset \mathcal{D}(x)$ and $\tau(e^{\perp}) < \varepsilon$ where $e^{\perp} = 1 - e$. We denote by $\widetilde{\mathcal{M}}$ the set of all τ-measurable operators affiliated with $\mathcal{M}$, which becomes a complete Hausdorff topological $*$-algebra in the measure topology [17, 21]. For $1 \le p < \infty$, the noncommutative L^p-spaces $L^p(\mathcal{M};\tau)$ is defined by $L^p(\mathcal{M};\tau) = \{x \in \widetilde{\mathcal{M}} \quad \tau(|x|^p) < \infty\}$, which becomes a Banach space with the L^p-norm $\|x\|_p = \tau(|x|^p)^{1/p}$ (see [4, 17, 19]). The trace τ extends to a linear functional on $L^1(\mathcal{M};\tau)$.

We denote by $\widetilde{\mathcal{M}}_{\rm sa}$ (resp. $\widetilde{\mathcal{M}}_+$) the set of all selfadjoint (resp. positive selfadjoint) operators in $\widetilde{\mathcal{M}}$. For $x \in \widetilde{\mathcal{M}}_{\rm sa}$ and a Borel subset E of $\mathbf{R}$, let $e_E(x)$ denote the spectral projection of x corresponding to E. The generalized s-numbers $\mu_t(x)$ of $x \in \widetilde{\mathcal{M}}$ are defined by

$$\mu_t(x) = \inf\{s \ge 0 : \tau(e_{(s,\infty)}(|x|)) \le t\}, \qquad t > 0.$$

A complete exposition on generalized s-numbers is found in [6]. We also define the spectral scale $\lambda_t(x)$ of $x \in \widetilde{\mathcal{M}}_{\rm sa}$ by

$$\lambda_t(x) = \inf\{s \in \mathbf{R} : \tau(e_{(s,\infty)}(x) \le t\}, \qquad 0 < t < \tau(1).$$

Properties of $\lambda_t(x)$ are analogous to those of $\mu_t(x)$ (see [18], [11, §6], [13, p. 5]).

The main result of this section is the following:

Theorem 1.1. *(1) If x, y are positive selfadjoint operators such that $\log x, \log y \in L^1(\mathcal{M};\tau)$, then $\log x^{1/2}yx^{1/2} \in L^1(\mathcal{M};\tau)$ and*

$$\tau(\log x^{1/2}yx^{1/2}) = \tau(\log x + \log y). \tag{1.1}$$

(2) If x, y are positive invertible operators in $\mathcal{M}$ such that $\log x, \log y \in L^1(\mathcal{M};\tau)$, then $\log xy, \log x^{1/2}yx^{1/2} \in L^1(\mathcal{M};\tau)$ and

$$\tau(\log xy) = \tau(\log x^{1/2}yx^{1/2}) = \tau(\log x + \log y). \tag{1.2}$$

Here let us note in (1) that the assumption $\log x, \log y \in L^1(\mathcal{M};\tau)$ contains $e_{\{0\}}(x) = e_{\{0\}}(y) = 0$. When x is a positive selfadjoint operator with $e_{\{0\}}(x) = 0$, we get $\log x \in L^1(\mathcal{M};\tau)$ if $x - 1, x^{-1} - 1 \in L^1(\mathcal{M};\tau)$. Also note in (2) that $\log xy$ can be defined by power series or by analytic functional calculus becuase the spectrum of xy is included in $[\alpha, \beta]$ for some $0 < \alpha \le \beta < \infty$.

In particular when $\mathcal{M} = \mathbf{B}(\mathcal{H})$ with the usual trace $\tau = Tr$, $L^1(\mathcal{M};\tau)$ is the space of all trace class operators on $\mathcal{H}$. Hence we have:

Corollary 1.2. *If $X, Y \in \mathbf{B}(\mathcal{H})$ are positive and $\log X, \log Y$ are in the trace class, then $\log XY$ and $\log X^{1/2}YX^{1/2}$ are in the trace class and*

$$Tr \log XY = Tr \log X^{1/2}YX^{1/2} = Tr(\log X + \log Y).$$

The proof of Theorem 1.1 will be presented in the rest of this section. The next lemma follows from [7] as mentioned in Introduction (see also [5]).

Lemma 1.3. *Assume $\tau(1) < \infty$. If $x, y \in \mathcal{M}$ are positive invertible, then equality (1.1) hold.*

Let us define for $x, y \in \widetilde{\mathcal{M}}_+$ and $n \in \mathbf{N}$

$$x_n = xe_{E_n}(x) + e_{E_n}(x)^\perp, \qquad y_n = ye_{E_n}(y) + e_{E_n}(y)^\perp, \tag{1.3}$$

where

$$E_n = \left[\frac{1}{n}, \frac{n}{n+1}\right] \cup \left[\frac{n+1}{n}, n\right].$$

Then:

Lemma 1.4. *If x, y are positive selfadjoint operators such that $\log x, \log y \in \widetilde{\mathcal{M}}$, then $x, y \in \widetilde{\mathcal{M}}_+$, $\log x^{1/2}yx^{1/2} \in \widetilde{\mathcal{M}}$, and $\log x_n^{1/2} y_n x_n^{1/2}$ converges in measure to $\log x^{1/2}yx^{1/2}$*

Proof. Since $x = \exp(\log x)$ and $x^{-1} = \exp(-\log x)$, we get $x, x^{-1} \in \widetilde{\mathcal{M}}_+$ by [22] and analogously $y, y^{-1} \in \widetilde{\mathcal{M}}_+$. Since

$$\tau(e_{[\frac{1}{n},n]}(x)^\perp) = \tau(e_{(n,\infty)}(|\log x|)) \to 0 \quad (n \to \infty)$$

and

$$\begin{aligned} \|(x_n^{1/2} - x^{1/2})e_{[\frac{1}{n},n]}(x)\| &= \|(1 - x^{1/2})e_{(\frac{n}{n+1},\frac{n+1}{n})}(x)\| \\ &\leq \left(\frac{n+1}{n}\right)^{1/2} - 1 \quad \to 0 \quad (n \to \infty), \end{aligned}$$

we have $x_n^{1/2} \to x^{1/2}$ in measure and similarly $y_n \to y$ in measure, so that $x_n^{1/2} y_n x_n^{1/2} - 1 \to x^{1/2}yx^{1/2} - 1$ in measure. Therefore it follows from [22] that

$$(\log x_n^{1/2} y_n x_n^{1/2})_+ = \log((x_n^{1/2} y_n x_n^{1/2} - 1)_+ + 1)$$

converges in measure to

$$(\log x^{1/2}yx^{1/2})_+ = \log((x^{1/2}yx^{1/2} - 1)_+ + 1).$$

Moreover

$$(\log x_n^{1/2} y_n x_n^{1/2})_- = (\log x_n^{-1/2} y_n^{-1} x_n^{-1/2})_+$$

converges in measure to

$$(\log x^{1/2}yx^{1/2})_- = (\log x^{-1/2}y^{-1}x^{-1/2})_+,$$

because x_n^{-1} and y_n^{-1} are defined in (1.3) for x^{-1} and y^{-1} in place of x and y, respectively. Hence we obtain the desired conclusions. ∎

Proof of Theorem 1.1. (1) Let x_n and y_n be as in (1.3) (note $x, y \in \widetilde{\mathcal{M}}_+$ by Lemma 1.4). Since

$$e_{E_n}(x) = e_{[\log \frac{n+1}{n}, \log n]}(|\log x|)$$

and

$$\log x - \log x_n = (\log x) e_{E_n}(x)^{\perp},$$

it follows that $\tau(e_{E_n}(x)) < \infty$ and $\tau(\log x_n) \to \tau(\log x)$. Analogously $\tau(e_{E_n}(y)) < \infty$ and $\tau(\log y_n) \to \tau(\log y)$. Now let

$$e_n = e_{E_n}(x) \vee e_{E_n}(y),$$

$$\tilde{x}_n = x e_{E_n}(x) + (e_n - e_{E_n}(x)), \qquad \tilde{y}_n = y e_{E_n}(y) + (e_n - e_{E_n}(y)).$$

Then $\tau(e_n) < \infty$ and $\tilde{x}_n, \tilde{y}_n$ are positive invertible in $e_n \mathcal{M} e_n$. Hence Lemma 1.3 implies that

$$\tau_n(\log \tilde{x}_n^{1/2} \tilde{y}_n \tilde{x}_n^{1/2}) = \tau_n(\log \tilde{x}_n + \log \tilde{y}_n)$$

where $\tau_n = \tau | e_n \mathcal{M} e_n$. Since

$$x_n^{1/2} y_n x_n^{1/2} = \tilde{x}_n^{1/2} \tilde{y}_n \tilde{x}_n^{1/2} + e_n^{\perp},$$

we get

$$\tau(\log x_n^{1/2} y_n x_n^{1/2}) = \tau_n(\log \tilde{x}_n^{1/2} \tilde{y}_n \tilde{x}_n^{1/2})$$

as well as $\tau(\log x_n) = \tau_n(\log \tilde{x}_n)$ and $\tau(\log y_n) = \tau_n(\log \tilde{y}_n)$. Therefore

$$\tau(\log x + \log y) = \lim_{n\to\infty} \tau(\log x_n + \log y_n) = \lim_{n\to\infty} \tau(\log x_n^{1/2} y_n x_n^{1/2}).$$

In the following let us prove that $\log x^{1/2} y x^{1/2} \in L^1(\mathcal{M}; \tau)$ and

$$\tau(\log x^{1/2} y x^{1/2}) = \lim_{n\to\infty} \tau(\log x_n^{1/2} y_n x_n^{1/2}),$$

which completes the proof of (1).

The case $\tau(1) < \infty$. For every $0 < t < \tau(1)$ we have by [11, §6]

$$\lambda_t(\log x_n^{1/2} y_n x_n^{1/2}) = \log \lambda_t(x_n^{1/2} y_n x_n^{1/2}) = \log \mu_t(x_n^{1/2} y_n x_n^{1/2})$$

and by [6, Lemma 2.5]

$$\begin{aligned}\mu_t(x_n^{1/2} y_n x_n^{1/2}) &= \mu_t(x_n^{1/2} y_n^{1/2})^2 \le (\mu_{t/2}(x_n^{1/2}) \mu_{t/2}(y_n^{1/2}))^2 \\ &= \mu_{t/2}(x_n) \mu_{t/2}(y_n) = \lambda_{t/2}(x_n) \lambda_{t/2}(y_n),\end{aligned}$$

so that

$$\lambda_t(\log x_n^{1/2} y_n x_n^{1/2}) \le \log \lambda_{t/2}(x_n) + \log \lambda_{t/2}(y_n) = \lambda_{t/2}(\log x_n) + \lambda_{t/2}(\log y_n).$$

Also

$$\begin{aligned}-\lambda_t(\log x_n^{1/2} y_n x_n^{1/2}) &= \lambda_{\tau(1)-t-0}(-\log x_n^{1/2} y_n x_n^{1/2}) = \lambda_{\tau(1)-t-0}(\log x_n^{-1/2} y_n^{-1} x_n^{-1/2})\\ &\le \lambda_{(\tau(1)-t)/2-0}(\log x_n^{-1}) + \lambda_{(\tau(1)-t)/2-0}(\log y_n^{-1})\\ &= -\lambda_{(\tau(1)+t)/2}(\log x_n) - \lambda_{(\tau(1)+t)/2}(\log y_n).\end{aligned}$$

Furthermore it is easy to see that

$$|\lambda_t(\log x_n)| \le |\lambda_t(\log x)|, \qquad 0 < t < \tau(1).$$

The above estimates show that

$$|\lambda_t(\log x_n^{1/2} y_n x_n^{1/2})| \le f(t), \qquad 0 < t < \tau(1),$$

where

$$f(t) = |\lambda_{t/2}(\log x)| + |\lambda_{t/2}(\log y)| + |\lambda_{(\tau(1)+t)/2}(\log x)| + |\lambda_{(\tau(1)+t)/2}(\log y)|.$$

Since by [6, Lemma 3.4]

$$\lambda_t(\log x_n^{1/2} y_n x_n^{1/2}) = \log \mu_t(x_n^{1/2} y_n x_n^{1/2})$$

converges a.e. to

$$\lambda_t(\log x^{1/2} y x^{1/2}) = \log \mu_t(x^{1/2} y x^{1/2})$$

and f is integrable on $(0, \tau(1))$ (see [12, Proposition 1.1]), the Lebesuge convergence theorem shows that $\lambda_t(\log x^{1/2} y x^{1/2})$ is integrable on $(0, \tau(1))$, i.e. $\log x^{1/2} y x^{1/2} \in L^1(\mathcal{M}; \tau)$, and that

$$\begin{aligned}\tau(\log x^{1/2} y x^{1/2}) &= \int_0^{\tau(1)} \lambda_t(\log x^{1/2} y x^{1/2}) dt\\ &= \lim_{n\to\infty} \int_0^{\tau(1)} \lambda_t(\log x_n^{1/2} y_n x_n^{1/2}) dt\\ &= \lim_{n\to\infty} \tau(\log x_n^{1/2} y_n x_n^{1/2}).\end{aligned}$$

The case $\tau(1) = \infty$. Since $\tau(e_{(0,\infty)}(|\log x_n^{1/2} y_n x_n^{1/2}|)) < \infty$, for every $t > 0$ we have by [13, p. 5]

$$\begin{aligned}\mu_t((\log x_n^{1/2} y_n x_n^{1/2})_+) &= \lambda_t(\log x_n^{1/2} y_n x_n^{1/2}) = \log \lambda_t(x_n^{1/2} y_n x_n^{1/2})\\ &= \log \mu_t(x_n^{1/2} y_n x_n^{1/2}) \le \log \mu_{t/2}(x_n)\mu_{t/2}(y_n)\\ &= \log \lambda_{t/2}(x_n) + \log \lambda_{t/2}(y_n)\\ &= \mu_{t/2}((\log x_n)_+) + \mu_{t/2}((\log y_n)_+)\\ &\le \mu_{t/2}((\log x)_+) + \mu_{t/2}((\log y)_+)\end{aligned}$$

and

$$\begin{aligned}\mu_t((\log x_n^{1/2}y_nx_n^{1/2})_-) &= \mu_t((\log x_n^{-1/2}y_n^{-1}x_n^{-1/2})_+)\\ &\le \mu_{t/2}((\log x^{-1})_+) + \mu_{t/2}((\log y^{-1})_+)\\ &= \mu_{t/2}((\log x)_-) + \mu_{t/2}((\log y)_-).\end{aligned}$$

Furthermore $\mu_{t/2}((\log x)_\pm)$ and $\mu_{t/2}(\log y)_\pm)$ are integrable on $(0,\infty)$. Since by Lemma 1.4 and [6, Lemma 3.4]

$$\mu_t((\log x_n^{1/2}y_nx_n^{1/2})_\pm) \to \mu_t((\log x^{1/2}yx^{1/2})_\pm) \text{ a.e.},$$

the Lebesgue convergence theorem shows that $\mu_t((\log x^{1/2}yx^{1/2})_\pm)$ are integrable on $(0,\infty)$, hence $\log x^{1/2}yx^{1/2} \in L^1(\mathcal{M};\tau)$, and that

$$\begin{aligned}&\tau(\log x^{1/2}yx^{1/2})\\ &= \int_0^\infty \mu_t((\log x^{1/2}yx^{1/2})_+)dt - \int_0^\infty \mu_t((\log x^{1/2}yx^{1/2})_-)dt\\ &= \lim_{n\to\infty}\left\{\int_0^\infty \mu_t((\log x_n^{1/2}y_nx_n^{1/2})_+)dt - \int_0^\infty \mu_t((\log x_n^{1/2}y_nx_n^{1/2})_-)dt\right\}\\ &= \lim_{n\to\infty} \tau(\log x_n^{1/2}y_nx_n^{1/2}).\end{aligned}$$

(2) When $x, y \in \mathcal{M}$ are positive invertible, since $(xy)^n = x^{1/2}(x^{1/2}yx^{1/2})^n x^{-1/2}$ for all $n \in \mathbf{N}$, computation by power series implies that

$$\log xy = x^{1/2}(\log x^{1/2}yx^{1/2})x^{-1/2}$$

Hence (2) immediately follows from (1). ∎

Theorem 1.1 leads us to extend the determinant theory in [7] to the semifinite case. Let us set the domain $\mathcal{D}(\Delta)$ by

$$\mathcal{D}(\Delta) = \{x \in \widetilde{\mathcal{M}} : e_{\{0\}}(|x|) = e_{\{0\}}(|x^*|) = 0 \text{ and } \log|x| \in L^1(\mathcal{M};\tau)\},$$

and define the (positive-valued) determinant $\Delta(x)$ of $x \in \mathcal{D}(\Delta)$ by

$$\Delta(x) = \exp\tau(\log|x|).$$

Then we have:

Proposition 1.5. (1) $\Delta(u_1 x u_2) = \Delta(x)$ *for every* $x \in \mathcal{D}(\Delta)$ *and unitaries* $u_1, u_2 \in \mathcal{M}$.
(2) $\Delta(x^*) = \Delta(x)$ *for every* $x \in \mathcal{D}(\Delta)$.
(3) $\Delta(x^{-1}) = \Delta(x)^{-1}$ *for every* $x \in \mathcal{D}(\Delta)$.
(4) *If* $x, y \in \mathcal{D}(\Delta)$, *then* $xy \in \mathcal{D}(\Delta)$ *and* $\Delta(xy) = \Delta(x)\Delta(y)$.
(5) $\Delta(x) \geq \Delta(y)$ *for every* $x, y \in \mathcal{D}(\Delta)$ *such that* $x \geq y \geq 0$.

Proof. (1) is easily seen. For $x, y \in \mathcal{D}(\Delta)$ let $x = v|x|$ and $y = w|y|$ be the polar decompositions of x and y, respectively. Then v and w are unitaries in $\mathcal{M}$. Since $x^* = |x|v^*$ and $x^{-1} = |x|^{-1}v^*$, (2) and (3) follow from (1). It is immediate that $e_{\{0\}}(|xy|) = e_{\{0\}}(|y^*x^*|) = 0$. Since $|xy|^2 = y^*x^*xy = |y|w^*|x|^2w|y|$, Theorem 1.1(1) implies that $\log|xy| \in L^1(\mathcal{M};\tau)$ and

$$\Delta(xy)^2 = \exp\tau(\log|xy|^2) = \exp\tau(\log|x|^2 + \log|y|^2) = \Delta(x)^2\Delta(y)^2.$$

Hence (4) is proved. For $x, y \in \mathcal{D}(\Delta)$ with $x \geq y \geq 0$, since $x^{-1/2}yx^{-1/2} \leq 1$, it follows from (3) and (4) that $\Delta(x)^{-1}\Delta(y) = \Delta(x^{-1/2}yx^{-1/2}) \leq 1$, implying (5). ∎

2. Inequality for relative entropy

As in §1 let $\mathcal{M}$ be a semifinite von Neumann algebra with a faithful normal semifinite trace τ.

Lemma 2.1. *Let* $x, y \in \widetilde{\mathcal{M}}_+$. *If* $x \leq \lambda y$ *for some* $\lambda \geq 0$, *then there exists a unique* $a \in \mathcal{M}$ *such that* $x^{1/2} = ay^{1/2}$ *and* $ae_{\{0\}}(y) = 0$. *Moreover* a *satisfies the following:*
(1) $\|a\| \leq \lambda^{1/2}$,
(2) $x^{1/2}(y+\varepsilon)^{-1/2} \to a$ *strongly** as $\varepsilon \downarrow 0$,
(3) $x^{1/2}(y+\varepsilon)^{-1}x^{1/2} \nearrow aa^*$ as $\varepsilon \downarrow 0$.

Proof. The assumption $x \leq \lambda y$ means that $\mathcal{D}(y^{1/2}) \subset \mathcal{D}(x^{1/2})$ and $\|x^{1/2}\xi\|^2 \leq \lambda\|y^{1/2}\xi\|^2$, $\xi \in \mathcal{D}(y^{1/2})$. So a bounded operator a is uniquely defined by $a(y^{1/2}\xi) = x^{1/2}\xi$, $\xi \in \mathcal{D}(y^{1/2})$, and $ae_{\{0\}}(y) = 0$. It is immediate to get $a \in \mathcal{M}$ and $\|a\| \leq \lambda^{1/2}$. We have as $\varepsilon \downarrow 0$

$$x^{1/2}(y+\varepsilon)^{-1/2} = ay^{1/2}(y+\varepsilon)^{-1/2} \to ae_{(0,\infty)}(y) = a \ \text{ strongly},$$

$$(x^{1/2}(y+\varepsilon)^{-1/2})^* = (y+\varepsilon)^{-1/2}y^{1/2}a^* \to e_{(0,\infty)}(y)a^* = a^* \ \text{ strongly},$$

$$x^{1/2}(y+\varepsilon)^{-1}x^{1/2} = ay(y+\varepsilon)^{-1}a^* \nearrow aa^*,$$

so that (2) and (3) hold. ∎

For $x, y \in \widetilde{\mathcal{M}}_+$ such that x is dominated by y (i.e. $x \leq \lambda y$ for some $\lambda \geq 0$), we write $aa^* = x^{1/2}y^{-1}x^{1/2}$ in view of (3) above.

Lemma 2.2. *Let $x, y \in L^1(\mathcal{M};\tau)_+$ $(= \widetilde{\mathcal{M}}_+ \cap L^1(\mathcal{M};\tau))$. If x is dominated by y, then $x^{1/2}(\log x^{1/2}y^{-1}x^{1/2})x^{1/2} \in L^1(\mathcal{M};\tau)$ and so $\tau(x^{1/2}(\log x^{1/2}y^{-1}x^{1/2})x^{1/2}) \in (-\infty, \infty)$.*

Proof. By definition,

$$x^{1/2}(\log x^{1/2}y^{-1}x^{1/2})x^{1/2} = y^{1/2}a^*(\log aa^*)ay^{1/2}$$

Using the polar decomposition of a we easily see that $a^*(\log aa^*)a = a^*a \log a^*a$. When $x \le \lambda y$ with $\lambda \ge 1$, since $a^*a - 1 \le a^*a \log a^*a \le \lambda \log \lambda$, we get

$$x - y = y^{1/2}(a^*a - 1)y^{1/2} \le x^{1/2}(\log x^{1/2}y^{-1}x^{1/2})x^{1/2} \le (\lambda \log \lambda)y,$$

implying the conclusion. ∎

Lemma 2.3. *Let $x, y, \tilde{y} \in L^1(\mathcal{M};\tau)_+$. If x is dominated by y and $y \le \tilde{y}$, then*

$$\tau(x^{1/2}(\log x^{1/2}y^{-1}x^{1/2})x^{1/2}) \ge \tau(x^{1/2}(\log x^{1/2}\tilde{y}^{-1}x^{1/2})x^{1/2}).$$

Proof. Let $\tilde{a}$ be defined for $x, \tilde{y}$ as a for x, y in Lemma 2.1. Since $x^{1/2}(y+\varepsilon)^{-1}x^{1/2} \ge x^{1/2}(\tilde{y}+\varepsilon)^{-1}x^{1/2}$ for $\varepsilon > 0$, Lemma 2.1(3) implies $aa^* \ge \tilde{a}\tilde{a}^*$. By the proof of Lemma 2.2 we have

$$\begin{aligned}
&\tau(x^{1/2}(\log x^{1/2}y^{-1}x^{1/2})x^{1/2}) \\
&= \tau(y^{1/2}(a^*a \log a^*a)y^{1/2}) = \lim_{\varepsilon \downarrow 0} \tau(y^{1/2}(a^*a \log(a^*a + \varepsilon))y^{1/2}) \\
&= \lim_{\varepsilon \downarrow 0} \tau(y^{1/2}a^*(\log(aa^* + \varepsilon))ay^{1/2}) = \lim_{\varepsilon \downarrow 0} \tau(x^{1/2}(\log(aa^* + \varepsilon))x^{1/2}) \\
&\ge \lim_{\varepsilon \downarrow 0} \tau(x^{1/2}(\log(\tilde{a}\tilde{a}^* + \varepsilon))x^{1/2}) = \tau(x^{1/2}(\log x^{1/2}\tilde{y}^{-1}x^{1/2})x^{1/2}).
\end{aligned}$$

The third equality in the above follows because $a^*a\log(a^*a + \varepsilon) = a^*(\log(aa^* + \varepsilon))a$. Also the inequality follows from the operator monotonity of $\log t$. ∎

Umegaki's relative entropy [25] for $x, y \in L^1(\mathcal{M};\tau)_+$ is defined by

$$S(x, y) = \tau(x(\log x - \log y)).$$

On the other hand, in view of Lemmas 2.2 and 2.3 let us define for $x, y \in L^1(\mathcal{M};\tau)_+$

$$\begin{aligned}
&S_{\mathrm{BS}}(x, y) \\
&\quad = \sup\{\tau(x^{1/2}(\log x^{1/2}\tilde{y}^{-1}x^{1/2})x^{1/2} : y \le \tilde{y} \in L^1(\mathcal{M};\tau), x \text{ is dominated by } \tilde{y}\},
\end{aligned}$$

which is in $(-\infty, \infty]$. This quantity is a version of that introduced in [3].

In fact, $S_{\mathrm{BS}}(x, y)$ is given as follows:

Proposition 2.4. *For every $x, y \in L^1(\mathcal{M};\tau)_+$,*

$$S_{\mathrm{BS}}(x,y) = \lim_{\varepsilon\downarrow 0} \tau(x^{1/2}(\log x^{1/2}(y+\varepsilon x)^{-1}x^{1/2})x^{1/2}). \tag{2.1}$$

Proof. Let α $(\in(-\infty,\infty])$ be the right-hand side of (2.1), which exists by Lemma 2.3. What we have to show is that

$$\alpha \geq \tau(x^{1/2}(\log x^{1/2}\tilde{y}^{-1}x^{1/2})x^{1/2}) \tag{2.2}$$

holds for every $\tilde{y} \in L^1(\mathcal{M};\tau)_+$ such that $y \leq \tilde{y}$ and $x \leq \lambda\tilde{y}$ for some $\lambda \geq 0$. For any $\varepsilon > 0$ we get by Lemma 2.3

$$\tau(x^{1/2}(\log x^{1/2}(y+\varepsilon x)^{-1}x^{1/2})x^{1/2}) \geq \tau(x^{1/2}(\log x^{1/2}(\tilde{y}+\varepsilon x)^{-1}x^{1/2})x^{1/2}),$$

so that

$$\alpha \geq \lim_{\varepsilon\downarrow 0} \tau(x^{1/2}(\log x^{1/2}(\tilde{y}+\varepsilon x)^{-1}x^{1/2})x^{1/2}). \tag{2.3}$$

Let $\tilde{a}$ and $\tilde{a}_\varepsilon$ be defined for $x, \tilde{y}$ and $x, \tilde{y}+\varepsilon x$, respectively, as in Lemma 2.1. Since $\tilde{y}+\varepsilon x \leq (1+\varepsilon\lambda)\tilde{y}$, we get $\tilde{a}_\varepsilon\tilde{a}_\varepsilon^* \geq (1+\varepsilon\lambda)^{-1}\tilde{a}\tilde{a}^*$ by Lemma 2.1(3). Hence as in the proof of Lemma 2.3

$$\begin{aligned}
&\tau(x^{1/2}(\log x^{1/2}(\tilde{y}+\varepsilon x)^{-1}x^{1/2})x^{1/2}) \\
&\quad = \lim_{\delta\downarrow 0} \tau(x^{1/2}(\log(\tilde{a}_\varepsilon\tilde{a}_\varepsilon^* + \delta))x^{1/2}) \\
&\quad \geq \lim_{\delta\downarrow 0} \tau(x^{1/2}(\log(1+\varepsilon\lambda)^{-1}(\tilde{a}\tilde{a}^* + \delta))x^{1/2}) \\
&\quad = -\tau(x)\log(1+\varepsilon\lambda) + \lim_{\delta\downarrow 0} \tau(x^{1/2}(\log(\tilde{a}\tilde{a}^* + \delta))x^{1/2}) \\
&\quad = -\tau(x)\log(1+\varepsilon\lambda) + \tau(x^{1/2}(\log x^{1/2}\tilde{y}^{-1}x^{1/2})x^{1/2}),
\end{aligned}$$

so that

$$\lim_{\varepsilon\downarrow 0} \tau(x^{1/2}(\log x^{1/2}(\tilde{y}+\varepsilon x)^{-1}x^{1/2})x^{1/2}) \geq \tau(x^{1/2}(\log x^{1/2}\tilde{y}^{-1}x^{1/2})x^{1/2}).$$

This together with (2.3) implies (2.2) required. ∎

By Proposition 2.4 and the proof of Lemma 2.2 we see that

$$S_{\mathrm{BS}}(x,y) \geq \tau(x-y)$$

holds for every $x, y \in L^1(\mathcal{M};\tau)_+$.

The main result of this section is the following:

Theorem 2.5. *Assume that $\mathcal{M}$ is hyperfinite. Then*

$$S(x,y) \le S_{\rm BS}(x,y) \tag{2.4}$$

holds for every $x, y \in L^1(\mathcal{M};\tau)_+$.

For the proof of Theorem 2.5 we first state the next lemma which was given in [14].

Lemma 2.6. *Assume that $\mathcal{M}$ is finite dimensional. Then inequality (2.4) holds for every $x, y \in \mathcal{M}_+$.*

Lemma 2.7. *Assume that $\mathcal{M}$ is hyperfinite and $\tau(1) < \infty$. If $x, y \in \mathcal{M}_+$ and y is invertible, then inequality (2.4) holds.*

Proof. Let $\{\mathcal{M}_j\}$ be an increasing net of finite dimensional subalgebras of $\mathcal{M}$ and ε_j be the conditional expectation of $\mathcal{M}$ onto $\mathcal{M}_j$ with respect to τ. By the martingale convergence of relative entropy (see [2, 16]) we have

$$S(x,y) = \lim_j S(\varepsilon_j(x), \varepsilon_j(y)). \tag{2.5}$$

Let us prove that

$$S_{\rm BS}(x,y) = \lim_j S_{\rm BS}(\varepsilon_j(x), \varepsilon_j(y)). \tag{2.6}$$

The martingale convergence theorem [24] says that $\varepsilon_j(x) \to x$ and $\varepsilon_j(y) \to y$ both strongly and in L^1-norm. Hence $a_j = \varepsilon_j(x)^{1/2}\varepsilon_j(y)^{-1/2} \to a = x^{1/2}y^{-1/2}$ strongly*, which implies that $a_j^* a_j \log a_j^* a_j \to a^* a \log a^* a$ strongly. Therefore (2.6) is seen because

$$\begin{aligned} S_{\rm BS}(\varepsilon_j(x), \varepsilon_j(y)) &= \tau(\varepsilon_j(y)^{1/2}(a_j^* a_j \log a_j^* a_j)\varepsilon_j(y)^{1/2}) \\ &= \tau(\varepsilon_j(y)(a_j^* a_j \log a_j^* a_j)) \\ &\to \tau(y(a^* a \log a^* a)) = S_{\rm BS}(x,y). \end{aligned}$$

Then inequality (2.4) follows from Lemma 2.6, (2.5) and (2.6). ∎

Proof of Theorem 2.5. By Proposition 2.4 and the lower semicontinuity of relative entropy (see [2]) it suffices to prove the case when $x \le \lambda y$ for some $\lambda \ge 0$. Furhter we may assume $e_{(0,\infty)}(y) = 1$. Let $e_n = e_{[1/n,n]}(y)$, $x_n = e_n x e_n$, and $y_n = y e_n$ for $n \in \mathbf{N}$. Then $\tau(e_n) < \infty$, $e_n \nearrow 1$, and $x_n \le \lambda y_n \le \lambda n e_n$. Hence $x_n, y_n \in (e_n \mathcal{M} e_n)_+$ and y_n is invertible in $e_n \mathcal{M} e_n$. Applying Lemma 2.7 in $e_n \mathcal{M} e_n$ with $\tau|e_n \mathcal{M} e_n$ we have

$$S(x_n, y_n) \le S_{\rm BS}(x_n, y_n).$$

Since $\|x_n - x\|_1 \to 0$ and $\|y_n - y\|_1 \to 0$ in $L^1(\mathcal{M};\tau)$, we have also

$$S(x,y) \le \liminf_{n\to\infty} S(x_n, y_n)$$

by the lower semicontinuity of relative entropy. Therefore, to show (2.4) it suffices to prove that

$$S_{\mathrm{BS}}(x,y) = \lim_{n\to\infty} S_{\mathrm{BS}}(x_n, y_n). \tag{2.7}$$

Let a be as in Lemma 2.1 and $a_n = x_n^{1/2} y_n^{-1/2} e_n$. Then for every $\xi \in e_n\mathcal{H}$ we get

$$\|a_n(y^{1/2}\xi)\|^2 = \|x_n^{1/2}\xi\|^2 = \langle e_n x e_n \xi, \xi\rangle = \|x^{1/2}\xi\|^2 = \|a(y^{1/2}\xi)\|^2,$$

so that $a_n^* a_n = e_n a^* a e_n$ because $y^{1/2} e_n \mathcal{H} = e_n \mathcal{H}$. This shows that $a_n^* a_n \to a^* a$ strongly and hence $a_n^* a_n \log a_n^* a_n \to a^* a \log a^* a$ strongly. Now we obtain (2.7) in the same way as (2.6). ∎

The relative operator entropy introduced in [9] is defined as follows: When $x, y \in \mathcal{M}$ are positive invertible,

$$\hat{S}(x,y) = x^{1/2}(\log x^{-1/2} y x^{-1/2}) x^{1/2} = -x^{1/2}(\log x^{1/2} y^{-1} x^{1/2}) x^{1/2}$$

For general $x, y \in \mathcal{M}_+$,

$$\hat{S}(x,y) = -\text{s-}\lim_{\varepsilon\downarrow 0} x^{1/2}(\log x^{1/2}(y+\varepsilon)^{-1} x^{1/2}) x^{1/2},$$

whenever the strong limit in the right-hand side exists.

See [8, 9] for properties of relative operator entropy. For convenience we here state the following:

1° If $x, y \in \mathcal{M}_+$ and x is dominated by y, then $\hat{S}(x,y)$ exists and

$$-\hat{S}(x,y) = x^{1/2}(\log x^{1/2} y^{-1} x^{1/2}) x^{1/2},$$

where $x^{1/2} y^{-1} x^{1/2}$ is in the sense defined after Lemma 2.1.

2° If $x, y, \tilde{y} \in \mathcal{M}_+$ with $y \le \tilde{y}$ and if $\hat{S}(x,y)$ exists, then $\hat{S}(x,\tilde{y})$ exists and

$$\hat{S}(x,y) \le \hat{S}(x,\tilde{y}).$$

Proposition 2.8. *If $x, y \in \mathcal{M}_+ \cap L^1(\mathcal{M};\tau)$ and $\hat{S}(x,y)$ exists, then*

$$S_{\mathrm{BS}}(x,y) = \tau(-\hat{S}(x,y)).$$

Proof. For every $\varepsilon > 0$, since $y \le y+\varepsilon x \le y+\varepsilon\|x\|$, it follows from 2° that $\hat{S}(x, y+\varepsilon x)$ exists and

$$x^{1/2}(\log x^{1/2}(y+\varepsilon\|x\|)^{-1}x^{1/2})x^{1/2} \le -\hat{S}(x, y+\varepsilon x) \le -\hat{S}(x, y).$$

Therefore $-\hat{S}(x, y+\varepsilon x) \nearrow -\hat{S}(x, y)$ as $\varepsilon \downarrow 0$. So the desired equality holds by 1°, Lemma 2.2 and Proposition 2.4. ∎

In particular when $\mathcal{M} = \mathbf{B}(\mathcal{H})$ with the usual trace $\tau = Tr$, we have:

Corollary 2.9. *Let X and Y be positive trace class operators. Then $S_{\mathrm{BS}}(X, Y) < \infty$ if and only if $\hat{S}(X, Y)$ exists and is in the trace class. In this case,*

$$S_{\mathrm{BS}}(X, Y) = Tr(-\hat{S}(X, Y)).$$

Proof. The 'if' part is seen by Proposition 2.8. Let us show the 'only if' part. By 1° and Proposition 2.4 we get

$$\infty > S_{\mathrm{BS}}(X, Y) = \lim_{\varepsilon \downarrow 0} Tr(-\hat{S}(X, Y+\varepsilon X)).$$

Moreover by 2°, $-\hat{S}(X, Y+\varepsilon X)$ is increasing as $\varepsilon \downarrow 0$. Hence there exists a trace class operator Z such that $-\hat{S}(X, Y+\varepsilon X) \nearrow Z$ as $\varepsilon \downarrow 0$. This shows that $\hat{S}(X, Y)$ exists and is in the trace class. ∎

References

1. H. Araki, Relative entropy of states of von Neumann algebras, *Publ. RIMS, Kyoto Univ.*, **11** (1976), 809–833.
2. H. Araki, Relative entropy for states of von Neumann algebras II, *Publ. RIMS, Kyoto Univ.*, **13** (1977), 173–192.
3. V. P. Belavkin and P. Staszewski, C^*-algebraic generalization of relative entropy and entropy, *Ann. Inst. H. Poincaré Phys. Théor.*, **37** (1982), 51–58.
4. J. Dixmier, Formes linéaires sur un anneau d'opérateurs, *Bull. Soc. Math. France*, **81** (1953), 9–39.
5. T. Fack, Sur la notion de valeur caractéristique, *J. Operator Theory*, **7** (1982), 307–333.
6. T. Fack and H. Kosaki, Generalized s-numbers of τ-measurable operators, *Pacific J. Math.*, **123** (1986), 269–300.
7. B. Fuglede and R. V. Kadison, Determinant theory in finite factors, *Ann. of Math.*, **55** (1952), 520–530.
8. J. I. Fujii, M. Fujii and Y. Seo, An extension of the Kubo-Ando theory: solidarities, *Math. Japon.*, **35** (1990), 387–396.

9. J. I. Fujii and E. Kamei, Relative operator entropy in noncommutative information theory, *Math. Japon.*, **34** (1989), 341–348.
10. I. C. Gohberg and M. G. Krein, Introduction to the Theory of Linear Nonselfadjoint Operators, *Transl. Math. Monographs*, Vol. 18, Amer. Math. Soc., 1969.
11. F. Hiai, Majorization and stochastic maps in von Neumann algebras, *J. Math. Anal. Appl.*, **109** (1985), 74–82.
12. F. Hiai and Y. Nakamura, Majorizations for generalized s-numbers in semifinite von Neumann algebras, *Math. Z.*, **195** (1987), 17–27.
13. F. Hiai and Y. Nakamura, Closed convex hulls of unitary orbits in von Neumann algebras, *Trans. Amer. Math. Soc.*, **323** (1991), 1–38.
14. F. Hiai and D. Petz, The proper formula for relative entropy and its asymptotics in quantum probability, *Comm. Math. Phys.*, **143** (1991), 99–114.
15. F. Hiai and D. Petz, The Golden-Thompson trace inequality is complemented, preprint.
16. H. Kosaki, Relative entropy of states: a variational expression, *J. Operator Theory*, **16** (1986), 335–348.
17. E. Nelson, Notes on non-commutative integration, *J. Funct. Anal.*, **15** (1974), 103–116.
18. D. Petz, Spectral scale of self-adjoint operators and trace inequalities, *J. Math. Anal. Appl.*, **109** (1985), 74–82.
19. I. Segal, A non-commutative extension of abstract integration, *Ann. of Math.*, **57** (1953), 401–457.
20. B. Simon, Trace Ideals and Their Applications, Cambridge Univ. Press, Cambridge, 1979.
21. M. Terp, L^p spaces associated with von Neumann algebras, Notes, Copenhagen Univ., 1981.
22. O. E. Tikhonov, Continuity of operator functions in topologies connected with a trace on a von Neumann algebra (Russian), *Izv. Vyssh. Uchebn. Zaved. Math.*, 1987, no. 1, 77-79; translated in *Soviet Math. (Iz. VUZ)*, **31** (1987), 110–114.
23. A. Uhlmann, Relative entropy and the Wigner-Yanase-Dyson-Lieb concavity in an interpolation theory, *Comm. Math. Phys.*, **54** (1977), 21–32.
24. H. Umegaki, Conditional expectation in an operator algebra, II, *Tôhoku Math. J.*, **8** (1956), 86–100.
25. H. Umegaki, Conditional expectation in an operator algebra, IV (entropy and information), *Kōdai Math. Sem. Rep.*, **14** (1962), 59–85.

Quantum Probability and Related Topics Vol. VIII (pp. 237–261)

Quantum Topological Entropy: First Steps of a "Pedestrian" Approach

Thomas Hudetz
Institut für Theoretische Physik
Universität Wien
Boltzmanngasse 5, A–1090 Wien

1 Introduction and Discussion

The purpose of this paper is to introduce a notion of topological entropy for automorphisms of arbitrary (non–commutative, but unital) *nuclear* C^*–algebras $\mathcal{A}$, generalizing the "classical" (see (2.1) below) topological entropy for a homeomorphism $T : X \to X$ of an arbitrary (possibly connected) compact Hausdorff space X, where the generalization is of course understood in the sense that the latter topological dynamical system (with $\mathbb{Z}$-action) is equivalently viewed as the C^*–dynamical system given by the T–induced automorphism of the *Abelian* C^*–algebra $\mathcal{A} = C(X)$ of (complex–valued) continuous functions on X.

In the same sense, also the Connes–Narnhofer–Thirring (CNT for short, [1]) entropy for a C^*–dynamical system (with respect to a single invariant state) is a generalization of the "classical" *measure–theoretic* (Kolmogorov–Sinai) dynamical entropy; and it was originally designed for this very same class of nuclear C^*–algebras, but only recently it has been extended to the class of quasi–local algebras in quantum statistical mechanics (by Park and Shin [2]). In both cases, the key concept is that of completely positive unital maps $\gamma : \mathcal{B} \to \mathcal{A}$ of finite rank (i.e. from any finite–dimensional unital C^*–algebra $\mathcal{B}$) into the given C^*–algebra $\mathcal{A}$, which is the "natural" extension of the special case of an inclusion $*$–homomorphism $\gamma = \imath_{\mathcal{B}} : \mathcal{B} \to \mathcal{A}$ of a finite–dimensional C^*–*sub*algebra $\mathcal{B} \subset \mathcal{A}$; and this extension is really necessary for a possible application of the CNT entropy theory to (nuclear) C^*–algebras $\mathcal{A}$ without sufficiently many finite–dimensional subalgebras, in particular the projectionless (or "connected", cf. [3] and e.g. [4]) C^*–algebras $\mathcal{A}$. The latter class of C^*–algebras contains as *Abelian* algebras $\mathcal{A}$ exactly the C^*–algebras $\mathcal{A} \cong C(X)$ of all possible *connected* compact spaces X; and even then every finite partition of unity $\alpha = \{A_i \in \mathcal{A}^+ \,|\, \sum_i A_i = \mathbb{1} \in \mathcal{A}\}$ determines a (completely) positive unital map $\gamma_\alpha : \mathcal{B}_\alpha \to \mathcal{A}$ from a certain finite–dimensional *Abelian* C^*–algebra $\mathcal{B}_\alpha$ into $\mathcal{A}$ (simply defining $\gamma_\alpha(e_i) = A_i$ with the minimal projections $e_i \in \mathcal{B}_\alpha$), but on the other hand this partition of unity α uniquely determines an open cover $\mathcal{U}(\alpha) = \{U_i \subset X \,|\, \bigcup_i U_i = X\}$ of X, simply defining the U_i to be the cozero–sets of $A_i \in \mathcal{A} = C(X)$, i.e. $X \setminus U_i = A_i^{-1}(0)$. In turn, by Urysohn's lemma, any finite open cover of X has always a *subordinate* partition α of the unit $\mathbb{1} \in C(X) = \mathcal{A}$, and thus the set of open covers of the form $\mathcal{U}(\alpha)$ is a *cofinal* subset of the set $\mathcal{O}(X)$ of all finite open covers of X, w.r.t. the natural partial order on $\mathcal{O}(X)$; cf. Sect. (2.2) below. – This correspondence $\gamma_\alpha \mapsto \mathcal{U}(\alpha)$ was used in [5] to "pull back" the classical definition of topological entropy (see still (2.1) below) to the category of Abelian C^*–algebras $\mathcal{A} \ni 1\!\!1$ (together with the "auxiliary" $\mathcal{B}_\alpha$) and positive unital maps (in particular the $\gamma_\alpha : \mathcal{B}_\alpha \to \mathcal{A}$ and the $*$–automorphisms $\theta : \mathcal{A} \to \mathcal{A}$), and we shall once more reformulate this simple redefinition in Section (2.3) below.

The challenging extension of this classically *equivalent* definition to the category of *non*–Abelian (nuclear) C^*–algebras $\mathcal{A} \ni 1\!\!1$ and (completely) positive unital maps is of course motivated by the above–mentioned CNT generalization of the measure–theoretic dynamical entropy: Our simple basic idea is to consider the (completely) positive unital maps $\gamma : \mathcal{B} \to \mathcal{A}$ of finite rank also as the "proper" non–commutative generalization of finite open *covers* for a compact space X; in much the same way as already the finite–dimensional C^*–*sub*algebras $\mathcal{B} \subset \mathcal{A}$ (if any exist) are viewed as the non–Abelian generalization of finite (measurable or clopen Borel–) *partitions* of X, originally in the CNT– and related work (see in particular [7, 6], and cf. e.g. [8, 5] for a review), but even so in Lindblad's later counter–proposal to it [9].

From the latter work, however, we take the essential idea of how to define the "joint topological entropy" $\bar{H}(\gamma_1, \ldots, \gamma_n)$ of several such positive unital maps $\gamma_k : \mathcal{B}_k \to \mathcal{A}$ ($k = 1, \ldots, n$) for non–commutative $\mathcal{A}$, generalizing the "topological" entropy $H(\mathcal{U}_1 \vee \ldots \vee \mathcal{U}_n)$ of the common refinement of several open covers $\mathcal{U}_k$ of a space X (see Def. (2.1) below for its classical definition) in the sense explained above: $\bar{H}(\gamma_1, \ldots, \gamma_n)$ will be defined (in Def. (3.8)) by a variational expression, formally similar to the CNT definition of $H_\omega(\gamma_1, \ldots, \gamma_n)$ for a particularly chosen state ω on $\mathcal{A}$ [1]; but now the supremum is taken over (finite) "square root" decompositions $\alpha_k \subset \mathcal{B}_k$ of the identity (in the sense of Lindblad's "operational partitions of unity" = OPUs [9]) in each of the preimage–algebras $\mathcal{B}_k$, respectively, and the functional to be maximized depends only on the n–tuple of "image–OPUs" $(\gamma_1[\alpha_1]_2, \ldots, \gamma_n[\alpha_n]_2) \subset \mathcal{A}^n$. Here $\gamma_k[\bullet]_2 : \mathcal{B}_k \to \mathcal{A}^+$ denotes the "quantum mechanically" *non*–linear application of the originally *linear* map $\gamma_k : \mathcal{B}_k \to \mathcal{A}$ (to be defined in Def.(3.6) below), and in the special case of the inclusion $*$–homomorphism $\gamma_k = \imath_{\mathcal{B}_k} : \mathcal{B}_k \to \mathcal{A}$ of a finite–dimensional *sub*algebra $\mathcal{B}_k \subset \mathcal{A}$ this coincides again with the usual *linear* application on the positive cone $\mathcal{B}_k^+$. In this case we thus have $\imath_{\mathcal{B}_k}[\alpha_k]_2 = \imath_{\mathcal{B}_k}(\alpha_k)$ for our OPUs $\alpha_k \subset \mathcal{B}_k^+$ with *positive* elements, which we can in turn identify with $\alpha_k = \imath_{\mathcal{B}_k}(\alpha_k)$ as OPUs in the *sub*algebras $\mathcal{B}_k \subset \mathcal{A}$; and then the "entropy" functional, maximized to give $\bar{H}(\imath_{\mathcal{B}_1}, \ldots, \imath_{\mathcal{B}_n}) \equiv \bar{H}(\mathcal{B}_1, \ldots, \mathcal{B}_n)$, depends more precisely only on the possible n–fold products (in $\mathcal{A}$) of the elements in α_k, always one for each k and *ordered* in k. This is exactly the same construction as used by Lindblad [9] to define the "ordered" refinement (or "composition", in his terminology) of the OPUs α_k "associated" with the resp. finite–dimensional subalgebras $\mathcal{B}_k \subset \mathcal{A}$; and evidently by this construction the functional $\bar{H}(\mathcal{B}_1, \ldots, \mathcal{B}_n)$, or more generally $\bar{H}(\gamma_1 \ldots, \gamma_n)$, is *not* necessarily symmetric in its arguments $\mathcal{B}_k$ resp. γ_k, in contradistinction to the CNT functional H_ω with state ω. In addition, we have to require in Def. (3.8) of $\bar{H}(\gamma_1, \ldots, \gamma_n)$ that the OPUs $\alpha_k \subset \mathcal{B}_k^+$ be determined by the respective *map* γ_k (not only by the counting index, i.e. for $\gamma_k = \gamma_\ell$ also $\alpha_k(\gamma_k) = \alpha_\ell(\gamma_\ell)$ is fixed in the optimization) in order to retain at least the invariance (3.11,iii) under repetitions of only *one* single map: $\bar{H}(\gamma_1, \ldots, \gamma_1) = \bar{H}(\gamma_1)$; which makes a further difference to the variational procedure in the CNT definition of $H_\omega(\gamma_1, \ldots, \gamma_n)$ where this (and even "better") invariance is obtained without such an *a priori* condition referring to repeated arguments.

On the other hand, however, the main advantage of our definition of $\tilde{H}(\gamma_1, \ldots, \gamma_n)$ seems to be that there is a very "natural" step-by-step physical interpretation for it, and thus also for the usual "dynamical" entropy $\hbar(\theta, \gamma)$ of Def. (3.17) below. We cannot discuss this interpretation here in an accessible and readable manner within the space allotted for the introduction, but we refer to a forthcoming "physical" study [11] (and also the "comprehensive" thesis [10]). We only want to announce here that we regard this physical interpretation for the quantum "topological" entropy $\hbar(\theta, \gamma)$ of Def. (3.17) as a "partially" satisfactory answer to the second part of "Problem 4" posed by Emch [12] (there for the quantum dynamical entropy w.r.t. a single θ-invariant state ω, e.g. the CNT entropy $h_\omega(\theta, \gamma)$ as comparatively used in Sect. (2.2),(2.3) below), namely to obtain at least a "sufficient characterization" of the entropy $\hbar(\theta, \gamma)$, "...in such a manner that the resulting entropy ..." $\hbar(\theta)$, here defined in Def. (3.20), "...be physically accessible to empirical testing". On the other hand, the "first" part of this "Problem 4" in [12], namely "to obtain a necessary and sufficient characterization of the ..." here approximating net τ in Def. (3.20) of $\hbar(\theta)$, "...in such a manner that the resulting entropy $\hbar(\theta)$ be mathematically adapted to the notion of K-flow" (in our case, of course that of C^*-algebraic K-flow as e.g. in "Axioms A" of [12]) seems still to be widely open, as of this writing (cf. Problem (3.22) below). Unfortunately, our definition of the entropy $\hbar(\theta, \gamma)$ as proposed here seems to be not even "mathematically adapted" to define a notion of *entropic* K-systems (and "memory loss", see [13] and cf. e.g. [12] but also [14]), in view of the "no-go" part of Prop. (3.18,iv) below (cf. also the *additional* "no-go" part of (3.21,ii), related to the main problem (3.22) and its "classical roots" in Prop. (2.10)). Taken together, this is what makes our present approach (only mathematically) "pedestrian"; but on the other hand, if one could succeed in finding a definition of quantum topological entropy which is closely "mimicking" the CNT construction [1] and presumably also related to the recent conceptual reformulation of the latter by Sauvageot and Thouvenot [15] (and thus mathematically much more "adapted" than the present approach), in turn the *second* part of the analogous "Problem 4" as posed by Emch [12] (and cited above) would pop up again, analogous also to Lindblad's explicit criticism of the CNT entropy as being "difficult to interpret in a physically meaningful way" [9].

2 Classical Theory and C^*–algebraic Reformulation

2.1 The classical theory of topological entropy

For the convenience of the reader and to fix our notation, we first recall the definitions and general properties of the "classical" topological entropy as in the original paper by Adler, Konheim and McAndrew [16], which contains already most of the essential parts of the classical theory. Throughout this section we use the standing notation (X, T) for a topological dynamical system, given by a compact Hausdorff space X (not necessarily metric) and a homeomorphism $T : X \to X$. By definition, any open cover of X possesses a *finite* subcover, and we can restrict ourselves to the latter from the very beginning, denoting by $\mathcal{O}(X)$ or simply $\mathcal{O}$ the set of finite open covers of X.

Definition (2.1): For $\mathcal{U}, \mathcal{V} \in \mathcal{O}(X)$ we define their join $\mathcal{U} \vee \mathcal{V} = \{U \cap V | U \in \mathcal{U}, V \in \mathcal{V}\} \in \mathcal{O}(X)$, and the action of T on $\mathcal{U} \in \mathcal{O}$ by $T^{-1}(\mathcal{U}) = \{T^{-1}(U) | U \in \mathcal{U}\}$.

(i) $N(\mathcal{U}) = \min\{\operatorname{card} \mathcal{U}' | \mathcal{U}' \subset \mathcal{U} : \mathcal{U}' \in \mathcal{O}(X)\}$ denotes the cardinality of a minimal subcover $\mathcal{U}'$ of $\mathcal{U} \in \mathcal{O}$. The "topological" *entropy* of $\mathcal{U} \in \mathcal{O}(X)$ is defined by $H(\mathcal{U}) = \log N(\mathcal{U})$.

(ii) The entropy of T w.r.t. $\mathcal{U} \in \mathcal{O}$ is defined as $h(T, \mathcal{U}) = \lim_{n\to\infty} \frac{1}{n} H(\mathcal{U} \vee T^{-1}\mathcal{U} \vee \ldots \vee T^{-n+1}\mathcal{U})$.

(iii) The *topological entropy* of T is $h(T) = \sup_{\mathcal{U}\in\mathcal{O}} h(T, \mathcal{U})$.

A cover $\mathcal{V} \in \mathcal{O}$ is said to be a *refinement* of a cover $\mathcal{U}$, $\mathcal{U} \prec \mathcal{V}$, if $\forall V \in \mathcal{V}\ \exists U \in \mathcal{U} : V \subset U$; and obviously $(\mathcal{O}, \succ)$ is a partially pre–ordered set, but even a (pre–)*directed* set with $\mathcal{U} \vee \mathcal{V} \succ \mathcal{U}, \mathcal{V}$.

Proposition (2.2): The entropy functionals (2.1,i–iii) have the following general properties: $\forall \mathcal{U}, \mathcal{V} \in \mathcal{O}(X)$,

ad (i)
(a) $\mathcal{U} \prec \mathcal{V} \Longrightarrow N(\mathcal{U}) \leq N(\mathcal{V}) \Longleftrightarrow H(\mathcal{U}) \leq H(\mathcal{V})$.
(b) $N(\mathcal{U}) \cdot N(\mathcal{V}) \geq N(\mathcal{U} \vee \mathcal{V}) \Longleftrightarrow H(\mathcal{U} \vee \mathcal{V}) \leq H(\mathcal{U}) + H(\mathcal{V})$.
(c) $N(T^{-1}\mathcal{U}) = N(\mathcal{U}) \Longleftrightarrow H(T^{-1}\mathcal{U}) = H(\mathcal{U})$.

ad (ii)
(d) $\mathcal{U} \prec \mathcal{V} \Longrightarrow h(T, \mathcal{U}) \leq h(T, \mathcal{V})$, follows from (a).
(e) $h(T, \mathcal{U}) \leq H(\mathcal{U})$, follows from (b) & (c).
(f) $h(T, \mathcal{U}) = h(T^{-1}, \mathcal{U})$, follows from (c).

ad (iii)
(g) $h(T) = \lim_{\mathcal{U}\in\mathcal{O}} h(T, \mathcal{U})$ as a *net* limit (corollary of d); and in particular for a *refining* sequence $\{\mathcal{U}_n \in \mathcal{O}\}_{n\in\mathbb{N}}$ of open covers, i.e. $\mathcal{U}_n \prec \mathcal{U}_{n+1}\ \forall n$ and such that $\forall \mathcal{V} \in \mathcal{O}\ \exists n \in \mathbb{N} : \mathcal{V} \prec \mathcal{U}_n$, as a *subnet*: $h(T) = \lim_{n\to\infty} h(T, \mathcal{U}_n)$.
(h) Let $\Phi : X \to X'$ be a homeomorphism onto a compact (Hausdorff) space X', then $h(T) = h(\Phi \circ T \circ \Phi^{-1})$, follows from (c).

(i) $h(T) = h(T^{-1})$, follows from (f).

As used already in [16], (g) simplifies the computation of $h(T)$ for *metric* spaces (X, d), where each sequence $\{\mathcal{V}_k \in \mathcal{O}(X)\}_{k\in\mathbb{N}}$ with diameters $d(\mathcal{V}_k) \equiv \max_{V_k\in\mathcal{V}_k} d(V_k)$ shrinking to zero, $d(\mathcal{V}_k) \to 0$, gives a *refining* sequence $\{\mathcal{U}_n | \mathcal{U}_n = \bigvee_{k=1}^n \mathcal{V}_k\}_{n\in\mathbb{N}}$ by Lebesgue's covering lemma.

Theorem (2.3): In addition, the topological entropy (2.1,iii) has the following special properties refering to the varying (non-conjugate) topological dynamical system (X, T):

(i) $h(T^k) = k \cdot h(T)\ \forall k \in \mathbb{N}$.

(ii) Let $T = T_1 \times T_2$ be the product homeomorphism of $T_i : X_i \to X_i$ $(i = 1, 2)$ acting on $X = X_1 \times X_2$ (with product topology). Then $h(T) = h(T_1) + h(T_2)$.

(iii) Denote (again generally) by $M(X) = M$ the convex compact space of (Borel) probability measures on X (with the weak-$*$ topology), and by T_M denote the induced affine homeomorphism $T_M : M \to M$ defined by $T_M(\mu) = \mu \circ T^{-1}\ \forall \mu \in M(X)$. If $h(T) > 0$, then $h(T_M) = \infty$ (cf. [17, 18]).

(iv) Denote by $M(X, T) \subset M(X)$ the (non-empty) convex compact set of T_M-fixed points $\mu = T_M(\mu)$, i.e. T-invariant probability measures μ on X. Then $h(T) = \sup_{\mu\in M(X,T)} h_\mu(T)$, where $h_\mu(T)$ denotes the measure-theoretic (Kolmogorov-Sinai, KS) entropy of T w.r.t. μ (cf. originally [19, 20]).

2.2 Motivations for the C^*-algebraic formulation

Now, this latter KS entropy was generalized finally by Connes, Narnhofer and Thirring (CNT for short, [1]) to the entropy $h_\omega(\theta)$ of a $*$-automorphism θ on a (unital) C^*-algebra $\mathcal{A}$ with invariant state $\omega = \omega \circ \theta \in S_{\mathcal{A}}$, the state space of $\mathcal{A}$ (cf. e.g. [8, 5] for a "historical" review and see in particular [7, 6]). Remember that, in the most general case, this CNT entropy is defined at first by $h_\omega(\theta) = \sup_{\gamma\in\mathcal{CP}_1(\mathcal{A})} h_\omega(\theta, \gamma)$, where $\mathcal{CP}_1(\mathcal{A})$ denotes the set of completely positive unital maps $\gamma : \mathcal{B} \to \mathcal{A}$ of finite rank (i.e. from any finite-dimensional unital C^*-algebra $\mathcal{B}$) into the given C^*-algebra $\mathcal{A}$, and where $h_\omega(\theta, \gamma)$ is the "entropy" [1] of $\theta \in * - \mathrm{Aut}(\mathcal{A})$ w.r.t. $\gamma \in \mathcal{CP}_1(\mathcal{A})$, *formally* analogous to (2.1,ii & iii) above. Actually, this supremum taken over $\mathcal{CP}_1(\mathcal{A})$ can already generally be restricted to the set $\mathcal{K}_1(\mathcal{A}) \subset \mathcal{CP}_1(\mathcal{A})$ of all completely positive unital maps $\gamma : M_n(\mathbb{C}) \to \mathcal{A}$ ($\forall n \in \mathbb{N}$!) from any matrix algebra into $\mathcal{A}$: $h_\omega(\theta) = \sup_{\gamma\in\mathcal{K}_1(\mathcal{A})} h_\omega(\theta, \gamma)$; and for an *Abelian* C^*-algebra $\mathcal{A} \cong C(X)$, X as above, and $\theta_T \in * - \mathrm{Aut}(\mathcal{A})$ induced by $T : X \to X$ as above (simply by $\theta_T(A) = A \circ T\ \forall A \in \mathcal{A} = C(X)$), we can restrict even further as stated in (i) of the following

Proposition (2.4)[5]: Let $\mathcal{A} = C(X)$ for some X and $\theta_T \in * - \mathrm{Aut}(\mathcal{A})$ for some $T : X \to X$ (as above) be fixed.

(i) Denote by $\mathcal{P}_1(\mathcal{A})$ the set of all *positive* (as well known, here "completely–" is redundant!) unital maps from finite–dimensional *Abelian* C^*–algebras into $\mathcal{A}$. Then the CNT entropy $h_\omega(\theta_T) = \sup_{\gamma\in\mathcal{P}_1(\mathcal{A})} h_\omega(\theta_T,\gamma)$, $\forall\omega \in S_{\mathcal{A}}$ with $\omega = \omega\circ\theta_T$.

(ii) For $\mu \in M(X)$ denote by $\omega_\mu \in S_{\mathcal{A}}$ the integral state $\omega_\mu = \int_X d\mu$. Then the CNT entropy $h_{\omega_\mu}(\theta_T) = h_\mu(T)$ $\forall\mu \in M(X,T)$, with then obviously invariant $\omega_\mu = \omega_\mu\circ\theta_T$ (and the KS entropy on the r.h.s.).

Both (i) and (ii) are not at all surprising, but also not quite "tautologically" trivial; and in particular (ii) merely points out that, for the *whole* category of Abelian C^*–algebras and their $*$–auto(or –iso)morphisms, the CNT entropy is really nothing but the "pullback" of the KS entropy (for the category of compact spaces X and homeomorphisms) via the Gelfand isomorphism $^\wedge : \mathcal{A} \to C(X)$ and Riesz' representation of states on $\mathcal{A}$. On the other hand, however, (i) & (ii) lead to a *new* "interpretation" (or equivalent redefinition) of this entropy $h_\mu(T)$ for $T: X \to X$ and $\mu \in M(X,T)$ above, in terms of "fuzzy partitions" (cf. [21, 22]) of the unit $\mathbb{1} \in C(X)$ and their "entropy" (cf. [5, Appendix] and see [23]), via the following simple construction:

Even for a general (non–commutative) C^*–algebra $\mathcal{A} \ni \mathbb{1}$, any such map $\gamma \in \mathcal{P}_1(\mathcal{A})$ with $\gamma : \mathcal{B} \to \mathcal{A}$ (and hence $\mathcal{B}$ Abelian!) as in (2.4.,i) *uniquely* determines a partition of the unit $\alpha_\gamma = \{A_i \in \mathcal{A}^+ | \sum_i A_i = \mathbb{1}\}$, simply defining $A_i = \gamma(e_i)$ with the minimal projectors $e_i \in \mathcal{B}$ ($i = 1,\ldots,\dim\mathcal{B}$); and this is the *natural* choice of α_γ for the calculation of the entropy $h_\omega(\theta_T,\gamma)$ in (i) (or of the "related", but much better calculable "fuzzy–entropy" [5, App.], [23]), as the latter depends (in the final effect, besides on ω and θ_T) only on the set $\{\gamma(e_i) \equiv A_i\}$, even when "ignoring" that γ is actually linear. – In turn, again generally, any such partition of the unit $\alpha = \{A_i \in \mathcal{A}^+ | i = 1,\ldots,n : \sum_{i=1}^n A_i = \mathbb{1}\}$ gives a positive unital map $\gamma_\alpha : \mathcal{B}_\alpha \to \mathcal{A}$ with $\mathcal{B}_\alpha = \mathbb{C}\oplus\mathbb{C}\oplus\ldots\oplus\mathbb{C}$ (n times; i.e. $\gamma_\alpha \in \mathcal{P}_1(\mathcal{A})$ as before), again simply defining $\gamma_\alpha(e_i) = A_i$ with the minimal projectors $e_i \in \mathcal{B}_\alpha$ (linearly extended), and thus γ_α is "uniquely" determined by the *set* (not n–tuple) α up to n–permutation automorphisms of $\mathcal{B}_\alpha$; which then allows to really equivalently rephrase the definition of $h_\omega(\theta_T)$ for $\mathcal{A} = C(X)$ as in (2.4,i & ii) in terms of these "fuzzy partitions" α of $\mathbb{1} \in \mathcal{A}$ (see [23]). We henceforth denote by $\mathcal{O}_1^+(\mathcal{A})$ generally the set of all such finite positive partitions α of the unit $\mathbb{1} \in \mathcal{A}$ in a C^*–algebra $\mathcal{A}$.

But now, on the other hand, for $\mathcal{A} = C(X)$ any such $\alpha \in \mathcal{O}_1^+(\mathcal{A})$, $\alpha = \{A_i \in \mathcal{A}^+ | \sum_i A_i = \mathbb{1}\}$, uniquely determines an open cover $\mathcal{U}(\alpha) \in \mathcal{O}(X)$ as used in (2.1 & 2) above, simply by $\mathcal{U}(\alpha) = \{U_i \subset X | X \setminus U_i = A_i^{-1}(0)\}$ (i.e. defining the U_i to be the *cozero*–sets of $A_i \in C(X)$). Of course, this map $\mathcal{O}_1^+(\mathcal{A}) \ni \alpha \mapsto \mathcal{U}(\alpha) \in \mathcal{O}(X)$ is highly non–injective, and we could define a suitable equivalence relation on $\mathcal{O}_1^+(\mathcal{A})$ and factor the map accordingly; but for our purpose we may let it be many–to–one. Even worse, for a *not* metrizable X (i.e. $\mathcal{A}$ not separable), the map $\alpha \mapsto \mathcal{U}(\alpha)$ is not even *surjective* into $\mathcal{O}(X)$ in general, whereas it evidently *is onto* $\mathcal{O}(X)$ for (X,d) metric (as then *every* open set $U \subset X$ is cozero–set of some $A \in \mathcal{A}^+$, e.g. take $A(x) = \inf_{y\in X\setminus U} d(x,y)$, $\forall x \in X$, with $X \setminus U = A^{-1}(0)$ as required; cf. also e.g. [24]). However, the subset $\{\mathcal{U}(\alpha) | \alpha \in \mathcal{O}_1^+(\mathcal{A})\} \subset \mathcal{O}(X)$ is always *cofinal* w.r.t. the partial order (2.1): $\forall\mathcal{V} \in \mathcal{O}(X)$ $\exists\alpha \in \mathcal{O}_1^+(\mathcal{A})$ such that $\mathcal{U}(\alpha) \succ \mathcal{V}$ (by Urysohn's lemma, cf. e.g. [24]); and thus we can reexpress the *topological* entropy (2.1,iii) generally as $h(T) = \sup_{\alpha \in \mathcal{O}_1^+(\mathcal{A})} h(T, \mathcal{U}(\alpha))$, again by (2.2.,d), also for X not necessarily metric.

By the same argument (2.2,d), we know that for $\gamma \in \mathcal{P}_1(\mathcal{A})$, $\gamma : \mathcal{B} \to \mathcal{A} = C(X)$ with "associated" $\alpha_\gamma \in \mathcal{O}_1^+(\mathcal{A})$ as before, and $\mathcal{U}(\alpha_\gamma) \in \mathcal{O}(X)$ as just above:

$$h\big(T, \mathcal{U}(\alpha_\gamma)\big) = \sup_{\beta \in \mathcal{O}_1^+(\mathcal{B})} h\big(T, \mathcal{U}(\gamma(\beta))\big),$$

which is clear from the fact that for $\delta = \{e_i \in \mathcal{B} | i = 1, \ldots, \dim \mathcal{B}\} \in \mathcal{O}_1^+(\mathcal{B})$ with the *minimal* projectors $e_i \in \mathcal{B}$ we have $\gamma(\delta) = \alpha_\gamma$ and then obviously $\mathcal{U}(\alpha_\gamma) \succ \mathcal{U}(\gamma(\beta))$ $\forall \beta \in \mathcal{O}_1^+(\mathcal{B})$; or "symbolically" $\mathcal{U}(\alpha_\gamma) = \bigvee_{\beta \in \mathcal{O}_1^+(\mathcal{B})} \mathcal{U}(\gamma(\beta))$ as a "supremum" (note that for the finite-dimensional *Abelian* preimage-algebra $\mathcal{B}$ there exist only finitely many *different* $\mathcal{U}(\gamma(\beta)) \in \mathcal{O}(X)$ as β varies over $\mathcal{O}_1^+(\mathcal{B})$, such that the supremum over $\mathcal{O}_1^+(\mathcal{B})$ is actually attained as a maximum). In this sense, α_γ is again the "natural" choice for computing the entropy (2.1,ii) of T w.r.t. an open cover "associated" with $\gamma \in \mathcal{P}_1(\mathcal{A})$; and by analogy with (2.4,i & ii) this makes it tempting to define the topological entropy (2.1,iii) of (X,T) *formally* "algebraically" for $\mathcal{A} = C(X)$ and $\theta_T \in * - \mathrm{Aut}(\mathcal{A})$ as before, using the above expression as a definition for "$\hbar(\theta_T, \gamma)$" and again defining $\hbar(\theta_T) = \sup_{\gamma \in \mathcal{P}_1(\mathcal{A})} \hbar(\theta_T, \gamma)$, such that (obviously from the above discussion) again $\hbar(\theta_T) = h(T)$ would be fulfilled (corresponding to (2.4,ii)).

However, for the desired generalization of this still only formally algebraic "definition" of $\hbar(\theta)$ to $\theta \in * - \mathrm{Aut}(\mathcal{A})$ for *non*-Abelian C^*-algebras $\mathcal{A}$ (see Sect. 3), where in addition for $\gamma \in C\mathcal{P}_1(\mathcal{A})$ with $\gamma : \mathcal{B} \to \mathcal{A}$ and also *non*-commutative preimage-algebra $\mathcal{B}$ there is *no* uniquely determined $\alpha_\gamma \in \mathcal{O}_1^+(\mathcal{A})$ any more (as there is for $\gamma \in \mathcal{P}_1(\mathcal{A})$ above), it seems to be necessary (see Sect. 3) to "mimick" the definition [1] of $h_\omega(\theta, \gamma)$ as in (2.4,i) even further in the algebraic definition of "$\hbar(\theta, \gamma)$" (although still equivalent to the above first attempt in the *Abelian* case $\mathcal{A} = C(X)$ and $\gamma \in \mathcal{P}_1(\mathcal{A})$); and consequently to consider the maps $\gamma \in (C)\mathcal{P}_1(\mathcal{A})$ as the "proper" (in Section 3, *non*-Abelian) algebraic generalization of the open covers $\mathcal{U} \in \mathcal{O}(X)$, instead of the more directly related partitions $\alpha \in \mathcal{O}_1^+(\mathcal{A})$ of $\mathbb{1} \in \mathcal{A}$ (as discussed before), using those maps γ already in the C^*-algebraic reformulation (and then non-Abelian generalization) of the entropy (2.1,i) for open covers $\mathcal{U} \in \mathcal{O}(X)$, as it then enters in the definition (2.1,ii) of $h(T, \mathcal{U})$.

First, we know from the above discussion of the latter entropy when restricted to all $\mathcal{U} = \mathcal{U}(\alpha)$ with $\alpha \in \mathcal{O}_1^+(\mathcal{A})$ "associated" to $\gamma \in \mathcal{P}_1(\mathcal{A})$, $\gamma : \mathcal{B} \to \mathcal{A} = C(X)$ and $\alpha = \gamma(\beta)$ for some $\beta \in \mathcal{O}_1^+(\mathcal{B})$, that *a fortiori* also for the former H-entropy (2.1,i):

$$H\big(\mathcal{U}(\alpha_\gamma)\big) = \max_{\beta \in \mathcal{O}_1^+(\mathcal{B})} H\big(\mathcal{U}(\gamma(\beta))\big)$$

is valid, with $\alpha_\gamma = \gamma(\delta)$ as before (follows from (2.2,a)). Thus the right hand side offers itself as the (again first only formally -) "algebraic" definition of an entropy $\bar{H}(\gamma)$ for $\gamma \in (C)\mathcal{P}_1(\mathcal{A})$; and it turns out that this definition can be made "purely" algebraic rather easily, but since in the definition (2.1,i) of $H(\mathcal{U}(\alpha)) = \log N(\mathcal{U}(\alpha))$ we possibly *omit* sets $U_i \in \mathcal{U}(\alpha)$ to get a minimal subcover $\mathcal{U}(\alpha)' \in \mathcal{O}(X)$, corresponding to omitting certain elements $A_i \in \alpha \in \mathcal{O}_1^+(\mathcal{A})$, we first have to weaken the requirement that α be a partition of unity $\mathbb{1} \in \mathcal{A}$.

2.3 C^*–algebraic reformulation of the classical theory

We do this using the notion of *strict* positivity in a (general, not necessarily Abelian) unital C^*-algebra:

Definition (2.5): Let $\mathcal{A} \ni 1\!\!1$ be a C^*-algebra, then $A \in \mathcal{A}^+$ is said to be *strictly* positive (denoted $A > 0$ in contradistinction to $A \geq 0$), if the following (rather obviously) *equivalent* conditions (i)–(iii) are fulfilled:

(i) $\phi(A) > 0 \;\; \forall \phi \in S_{\mathcal{A}}$

(ii) $\exists A^{-1} \in \mathcal{A} : AA^{-1} = A^{-1}A = 1\!\!1$

(iii) $\exists \varepsilon > 0 : A \geq \varepsilon \cdot 1\!\!1$.

Note that in particular for an *Abelian* C^*-algebra $\mathcal{A}$, again isomorphic to $C(X)$ for some X as before via Gelfand's isomorphism $\wedge : \mathcal{A} \to C(X)$, we have that $A > 0$ in $\mathcal{A}$ iff $\hat{A}(x) > 0 \;\; \forall x \in X$.

Again for a *general* C^*-algebra $\mathcal{A} \ni 1\!\!1$, clearly $\mathcal{A}^+ \ni A > 0 \iff A^2 > 0$, but as an easy exercise (in elementary C^*-theory) one sees that for a set $\alpha = \{A_i \in \mathcal{A}^+ | i = 1, \ldots, n\}$ also even

$$\left(\sum_{i=1}^{n} A_i\right) > 0 \iff \left(\sum_{i=1}^{n} A_i^2\right) > 0;$$

and we call such a set α fulfilling these equivalent conditions a "positive operator cover" for $\mathcal{A}$, and $\mathcal{O}^+(\mathcal{A}) \supset \mathcal{O}_1^+(\mathcal{A})$ denotes the set of all these α.

Now in turn again for *Abelian* $\mathcal{A} = C(X)$, we can directly extend our "definition" (in Sect. (2.2) above) of the map $\mathcal{O}_1^+(\mathcal{A}) \ni \alpha \mapsto \mathcal{U}(\alpha)[\in \mathcal{O}(X)]$ also to $\alpha \in \mathcal{O}^+(\mathcal{A})$ but even to $\alpha = \{A_i \in \mathcal{A}^+\} \notin \mathcal{O}^+(\mathcal{A})$, and obviously from above $\alpha \in \mathcal{O}^+(\mathcal{A}) \iff \mathcal{U}(\alpha) \in \mathcal{O}(X)$. Furthermore, for $\mathcal{A} = C(X)$ and $\alpha = \{A_i\}, \beta = \{B_j\} \in \mathcal{O}^+(\mathcal{A})$ we can define their "refinement" $\alpha \vee \beta = \{A_i \cdot B_j \in \mathcal{A}^+\}$, such that obviously $\alpha \vee \beta \in \mathcal{O}^+(\mathcal{A})$ and $\mathcal{U}(\alpha \vee \beta) = \mathcal{U}(\alpha) \vee \mathcal{U}(\beta)$ with the join of Def. (2.1) in $\mathcal{O}(X)$. And for $T : X \to X$ with induced $\theta_T \in * - \mathrm{Aut}(\mathcal{A})$ as in (2.4), we have again obviously: $T^{-1}\mathcal{U}(\alpha) = \mathcal{U}(\theta_T(\alpha)) \;\; \forall \alpha \in \mathcal{O}^+(\mathcal{A})$. Now, we can first "algebraically translate" the classical definition of topological entropy in Def. (2.1) as follows:

Definition (2.6): Let $\mathcal{A}$ be an *Abelian* C^*-algebra with $\theta \in * - \mathrm{Aut}(\mathcal{A})$. For $\alpha, \beta \in \mathcal{O}^+(\mathcal{A})$ we define their common refinement $\alpha \vee \beta \in \mathcal{O}^+(\mathcal{A})$ as just above.

(i) $\bar{N}(\alpha) = \min\{\mathrm{card}\, \alpha' | \alpha' \subset \alpha, \;\; \alpha' \in \mathcal{O}^+(\mathcal{A})\} \in \mathbb{N}$ denotes the cardinality of a minimal "operator subcover" α' of $\alpha \in \mathcal{O}^+(\mathcal{A})$.

For $\gamma_1, \ldots, \gamma_n \in \mathcal{P}_1(\mathcal{A})$ with $\gamma_k : \mathcal{B}_k \to \mathcal{A}$ ($\mathcal{B}_k$ hence a finite-dimensional Abelian C^*-algebra; $k = 1, \ldots, n$), their *entropy* is defined by

$$\bar{H}(\gamma_1, \ldots, \gamma_n) = \max_{\{(\beta_1(\gamma_1), \ldots, \beta_n(\gamma_n))\}} \log \bar{N}(\gamma_1(\beta_1) \vee \ldots \vee \gamma_n(\beta_n)),$$

where the maximum is taken (first as supremum) over the set of n–tuples with elements $\beta_k(\gamma_k) \in \mathcal{O}_1^+(\mathcal{B}_k)$ determined by the resp. *map* γ_k ($\forall k = 1, \ldots, n$; i.e. for $\gamma_k = \gamma_\ell$ also $\beta_k(\gamma_k) = \beta_\ell(\gamma_\ell)$), and the maximum rather obviously exists (cf. the discussion in Sect. (2.2), in particular before Sect. (2.3), and remember Prop.(2.2,b); see also Sect. 3 below).

(ii) The entropy of θ w.r.t. $\gamma \in \mathcal{P}_1(\mathcal{A})$ is defined as $\hbar(\theta,\gamma) = \lim_{n\to\infty} \frac{1}{n}\bar{H}(\gamma, \theta\circ\gamma, \ldots, \theta^{n-1}\circ\gamma)$, where again the limit obviously exists (cf. (2.2,b & c)).

(iii) The "topological" *entropy* of θ is $\hbar(\theta) = \sup_{\gamma\in\mathcal{P}_1(\mathcal{A})} \hbar(\theta,\gamma)$.

Summarizing our findings thus far, we have the

Theorem (2.7):(cf. [5]) Let (X,T) be a topological dynamical system with "associated" C^*–dynamical system $\mathcal{A} = C(X)$ and $\theta_T \in * - \mathrm{Aut}(\mathcal{A})$, then the definitions (2.1) and (2.6) coincide: $h(T) = \hbar(\theta_T)$.

Corollary (2.8): Let $\mathcal{A}$ be a (separable) Abelian C^*-algebra with $\theta \in * - \mathrm{Aut}(\mathcal{A})$, and denote by $S^\theta_{\mathcal{A}}$ the set of θ–invariant states $\omega = \omega\circ\theta \in S_{\mathcal{A}}$, then obviously by (2.3,iv), (2,4,ii) and (2.7): $\hbar(\theta) = \sup_{\omega\in S^\theta_{\mathcal{A}}} h_\omega(\theta)$, with the CNT entropy on the right hand side.

In view of the definition (2.6,iii) of $\hbar(\theta) = \sup_{\gamma\in\mathcal{P}_1(\mathcal{A})} \hbar(\theta,\gamma)$, resp. the result $h_\omega(\theta) = \sup_{\gamma\in\mathcal{P}_1(\mathcal{A})} h_\omega(\theta,\gamma)$ of (2.4,i) for *Abelian* $\mathcal{A}$, it is tempting to think (cf. [5]) of a *direct* "purely algebraic" proof of (2.8), simply by showing that $\hbar(\theta,\gamma) = \sup_{\omega\in S^\theta_{\mathcal{A}}} h_\omega(\theta,\gamma)$, or at least $\hbar(\theta,\gamma) \geq h_\omega(\theta,\gamma)$ $\forall\omega \in S^\theta_{\mathcal{A}}$. But this can certainly *not* be true for *all* $\gamma \in \mathcal{P}_1(\mathcal{A})$, if ever, for the following reason: As shown in [1], $h_\omega(\theta,\gamma)$ is norm–continuous in γ even for *non–Abelian* $\mathcal{A}$ and $\gamma \in \mathcal{CP}_1(\mathcal{A})$ with our notation from Sect. (2.2), what then implies the "non–Abelian KS theorem" for *nuclear* $\mathcal{A}$ (recently generalized for quasi–local C^*–algebras by Park and Shin [2]). Just to remind the reader and again fix the notation, we state

Proposition (2.9)[1]: Let $\mathcal{A} \ni 1\!\!1$ be a (general) unital C^*–algebra with $\theta \in * - \mathrm{Aut}(\mathcal{A})$, $\omega \in S^\theta_{\mathcal{A}}$.

(i) $\forall\gamma \in \mathcal{CP}_1(\mathcal{A})$, $\gamma : \mathcal{B} \to \mathcal{A}$, and $\forall\varepsilon > 0$ $\exists\delta(\dim\mathcal{B},\varepsilon) > 0$: $\forall\gamma' \in \mathcal{CP}_1(\mathcal{A})$, $\gamma' : \mathcal{B} \to \mathcal{A}$ with $\|\gamma - \gamma'\| < \delta$, the CNT entropies $|h_\omega(\theta,\gamma) - h_\omega(\theta,\gamma')| < \varepsilon$.

(ii) If $\mathcal{A}$ is *nuclear* [25], there exists a net of completely positive unital maps $\sigma_\nu : \mathcal{A} \to \mathcal{B}_\nu$, $\tau_\nu : \mathcal{B}_\nu \to \mathcal{A}$ with finite–dimensional C^*–algebras (or full matrix algebras) $\mathcal{B}_\nu$ ($\nu \in N$, a directed set), such that $\lim_{\nu\in N} \|\tau_\nu\circ\sigma_\nu(A) - A\| = 0$, $\forall A \in \mathcal{A}$. Then as a corollary of (i), quite obviously the full CNT entropy $h_\omega(\theta) = \lim_{\nu\in N} h_\omega(\theta,\tau_\nu)$, although $\{\tau_\nu|\nu \in N\} \subset \mathcal{CP}_1(\mathcal{A})$. As is well known (and rather easily seen), there exists an "approximating" *sequence* $\sigma_n : \mathcal{A} \to \mathcal{B}_n$, $\tau_n : \mathcal{B}_n \to \mathcal{A}$ ($n \in \mathbb{N}$) *iff* $\mathcal{A}$ is *separable*; and in this case $h_\omega(\theta) = \lim_{n\to\infty} h_\omega(\theta,\tau_n)$.

But for our "topological" entropy $\hbar(\theta,\gamma)$ of (2.6,ii) with $\mathcal{A}$ *Abelian* and $\gamma \in \mathcal{P}_1(\mathcal{A})$ as before, we have unfortunately the following continuity–"no go" result:

Proposition (2.10): Let $\mathcal{A} \ni 1\!\!1$ be an Abelian C^*–algebra with $\theta \in * - \mathrm{Aut}(\mathcal{A})$.

(i) For $\gamma \in \mathcal{P}_1(\mathcal{A})$, $\gamma : \mathcal{B} \to \mathcal{A}$ ($\mathcal{B}$ finite–dimensional Abelian), and $\forall \delta > 0$, $\exists \gamma' \in \mathcal{P}_1(\mathcal{A})$, $\gamma' : \mathcal{B} \to \mathcal{A}$ such that $\|\gamma - \gamma'\| < \delta$, but $\hbar(\theta,\gamma') = 0$.

(ii) For any $\gamma' \in \mathcal{P}_1(\mathcal{A})$ with $\hbar(\theta,\gamma') = 0$ as above, $\exists \varepsilon(\gamma') > 0$ such that $\hbar(\theta,\gamma'') = 0$ $\forall \gamma'' \in \mathcal{P}_1(\mathcal{A})$, $\gamma'' : \mathcal{B} \to \mathcal{A}$ with $\|\gamma' - \gamma''\| < \varepsilon$.

We leave the proof to the reader as an easy but tedious exercise (going through the above definitions (2.6,i & ii) and using the correspondence $\mathcal{P}_1(\mathcal{A}) \ni \gamma \mapsto \alpha_\gamma \in \mathcal{O}_1^+(\mathcal{A})$ as discussed in Sect. (2.2) for computing $\bar{H}(\gamma_1,\ldots,\gamma_n) = \log \bar{N}(\alpha_{\gamma_1} \vee \ldots \vee \alpha_{\gamma_n}) = 0$ whenever $\exists A_k \in \alpha_{\gamma_k} : A_k > 0$ $\forall k = 1,\ldots,n$). – Thus the $\gamma' : \mathcal{B} \to \mathcal{A}$ with $\hbar(\theta,\gamma') = 0$ are an open dense subset of $\{\gamma \in \mathcal{P}_1(\mathcal{A}) | \gamma . \mathcal{B} \to \mathcal{A}\} \ni \gamma'$ in norm; and hence, also in view of the proof [1] of the "non–Abelian KS theorem" (2.9,ii) for $h_\omega(\theta)$, it is clear that the mere existence of an approximating sequence of maps $\tau_n : \mathcal{A}_n \to \mathcal{A}$, $\sigma_n : \mathcal{A} \to \mathcal{A}_n$ with $\|\tau_n \circ \sigma_n(A) - A\| \to 0$ $(\forall A \in \mathcal{A})$ is not at all *sufficient* for proving that $\hbar(\theta,\tau_n) \to \hbar(\theta)$ by the same (norm–approximation) argument. Surprisingly, however, it turns out that in this Abelian case $\mathcal{A} = C(X)$ with (X,d) metric (as $\mathcal{A}$ is necessarily separable by (2.9,ii)), any "refining" sequence of maps $\{\tau_n \in \mathcal{P}_1(\mathcal{A})\}_{n\in\mathbb{N}}$ with $\hbar(\theta,\tau_n) \to \hbar(\theta)$ has to be *necessarily* of this form (together with the appropriate maps σ_n): Note first that we can define a partial pre–order $\succ$ on $\mathcal{O}^+(\mathcal{A})$ by $\alpha \succ \beta$ for $\alpha,\beta \in \mathcal{O}^+(\mathcal{A})$, if $\forall A_i \in \alpha$ $\exists B_{j(i)} \in \beta$ and $\lambda_i \in (0,1]$: $\lambda_i A_i \le B_{j(i)}$; such that obviously $\alpha \succ \beta \iff \mathcal{U}(\alpha) \succ \mathcal{U}(\beta) \in \mathcal{O}(X)$ as defined in Def. (2.1), with our notation from Sect. (2.2).

Lemma (2.11): Let (X,d) be a *metric* compact space and $\mathcal{A} = C(X)$. Choose a refining sequence $\{\mathcal{U}_n \in \mathcal{O}(X) | d(\mathcal{U}_n) \to 0\}_{n\in\mathbb{N}}$ of open covers as in (2.2,g) and corresponding partitions of unity $\{\beta_n \in \mathcal{O}_1^+(\mathcal{A})\}_{n\in\mathbb{N}}$ such that $\mathcal{U}_n = \mathcal{U}(\beta_n)$, with our notation from Sect. (2.2). For $\beta_n = \{f_{i_n}^{(n)} \in \mathcal{A}^+ | i_n = 1,\ldots,N_n : \sum_{i_n=1}^{N_n} f_{i_n}^{(n)} = 1\!\!1\}$ define maps $\tau_n : \mathcal{B}_n \equiv \bigoplus_{k=1}^{N_n}(\mathbb{C})_k \to \mathcal{A}$ by $\tau_n(e_{i_n}^{(n)}) = f_{i_n}^{(n)}$ with the minimal projections $e_{i_n}^{(n)} \in \mathcal{B}_n$, such that for $\tau_n \in \mathcal{P}_1(\mathcal{A})$ again with our notation from Sect. (2.2): $\tau_n = \gamma_{\beta_n}$ resp. $\beta_n = \alpha_{\tau_n}$. Then there exists a sequence $\{\sigma_n : \mathcal{A} \to \mathcal{B}_n\}_{n\in\mathbb{N}}$ of positive unital maps σ_n with the approximation property (2.9), i.e. $\tau_n \circ \sigma_n \to \mathrm{Id}_{\mathcal{A}}$ in pointwise norm; and for any $\theta \in * - \mathrm{Aut}(\mathcal{A})$ we have $\hbar(\theta) = \lim_{n\to\infty} \hbar(\theta,\tau_n)$.

Proof: Choose points $x_{i_n}^{(n)} \in X$ such that $f_{i_n}^{(n)}(x_{i_n}^{(n)}) > 0$ $\forall i_n = 1,\ldots,N_n$, $n \in \mathbb{N}$, and define $\sigma_n(A) = \sum_{i_n=1}^{N_n} e_{i_n}^{(n)} \cdot A(x_{i_n}^{(n)})$, $\forall A \in \mathcal{A}$, $n \in \mathbb{N}$. Then obviously $\tau_n \circ \sigma_n(A) = \sum_{i_n} f_{i_n}^{(n)} \cdot A(x_{i_n}^{(n)}) \to A$ pointwise on X (i.e. $\tau_n \circ \sigma_n(A)|_x \to A|_x$, $\forall A \in \mathcal{A}$, $x \in X$) as $\bigcup_{n\in\mathbb{N}} \beta_n$ separates the points of X (and actually β_n separates "arbitrarily well" for $n \to \infty$, because obviously $\beta_n \prec \beta_{n+1}$, $\forall n \in \mathbb{N}$, as defined before), which implies by a "Stone–Weierstrass" type argument (actually by a variant of Dini's theorem, although the function sequence $\tau_n \circ \sigma_n(A)$ is not monotonic in general) that $\|\tau_n \circ \sigma_n(A) - A\| \to 0$ $\forall A \in \mathcal{A}$.

In addition, it is evident by (2.2,g) and definition (2.6,ii) of $\hbar(\theta_T, \tau_n) = h(T, \mathcal{U}(\beta_n))$ for any homeomorphism $T : X \to X$, that $\hbar(\theta_T) = \lim_{n\to\infty} \hbar(\theta_T, \tau_n)$.

This detailed simple proof is an exception of our rule to skip *all* of them in this paper; but actually (2.11) paves the way out of the "no-go" impasse (2.10) via the following more general sublemma (with essentially the same proof):

Sublemma (2.11,1): Let $\mathcal{A} \ni \mathbb{1}$ be a non-separable Abelian C^*-algebra, and note that $\mathcal{O}_1^+(\mathcal{A})$ with the partial pre-order $\succ$ as introduced before is a (pre-)directed set with the common refinement (as in (2.6)) $\alpha \vee \beta \succ \alpha, \beta \in \mathcal{O}_1^+(\mathcal{A})$. Considering $\mathcal{P}_1(\mathcal{A}) \ni \gamma_\alpha$ as a net indexed by $\alpha \in \mathcal{O}_1^+(\mathcal{A})$ with our notation from Sect. (2.2), there exists a net $\{\sigma_\alpha | \alpha \in \mathcal{O}_1^+(\mathcal{A})\}$ of positive unital maps σ_α from $\mathcal{A}$ into the resp. preimage-algebra of γ_α, such that $\lim_{\alpha\in\mathcal{O}_1^+(\mathcal{A})} \|\gamma_\alpha \circ \sigma_\alpha(A) - A\| = 0$, $\forall A \in \mathcal{A}$; and for any $\theta \in * - \mathrm{Aut}(\mathcal{A})$ we have $\hbar(\theta) = \lim_{\alpha\in\mathcal{O}_1^+(\mathcal{A})} \hbar(\theta, \gamma_\alpha)$ (as a net limit).

Definition (2.12): Let $\mathcal{A} \ni \mathbb{1}$ be an Abelian C^*-algebra and $\theta \in * - \mathrm{Aut}(\mathcal{A})$. For an approximating sequence or net $\tau = \{\tau_\nu \in \mathcal{P}_1(\mathcal{A}) | \nu \in N\}$ such that $\tau_\nu \circ \sigma_\nu \to \mathrm{Id}_\mathcal{A}$ in pointwise norm (with positive unital maps σ_ν as usual), we define the "τ-topological" entropy $\hbar_\tau(\theta) = \limsup_{\nu\in N} \hbar(\theta, \tau_\nu)$. - Note that obviously by definition (2.6,iii) we have generally $\hbar_\tau(\theta) \leq \hbar(\theta)$, and in particular $\hbar_\tau(\theta) = \hbar(\theta)$ if τ forms also a *subnet* of $\mathcal{P}_1(\mathcal{A}) \ni \gamma_\alpha$ w.r.t. the net structure from the directed index set $\mathcal{O}_1^+(\mathcal{A}) \ni \alpha$ as in (2.11 & ,1) above (i.e. if the directed set N is order-isomorphic to a cofinal sub-poset of $\mathcal{O}_1^+(\mathcal{A})$).

3 Quantum Topological Entropy for C^*-dynamical Systems

3.1 The general theory for nuclear C^*-algebras

Definition (3.1): Let $\mathcal{A} \ni \mathbb{1}$ be a (first general) non-Abelian C^*-algebra. Using the general definition (2.5), we introduce the following notations: $\mathcal{O}(\mathcal{A})$ denotes the set of all finite "operator covers" $\alpha = \{A_i \in \mathcal{A} | \sum_i A_i^* A_i > 0\}$, $\mathcal{O}_2(\mathcal{A})$ is the set of (not necessarily positive) "operator partitions of unity" (OPUs) $\beta = \{B_j \in \mathcal{A} | \sum_j B_j^* B_j = \mathbb{1}\}$, and $\mathcal{O}_2^+(\mathcal{A})$ denotes the set of *positive* "square root" OPUs $\beta = \{B_j \in \mathcal{A}^+ | \sum_j B_j^2 = \mathbb{1}\}$. Note that obviously $\mathcal{O}_2^+(\mathcal{A}) \subset \mathcal{O}_2(\mathcal{A}) \subset \mathcal{O}(\mathcal{A})$ resp. $\mathcal{O}_2^+(\mathcal{A}) \subset \mathcal{O}^+(\mathcal{A}) \subset \mathcal{O}(\mathcal{A})$ with $\mathcal{O}^+$ as in Sect. (2.3), and "symbolically" for $\beta \in \mathcal{O}_2^+(\mathcal{A})$ element-wise $(\beta^2) \in \mathcal{O}_1^+(\mathcal{A})$, resp. for $\alpha \in \mathcal{O}_1^+(\mathcal{A})$ in turn $(\alpha^{\frac{1}{2}}) \in \mathcal{O}_2^+(\mathcal{A})$.

For $\alpha, \beta \in \mathcal{O}(\mathcal{A})$, $\alpha = \{A_i\}$ and $\beta = \{B_j\}$, we can well define their "ordered refinement" $\alpha \vec{\vee} \beta = \{B_j A_i\} \in \mathcal{O}(\mathcal{A})$, as we have for $\sum_j B_j^* B_j \geq \varepsilon \cdot \mathbb{1}$ (for some $\varepsilon > 0$, using (2.5,iii)) that $\sum_{i,j} A_i^* B_j^* B_j A_i \geq \varepsilon \sum_i A_i^* A_i > 0$.

Obviously, $\vec{\vee} : \mathcal{O} \times \mathcal{O} \to \mathcal{O}$ is associative (because $\mathcal{A}$ is so), but *not* commutative for *non*-Abelian $\mathcal{A}$, and $\vec{\vee} \cdot \mathcal{O}_2 \times \mathcal{O}_2 \to \mathcal{O}_2$ (but $\vec{\vee} : \mathcal{O}^+ \times \mathcal{O}^+ \to \mathcal{O}$, with image more than $\mathcal{O}^+$!). On the other hand, for *Abelian* $\mathcal{A}$ and $\alpha, \beta \in \mathcal{O}^+(\mathcal{A})$ of course $\alpha \vec{\vee} \beta = \alpha \vee \beta$ with the "classical" refinement as used in Def. (2.6).

Now we can first generalize for non-Abelian $\mathcal{A}$ only the *first* part of Def. (2.6,i), and also the partial pre-order $\succ$ on $\mathcal{O}^+(\mathcal{A})$ as in (2.11 & ,1), but necessarily also extended to all of $\mathcal{O}(\mathcal{A})$:

Definition (3.2): Let again $\mathcal{A} \ni 1\!\!1$ be a general C^*-algebra.

(i) For $\alpha \in \mathcal{O}(\mathcal{A})$, $\bar{N}(\alpha) = \min\{\text{card}\,\alpha' | \alpha' \subset \alpha, \alpha' \in \mathcal{O}(\mathcal{A})\} \in \mathbb{N}$ denotes the cardinality of a *minimal* "operator subcover" α' of α.

(ii) We define a partial pre-order $\succ$ on $\mathcal{O}(\mathcal{A})$ by $\alpha \succ \beta$ for $\alpha, \beta \in \mathcal{O}(\mathcal{A})$, if $\forall A_i \in \alpha\ \exists B_{j(i)} \in \beta$ and $\lambda_i \in (0,1]$: $\lambda_i A_i^* A_i \leq B_{j(i)}^* B_{j(i)}$.

Note that these are really generalizations of the "classical" definitions, although formally different for *Abelian* $\mathcal{A}$; remember that generally $\alpha \in \mathcal{O}^+(\mathcal{A}) \Longleftrightarrow \alpha^2 \in \mathcal{O}^+(\mathcal{A})$ (element-wise!), resp. for $A, B \in \mathcal{A}^+ : \lambda A \leq B \Longrightarrow \lambda^{\frac{1}{2}} A^{\frac{1}{2}} \leq B^{\frac{1}{2}}$ (being equivalent for Abelian $\mathcal{A}$).

Proposition (3.3): For these preceding definitions, we have the following rather obvious properties: $\forall \alpha, \beta, \gamma \in \mathcal{O}(\mathcal{A})$,

ad (ii) (a) $\alpha \prec \alpha \vec{\vee} \beta$, but *not* necessarily $\beta \prec \alpha \vec{\vee} \beta$ (i.e. the "directed join" $\vec{\vee}$ does *not* render $(\mathcal{O}, \succ)$ a *directed* set any more).

(b) $\beta \succ \gamma \Longrightarrow \alpha \vec{\vee} \beta \succ \alpha \vec{\vee} \gamma$, but *not* necessarily also $\beta \vec{\vee} \alpha \succ \gamma \vec{\vee} \alpha$!

ad (i) (c) $\bar{N}(\alpha) \leq \bar{N}(\alpha \vec{\vee} \beta) \leq \bar{N}(\alpha) \cdot \bar{N}(\beta)$, but *not* necessarily also $\bar{N}(\beta) \leq \bar{N}(\alpha \vec{\vee} \beta)$, as follows from (a) together with (f) below (this generalizes (2.2,a & b) for N partly).

(d) $\bar{N}(\theta(\alpha)) = \bar{N}(\alpha)\ \forall \theta \in * - \text{Aut}(\mathcal{A})$ (generalizes (2.2,c) for N).

(e) $\bar{N}(\alpha \vec{\vee} \alpha \vec{\vee} \ldots \vec{\vee} \alpha) = \bar{N}(\alpha)\ \forall \alpha \in \mathcal{O}^+(\mathcal{A})$ and for arbitrary $\vec{\vee}$-"powers" of α (generalizes $N(\mathcal{U}) = N(\mathcal{U} \vee \mathcal{U})$ in (2.1,i), as follows from $\mathcal{U} \vee \mathcal{U} \succ \mathcal{U}$ *and* $\mathcal{U} \vee \mathcal{U} \prec \mathcal{U}$ for $\mathcal{U} \in \mathcal{O}(X)$ by (2.2,a)), but *not* necessarily also $\bar{N}(\alpha \vec{\vee} \alpha) \leq \bar{N}(\alpha)$ for $\alpha \in \mathcal{O}(\mathcal{A}) \setminus \mathcal{O}^+(\mathcal{A})$.

(f) $\alpha \prec \beta \Longrightarrow \bar{N}(\alpha) \leq \bar{N}(\beta)$ (generalizes (2.2,a) for N).

Lemma (3.4): Let $\mathcal{B}$ be a *finite*-dimensional C^*-algebra, i.e. $*$-isomorphic to $\bigoplus_{k=1}^n M_{d_k} \cong \mathcal{B}$ with full $d_k \times d_k$-matrix algebras M_{d_k}, and define $D(\mathcal{B})$ to be the dimension of a maximal Abelian subalgebra $\mathcal{C} \subset \mathcal{B}$: $D(\mathcal{B}) = \sum_{k=1}^n d_k$. Then for any $\alpha \in \mathcal{O}(\mathcal{B})$ there exists a $\beta \in \mathcal{O}^+(\mathcal{B})$, $\beta = \{e_i \in \mathcal{B}^+ | i = 1, \ldots, D(\mathcal{B}) \cdot \sum_{i=1}^D e_i > 0\}$ with *minimal* projectors $e_i \in \mathcal{B}$, such that $\beta \succ \alpha$ as defined in (3.2,ii).

Corollary (3.5): Let $\mathcal{B}$ again be a finite-dimensional C^*-algebra with $D(\mathcal{B})$ defined as in (3.4), then $\forall \alpha \in \mathcal{O}(\mathcal{B}) : \bar{N}(\alpha) \leq D(\mathcal{B})$, as follows from (3.3,f) & (3.4). Obviously, there exists a $\beta \in \mathcal{O}_1^+(\mathcal{B}) \cap \mathcal{O}_2^+(\mathcal{B})$ with again *minimal* projectors e_i, $\beta = \{e_i \in \mathcal{B}^+ | i = 1, \ldots, D(\mathcal{B}) : \sum_{i=1}^D e_i = 1\!\!1\}$, such that $\bar{N}(\beta) = D(\mathcal{B})$; and together we have that $\max_{\alpha \in \mathcal{O}(\mathcal{B})} \bar{N}(\alpha) = D(\mathcal{B}) = \max_{\beta \in \mathcal{O}_2^+(\mathcal{B})} \bar{N}(\beta)$.

Definition (3.6): Let $\gamma : \mathcal{B} \to \mathcal{A}$ be a positive linear map between C^*–algebras $\mathcal{B}$ and $\mathcal{A}$, then we define the "2-application" $\gamma[\bullet]_2 : \mathcal{B} \to \mathcal{A}^+$ (as a *map*, also denoted by "absolute value" $|\gamma|$) by $\gamma[B]_2 = [\gamma(B^*B)]^{\frac{1}{2}}$. Obviously, $|\gamma| : \mathcal{B} \to \mathcal{A}^+$ is still positive but *not* linear any more (only still positive homogeneous: $\gamma[\lambda B]_2 = \lambda \cdot \gamma[B]_2, \ \forall \lambda \geq 0, \ B \in \mathcal{B}$); and if γ is a $*$–homomorphism, then $|\gamma|_{\mathcal{B}^+} = \gamma_{\mathcal{B}^+}$. (Less trivially, this latter statement has a partial converse: If γ is even *completely* positive, then $\gamma[B]_2 = \gamma(B)$ for a particular $B \in \mathcal{B}$ implies that B is in the "left multiplicative domain" of γ, i.e. $\gamma(C) \cdot \gamma(B) = \gamma(CB)\ \forall C \in \mathcal{B}$; see [26].) Note that generally $\gamma[B]_2 = \gamma(B)$ at least whenever $B = B^2 \in \mathcal{B}^+$ *and* $\gamma(B) = \gamma(B)^2 \in \mathcal{A}^+$ are *projections.*

Lemma (3.7):

(i) Let $\gamma : \mathcal{B} \to \mathcal{A}_1$ and $\theta : \mathcal{A}_1 \to \mathcal{A}_2$ be positive linear maps between C^*–algebras $\mathcal{B}, \mathcal{A}_1, \mathcal{A}_2$, then $\theta[\gamma[B]_2]_2 = (\theta \circ \gamma)[B]_2\ \forall B \in \mathcal{B}$ (or alternatively, symbolically: $|\theta| \circ |\gamma| = |\theta \circ \gamma|$).

(ii) If θ is even a $*$–homomorphism, then also $\theta(\gamma[B]_2) = (\theta \circ \gamma)[B]_2\ \ \forall B \in \mathcal{B}$ (or $\theta \circ |\gamma| = |\theta \circ \gamma|$; note that also for γ a $*$–homomorphism and θ general: $|\theta \circ \gamma| = |\theta| \circ \gamma$).

Definition (3.8): Let $\gamma_1, \ldots, \gamma_n$ be positive unital (linear) maps $\gamma_k : \mathcal{B}_k \to \mathcal{A}$ from finite–dimensional C^*–algebras $\mathcal{B}_k$ into the C^*–algebra $\mathcal{A} \ni 1\!\!1$, then their *entropy* is defined by

$$\bar{H}(\gamma_1, \ldots, \gamma_n) = \max_{\{(\alpha_1(\gamma_1), \ldots, \alpha_n(\gamma_n))\}} \log \bar{N}(\gamma_1[\alpha_1]_2 \vec{\vee} \ldots \vec{\vee} \gamma_n[\alpha_n]_2),$$

where the maximum is taken (first as supremum) over the set of n–tuples with elements $\alpha_k \in \mathcal{O}_2^+(\mathcal{B}_k)$ determined by the respective *map* γ_k (i.e. for $\gamma_k = \gamma_\ell$ also $\alpha_k(\gamma_k) = \alpha_\ell(\gamma_\ell)$), and the maximum exists because of (3.11,ii) and (3.10) below; and where of course $\gamma_k[\alpha_k]_2 \in \mathcal{O}_2^+(\mathcal{A})$ is understood element–wise.

For *Abelian* $\mathcal{A}$ and $\gamma_1, \ldots, \gamma_n \in \mathcal{P}_1(\mathcal{A})$, i.e. also $\mathcal{B}_k$ Abelian $\forall k$, this definition rather obviously coincides again with (2.6,i); and on the other hand for non–Abelian $\mathcal{A}$ but the *inclusion* $*$–homomorphism $\gamma_k = \imath_{\mathcal{B}_k} : \mathcal{B}_k \to \mathcal{A}$ of finite–dimensional $*$–*sub*algebras $\mathcal{B}_k \subset \mathcal{A}$, we see from (3.6) that $\imath_{\mathcal{B}_k}[\alpha_k]_2 = \imath_{\mathcal{B}_k}(\alpha_k)\ \forall \alpha_k \in \mathcal{O}_2^+(\mathcal{B}_k)$, and thus we can denote

$$\bar{H}(\mathcal{B}_1, \ldots, \mathcal{B}_n) \equiv \bar{H}(\imath_{\mathcal{B}_1}, \ldots, \imath_{\mathcal{B}_n}) = \max_{\{(\alpha_1(\mathcal{B}_1), \ldots, \alpha_n(\mathcal{B}_n))\}} \log \bar{N}(\alpha_1 \vec{\vee} \ldots \vec{\vee} \alpha_n).$$

Lemma (3.9): Let $\gamma : \mathcal{B} \to \mathcal{A}$ be a positive unital (linear) map from a finite–dimensional C^*–algebra $\mathcal{B}$ into the C^*-algebra $\mathcal{A} \ni 1\!\!1$, then $\forall \alpha \in \mathcal{O}^+(\mathcal{B})$:

(i) $\bar{N}(\gamma[\alpha]_2) \leq \bar{N}(\alpha)$, where we should emphasize at this occasion that we understand it as defined by the arguments of $\bar{N}$ (or $\bar{H}$) to which algebra they refer (here $\mathcal{A}$ on l.h.s., $\mathcal{B}$ on r.h.s.), and

(ii) also $\gamma(\alpha) \in \mathcal{O}^+(\mathcal{A})$ with $\bar{N}(\gamma[\alpha]_2) \leq \bar{N}(\gamma(\alpha))$. In particular, for a positive operator cover β of $\mathcal{B}$, $\beta = \{p_i = p_i^2 \in \mathcal{B}^+ | \sum_i p_i > 0\}$ with *projections* p_i even $\bar{N}(\gamma[\beta]_2) = \bar{N}(\gamma(\beta))$.

Proof: The inequalities (i) resp. (ii) follow from the implications

(i) $\sum_i A_i^2 > 0 \Longrightarrow \sum_i \gamma(A_i^2) > 0$ (obvious by (2.5,iii)), resp.

(ii) $\sum_i \gamma(A_i)^2 > 0 \Longrightarrow \sum_i \gamma(A_i^2) > 0$ (for which Schwarz positivity of γ is not even necessary, as $\gamma(A^2) \geq \gamma(A)^2 \ \forall A = A^*$ by Kadison's ancient result [27]);

which obviously both cannot be reversed in general. But in (ii), somewhat "paradoxically", always $\gamma(\alpha) \in \mathcal{O}^+(\mathcal{A})$, as follows from our well-known equivalence (cf. after (2.5)): $\sum_i A_i^2 > 0 \Longleftrightarrow \sum_i A_i > 0 \Longrightarrow \sum_i \gamma(A_i) > 0 \Longleftrightarrow \sum_i \gamma(A_i)^2 > 0$.
And if even $A_i^2 = A_i \ \forall i$, we can use the latter equivalence to reverse the inequality (ii).

Corollary (3.10): Let $\gamma : \mathcal{B} \to \mathcal{A}$ be as in (3.9), then $\bar{H}(\gamma) \leq \log D(\mathcal{B}) = \bar{H}(\mathcal{B})$, as follows from (3.9,i),(3.5) and the definitions. Thus $\bar{H}(\gamma) = \max_{\alpha \in \mathcal{O}_2^+(\mathcal{B})} \log \bar{N}(\gamma[\alpha]_2)$ is really a maximum; but even $\bar{H}(\gamma) = \max_{\alpha \in \mathcal{O}_2(\mathcal{B})} \log \bar{N}(\gamma[\alpha]_2)$ does not give more, because for $\alpha = \{A_i \in \mathcal{B} \mid \sum_i A_i^* A_i = \mathbb{1}\} \in \mathcal{O}_2(\mathcal{B})$ we have that (symbolically) $|\alpha| = \{(A_i^* A_i)^{\frac{1}{2}} \in \mathcal{B}^+\} \in \mathcal{O}_2^+(\mathcal{B})$, and obviously by Def. (3.6): $\gamma[\alpha]_2 = \gamma[|\alpha|]_2$. By the same argument, we can take $\alpha_n(\gamma_n) \in \mathcal{O}_2(\mathcal{B}_n)$ in Def. (3.8), but *not* for $k = 1, \ldots, (n-1)$ if we want to keep to the equivalent expression for *sub*algebras $\mathcal{B}_k \subset \mathcal{A}$ in (3.8).

Proposition (3.11): Let $\mathcal{A} \ni \mathbb{1}$ be a (first general) C^*-algebra, then the entropy functional (3.8) has the following general properties:

(i) For $\theta_k : \mathcal{B}_k \to \mathcal{A}_k$ and $\gamma_k : \mathcal{A}_k \to \mathcal{A}$ positive unital (linear) maps with finite-dimensional C^*-algebras $\mathcal{A}_k, \mathcal{B}_k$ ($\forall k = 1, \ldots, n$), such that (cf. Def. (3.8)!) for $\gamma_i = \gamma_j$ also $\gamma_i \circ \theta_i = \gamma_j \circ \theta_j$ ($\forall i, j = 1, \ldots, n$): $\bar{H}(\gamma_1 \circ \theta_1, \ldots, \gamma_n \circ \theta_n) \leq \bar{H}(\gamma_1, \ldots, \gamma_n)$, which follows from $\theta_k[\mathcal{O}_2^+(\mathcal{B}_k)]_2 \subset \mathcal{O}_2^+(\mathcal{A}_k)$ using (3.7,i) (and generalizes (2.2,a) for H).

(ii) For $\gamma_k : \mathcal{A}_k \to \mathcal{A}$ positive unital maps with finite-dimensional C^*-algebras $\mathcal{A}_k$,

$$\bar{H}(\gamma_1, \ldots, \gamma_m) \leq \bar{H}(\gamma_1, \ldots, \gamma_n) \leq \bar{H}(\gamma_1 \ldots, \gamma_m) + \bar{H}(\gamma_{m+1}, \ldots, \gamma_n)$$

$\forall m < n$ (where literally the order of $(1, \ldots, n)$ has to be respected), which follows from (3.3,c) and generalizes (2.2,a & b) for H partly.
If in addition $\gamma_i \neq \gamma_j \ \forall i \neq j \in \{1, \ldots, n\}$, then even $\bar{H}(\gamma_{i_1}, \ldots, \gamma_{i_\ell}) \leq \bar{H}(\gamma_1, \ldots, \gamma_n) \ \forall \ell \leq n$ and $1 \leq i_1 < i_2 < \ldots < i_\ell \leq n$, as follows directly from Def. (3.8).

(iii) For $\gamma_1 : \mathcal{A}_1 \to \mathcal{A}$ as before, $\bar{H}(\gamma_1, \ldots, \gamma_1) = \bar{H}(\gamma_1)$ for arbitrarily many repetitions of only *one* single γ_1 (as follows from (3.3,e) because of the definition (3.8) for $\bar{H}$, and generalizes $H(\mathcal{U} \vee \mathcal{U}) = H(\mathcal{U})$ in (2.1,i) as noted in (3.3,e)).

(iv) For $\theta \in * - \mathrm{Aut}(\mathcal{A})$ and $\gamma_k : \mathcal{A}_k \to \mathcal{A}$ as before, $\bar{H}(\theta \circ \gamma_1, \ldots, \theta \circ \gamma_n) = \bar{H}(\gamma_1, \ldots, \gamma_n)$, which follows from (3.3,d) using (3.7,ii) (and generalizes (2.2,c) for H).

(v) If there exists a finite-dimensional C^*-subalgebra $\mathcal{B} \subset \mathcal{A}$ such that $\mathrm{Im}(\gamma_k) \subseteq \mathcal{B}$ $\forall k = 1, \ldots, n$, then $\bar{H}(\gamma_1, \ldots, \gamma_n) \leq \log D(\mathcal{B}) \equiv \bar{H}(\mathcal{B})$ with the "Abelian dimension" $D(\mathcal{B})$ as introduced in (3.4); which follows again from (3.4) applied to $\gamma_1[\alpha_1]_2\vec{\vee}\ldots\vec{\vee}\gamma_n[\alpha_n]_2 \in \mathcal{O}_2(\mathcal{B})$ in Def. (3.8), and (3.3,f).

(vi) More generally (but much less applicable): If there exists a positive unital *injective* (linear) map $\gamma : \mathcal{B} \to \mathcal{A}$ with a finite-dimensional C^*-algebra $\mathcal{B}$ such that for $\gamma_k : \mathcal{A}_k \to \mathcal{A}$ as before, the set

$$\{\gamma_1(B_1)^{\frac{1}{2}} \cdot \gamma_2(B_2)^{\frac{1}{2}} \cdot \ldots \cdot \gamma_n(B_n) \cdot \ldots \cdot \gamma_2(B_2)^{\frac{1}{2}} \cdot \gamma_1(B_1)^{\frac{1}{2}} | \forall B_k \in \mathcal{A}_k^+, \ \|B_k\| \leq 1\} \subseteq \gamma(\mathcal{B}^+),$$

then $\bar{H}(\gamma_1, \ldots, \gamma_n) \leq \bar{H}(\gamma)$ ($\leq \bar{H}(\mathcal{B})$ by (3.10) above); which follows directly from Def. (3.8).

Corollary (3.12): ad (i) If $\gamma_k : \mathcal{A}_k \to \mathcal{A}$ and $\mathcal{A}_k \subset \mathcal{B}_k$ with conditional expectations $E_k : \mathcal{B}_k \to \mathcal{A}_k$ (such that the inclusions $\imath_{\mathcal{A}_k} : \mathcal{A}_k \to \mathcal{B}_k$ have E_k as left $\circ$-inverse: $E_k \circ \imath_{\mathcal{A}_k} = \mathrm{Id}_{\mathcal{A}_k}$; and assuming that for $\mathcal{A}_i = \mathcal{A}_j$ also $E_i = E_j$, $\forall i, j = 1, \ldots, n$), then $\bar{H}(\gamma_1 \circ E_1, \ldots, \gamma_n \circ E_n) = \bar{H}(\gamma_1, \ldots, \gamma_n)$. Thus we may again assume (as in [1, (III.5,1)]) for the computation of $\bar{H}(\gamma_1, \ldots, \gamma_n)$ that the γ_k are maps from resp. full matrix algebras M_{d_k} into $\mathcal{A}$; or $\gamma_k \in \mathcal{K}_1(\mathcal{A})$ (with our notation at the beginning of Sect. (2.2)) if we restrict to *completely* positive unital maps $\gamma_k \in \mathcal{CP}_1(\mathcal{A})$.

Definition (3.13): We define n positive unital (linear) maps $\gamma_k : \mathcal{B}_k \to \mathcal{A}$ ($k = 1, \ldots, n$) to be *independently covering* (for the C^*-algebra $\mathcal{A} \ni \mathbb{1}$), if for $B_k \in \mathcal{B}_k^+$ but $\gamma_k(B_k)$ *not* invertible ($\gamma_k(B_k) \not> 0$, $\forall k$) $\exists \omega \in S_{\mathcal{A}} : \ \omega \circ \gamma_k(B_k) = 0$ $\forall k = 1, \ldots, n$. By (2.5,i), this condition is equivalent to the following reformulations:

$$B_k \in \mathcal{B}_k^+ \ (k = 1, \ldots, n) : \gamma_k(B_k) \not> 0 \ \forall k \implies \sum_{k=1}^{n} \gamma_k(B_k) \not> 0$$

$$\Leftrightarrow: \ B_k \in \mathcal{B}_k^+ \ (k = 1, \ldots, n) : \sum_{k=1}^{n} \gamma_k(B_k) > 0 \implies \exists k : \gamma_k(B_k) > 0.$$

We say that two maps γ_1 and γ_2 as before are "commuting" (denoted symbolically $[\gamma_1, \gamma_2] = 0$) if $[\gamma_1(B_1), \gamma_2(B_2)] = 0$ $\forall B_k \in \mathcal{B}_k$ ($k = 1, 2$).

Proposition (3.14): If n positive unital (linear) maps $\gamma_k : \mathcal{B}_k \to \mathcal{A}$ from finite-dimensional C^*-algebras $\mathcal{B}_k$ into $\mathcal{A} \ni \mathbb{1}$ ($k = 1, \ldots, n$) are independently covering *and* pairwise commuting, $[\gamma_i, \gamma_j] = 0$ $\forall i \neq j \in \{1, \ldots, n\}$, then $\bar{H}(\gamma_1, \ldots, \gamma_n) = \sum_{k=1}^{n} \bar{H}(\gamma_k)$; i.e. the optimal upper bound of (3.11,ii) is attained.
In particular, this is the case if $\mathcal{A} = \mathcal{A}_1 \otimes \ldots \otimes \mathcal{A}_n$ with (e.g. nuclear) unital C^*-algebras $\mathcal{A}_k \ni \mathbb{1}_k$ and $\gamma_k : \mathcal{B}_k \to \mathbb{1}_1 \otimes \ldots \otimes \mathcal{A}_k \otimes \ldots \otimes \mathbb{1}_n$ ($\forall k = 1, \ldots, n$).

Sketch of Proof: Let $\alpha_k \in \mathcal{O}_2^+(\mathcal{B}_k)$ be such that $\bar{H}(\gamma_k) = \log \bar{N}(\gamma_k[\alpha_k]_2)$ $\forall k = 1, \ldots, n$, by (3.10). Then, for the corresponding "γ_k-minimal" subcovers $\gamma_k[\alpha_k']_2 \in \mathcal{O}^+(\mathcal{A})$ ($\alpha_k' \subset \alpha_k$ with $\mathrm{card}\,\alpha_k' = \exp \bar{H}(\gamma_k)$), clearly $\gamma_1[\alpha_1']_2\vec{\vee}\ldots\vec{\vee}\gamma_k[\alpha_n']_2$ is always a subcover of the "unprimed" n-fold $\vec{\vee}$-refinement of all $\gamma_k[\alpha_k]_2$; but it is even *minimal* by the assumption

(3.13) for $\{\gamma_k | k = 1, \ldots, n\}$, as otherwise at least one α'_k could *not* be γ_k–minimal (an elementary C^*–exercise for the reader). Thus by Def. (3.8), $\bar{H}(\gamma_1, \ldots, \gamma_n) \geq \sum_{k=1}^n \bar{H}(\gamma_k)$; and (3.11,ii) gives the converse. The rest is obvious.

Lemma (3.15): Let $\mathcal{A}_1, \mathcal{A}_2 \subset \mathcal{A}$ be unital C^*–*sub*algebras of $\mathcal{A} \ni \mathbb{1}$ which are commuting ($[\mathcal{A}_1, \mathcal{A}_2] = 0$), then $\imath_{\mathcal{A}_1}$ and $\imath_{\mathcal{A}_2}$ (the $*$–homomorphic inclusions) are independently covering *iff* $\mathcal{A}_1$ and $\mathcal{A}_2$ are C^*–independent in the sense of [28]; with proof of the nontrivial implication similar to that of (3.2,(1)⇒(2)) in [28].

Remark (3.16): For an n–tuple of positive unital (linear) maps $\gamma_k : \mathcal{B}_k \to \mathcal{A}$ ($k = 1, \ldots, n$), an "Abelian model" is determined by an n–tuple $(\alpha_1(\gamma_1), \ldots, \alpha_n(\gamma_n))$ of $\alpha_k(\gamma_k) \in \mathcal{O}_2^+(\mathcal{B}_k)$ as in Def. (3.8), and is defined as follows:

For $\alpha_k = \{B_{i_k}^{(k)} \in \mathcal{B}_k^+ | i_k = 1, \ldots, N_k : \sum_{i_k=1}^{N_k} (B_{i_k}^{(k)})^2 = \mathbb{1}_k\}$, choose the Abelian C^*–algebra $\mathcal{C} = \bigotimes_{k=1}^n (\bigoplus_{i_k=1}^{N_k} (\mathbb{C})_{i_k})$, with corresponding minimal projectors $e_{(i_1, \ldots, i_n)} \in \mathcal{C}$. Using the notations $I_n = (i_1, \ldots, i_n) \in \mathbb{N}^n$ and $E_{I_n} = \gamma_n[B_{i_n}]_2 \cdot \ldots \cdot \gamma_1[B_{i_1}]_2$, define a positive unital linear map $P_\omega : \mathcal{A} \to \mathcal{C}$ (for fixed state $\omega \in S_\mathcal{A}$) by

$$P_\omega(A) = \sum_{I_n} \omega(E_{I_n}^* \cdot A \cdot E_{I_n}) \cdot \omega(E_{I_n}^* E_{I_n})^{-1} \cdot e_{I_n},$$

and a state $\mu_\omega \in S_\mathcal{C}$ by $\mu_\omega(e_{I_n}) = \omega(E_{I_n}^* E_{I_n})$. Then, for fixed $(\alpha_1, \ldots, \alpha_n)$, $\omega \mapsto P_\omega$ is a map from the state space $S_\mathcal{A}$ of $\mathcal{A}$ into the set $\mathcal{P}_1(\mathcal{A}, \mathcal{C})$ of positive unital maps from $\mathcal{A}$ to $\mathcal{C}$; and $\omega \mapsto \mu_\omega$ is a weak–$*$ continuous affine map $M : S_\mathcal{A} \to S_\mathcal{C}$, whose adjoint $M^* : \mathcal{C} \to \mathcal{A}$ is given by the positive unital (linear) map $M^*(e_{I_n}) = E_{I_n}^* E_{I_n}$. Now, it is not true in general that $\mu_\omega \circ P_\omega = \omega \ \forall \omega \in S_\mathcal{A}$ (only if ω is "tracial" at least for E_{I_n}, $\forall I_n$); but $C_{I_n} : \omega \mapsto \mu_\omega(e_{I_n}) = \omega \circ M^*(e_{I_n})$ is a weak–$*$ continuous (affine) function on $S_\mathcal{A}$, $C_{I_n} \in C(S_\mathcal{A})$, and for $\beta_M = \{C_{I_n} | I_n\} \in \mathcal{O}_1^+(C(S_\mathcal{A}))$ we have $\bar{H}(\gamma_1, \ldots, \gamma_n) = \max_{\{M = M(\alpha_1, \ldots, \alpha_n)\}} \log \bar{N}(\beta_M)$, by definitions (3.8), (3.2,i) resp. (2.6,i), and (2.5,i).
Note, finally, that for the *pure* states $e_{I_n}^* \in S_\mathcal{C} \subset \mathcal{C}^*$, defined by $e_{I_n}^*(e_{J_n}) = \delta_{I_n J_n}$, we get the "reduced" states $e_{I_n}^* \circ P_\omega(A) = \omega(E_{I_n}^* \cdot A \cdot E_{I_n}) \cdot \mu_\omega(e_{I_n})^{-1}$, $\forall A \in \mathcal{A}$.

Definition (3.17): For a positive unital map of finite rank $\gamma : \mathcal{B} \to \mathcal{A}$ into a C^*–algebra $\mathcal{A} \ni \mathbb{1}$, and for $\theta \in * - \mathrm{Aut}(\mathcal{A})$, the "entropy" of θ w.r.t γ is defined by $\hbar(\theta, \gamma) = \lim_{n \to \infty} \frac{1}{n} \bar{H}(\gamma, \theta \circ \gamma, \ldots, \theta^{n-1} \circ \gamma)$, where the limit exists by (3.11,ii & iv), and actually is the *infimum*.

Proposition (3.18): The "entropy" (3.17) has the following easily provable properties (as corollaries of (3.11)):

(i) $\hbar(\theta, \gamma) \leq \bar{H}(\gamma)$, generalizes (2.2,e).

(ii) For $\gamma_2 : \mathcal{B}_2 \to \mathcal{A}$ and $\gamma_1 : \mathcal{B}_1 \to \mathcal{B}_2$ with finite–dimensional $\mathcal{B}_k$ ($k = 1, 2$): $\hbar(\theta, \gamma_2 \circ \gamma_1) \leq \hbar(\theta, \gamma_2)$; generalizes (2.2,d).

(iii) $\hbar(\sigma \circ \theta \circ \sigma^{-1}, \sigma \circ \gamma) = \hbar(\theta, \gamma) \ \forall \sigma \in * - \mathrm{Aut}(\mathcal{A})$.

(iv) $\hbar(\theta^n,\gamma) \leq |n| \cdot \hbar(\theta,\gamma)$ $\forall n \in \mathbb{Z}$ (generalizes (2.2,f), if $n = -1$); but $\hbar(\theta,\gamma) \leq \hbar(\theta^n,\gamma)$ is not provable here.

(v) For $\mathcal{A} = \mathcal{A}_1 \otimes \mathcal{A}_2$, $\gamma_k : \mathcal{B}_k \to \mathcal{A}_k$ (e.g. with nuclear $\mathcal{A}_k$, $k = 1,2$), and $\theta = \theta_1 \otimes \theta_2$ with $\theta_k \in * - \mathrm{Aut}(\mathcal{A}_k)$: $\hbar(\theta,\gamma_1 \otimes \gamma_2) \geq \hbar(\theta_1,\gamma_1) + \hbar(\theta_2,\gamma_2)$, as follows from Lemma (3.19) below; but the converse inequality is unfortunately improvable thus far (cf. [29, Lemma(3.4)] for the CNT entropy).

(vi) If there exists a sequence of positive unital injective maps $\gamma_n : \mathcal{B}_n \to \mathcal{A}$ of finite rank, such that for given $\gamma : \mathcal{B} \to \mathcal{A}$:

$$\{\gamma(B_1)^{\frac{1}{2}} \cdot \theta \circ \gamma(B_2)^{\frac{1}{2}} \cdot \ldots \cdot \theta^{n-1} \circ \gamma(B_n) \cdot \ldots \cdot \gamma(B_1)^{\frac{1}{2}} | \forall B_k \in \mathcal{B}^+, \|B_k\| \leq 1\} \subseteq \gamma_n(\mathcal{B}_n^+),$$

$\forall n \in \mathbb{N}$, then $\hbar(\theta,\gamma) \leq \liminf_{n\in\mathbb{N}} \frac{1}{n}\bar{H}(\gamma_n)$.

Lemma (3.19): Actually, we need for the proof of (v) the following corollary of the proof of (3.14): For $\mathcal{A} = \mathcal{A}_1 \otimes \mathcal{A}_2$, let $\beta_k \in \mathcal{O}_2(\mathcal{A}_k)$ (e.g. with nuclear $\mathcal{A}_k$ $k = 1,2$), and $\alpha_1 = \beta_1 \otimes \mathbb{1}_2$, $\alpha_2 = \mathbb{1}_1 \otimes \beta_2 \in \mathcal{O}_2(\mathcal{A})$, then $\bar{N}(\alpha_1 \vec{\vee} \alpha_2) = \bar{N}(\alpha_1) \cdot \bar{N}(\alpha_2)$ (again by much the classical argument as implicitely used in [16], using (3.13) for $\mathcal{A}_k$).

Definition (3.20): Let $\mathcal{A} \ni \mathbb{1}$ be a *nuclear* C^*-algebra and *choose* an approximating net of completely positive unital maps $\tau = \{\tau_\nu \in \mathcal{CP}_1(\mathcal{A}) | \nu \in N\}$, $\tau_\nu : \mathcal{A}_\nu \to \mathcal{A}$ (with corresponding $\sigma_\nu : \mathcal{A} \to \mathcal{A}_\nu$) as defined in (2.9,ii). We define the "τ-topological" *entropy* of $\theta \in * - \mathrm{Aut}(\mathcal{A})$ by repeating (2.12): $\hbar_\tau(\theta) = \limsup_{\nu\in N} \hbar(\theta,\tau_\nu)$.
For $\sigma \in * - \mathrm{Aut}(\mathcal{A})$, we use the notation $\sigma(\tau) = \{\sigma \circ \tau_\nu \in \mathcal{CP}_1(\mathcal{A}) | \nu \in N\}$, which is again an approximating net (with corresponding $\sigma_\nu \circ \sigma^{-1} : \mathcal{A} \to \mathcal{A}_\nu$).

Corollary (3.21): The τ-topological entropy has the following easily provable properties (by (3.18) before):

(i) $\hbar_{\sigma(\tau)}(\sigma \circ \theta \circ \sigma^{-1}) = \hbar_\tau(\theta)$, $\forall \sigma \in * - \mathrm{Aut}(\mathcal{A})$; generalizes (2.2,h).

(ii) $\hbar_\tau(\theta^n) \leq |n| \cdot \hbar_\tau(\theta)$, $\forall n \in \mathbb{Z}$ (unfortunately, the converse inequality is unprovable thus far; cf. [30, Corollary(2.9)] for the CNT entropy); generalizes (2.3,i) partly.

(iii) For $\mathcal{A} = \mathcal{A}_1 \otimes \mathcal{A}_2$ with nuclear $\mathcal{A}_k$ and $\theta_k \in * - \mathrm{Aut}(\mathcal{A}_k)$, $\theta = \theta_1 \otimes \theta_2 \in * - \mathrm{Aut}(\mathcal{A})$; and for approximating nets $\tau_k = \{\tau_{\nu_k} . \mathcal{A}_{\nu_k} \to \mathcal{A}_k | \nu_k \in N_k\}$ of $\mathcal{A}_k$, also $\tau = \{\tau_{\nu_1} \otimes \tau_{\nu_2} : \mathcal{A}_{\nu_1} \otimes \mathcal{A}_{\nu_2} \to \mathcal{A} | (\nu_1,\nu_2) \in N_1 \times N_2\}$ is an approximating net of $\mathcal{A}$. Then by (3.18,v): $\hbar_\tau(\theta) \geq \hbar_{\tau_1}(\theta_1) + \hbar_{\tau_2}(\theta_2)$; generalizes (2.3,ii) partly.

Problem (3.22): Characterize those approximating nets τ of $\mathcal{A}$ as in (3.20), for which the (first "formal") supremum $\hbar(\theta) \equiv \sup_{\gamma\in\mathcal{CP}_1(\mathcal{A})} \hbar(\theta,\gamma) = \sup_{\gamma\in\mathcal{K}_1(\mathcal{A})} \hbar(\theta,\gamma)$ (cf. (3.12)) is attained: $\hbar_\tau(\theta) = \hbar(\theta)$ $\forall \theta \in * - \mathrm{Aut}(\mathcal{A})$; cf. (2.12) for Abelian $\mathcal{A}$. Is there any relation still to the partial pre-order (3.2,ii) for *non*-Abelian $\mathcal{A}$? Note that it is possibly not true in general that $\hbar_{\sigma(\tau)}(\theta) = \hbar(\theta)$, contrasting (3.21,i) above.

3.2 AF algebra theory and first applications

Let $\mathcal{A} = \overline{\bigcup_{n\in\mathbb{N}} \mathcal{A}_n}$ be a (separable) unital AF algebra, i.e. $\mathcal{A}_n \subset \mathcal{A}_{n+1}$ ($\forall n \in \mathbb{N}$) with unital inclusions of finite-dimensional C^*-algebras $\mathcal{A}_n$, and $\mathcal{A} = \overline{\mathcal{A}_\infty}$ is the norm completion of the inductive limit $\mathcal{A}_\infty = \bigcup_{n\in\mathbb{N}} \mathcal{A}_n$, which is a locally semisimple $*$-algebra (with pre-C^*-norm). $\mathcal{A}$ is nuclear with approximating maps (2.9,ii) given by $\tau_n = \imath_{\mathcal{A}_n} . \mathcal{A}_n \to \mathcal{A}$ ($*$-homomorphic inclusions) and with any chosen norm-one projections (conditional expectations) $\sigma_n : \mathcal{A} \to \mathcal{A}_n$; and we can reformulate our general definition (3.20) of "quantum topological" entropy in this AF case as follows:

Definition (3.23): Let $\mathcal{A} = \overline{\mathcal{A}_\infty}$ be a unital AF algebra as above, then the "$\mathcal{A}_\infty$-topological" entropy of $\theta \in * - \mathrm{Aut}(\mathcal{A})$ is defined by: $\hbar_{\mathcal{A}_\infty}(\theta) = \sup_{\mathcal{B}\subset\mathcal{A}_\infty} \hbar(\theta, \mathcal{B})$, where the supremum is taken over all finite-dimensional C^*-subalgebras $\mathcal{B} \subset \mathcal{A}_\infty$.
This is really a special case of our general definition (3.20) because of (i) of the following concretization of (3.18):

Corollary (3.24): The entropy $\hbar(\theta, \mathcal{B}) \equiv \hbar(\theta, \imath_{\mathcal{B}})$ with (3.17) for a finite-dimensional *sub*algebra $\mathcal{B} \subset \mathcal{A}$ has the following properties:

(i) For $\mathcal{B}_1 \subset \mathcal{B}_2 \subset \mathcal{A}$, $\hbar(\theta, \mathcal{B}_1) \leq \hbar(\theta, \mathcal{B}_2)$.

(ii) $\hbar(\theta^n, \mathcal{B}) \leq |n| \cdot \hbar(\theta, \mathcal{B})$ $\forall n \in \mathbb{Z}$.

(iii) If there exists an increasing sequence $\mathcal{B}_n \subseteq \mathcal{B}_{n+1}$ of finite-dimensional subalgebras $\mathcal{B}_n \subset \mathcal{A}$ such that $\mathcal{B}, \theta(\mathcal{B}), \ldots, \theta^{n-1}(\mathcal{B}) \subset \mathcal{B}_n$ (as not necessarily commuting subalgebras) $\forall n \in \mathbb{N}$, then $\hbar(\theta, \mathcal{B}) \leq \liminf_{n\to\infty} \frac{1}{n} \bar{H}(\mathcal{B}_n)$.

Subcorollary (3.25): For an AF algebra $\mathcal{A} = \overline{\mathcal{A}_\infty}$ as in Def. (3.23), $\hbar_{\mathcal{A}_\infty}(\theta) = \lim_{n\to\infty} \hbar(\theta, \mathcal{A}_n)$; follows from (i).

Again generally, we can concretize also (3.11) by
Corollary (3.26): The entropy $\bar{H}(\mathcal{B}_1, \ldots, \mathcal{B}_n)$ with (3.8) for finite-dimensional *sub*algebras $\mathcal{B}_k \subset \mathcal{A}$ ($k = 1, \ldots, n$) has the following general properties (from (3.11)):

(i) For $\mathcal{B}_k \subset \mathcal{A}_k \subset \mathcal{A}$, such that for $\mathcal{A}_i = \mathcal{A}_j$ also $\mathcal{B}_i = \mathcal{B}_j$ ($\forall i, j = 1, \ldots, n$), $\bar{H}(\mathcal{B}_1, \ldots, \mathcal{B}_n) \leq \bar{H}(\mathcal{A}_1, \ldots, \mathcal{A}_n)$.

(ii) $\bar{H}(\mathcal{B}_1, \ldots, \mathcal{B}_m) \leq \bar{H}(\mathcal{B}_1, \ldots, \mathcal{B}_n) \leq \bar{H}(\mathcal{B}_1, \ldots, \mathcal{B}_m) + \bar{H}(\mathcal{B}_{m+1}, \ldots, \mathcal{B}_n)$, $\forall m < n$.
If in addition $\mathcal{B}_i \neq \mathcal{B}_j$ $\forall i \neq j \in \{1, \ldots, n\}$, then even $\bar{H}(\mathcal{B}_{i_1}, \ldots, \mathcal{B}_{i_\ell}) \leq \bar{H}(\mathcal{B}_1, \ldots, \mathcal{B}_n)$ $\forall \ell \leq n$ and $1 \leq i_1 < i_2 < \ldots < i_\ell \leq n$.

(iii) $\bar{H}(\mathcal{B}, \ldots, \mathcal{B}) = \bar{H}(\mathcal{B})$ with arbitrarily many repetitions of only *one* single $\mathcal{B}$.

(iv) For $\theta \in * - \mathrm{Aut}(\mathcal{A})$, $\bar{H}(\theta(\mathcal{B}_1), \ldots, \theta(\mathcal{B}_n)) = \bar{H}(\mathcal{B}_1, \ldots, \mathcal{B}_n)$.

(v) If there exists a finite-dimensional $\mathcal{B} \subset \mathcal{A}$ such that $\mathcal{B}_1, \ldots, \mathcal{B}_n \subset \mathcal{B}$, then $\bar{H}(\mathcal{B}_1, \ldots, \mathcal{B}_n) \leq \bar{H}(\mathcal{B})$.

(vi) If $\mathcal{B}_1, \ldots, \mathcal{B}_n$ are pairwise commuting ($[\mathcal{B}_i, \mathcal{B}_j] = 0 \ \forall i \neq j$) *and* independently covering in the sense of (3.14),(3.15), then $\bar{H}(\mathcal{B}_1, \ldots, \mathcal{B}_n) = \sum_{k=1}^{n} \bar{H}(\mathcal{B}_k)$.

Note that in (vi) the condition (3.13) for $\{\mathcal{B}_k | k = 1, \ldots, n\}$ is seemingly stronger than only *pairwise* C^*–independence of $(\mathcal{B}_i, \mathcal{B}_j)$ $\forall i \neq j$ in the sense of [28], as it means that also the "multiple correlations" between $\mathcal{B}_1, \ldots, \mathcal{B}_n$ vanish in the sense of [28, (3.2,4)].

Proposition (3.27): In addition, the entropy (3.8) has the following special properties referring to resp. only *one* (particular) finite–dimensional *sub*algebra of $\mathcal{A}$:

(i) For $\mathcal{B}_1, \ldots, \mathcal{B}_n \subset \mathcal{A}$ as in (3.26), $\bar{H}(\mathcal{B}_1, \ldots, \mathcal{B}_{n-1}, \mathcal{B}_n, \ldots, \mathcal{B}_n) = \bar{H}(\mathcal{B}_1, \ldots, \mathcal{B}_n)$ with arbitrarily many repetitions of the *last* entry $\mathcal{B}_n$ in $\bar{H}$ (but not provable for the other entries $\mathcal{B}_1, \ldots, \mathcal{B}_{n-1}$ analogously!); generalizes (3.26,iii) a tiny bit further.

(ii) For $\mathcal{B}_1, \mathcal{B}_2 \subset \mathcal{A}$, we define $\mathcal{B}_1 \overset{\delta}{\subset} \mathcal{B}_2 \iff \forall B_1 \in \mathcal{B}_1 \ \exists B_2 \in \mathcal{B}_2 : \|B_1 - B_2\| \leq \delta \|B_1\|$. Then if $\mathcal{B}_1 \overset{\delta}{\subset} \mathcal{B}_2 \subset \mathcal{A}$ with $\delta < 10^{-4}$, always $\bar{H}(\mathcal{B}_1) \leq \bar{H}(\mathcal{B}_2)$ (but *not* provable for more than *one* argument analogously!).

Corollary (3.28): Defining in (ii): $\|\mathcal{B}_1 - \mathcal{B}_2\| = \inf\{\delta > 0 | \mathcal{B}_1 \overset{\delta}{\subset} \mathcal{B}_2, \mathcal{B}_2 \overset{\delta}{\subset} \mathcal{B}_1\}$, we have obviously: $\|\mathcal{B}_1 - \mathcal{B}_2\| = \delta$ with $\delta < 10^{-4} \Longrightarrow \bar{H}(\mathcal{B}_1) = \bar{H}(\mathcal{B}_2)$.

Although this looks like a "first step" towards a norm–continuity of $\hbar(\theta, \mathcal{B})$ analogous to (2.9,i), it seems to be already the "last" step (into an impasse); not to speak about the generalization for $\hbar(\theta, \gamma)$ (for $\gamma \in \mathcal{CP}_1(\mathcal{A})$), in view of our "no–go" Proposition (2.10) already for *Abelian* $\mathcal{A}$. Note that the premise of (3.28) is actually *nonsense* for Abelian $\mathcal{A} \supset \mathcal{B}_1, \mathcal{B}_2$, as then rather obviously $\|\mathcal{B}_1 - \mathcal{B}_2\| = 1 \ \forall \mathcal{B}_1 \neq \mathcal{B}_2$; which could be viewed as rendering any non–commutative analogue of (2.9,i) "unnecessary" even for $\hbar(\theta, \mathcal{B})$ (cf. however [31, Prop.1] and its discussion in (3.35,2) below).

Proof of (3.27):

(i) By Def. (3.8), we have to consider for the l.h.s. $\bar{N}(\alpha_1 \vec{\nabla} \alpha_2 \vec{\nabla} \ldots \vec{\nabla} \alpha_n \vec{\nabla} \alpha_n \vec{\nabla} \ldots \vec{\nabla} \alpha_n)$ with $\alpha_k \in \mathcal{O}_2^+(\mathcal{B}_k)$. Now $\alpha_n \vec{\nabla} \alpha_n \vec{\nabla} \ldots \vec{\nabla} \alpha_n \in \mathcal{O}_2(\mathcal{B}_n)$ only, but as in (3.10) $\beta_n = |\alpha_n \vec{\nabla} \alpha_n \vec{\nabla} \ldots \vec{\nabla} \alpha_n| \in \mathcal{O}_2^+(\mathcal{B}_n)$, and obviously $\bar{N}(\alpha_1 \vec{\nabla} \ldots \vec{\nabla} \alpha_{n-1} \vec{\nabla} \beta_n)$ is the same as the above expression. Thus $\bar{H}(\mathcal{B}_1, \ldots, \mathcal{B}_n, \mathcal{B}_n, \ldots, \mathcal{B}_n) \leq \bar{H}(\mathcal{B}_1, \ldots, \mathcal{B}_{n-1}, \mathcal{B}_n)$, and (3.26,ii) gives the converse.

(ii) By Thm. (5.3) of [32], there exists a unitary $U \in \mathcal{A}$ (such that $\|\mathbb{1} - U\| \leq 120\sqrt{\delta}$) with $U\mathcal{B}_1 U^* \subset \mathcal{B}_2$. Thus by (3.26,i & iv): $\bar{H}(\mathcal{B}_2) \geq \bar{H}(U\mathcal{B}_1 U^*) = \bar{H}(\mathcal{B}_1)$.

Now let again $\mathcal{A} = \overline{\bigcup_{n \in \mathbb{N}} \mathcal{A}_n}$ be AF, then $\mathcal{A}$ is *uniquely* determined up to $*$–isomorphism by the norm–dense, locally semisimple $*$–subalgebra $\mathcal{A}_\infty = \bigcup_{n \in \mathbb{N}} \mathcal{A}_n$, or equivalently by a corresponding Bratteli diagram [33], which is in turn equivalent to the sequence of inclusion matrices $\{[\mathcal{A}_n \to \mathcal{A}_{n+1}] | \forall n \in \mathbb{N}\}$ (cf. [34] for the notation and terminology in the following).

Definition (3.29): We say with Choda [34] that the sequence $\{\mathcal{A}_n\}_{n \in \mathbb{N}}$ as above is *periodic* with period p, if $\exists n_0 \in \mathbb{N}$ such that $\forall j \geq n_0$:

(i) $[\mathcal{A}_j \to \mathcal{A}_{j+1}] = [\mathcal{A}_{j+p} \to \mathcal{A}_{j+p+1}]$

(ii) The (hence necessarily square–) matrix $T_j = [\mathcal{A}_j \to \mathcal{A}_{j+p}]$ is *primitive* ($\Longleftrightarrow \exists \ell \in \mathbb{N}$: $(T_j^\ell)_{ik} > 0\ \forall i, k \Longleftrightarrow$ by the inclusion $\mathcal{A}_j \subset \mathcal{A}_{j+\ell p}$ each simple direct summand of $\mathcal{A}_j$ is "contained" in every simple direct summand of $\mathcal{A}_{j+\ell p}$), which implies that T_j has a unique Perron–Frobenius eigenvalue $\beta_j > 0$ (which, together with (i), is actually independent of $j \geq n_0$!).

Lemma (3.30): For a periodic sequence $\{\mathcal{A}_n\}_{n\in\mathbb{N}}$ as above with period p and Perron–Frobenius eigenvalue β of $T_j = [\mathcal{A}_j \to \mathcal{A}_{j+p}]$ for $j \geq n_0$, we have $\lim_{n\to\infty} \frac{1}{n}\bar{H}(\mathcal{A}_n) = \frac{1}{p}\log\beta$ (cf. [34] for the Connes–Størmer entropy [6]).

Proof: Denote the resp. dimension vector of $\mathcal{A}_n \cong \bigoplus_{k(n)=1}^{N_n} M_{d_{k(n)}}$ by $\vec{a}_n = (d_{k(n)}^{(n)})$, and the resp. trace vector by $\vec{t_n} = (t_{k(n)}^{(n)})$ corresponding to the trace $\tau \in S_\mathcal{A}$ by $t_{k(n)}^{(n)} = \tau(\mathbb{1}_{k(n)}^{(n)})/d_{k(n)}^{(n)}$, where $\mathbb{1}_{k(n)}^{(n)}$ are the minimal central projections of $\mathcal{A}_n$ ($\forall k(n) = 1, \ldots, N_n$, $\forall n \in \mathbb{N}$). As indicated in [34], we have $\vec{t_j} = \beta \cdot \vec{t}_{j+p}$ $\forall j \geq n_0$. Fix $j \geq n_0$ and let $r_j = \min_{k(j)} t_{k(j)}^{(j)}$ and $s_j = \max_{k(j)} t_{k(j)}^{(j)}$. Then since

$$1 = \sum_{k(j+np)} t_k^{(j+np)} d_k^{(j+np)} = \sum_{k(j)} \beta^{-n} t_k^{(j)} d_k^{(j+np)}$$

(note that $N_j = N_{j+np}$ $\forall n \in \mathbb{N}$), we get $\beta^n s_j^{-1} \leq \sum_k d_k^{(j+np)} \leq \beta^n r_j^{-1}$, $\forall n \in \mathbb{N}$.
By Def. (3.4), $\sum_k d_k^{(j+np)} = D(\mathcal{A}_{j+np})$, so that by (3.5) and Def. (3.8): $n\log\beta - \log s_j \leq \bar{H}(\mathcal{A}_{j+np}) \leq n\log\beta - \log r_j$, $\forall n \in \mathbb{N}$, and $\lim_{n\to\infty}\frac{1}{n}\bar{H}(\mathcal{A}_n) = \lim_{n\to\infty}\frac{1}{np}\bar{H}(\mathcal{A}_{j+np}) = \frac{1}{p}\log\beta$.

Definition (3.31): For an AF algebra $\mathcal{A} = \overline{\bigcup_{n\in\mathbb{N}}\mathcal{A}_n} = \overline{\mathcal{A}_\infty}$, $\theta \in * - \mathrm{Aut}(\mathcal{A})$ is said to be "$\mathcal{A}_\infty$-shifty" if the following conditions are fulfilled (cf. [34]):

(i) $\forall j, m \in \mathbb{N}$ $\exists \sigma \in * - \mathrm{Aut}(\mathcal{A})$ such that $\sigma(\mathcal{A}_{j+m}) \supset \mathcal{A}_j, \theta(\mathcal{A}_j), \ldots, \theta^{m-1}(\mathcal{A}_j)$ as (not necessarily commuting) subalgebras, and

(ii) there is a sequence $\{n_j \in \mathbb{N} | n_{j+1} > n_j\}_{j\in\mathbb{N}}$ such that $\mathcal{A}_j, \theta^{n_j}(\mathcal{A}_j), \ldots, \theta^{kn_j}(\mathcal{A}_j)$ are pairwise commuting *and* independently covering as in (3.26,v), $\forall k \in \mathbb{N}$; and such that $\lim_{j\to\infty}\frac{n_j - j}{j} = 0$.

Theorem (3.32): Let $\mathcal{A} = \overline{\bigcup_{n\in\mathbb{N}}\mathcal{A}_n}$ be an AF algebra with an "$\mathcal{A}_\infty$-shifty" automorphism $\theta \in * - \mathrm{Aut}(\mathcal{A})$, then $\hbar_{\mathcal{A}_\infty}(\theta) = \lim_{n\to\infty}\frac{1}{n}\bar{H}(\mathcal{A}_n)$.

Proof:[34] By (3.25), (3.26;ii, iv & v) (cf. also (3.24,iii)!),

$$\hbar_{\mathcal{A}_\infty}(\theta) = \lim_{j\to\infty}\hbar(\theta, \mathcal{A}_j) = \lim_{j\to\infty}\lim_{n\to\infty}\frac{1}{n}\bar{H}(\mathcal{A}_j, \theta(\mathcal{A}_j), \ldots, \theta^{n-1}(\mathcal{A}_j)) \leq$$

$$\leq \lim_{j\to\infty}\lim_{n\to\infty}\frac{1}{n}[\bar{H}(\mathcal{A}_j, \ldots, \theta^{n-j}(\mathcal{A}_j)) + \bar{H}(\theta^{n-j+1}(\mathcal{A}_j), \ldots, \theta^{n-1}(\mathcal{A}_j))] \leq$$

$$\leq \lim_{j\to\infty}\liminf_{n\to\infty}\frac{1}{n}[\bar{H}(\sigma(\mathcal{A}_n)) + \bar{H}(\theta^{n-j+1}\circ\sigma(\mathcal{A}_{2j-1}))] =$$

$$= \lim_{j\to\infty}\liminf_{n\to\infty}\frac{1}{n}[\bar{H}(\mathcal{A}_n) + \bar{H}(\mathcal{A}_{2j-1})] = \liminf_{n\to\infty}\frac{\bar{H}(\mathcal{A}_n)}{n}.$$

On the other hand, by (3.24,ii) resp. (3.21,ii):

$$n_j \cdot \hbar_{\mathcal{A}_\infty}(\theta) \geq \hbar_{\mathcal{A}_\infty}(\theta^{n_j}) \geq \hbar(\theta^{n_j}, \mathcal{A}_j) = \bar{H}(\mathcal{A}_j)$$

by assumption (ii) and (3.26,vi). Hence

$$\hbar_{\mathcal{A}_\infty}(\theta) \geq \frac{\bar{H}(\mathcal{A}_j)}{n_j} = \frac{\bar{H}(\mathcal{A}_j)}{j} - \frac{n_j - j}{j}\frac{\bar{H}(\mathcal{A}_j)}{n_j}$$

which implies $\hbar_{\mathcal{A}_\infty}(\theta) \geq \limsup_{n\to\infty} \frac{1}{n}\bar{H}(\mathcal{A}_n)$ (again using assumption (ii) on $\{n_j\}$).

Corollary (3.33): Let $\mathcal{A} = \overline{\bigcup_{n\in\mathbb{N}} \mathcal{A}_n}$ be an AF algebra with a *periodic* sequence $\{\mathcal{A}_n\}_{n\in\mathbb{N}}$ (with period p and Perron–Frobenius eigenvalue β of the inclusion matrix $[\mathcal{A}_j \to \mathcal{A}_{j+p}]$ $\forall j \geq n_0$), then for any "$\mathcal{A}_\infty$-shifty" $\theta \in * - \mathrm{Aut}(\mathcal{A})$: $\hbar_{\mathcal{A}_\infty}(\theta) = \frac{1}{p}\log\beta$.

Examples (3.34):

(a) Let $\mathcal{A}(n) = \overline{\bigotimes_{k\in\mathbb{Z}}(M_n)_k}$ be the n^∞-UHF algebra and $\mathcal{A}_N(n) = \bigotimes_{k=-N}^{N}(M_n)_k$, $\mathcal{A}_\infty(n) = \bigcup_{N\in\mathbb{N}} \mathcal{A}_N(n)$; then for $\theta_n \in * - \mathrm{Aut}(\mathcal{A}(n))$ determined by the unit shift ("n-shift"?) $\theta_n : (M_n)_k \to (M_n)_{k+1}$ ($\forall k \in \mathbb{N}$), we get $\hbar_{\mathcal{A}_\infty(n)}(\theta_n) = \log n$ (follows from (3.33) with $p = 1$ and $[\mathcal{A}_N(n) \to \mathcal{A}_{N+1}(n)] = n \in \mathbb{N}$).

(b) $\mathcal{A}_N(g) = C^*(\{e_i | i = -N, \ldots, N\})$ with the relations $e_i^* = e_i$, $e_i^2 = \mathbb{1}$ and $e_i e_j = e_j e_i (-1)^{g(|i-j|)}$ $(i \neq j)$ with $g : \mathbb{N} \to \{0,1\}$. Again $\mathcal{A}_\infty(g) = \bigcup_{N\in\mathbb{N}} \mathcal{A}_N(g)$ and $\mathcal{A}_g = \overline{\mathcal{A}_\infty(g)}$; then for $\theta_g \in * - \mathrm{Aut}(\mathcal{A}_g)$ determined by $\theta_g(e_i) = e_{i+1}$ $\forall i \in \mathbb{Z}$, we get $\hbar_{\mathcal{A}_\infty(g)}(\theta_g) = \frac{1}{2}\log 2$ if $g(n) = 1$ $\forall n \in \mathbb{N}$ *or* if $\exists n_0 : g(n) = 0$ $\forall n \geq n_0$ (but *not* with $g \equiv 0$, in which case we would have the classical "2-shift" with $\hbar(\theta) = \log 2$, corresponding to example (a) with θ restricted to a maximally Abelian C^*-subalgebra $C(\{0,1\}^{\mathbb{Z}}) \subset \mathcal{A}(2)$); see [14] for the proof.

(c) $\mathcal{A}_N(\lambda) = C^*(\{p_i | i = -N, \ldots, N\})$ with the relations $p_i = p_i^* = p_i^2$, $p_i p_{i\pm1} p_i = \lambda p_i$ and $[p_i, p_j] = 0$ $\forall i,j : |i-j| \geq 2$; for $\lambda \in \{(4\cos^2\frac{\pi}{m})^{-1} | m \in \mathbb{N} \setminus \{1,2\}\}$. Again $\mathcal{A}_\infty(\lambda) = \bigcup_{N\in\mathbb{N}} \mathcal{A}_N(\lambda)$ and $\mathcal{A}_\lambda = \overline{\mathcal{A}_\infty(\lambda)}$. Then, for $\theta_\lambda \in * - \mathrm{Aut}(\mathcal{A}_\lambda)$ determined by $\theta_\lambda(p_i) = p_{i+1}$ $\forall i \in \mathbb{Z}$, we get $\hbar_{\mathcal{A}_\infty(\lambda)}(\theta_\lambda) = -\frac{1}{2}\log\lambda$. (Note that (c) for $\lambda = \frac{1}{2}$ is identical with (b) for $g(n) = \delta_{n1}$ by $p_i = \frac{1}{2}(e_i + \mathbb{1})$; and for $\lambda \neq \frac{1}{2}$, (c) follows again from (3.33) by the same argument as in [34]).

(d) Let $(\mathcal{A}_A, \theta_A)$ be the shift θ_A on an AF algebra $\mathcal{A}_A$ associated with a topological Markov chain as treated by Evans [31] with his "AF–imitation" of the topological entropy (via the Connes–Størmer entropy [6]); but without repeating this lengthy example here, we just compare with Evans' notation: Our $(\mathcal{A}_A, \theta_A) = (C^A, \sigma_0)$ of Evans, where A is an aperiodic $(n \times n)$–matrix with entries in $\{0,1\}$, and our $\mathcal{A}_s = N_s$ (with $s \in \mathbb{N}$) resp. $\mathcal{A}_\infty(A) = \bigcup_{s\in\mathbb{N}} N_s$ of Evans. Then, repeating the proof of the main theorem in [31] with our $\bar{H}$ instead of Evans' H and using properties (3.26,i & iii) and (3.10) of $\bar{H}$ (resp. its definition (3.8)), we immediately get the same result:

$\hbar_{\mathcal{A}_\infty(A)}(\theta_A) = \log\lambda$, where λ is the spectral radius of A (note that we do not need at all even the *ingredients* of Evans' Prop. 2 for the second part of the rewritten proof, as it amounts in our case to the *same* estimate by "$\log|\mathcal{M}_{s,s+k}|$" in Evans' notation as the first part, only from below instead from above).

Concluding Remarks (3.35):

1. Possibly, the main problem (3.22) of our thus "pedestrian" approach could be solved more easily in this special AF case as above, i.e. the problem to characterize (or identify) the norm-generating locally semisimple $*$-subalgebras $\mathcal{A}_\infty$ of the AF algebra $\mathcal{A}$ such that the (first "formal") supremum $\hbar(\theta) = \sup_{\gamma\in\mathcal{CP}_1(\mathcal{A})}\hbar(\theta,\gamma)$ is attained: $\hbar_{\mathcal{A}_\infty}(\theta) = \hbar(\theta), \ \forall\theta\in * - \mathrm{Aut}(\mathcal{A})$.

2. The advantage of Evans' approach [31] cited in (d) is to "circumvent" this problem (1.) by using the norm continuity (2.9,i) of the Connes–Størmer entropy ([6], for finite-dimensional subalgebras $\mathcal{B}\subset\mathcal{A}$); but on the other hand, this "AF-imitation" of the topological entropy has the serious drawback of an "AF-limitation" in the following sense: To extend Evans' definition further for non-AF (nuclear) C^*-algebras, one would have to repeat his construction with the CNT entropy as in (2.9,i) for $\gamma\in\mathcal{CP}_1(\mathcal{A})$ instead of the Connes-Størmer entropy; and then it seems to be impossible (to show?) that this analogously extended definition of Evans' entropy $h_E(\theta)$ coincides again with the "classical" topological entropy (2.1,iii) for a homeomorphism $T: X \to X$ of a *connected* compact space X: $h_E(\theta_T) = h(T)$? (Cf. [5, App.], and contrast with Thm. (2.7) here.)

3. Naively, one could think of a *direct* "classical" definition of a "quantum" topological entropy $h_{cl}(\theta)$ for a C^*-dynamical system $(\mathcal{A},\theta)$, simply defining $h_{cl}(\theta) = h(T_\theta)$ with the topological entropy (2.1,iii) for the adjoint (affine) weak-$*$ homeomorphism $T_\theta: S_\mathcal{A}\to S_\mathcal{A}$ (defined by $T_\theta(\omega) = \omega\circ\theta, \ \forall\omega\in S_\mathcal{A}$). But by Thm. (2.3,iii) this is no *generalization* of the topological entropy $h(T)\neq h(T_M)$, and one could only think to regard it as a non-Abelian *analogue*; but also that is definitely ruled out by the following most simple "counter-example": In (3.34,a) already with $n = 2$, $h_{cl}(\theta_2) = \infty$; which we leave to the reader as a not quite trivial exercise (see [10]).

4. In all the AF examples (3.34), we always have the desirable generalization of (2.8): $\hbar_{\mathcal{A}_\infty}(\theta) = \sup_{\omega\in S^\theta_\mathcal{A}} h_\omega(\theta)$, with the CNT entropy on the r.h.s.; but without having computed $h_\tau(\theta)$ for any non-AF nuclear C^*-algebra with ("optimal") approximating net τ (and also not yet the CNT entropy, for non-trivial ω and θ, by the way!), we are far from a conjecture (not to speak about a general proof, cf. already [5] for *Abelian* $\mathcal{A}$).

Acknowledgements: I have to thank the organizers of "QP VIII", Profs. L. Accardi and W. von Waldenfels, for inviting me to this meeting in Oberwolfach with its included financial support by the institute; and thanks go to the referee for pointing out a (terrible) mistake in the original proof of the nevertheless correct Lemma (3.30).

Financial support by "Fonds zur Förderung der wissenschaftlichen Forschung in Österreich" (Proj. P7101–Phy) is gratefully acknowledged.

Remark: Two earlier versions of this contribution have been previously distributed as preprints UWThPh–1991–62 (Vienna) resp. NTZ–37/1991 (Leipzig).

References

[1] Connes, A., Narnhofer, H. and Thirring, W.: "Dynamical Entropy for C^* Algebras and von Neumann Algebras", Commun. Math. Phys. **112**, 691–719 (1987).

[2] Park, Y. M. and Shin, H. H.: "Dynamical Entropy of Quasi-local Algebras in Quantum Statistical Mechanics", Commun. Math. Phys. **144**, 149–161 (1992).

[3] Effros, E. G.: "Why the Circle is Connected: An Introduction to Quantized Topology", Mathematical Intelligencer **11**, 27–34 (1989).

[4] Blackadar, B. E.: "A simple unital projectionless C^*-algebra", J. Operator Theory **5**, 63–71 (1981).

[5] Hudetz, T.: "Algebraic Topological Entropy", in *Nonlinear Dynamics and Quantum Dynamical Systems*, G. A. Leonov et al. (eds.), Akademie–Verlag Berlin (1990), p. 27–41.

[6] Connes, A. and Størmer, E.: "Entropy for Automorphisms of II_1 von Neumann Algebras", Acta Math. **134**, 289–306 (1975).

[7] Connes, A.: "Entropie de Kolmogoroff–Sinai et mécanique statistique quantique", C. R. Acad. Sci. Paris Ser. I Math. **301**, 1–6 (1985).

[8] Hudetz, T.: "Non–linear Entropy Functionals and a Characteristic Invariant of Symmetry Group Actions on Infinite Quantum Systems", in *Selected Topics in QFT and Mathematical Physics*, J. Niederle and J. Fischer (eds.), World Scientific, Singapore (1990), p. 110–124.

[9] Lindblad, G.: "Dynamical Entropy for Quantum Systems", in *Quantum Probability and Applications III*, L. Accardi and W. von Waldenfels (eds.), Springer LNM 1303, Berlin (1988), p. 183–191.

[10] Hudetz, T.: "Quantum Topological Entropy: The 'Canonical' Approach", in preparation (1992).

[11] Hudetz, T.: Ph.D. Thesis, Univ. Vienna (in German), unpublished (1992).

[12] Emch, G. G.: "Kolmogorov Flows, Dynamical Entropies and Mechanics", to appear in *Quantum Probability VII (New Delhi)*, L. Accardi et al. (eds.), World Scientific, Singapore (1992).

[13] Narnhofer, H. and Thirring, W.: "Quantum K-Systems", Commun. Math. Phys. **125**, 565–577 (1989).

[14] Narnhofer, H. and Thirring, W.: "Chaotic Properties of the Noncommutative 2-shift", to appear in *From Phase Transitions to Chaos, Topics in Modern Statistical Physics*, G. Györgyi et al. (eds.), World Scientific, Singapore (1992).

[15] Sauvageot, J.-L. and Thouvenot, J.-P.: "Une nouvelle définition de l'entropie dynamique des systemes non commutatifs", Commun. Math. Phys. **145**, 411–423 (1992).

[16] Adler, R. L., Konheim, A. G. and McAndrew, M. H.: "Topological Entropy", Trans. Am. Math. Soc. **114**, 309–319 (1965).

[17] Bauer, W. and Sigmund, K.: "Topological Dynamics of Transformations Induced on the Space of Probability Measures", Monatshefte Math. **79**, 81–92 (1975).

[18] Sigmund, K.: "Affine Transformations on the Space of Probability Measures", Societé Mathématique de France, Astérisque **51**, 415–427 (1978).

[19] Dinaburg, E. I.: "On the Relations Among Various Entropy Characteristics of Dynamical Systems", Mathematics of the USSR - Izvestiya **5**, 337–378 (1971); reprinted in *Dynamical Systems*, Ya. G. Sinai (ed.), World Scientific, Singapore (1991), p. 97–138.

[20] Goodman, T. N. T.:"Relating Topological Entropy and Measure Entropy", Bull. London Math. Soc. **3**, 176–180 (1971).

[21] Kuriyama, K.: "Entropy of a Finite Partition of Fuzzy Sets", J. Math. Anal. Appl. **94**, 38–43 (1983).

[22] Maličký, P. and Riečan, B.: "On the Entropy of Dynamical Systems", in *Ergodic Theory and Related Topics II*, H. Michel et al. (eds.), Teubner, Leipzig (1987), p. 135–138.

[23] Hudetz, T.: "Entropy and Dynamical Entropy for Continuous Fuzzy Partitions", in preparation (1992).

[24] Gillmann, L. and Jerison, M.: *Rings of Continuous Functions*, Springer GTM 43, New York (1960).

[25] Choi, M.-D. and Effros, E. G.: "Nuclear C^*-algebras and the Approximation Property", Am. J. Math. **100**, 61–79 (1978).

[26] Choi, M.-D.: "A Schwarz Inequality for Positive Linear Maps on C^*-algebras", Illinois J. Math. 18, 565–574 (1974).

[27] Kadison, R. V.: "A Generalized Schwarz Inequality and Algebraic Invariants for Operator Algebras", Ann. Math. **56**, 494–503 (1952).

[28] Summers, S. J.: "On the Independence of Local Algebras in Quantum Field Theory", Rev. Math. Phys. **2**, 201–247 (1990).

[29] Størmer, E. and Voiculescu, D.: "Entropy of Bogoliubov Automorphisms of the Canonical Anticommutation Relations", Commun. Math. Phys. **133**, 521–542 (1990).

[30] Hudetz, T.: "Spacetime Dynamical Entropy of Quantum Systems", Lett. Math. Phys. **16**, 151–161 (1988).

[31] Evans, D. E.: "Entropy of Automorphisms of AF Algebras", Publ. RIMS Kyoto Univ. **18**, 1045–1051 (1982).

[32] Christensen, E.: "Near Inclusions of C^*-algebras", Acta Math. **144**, 249–265 (1980).

[33] Bratteli, O.: "Inductive Limits of Finite Dimensional C^*-algebras", Trans. Am. Math. Soc. **171**, 195–234 (1972).

[34] Choda, M.: "Entropy for $*$-Endomorphisms and Relative Entropy for Subalgebras", J. Operator Theory (to appear).

Quantum Probability and Related Topics Vol. VIII (pp. 263–270)

ON FEYNMAN–KAC COCYCLES

R L Hudson

Mathematics Department
University of Nottingham
Nottingham NG7 2RD
UK

Abstract. We consider the operator Feynman–Kac formula in which the unperturbed semigroup is got by contracting a quantum stochastic evolution, and ask when is the perturbation cocycle of exponential integral form.

§1. Statement of the problem

An analysis of the Feynman–Kac formula [HIP] at operator rather than super-operator [Ac] level, reveals that the unperturbed semigroup is the expectation of a random unitary evolution and the perturbed semigroup is the expectation of a perturbation of this evolution which is effected by a cocycle possessing covariance properties for the one-dimensional Euclidean group of translations and reflections of the real line.

Such random evolutions are readily constructed using quantum stochastic calculus. Let $U_{s,t}$ denote the solution at time s of the quantum stochastic differential equation

$$dU = U(l dA^\dagger - l^* dA + (ih - \tfrac{1}{2}l^* l)ds), \quad U(t) = 1 \tag{1.1}$$

in which l and $h = h^*$ are ampliations to the Fock space $\Gamma(L^2(\mathbb{R}))$ of operators in an initial space $\mathcal{H}_0$. Then $(U_{s,t} : s \geqslant t)$ is a covariantly adapted evolution in the sense of [HIP] in which the filtration of von Neumann algebras $(N, (N_{s,t} : s \geqslant t))$ is got by taking

$$N = B(\mathcal{H}_0) \otimes B(\Gamma(L^2(\mathbb{R}))), \qquad N_{s,t} = B(\mathcal{H}_0) \otimes 1 \otimes B(\Gamma(L^2[t,s])) \otimes 1$$

and the Euclidean group acts by second quantising shifts and reflections on $L^2(\mathbb{R})$. If $h = 0$ $(U_{s,t})$ is reflectively covariantly adapted. The vacuum conditional expectation onto $B(\mathcal{H}_0)$ is a reflectively covariant reducing map. The reduced evolution

$$T_t = \mathbb{E}_0[U_{s+t,s}]$$

is the contraction semigroup on $\mathcal{H}_0$ generated by $ih - \frac{1}{2}l^* l$. Given $V \in B(\mathcal{H}_0)$ identified with its ampliation to Fock space we may now construct the cocycle $M^V = (M^V_{s,t}, s \geq t)$ defined by the ordinary differential equation

$$\partial_s M^V_{s,t} = -M^V_{s,t} U_{s,t} V U^{-1}_{s,t}, \qquad M^V_{t,t} = 1,$$

the perturbed evolution

$$U^V_{s,t} = M^V_{s,t} U_{s,t}$$

and the corresponding semigroup

$$T^V_t = \mathbb{E}_0[U^V_{s+t,s}]$$

whose generator is $ih - \frac{1}{2}l^*l + V$. The formula

$$e^{t(ih-\frac{1}{2}l^*l-V)} = \mathbb{E}_0[M^V_{s+t,s} U_{s+t,s}] \tag{1.2}$$

is the abstract form of Feynman–Kac formula of [HIP]. The standard Feynman–Kac formula of Brownian motion is obtained by taking $h = 0$, $l = l^* = p$, the momentum operator in $\mathcal{H}_0 = L^2(\mathbb{R})$, identifying $i(A^\dagger - A)$ as Brownian motion, and taking the perturbation potential V to be a function of q. The 'oscillator process' Feynman–Kac formula of the Ornstein–Uhlenbeck process [Si] is got by taking $l = a$, the annihilation operator $q + ip$ [HIP]. In both these cases, and also [HP1] in the case of the 'anti Ornstein–Uhlenbeck process' got by taking $l = a^\dagger$, provided that V is a function of q (or more generally, of a fixed real linear combination

$$V = v(xp + yq)$$

of p and q), the process $(U_{s,t} V U^{-1}_{s,t} : s \geq t)$ is commutative. (Note that in all three cases l is unbounded.) Consequently [HIP] the cocycle M^V can be expressed in exponential integral form

$$M^V_{s,t} = \exp\left(-\int_t^s U_{s,\tau} V U^{-1}_{s,\tau}\, d\tau\right).$$

The purpose of this paper is to investigate conditions on V, h and l under which this is the case. By covariance, the problem amounts to the following.

When is the process $(V(t): t \geq 0)$ commutative, where $V(t) = U(t) V U(t)^{-1}$, and $U(t) = U_{t,0}$?

Acknowledgement

I am grateful to J M Lindsay for pointing out an error in an earlier version of this work and for suggesting numerous other improvements.

§2. The bounded case

Assuming that V, h and l are bounded operators on the initial space $\mathcal{H}_0$ we denote by j the flow

$$j_t(x) = U(t)xU(t)^{-1}, \quad x \in B(\mathcal{H}_0)$$

on $B(\mathcal{H}_0)$ and by α, $\alpha^\dagger$ and τ its structure maps

$$\alpha(x) = [l, x], \qquad \alpha^\dagger(x) = [x, l^*],$$

$$\tau(x) = i[h, x] - \tfrac{1}{2}(l^*lx - 2l^*xl + xl^*l),$$

so that

$$dj_t(x) = x + \int_0^t \big(j_s(\alpha(x))\, dA^\dagger + j_s(\alpha^\dagger(x))\, dA + j_s(\tau(x))\, ds\big).$$

Theorem 2.1. Under these assumptions, the process $V(t)$ is commutative if and only if, for arbitrary $n \in \mathbb{N}$,

$$[V, \kappa_n \ldots \kappa_1(V)] = 0, \tag{2.1}$$

where $\kappa_1, \ldots, \kappa_n$ are arbitrary elements of the set of structure maps $\mathcal{S} = \{\alpha, \alpha^\dagger, \tau\}$.

Proof. Assuming that

$$[V, j_t(V)] = 0, \quad t \geqslant 0$$

we have, since $dV = 0$ and V commutes with $dA^\dagger$, dA and dt

$$\begin{aligned} 0 &= [V, dj_t(V)] \\ &= \sum_{\kappa \in \mathcal{S}} [V, j_t(\kappa(V))]\, dK, \end{aligned}$$

where dK is the stochastic differential corresponding to $\kappa \in \mathcal{S}$. By independence of the stochastic differentials [Li] we must have

$$[V, j_t(\kappa(V))] = 0, \quad t \geqslant 0$$

for each $\kappa \in \mathcal{S}$. Repeating the argument, differentiating n times in total, and finally setting $t = 0$, we obtain (2.1).

For the converse we use an adapted form of the argument of Theorem 7.1 of [FS]. We have

$$[V, j_t(V)] = \sum_{\kappa \in \mathcal{S}} \int_0^t [V, j_s(\kappa(V))]\, dK(s).$$

Since (2.1) holds, we may iterate, obtaining

$$[V, j_t(V)] = \sum_{\kappa_1,\ldots,\kappa_n \in \mathcal{I}} \int_{0<s_n<\ldots<s_1<t} [V, j_{s_n}(\kappa_n \ldots \kappa_1(V))]\, dK_n(s_n)\ldots dK_1(s_1).$$

Using the basic estimate of quantum stochastic calculus in the form

$$\left\| \int_0^t E\, dK u \otimes \psi(f) \right\|^2 \leq c(f,T) \int_0^t \| E(s) u \otimes \psi(f) \|^2\, ds$$

for $t \leq T$ and locally bounded f, we find that for arbitrary such f and $u \in \mathcal{H}_0$, using the contractivity of j,

$$\| [V, j_t(V)] u \otimes \psi(f) \|^2 \leq \frac{(c(f,T)MT)^n}{n!} \|V\| \, \|u \otimes \psi(f)\|^2 \underset{n}{\longrightarrow} 0,$$

where $M = \max(\|\alpha\|, \|\alpha^\dagger\| \, \|\tau\|)$ and α, $\alpha^\dagger$ and τ are regarded as elements of the Banach space $B(B(\mathcal{H}_0))$. Hence $[V, j_t(V)] = 0$ as required. □

We consider the case when $\mathcal{H}_0 = L^2(\mathbb{R})$ and seek evolutions U_t for which the Feynman–Kac cocycle M^V will have exponential integral form for every potential of the form $V = v(q)$ of a bounded measurable function of the position operator q. By Theorem 2.1 we must have, for all such V,

$$[V, [l, V]] = 0. \tag{2.2}$$

If v is a continuous bijective function, for example if $v(x) = \tanh \lambda x$ for fixed $\lambda \in \mathbb{R}$, operators which commute with V are themselves multiplications so we can write

$$[l, V] = \hat{V}, \tag{2.3}$$

where $\hat{V}$ is multiplication by a function $\hat{v}$. Since $ad_l : V \mapsto [l, V]$ is a derivation, the same is true for products of such functions v and hence, by the Stone–Weierstrass theorem and boundedness of the map ad_l, for arbitrary V belonging to the commutative C*-algebra of multiplications by continuous functions on $\mathbb{R}$ possessing limits at $\mp\infty$. Passing to weak operator limits using the weak operator continuity of ad_l on bounded subsets of the von Neumann algebra $B(\mathcal{H}_0)$ we conclude that (2.3) holds for multiplications V by arbitrary bounded measurable functions and hence that ad_l is a derivation of the von Neumann algebra M of such multiplications. Since Abelian von Neumann algebras have no non-trivial derivations we see that $\alpha = ad_l$ vanishes on M and hence, since M is maximal Abelian, that $l \in M$. In particular, for all $V \in M$, $\alpha(V) = 0$ and

$$\tau(V) = i[h, V] - \tfrac{1}{2}(l^* l V - 2 l^* V l + V l^* l) = i[h, V].$$

We may now repeat the argument with l replaced by ih to conclude that $h \in M$ and $\tau(V) = 0$ for all $V \in M$. Thus, for $V \in M$, $j_t(V) = V$ for all t and the Feynman–Kac formula (1.2) takes the trivial form

$$
\begin{aligned}
e^{t(ih-\frac{1}{2}l^*l-V)} &= \mathbb{E}_0[e^{-tV}U_{s+t,s}] \\
&= e^{-tV}\mathbb{E}_0[U_{s+t,s}] \\
&= e^{-tV}e^{t(ih-\frac{1}{2}l^*l)}
\end{aligned}
$$

of a product of commuting semigroups of elements of M. We have proved

Theorem 2.2. It is not possible to find a unitary process, driven by a stochastic differential equation (1.1) with bounded coefficients l and h, for which there is a non-trivial perturbation cocycle of exponential integral form for arbitrary V of the form $v(q)$ where v is a bounded measurable function.

§3. Some remarks about Feynman–Kac cocycles for the Weyl algebra

We denoted by $\mathcal{W}$ the *Weyl algebra* of polynomials in the indeterminate p and q satisfying

$$[p,q] = -i,$$

equipped with the involution $\dagger$ which makes p and q self-adjoint.

Theorem 3.1. Let α, $\alpha^\dagger$ and τ be linear maps $\mathcal{W} \to \mathcal{W}$ satisfying the structure relations of a quantum flow

$$\alpha(xy) = \alpha(x)y + x\alpha(y), \qquad \alpha^\dagger(x) = \alpha(x^\dagger)^\dagger \tag{3.1}$$

$$\tau(xy) = \tau(x)y + x\tau(y) + \alpha^\dagger(x)\alpha(y), \quad \tau = \tau^\dagger. \tag{3.2}$$

In order that for every element $V \in \mathcal{W}$ of form

$$V = v(xp+yq), \quad x, y \in \mathbb{R}$$

that is a polynomial in a real linear combination of p and q, the condition (2.1) holds, it is necessary and sufficient that α and τ take the form

$$\alpha(x) = [l,x], \qquad \tau(x) = i[h,x] - \tfrac{1}{2}(l^\dagger lx - 2l^\dagger xl + xl^\dagger l), \tag{3.3}$$

where l and h are, respectively, complex and real linear polynomials in p and q.

Proof. To prove the necessity, we note first that, since every derivation of $\mathcal{W}$ is inner [Se], [HP1], α is of form $\alpha(x) = [l,x]$ for some $l \in \mathcal{W}$, and furthermore since τ_0 defined by

$$\tau_0(x) = -\tfrac{1}{2}(l^\dagger lx - 2l^* xl + xl^\dagger l)$$

satisfies (3.2), τ must be of form

$$\tau(x) = i[h, x] - \tfrac{1}{2}(l^{\dagger}lx - 2l^{\dagger}xl + xl^{\dagger}l),$$

where $h = h^{\dagger} \in \mathcal{W}$. To see that l is linear in p and q we first take $V = q$. From (2.1) we have

$$[q, [l, q]] = 0$$

from which it follows that $[l, q]$ must be a polynomial in q and hence that l is of the form $pf(q) + g(q)$ where f and g are polynomials. Similarly, taking $V = p$, we find that l must be of the form $qh(p) + k(p)$ where h and k are polynomials. Combining these statements we conclude that l must be of the form

$$l = bpq + cp + dq + e,$$

where $b, c, d, e \in \mathbb{C}$. Now take $V = p + q$. Then

$$\alpha(V) = [l, V] = ib(-q+p) - ic + id.$$

It follows from $[V, \alpha(V)] = 0$ that $b = 0$ and hence that l is linear in p and q as claimed. It can now be verified that each $V = v(xp + yq)$ commutes with $\tau_0(V) = -\frac{1}{2}(l^{\dagger}lV - 2l^{\dagger}Vl + Vl^{\dagger}l)$ as well as with $\tau(V)$ and must thus commute with the difference $i[h, V]$. Repetition of the argument for l shows that h must also be linear in p and q; it is a real linear combination since $h = h^{\dagger}$

Conversely suppose that α and τ have the forms (3.3). Fix $x, y \in \mathbb{R}$. An element w of $\mathcal{W}$ commutes with all polynomials $v(xp + yq)$, in particular with $xp + yq$ itself, if and only if it is itself such a polynomial. The proof will therefore be completed by showing that, if w is such a polynomial then so too are $\alpha(w)$, $\alpha^{\dagger}(w)$, $\tau(w)$, since then repeated action of the maps α, $\alpha^{\dagger}$ and τ on the polynomial V can only yield further polynomials, which will commute with V. Assuming that $l = cp + dq + e$ we have $l^{\dagger} = \bar{c}p + \bar{d}q + \bar{e}$ and hence

$$\alpha(w) = i(dx - cy)w', \qquad \alpha^{\dagger}(w) = -i(\bar{d}x - \bar{c}y)w', \tag{3.4}$$

where w' is the derivative of the polynomial w. Writing $\tau(w)$ in the form

$$\tau(w) = -\tfrac{1}{2}l^{\dagger}\alpha(w) - \tfrac{1}{2}\alpha^{\dagger}(w)l$$

we have, using (3.4), that

$$\begin{aligned}[xp + yq, \tau(w)] &= -\tfrac{1}{2}i(y\bar{c} - x\bar{d})(dx - cy)w' \\ &\quad + \tfrac{1}{2}i(\bar{d}x - \bar{c}y)(yc - xd)w' \\ &= 0.\end{aligned}$$

Hence $\tau(w)$ is also a polynomial as required. □

§4. Conclusion

Theorem 3.2 indicates that the only Feynman–Kac cocycles constructed by means of quantum stochastic calculus for one-dimensional quantum particle mechanics for which the perturbation cocycle has exponential integral form correspond to the linear quantum flows of [HP1], in which creation and annihilation operators satisfy linear quantum stochastic differential equations with constant coefficients. (The possibility of including the term $i[h,x]$ in (3.3) was not considered in [HP1], but the new freedom generated thereby amounts only to transforming to new canonical coordinates related to the old by a phase-space translation.) In [HP1] three canonical forms for such flows were found under the actions of linear canonical transformations and gauge transformations in Fock space, for which the corresponding Feynman–Kac formulae are, respectively, the usual one for Brownian motion, the 'oscillator process' [Si] one corresponding to the Ornstein–Uhlenbeck process, and that for the 'anti-Ornstein–Uhlenbeck process' of [HP1]. Thus it may be asserted that these three are the only 'useful' Feynman–Kac formulae. However, to reach this conclusion we have had to assume exponential integral form for the cocycle whenever V is a function of a fixed linear combination $xp+yq$, and our analysis evidently exploits the resulting canonical symmetry. The question of classifying Feynman–Kac cocycles of exponential-integral type only for potentials which are functions of q remains open.

References

[Ac] Accardi L, A quantum formulation of the Feynman–Kac formula, Random Fields, Vol I, Proceedings Esztergom 1979, ed J Fritz *et al*, North Holland (1981).

[FS] Fagnola F and Sinha K B, On quantum flows with unbounded structure maps, to appear in *J London Math Soc*.

[HIP] Hudson R L, Ion P D F and Parthasarathy K R, Time-orthogonal unitary dilations and non-commutative Feynman–Kac formulae, *Commun Math Phys* **83** 761–80 (1982).

[HP1] Hudson R L and Parthasarathy K R, Construction of quantum diffusions, pp 173–198, in *Quantum Probability*, proceedings, Rome 1982, ed L Accardi *et al*, Springer LNM **1055**.

[HP2] Hudson R L and Parthasarathy K R, Quantum Ito's formula and stochastic evolutions, *Commun Math Phys* **93** 301–323 (1984).

[Li] Lindsay J M, Independence for quantum stochastic integrators, pp 325–332 in *Quantum Probability VI*, eds L Accardi *et al*, World Scientific (1991).

[Se] Segal I E, Quantized differential forms, *Topology* **7** 147–172 (1968).

[Si] Simon B, *Functional integration and quantum physics*, Academic Press, New York (1979).

Quantum Probability and Related Topics Vol. VIII (pp. 271–280)

THE KERNEL OF A FOCK SPACE OPERATOR I

J M Lindsay

§0. Introduction

Symmetric Fock space over $L^2(M)$ is usually viewed as consisting of square-summable sequences (f_n) of symmetric square-integrable functions (f_n on M^n). When M is non-atomic however, it is often convenient to replace the sequence (f_n) by a single function f on Γ, the *finite power set* of M. In this way Fock space becomes an L^2-space itself — the measure on Γ being composed from the product measures on M^n in a natural fashion ([Gui]).

Exploiting this viewpoint, a class of *integral-sum* operators has been introduced for describing solutions of quantum stochastic differential equations ([Maa], [Me 1] p.305, [Me 2] p.39, [L 1], [LP]). The operators have a formal similarity with the Hilbert–Schmidt class, they are defined however in terms of the *union and partition operations* which are closely connected with the gradient operator and Hitsuda-Skorohod integral of Malliavin calculus (see [L 2]). Here we are concerned with the 3-argument integral-sum operators introduced by Meyer. The operator X with kernel x is formally the multiple quantum stochastic integral

$$\iiint_{\Gamma^3} x(\rho, \sigma, \tau)\, dA^*_\rho dA_\sigma dA_\tau, \tag{0.1}$$

where $dA^*_{\{r_1,\dots,r_n\}} = dA^*_{r_1} \dots dA^*_{r_n}$ etc and A^*, Λ, A are respectively the creation, preservation and annihilation processes ([HuP], [Par]). Indeed the action of X on $k \in L^2(\Gamma)$ may be deduced by using the representation

$$k = \int_\Gamma k(\omega)\, dA^*_\omega\, \delta_\varnothing$$

and applying the formal quantum Itô relations (see [Me 1] p.81, [LM]).

In this note the uniqueness of the kernel of such operators is demonstrated. More specifically, it is shown that if an integral-sum operator annihilates a *finite particle domain* or, equally, an *exponential domain*, then the kernel must be zero. This in turn indicates a limitation of this class: if an integral-sum operator has domain including either of these domains, then it is *determined* by its action on that domain. Precisely *how* the kernel is determined is a little delicate — this will be described in a companion paper ([BL]). Operators of the form (0.1) but with a fourth variable, integrated over in the ordinary sense, have also been considered ([L 1]). They have the attractive property that the kernel of a product of two such operators is given simply by a sum, over partitions, of products of the two kernels evaluated at appropriate sets. Thus, unlike the convolution of 3-argument kernels, no integration is involved. However, uniqueness clearly fails here. In contrast, if one allows *distributions* as kernels, then 2-arguments suffice since (0.1) coincides with

$$\iint_{\Gamma^2} \kappa(\alpha,\gamma)\, \mathrm{d}A^*_\alpha \mathrm{d}A_\gamma = \iint_{\Gamma^2} \kappa(\alpha,\gamma)\, \partial^*_\alpha \partial_\gamma \, \mathrm{d}\alpha\, \mathrm{d}\gamma$$

where κ is the distribution

$$\varphi \mapsto \iiint_{\Gamma^3} x(\alpha,\omega,\gamma)\varphi(\alpha\cup\omega,\omega\cup\gamma)\, \mathrm{d}\alpha \mathrm{d}\omega \mathrm{d}\gamma$$

and $\partial_\gamma = \prod_{c\in\gamma} \partial_c$ is a product of Hida differential operators. This white noise calculus for Fock space operators is currently being developed by Huang, Obata and others (see [Hua], [HOS]). The importance of preservation integrals, which are in a sense lost in the distribution-theoretic view, has been re-affirmed by recent results of Attal (see [At 1]).

§1. Notations and Preliminaries

For any set S let Γ_S be its finite power set : $\{\sigma \subset S : \#\sigma < \infty\}$. Γ_S has the countable partition $\bigcup_{n\geqslant 0}\Gamma_n(S)$ where $\Gamma_n(S) := \{\sigma \subset S : \#\sigma = n\}$. $\{\sigma \subset S : \#\sigma \leqslant n\}$ will be denoted $\Gamma_{\leqslant n}(S)$, and $\Gamma_1(S)$ will be identified with S itself, so that $S \subset \Gamma_S$. Let $S^{(n)} = \{s \in S^n : s_i \neq s_j \text{ if } i \neq j\}$, $S^{(0)} = \{0\}$ and let

$$\Phi : \bigcup_{n\geqslant 0} S^{(n)} \to \Gamma_S$$

be the map which takes 0 to $\varnothing$, and reduces $s \in S^{(n)}$ to the collection of its coordinates: $\{s_1,\ldots,s_n\}$.

Γ_S inherits a measurable structure from S as follows. Let $\mathcal{F}$ be a σ-algebra on S, then $U \subset \Gamma$ is measurable if, for each n, $\Phi^{-1}(U \cap \Gamma_n) \in \mathcal{F}^n \cap S^{(n)}$ If $\mathcal{F}$ contains all the singleton sets $\{s\}$, $s \in S$, then the union maps $\sigma \in \Gamma^d \mapsto |\sigma| := \sigma_1 \cup \ldots \cup \sigma_d$ $(d \geqslant 2)$ are measurable with respect to the product σ-algebras.

Finally, a non-atomic measure space m on $(S, \mathcal{F})$ determines a measure Γ_m on $(\Gamma_S, \Gamma_{\mathcal{F}})$ by

$$U \mapsto \iota_\varnothing(U) + \sum\nolimits_{n \geqslant 1} (n!)^{-1} \overline{m^n}\left(\Phi^{-1}(U \cap \Gamma_n)\right),$$

where $\iota_\varnothing(U) = 1$ if $\varnothing \in U$, and 0 otherwise, and $\overline{m^n}$ is the completion of the product measure m^n. One consequence of non-atomicity that is used all the time is that, for $d \geqslant 2$, $\left\{\omega \in (\Gamma_S)^d : \omega_i \cap \omega_j \neq \varnothing \text{ for some } i \neq j\right\}$ is $(\Gamma_m)^d$-null.

If $M = (S, \mathcal{F}, m)$ then $\Gamma_M := (\Gamma_S, \Gamma_{\mathcal{F}}, \Gamma_m)$ is called the *symmetric measure space* of M and $L^2(\Gamma_M)$ will be called *Guichardet space* over M. By construction Guichardet space is naturally isomorphic to symmetric Fock space: $\bigoplus_{n \geqslant 0} L^2_{\text{sym.}}(M^n)$.

For a function $\varphi: S \to \mathbb{C}$, let $\pi_\varphi: \Gamma_S \to \mathbb{C}$ denote the *product function*

$$\sigma \mapsto \prod\nolimits_{s \in \sigma} \varphi(s).$$

The map $\varphi \in L^2(M) \mapsto \pi_\varphi \in L^2(\Gamma_M)$ is continuous and

$$\left\{\pi_\varphi : \varphi \in L^2(M),\ \|\varphi\|_\infty \leqslant 1\right\}$$

is total in $L^2(\Gamma_M)$. The fundamental property of the symmetric measure is:

$\not{\Sigma}$-Lemma: For $d \geqslant 2$ let $g : \Gamma^d \to \mathbb{C}$ be non-negative, or integrable, then

$$\int_\Gamma \cdots \int_\Gamma g(\sigma_1, \ldots, \sigma_d)\, d\sigma_1 \ldots d\sigma_d = \int_\Gamma \sum\nolimits_{|\alpha| = \sigma} g(\alpha_1, \ldots, \alpha_d)\, d\sigma$$

where the sum is over partitions of σ into d parts: $(\alpha_1, \ldots, \alpha_d)$, and $d\Gamma_m(\sigma)$ is abbreviated to $d\sigma$.

A proof may be found in [LP]. As noted by Meyer ([Me 3]), the $\not{\Sigma}$-relation is linear in g, and is trivial for functions of the form $\pi_{\varphi_1} \otimes \cdots \otimes \pi_{\varphi_d}$, $\varphi_i \in L^1(M)$.

§2. Integral-sum operators

Fix a σ-finite, non-atomic, separable measure space M in which each singleton set is measurable, and let $\Gamma = \Gamma_M$ be its symmetric measure space. For a function $x : \Gamma^3 \to \mathbb{C}$, x' will denote the following transform of x:

$$x' : (\alpha, \beta, \gamma) \mapsto \sum_{\omega \subset \beta} x(\alpha, \omega, \gamma). \tag{2.1}$$

x may be recovered from x' by the Möbius-type inversion:

$$x(\alpha, \beta, \gamma) = \sum_{\omega \subset \beta} (-1)^{\#\bar{\omega}} x'(\alpha, \omega, \gamma), \tag*{(2.1)$'$}$$

where the sum is over partitions of β into two parts: $(\omega, \bar{\omega})$. Measurability of x (in the product σ-algebra of Γ^3) is equivalent to measurability of x'

Definition 2.1: Let $x : \Gamma^3 \to \mathbb{C}$ and $k : \Gamma \to \mathbb{C}$ be measurable, and consider the following condition on the pair (x, k):

$$\int_\Gamma |x(\alpha, \beta, \omega) k(\omega \cup \beta \cup \gamma)| \, d\omega < \infty \quad \text{for a.a. } (\alpha, \beta, \gamma) \in \Gamma^3 \tag{2.2}$$

If (x, k) satisfies (2.2), then write $X_3 k$ for the a.e. defined function

$$(X_3 k)(\sigma) = \int_\Gamma \sum_{|\alpha| = \sigma} x(\alpha_1, \alpha_2, \omega) k(\omega \cup \alpha_2 \cup \alpha_3) \, d\omega \tag{2.3}$$

the sum being over all partitions of σ into three: $(\alpha_1, \alpha_2, \alpha_3)$.

The condition (2.2) on x, for fixed k, is equivalent to the following condition on its transform x':

$$\int_\Gamma |x'(\alpha, \beta, \omega) \, k(\omega \cup \beta)| \, d\omega < \infty \quad \text{for a.a. } (\alpha, \beta) \in \Gamma^2, \tag*{(2.2)$'$}$$

moreover, when satisfied,

$$(X_3 k) k(\sigma) = \int_\Gamma \sum_{\alpha \subset \sigma} x'(\alpha, \bar{\alpha}, \omega) k(\omega \cup \bar{\alpha}) \, d\omega \tag*{(2.3)$'$}$$

for a.a. σ. One advantage of this alternative representation is the identity

$$\iiint_{\Gamma^3} h(\alpha \cup \beta) x'(\alpha, \beta, \gamma) k(\beta \cup \gamma) \, d\alpha \, d\beta \, d\gamma = \int_\Gamma h(\sigma)(X_3 k)(\sigma) \, d\sigma \tag{2.4}$$

— valid when either side is defined — which is an immediate consequence of the $\int\!\!\!\!\sum$-Lemma. Some other routine consequences of the $\int\!\!\!\!\sum$-Lemma are collected next.

Lemma 2.2: Let $x : \Gamma \to \mathbb{C}$ and $k : \Gamma \to \mathbb{C}$ be measurable. Then

(i) $x = 0$ a.e. if and only if $x' = 0$ a.e.;

(ii) if either $x = 0$ a.e. or $k = 0$ a.e. then $X_3 k$ is well-defined and vanishes a.e.

Thus each *kernel* x determines a partially defined linear operator X_3 on $L^0(\Gamma)$ — the linear space of measurable functions modulo null functions. Sufficient conditions for X_3 to be a densely defined operator on $L^2(\Gamma)$ are given in [Me 2], [L 1] and [BL].

The following classes of functions are relevant:

$$\mathcal{H}_a := \left\{ k \in L^0(\Gamma) : \|k\|_{(a)} := \left\{ \int_\Gamma a^{\#\sigma} |k(\sigma)|^2 \, d\sigma \right\}^{1/2} < \infty \right\} \qquad (a>0) \tag{2.5}$$

$$\mathcal{K} := \bigcap_{a>0} \mathcal{H}_a$$

$$\mathcal{K}_b^3 := \left\{ x \in L^0(\Gamma) : \exists K > 0 \text{ s.t. } \iint_{\Gamma^2} \sup_\beta |a^{\#\alpha} c^{\#\gamma} K^{\#\beta} x(\alpha,\beta,\gamma)|^2 \, d\alpha \, d\gamma < \infty \quad \forall a, c > 0 \right\}.$$

Kernels from $\mathcal{K}_b^3$ determine an algebra of operators each leaving $\mathcal{K}$ invariant, and the kernel of the product is given by a convolution-like product of the corresponding kernels ([Me 1] p309, [L 1]). X_3 is bounded as an operator from $\mathcal{H}_a$ to $\mathcal{H}_b$ for values of (a, b) depending on x, moreover the bound is given by an $L^2 \times L^\infty \times L^2$ norm on the kernel — see [Bel] and [BL] Proposition 2.2.

§3. Uniqueness of the kernel

Injectivity of the map from kernel to integral-sum operator will be demonstrated in two senses: the operator is first viewed on finite particle vectors, and secondly on an exponential domain. Let M and Γ be as in §2 and fix a countable generating ring $\mathcal{R}$, consisting of sets with finite measure. Let $\mathcal{F}_\mathcal{R}$ and $\mathcal{E}_\mathcal{R}$ denote respectively the linear spans of

$$\left\{ \chi_{\Gamma_n(E)} \pi_\varphi : n \in \mathbb{N},\ E \in \mathcal{R},\ \varphi \in \mathcal{D} \right\}$$

$$\{\pi_\varphi : \varphi \in \mathcal{D}, \|\varphi\|_\infty \leqslant 1\}$$

where $\mathcal{D}$ is the complex rational span of the collection of indicator functions of sets in $\mathcal{R}$. The proofs exploit the fact that for almost all $\tau \in \Gamma$,

$$\mathcal{D}_{(\tau, u)} := \{\varphi \in \mathcal{D} : \varphi(t) = u(t) \, \forall t \in \tau\}$$

is dense in $L^2(M)$ for each $u \in [\mathbb{Q} + i\mathbb{Q}]^\tau$, and also the totality of product functions in $L^2(\Gamma_F)$ for any *finite* set F, where the measure on Γ_F is the counting measure: $\mu(\{\alpha\}) = 1 \; \forall \alpha \in \Gamma_F$.

Proposition 3.1 Let $x : \Gamma_M^3 \to \mathbb{C}$. Suppose that x is locally integrable in the third variable, in the sense that for each $n \in \mathbb{N}, E \in \mathcal{R}$,

$$\int_{\Gamma_n(E)} |x(\alpha, \beta, \omega)| \, d\omega < \infty \quad \text{for a.a. } (\alpha, \beta) \in \Gamma_M^2 .$$

Then

(i) $X_3 k$ is defined for each $k \in \mathcal{F}_{\mathcal{R}}$;

(ii) if X_3 annihilates $\mathcal{F}_{\mathcal{R}}$, then $x = 0$ a.e.

Proof: If $k \in \mathcal{F}_{\mathcal{R}}$ then for some $n \in \mathbb{N}$, $K > 0$ and $E \in \mathcal{R}$,

$$|k| \leqslant K \chi_{\Gamma_{\leqslant n}(E)} ,$$

so (x, k) satisfies (2.2), and $X_3 k$ is defined.
Suppose X_3 annihilates $\mathcal{F}_{\mathcal{R}}$. Fix $j, k, l \in \mathbb{N}$, put $n = j + k$, $m = k + l$ and $p = \min\{n, m\}$, and let $x'_{j,k,l}$ denote x' times the indicator function of $\Gamma_j \times \Gamma_k \times \Gamma_l$. Then, since $\mathcal{D}$ is countable, there is a null set N such that for $\sigma \notin N \cap \Gamma_m$,

$$\int_\Gamma \sum_{\alpha \subset \sigma} x'(\alpha, \bar{\alpha}, \omega)(\chi_{\Gamma_n} \pi_\varphi)(\omega \cup \bar{\alpha}) \, d\omega = 0$$

for each $\varphi \in \mathcal{D}_{(\sigma, su)}$, $u \in [\mathbb{Q} + i\mathbb{Q}]^\sigma$ and $s \in \mathbb{Q}$. Thus

$$\int_\Gamma d\omega \, \pi_\varphi(\omega) \sum_{\alpha \subset \sigma} \pi_u(\bar{\alpha}) \sum_{i=0}^{p} s^i \, x'_{m-i, i, n-i} (\alpha, \bar{\alpha}, \omega) = 0$$

for φ, u and s as above. Successively using the totality of $\{\pi_\varphi : \varphi \in \mathcal{D}_{(\sigma, su)}\}$ in $L^2(\Gamma)$, for each (s,u); the totality of $\{\pi_u : u \in [\mathbb{Q}+i\mathbb{Q}]^\sigma\}$ in $L^2(\Gamma(\sigma))$; and the totality of $\{(1,s,\ldots,s^p) : s \in \mathbb{Q}\}$ in $\mathbb{C}^{p+1}$; and Fubini's Theorem, we see that for a.a. $(\sigma,\omega) \in \Gamma^2$,

$$x'_{m-i,i,n-i}(\alpha,\bar{\alpha},\omega) = 0 \quad \forall \alpha \subset \sigma,\ i = 0,\ldots,p.$$

Putting $i = k$ and applying the $\oint$-Lemma gives

$$0 = \iint_{\Gamma^2} \sum_{\alpha \subset \sigma} |x'_{j,k,l}(\alpha,\bar{\alpha},\omega)|\ \mathrm{d}\sigma\, \mathrm{d}\omega$$

$$= \iiint_{\Gamma^3} |x'_{j,k,l}(\alpha,\beta,\omega)|\ \mathrm{d}\alpha\, \mathrm{d}\beta\, \mathrm{d}\omega.$$

Since j, k and l were arbitrary, $x' = 0$ a.e., and so by Lemma 2.2 $x = 0$ a.e. also. ∎

Proposition 3.2: Let $x : \Gamma_M^3 \to \mathbb{C}$ be locally integrable in its third variable in the sense that for each $E \in \mathcal{R}$,

$$\int_{\Gamma(E)} |x(\alpha,\beta,\omega)|\ \mathrm{d}\omega < \infty \quad \text{for a.a. } (\alpha,\beta) \in \Gamma_M^2.$$

Then

(i) $X_3 k$ is defined for each $k \in \mathcal{E}_{\mathcal{R}}$;

(ii) if X_3 annihilates $\mathcal{E}_{\mathcal{R}}$, then $x = 0$ a.e.

Proof: If $k \in \mathcal{E}_{\mathcal{R}}$, then $|k| \leqslant K\chi_{\Gamma(E)}$ for some $K > 0$ and $E \in \mathcal{R}$, so (x,k) satisfies (2.2) and $X_3 k$ is defined.
Suppose that X_3 annihilates $\mathcal{E}_{\mathcal{R}}$. Since $\mathcal{D}$ is countable there is a null set N such that for $\sigma \notin N$,

$$\int_\Gamma \mathrm{d}\omega\, \pi_\varphi(\omega) \sum_{\alpha \subset \sigma} \pi_\varphi(\bar{\alpha})\, x'(\alpha,\bar{\alpha},\omega) = 0$$

for each $\varphi \in \mathcal{D}$ for which $\|\varphi\|_\infty \leqslant 1$. Letting φ run through $\{\varphi \in \mathcal{D}_{(\sigma,u)} : \|\varphi\|_\infty \leqslant 1\}$, and u run through $[\mathbb{Q}+i\mathbb{Q}]^\sigma$, gives

$$x'(\alpha,\bar{\alpha},\omega) = 0 \quad \text{for a.a. } \omega \text{ and each } \alpha \subset \sigma.$$

Again, by the $\oint$-Lemma and Fubini's Theorem, $x' = 0$ a.e., and so also $x = 0$ a.e.

In the context of distributions, the map

$$(\varphi,\psi) \mapsto \langle \pi_\varphi \otimes \pi_{\varphi\psi} \otimes \pi_\psi, x' \rangle = \langle \pi_\varphi, X_3 \pi_\psi \rangle$$

is called the symbol of the operator X_3. We have shown that 3-argument integral-sum operators are determined by their symbols. Another consequence which may be of use in Fock space analysis is:

Corollary 3.3: The following collection is total in $L^2(\Gamma^3)$:

$$\mathscr{C} := \left\{ \pi_{\varphi_1} \otimes \pi_{\varphi_1\varphi_2} \otimes \pi_{\varphi_2} \colon \varphi_i \in \mathscr{D},\ \|\varphi_i\| \leqslant 1 \quad i=1,2 \right\}$$

Proof: Suppose that $y \in L^2(\Gamma^3)$ is orthogonal to $\mathscr{C}$. Let x be the inverse transform of y, given by (2.1)$'$. Then for each $\pi_\psi \in \mathscr{E}_{\mathfrak{R}}$, ($\psi$ real-valued),

$$0 = \langle \pi_\varphi \otimes \pi_{\varphi\psi} \otimes \pi_\psi, y \rangle = \langle \pi_\varphi, X_3 \pi_\psi \rangle \qquad \forall\, \varphi \in L^2(M)$$

and so $X_3 \pi_\psi = 0$ a.e. Since X_3 annihilates $\mathscr{E}_{\mathfrak{R}}$, x and therefore also $y = x'$, vanishes almost everywhere. The totality of $\mathscr{C}$ follows.

■

The proof of Proposition 3.1 should have appeared in [L 1]. Unfortunately it did not survive the transition from preprint to publication. An indication of the result was given by Meyer, who examined the matrix elements of X_3 with respect to 3- and 2-particle functions ([Me 2] p.46). Attal is considering the multidimensional case, and has results on the uniqueness of (n^2+2n)-argument kernels ([At 2]).

References

[At 1] S Attal: Characterisations of some operators on Fock space, *Quantum Probability & Related Topics VIII (this volume).*

[At 2] S Attal: Problèmes d'unicité dans les représentations d'opérateurs sur l'espace de Fock. *Sém. de Prob. XXVI (to appear).*

[Bel] V P Belavkin: A quantum non-adapted Itô formula and stochastic analysis in Fock scale, *J. Funct. Anal.* **102** no. 2 (1991) 414–447.

[BL] V P Belavkin and J M Lindsay: The kernel of a Fock space operator II, *Quantum Probability & Related Topics VIII (this volume).*

[Gui] A Guichardet: "Symmetric Hilbert spaces and related topics". Springer LNM **261** (1972).

[HOS] T Hida, N Obata and K Saitô: Infinite dimensional rotations and Laplacians in terms of white noise calculus, *Nagoya Math J.* **128** (1992) 65–93.

[Hua] Z-Y Huang: Quantum white noises — white noise approach to quantum stochastic calculus, *Nagoya Math. J.* **129** (1993) 1–20.

[HuP] R L Hudson and K R Parthasarathy: Quantum Itô's formula and stochastic evolutions, *Commun. Math. Phys.* **93** (1984) 301–323.

[L 1] J M Lindsay: On set convolutions and integral-sum kernel operators, *in* "Proceedings of Vth International Vilnius Conference on Probability and Mathematical Statistics, 1989" (ed. B Grigelionis *et al*) Volume II, 105–123, VSP, Utrecht 1990.

[L 2] J M Lindsay: Quantum and non-causal stochastic calculus, *Probab. Th. Rel. Fields (to appear).*

[LM] J M Lindsay and H Maassen: An integral kernel approach to noise, *in* "Quantum Probability and Applications III: Proceedings, Oberwolfach 1987" (ed. L Accardi and W von Waldenfels). Springer LNM **1303** (1988) 192–208.

[LP] J M Lindsay and K R Parthasarathy: Cohomology of power sets with applications in quantum probability, *Commun. Math. Phys.* **124** (1989) 337–364.

[Maa] H Maassen: Quantum Markov processes in Fock space described by integral kernels, *in* "Quantum Probability and Applications II: Proceedings, Heidelberg 1984", (ed. L Accardi and W von Waldenfels). Springer LNM **1136** (1985) 361–374.

[Me 1] P-A Meyer: Eléments de probabilités quantiques I-V, *in* "Séminaire de Probabilités XX" (ed. J Azéma and M Yor). Springer LNM **1204** (1986) 186–312.

[Me 2] P-A Meyer: Eléments de probabilités quantiques VI-VIII, *in* "Séminaire de Probabilités XXI" (ed. J Azéma, P-A Meyer and M Yor). Springer LNM **1247** (1987) 34–80.

[Me 3] P-A Meyer: Fock spaces in classical and non-commutative probability, Ch IV, Strasbourg (1989).

[Par] K R Parthasarathy: "An Introduction to Quantum Stochastic Calculus", Birkhaüser Verlag, Basel/Boston/Berlin, (1992).

Mathematics Department
University Park
Nottingham NG7 2RD
England

Quantum Probability and Related Topics Vol. VIII (pp. 281–295)

BRAIDED GROUPS AND BRAID STATISTICS

Shahn Majid[1]

Department of Applied Mathematics
& Theoretical Physics
University of Cambridge
Cambridge CB3 9EW, U.K.

ABSTRACT We report on a generalization of groups and supergroups relevant to particles of braid statistics. This is based on the notion of a braided group, for which the ring of co-ordinate functions is acted upon by a braid group action Ψ, generalizing $\Psi = \pm 1$ familiar for super-groups. We explain the physical motivation, the current status of the programme, and some problems that must be solved for a systematic approach to braided increment processes.

1 INTRODUCTION

The physical motivation for the present work comes from a recent proof of a bose-fermi statistics theorem of Doplicher and Roberts in the framework of algebraic quantum field theory, as explained in [5]. The interesting thing about this proof is that it works only in four space-time dimensions. In three or two dimensions the superselection structure of the theory need not correspond to that of bose or fermi particles but rather to particles of braid statistics[8][7][12]. It is well known that the processes that arise as central limits of baths of particles depend only on the statistics of the particles, rather than on their internal symmetries. In general, this means that it is hard to obtain new and yet physical kinds of processes. Because the situation in two and three dimensions is different, with new more general braid statistics allowed, it suggests that we should look there for new physical kinds of noise processes.

Briefly, the results of [5] can be outlined as follows. Let O denote the "diamond-shaped" intersection between the forward light cone of a point in space-time and the rear light cone of a future point (in two space-time dimensions these are rectangles). Let $\mathcal{A}(O)$ denote the algebra of observables of the quantum theory that are localized in O. Thus the entire C^* algebra of observables is the completion

$$\mathcal{A} = \overline{\cup_O \mathcal{A}(O)}.$$

If a region is not of the type O we define its algebra of localized observables in the same way as the completion over all O in the region. For an algebraic quantum field theory one makes various postulates. Among the most important is Haag duality

$$\mathcal{A}(O_1) \subset \mathcal{A}(O_2)', \qquad \forall O_1 \subset O_2'$$

where the first prime denotes commutant and the second denotes spacelike complement. Of course, there should be an action α of the Poincaré group P such that $\alpha_L(\mathcal{A}(O)) = \mathcal{A}(L(O))$

[1]SERC Fellow and Drapers Fellow of Pembroke College, Cambridge

for all $L \in P$. Finally, there should be a vacuum representation π_0 of $\mathcal{A}$ such that $\pi_0(\mathcal{A}(O)) = \pi_0(\mathcal{A}(O'))'$.

A representation π of $\mathcal{A}$ is localized at O if it coincides with the vacuum when restricted to $\mathcal{A}(O')$. Such representations are "close" to the vacuum. One of the results in [5] is that such localized representations are of the form $\pi = \pi_0 \circ \rho$ where $\rho : \mathcal{A} \to \mathcal{A}$ is an endomorphism and is the identity when restricted to $\mathcal{A}(O')$ (an endomorphism localized at O). If π_1, π_2 correspond to local endomorphisms ρ_1, ρ_2, then we can obtain a new representation $\pi_1 \otimes \pi_2$ corresponding to a local endomorphism obtained from $\rho_1 \circ \rho_2$. In constructing the latter as a local endomorphism, the local region of ρ_1 has to be transported round to that of ρ_2. Note that this fails in spacetime dimension 2 where the causal complement of a point is not connected. There are also some more subtle problems in dimension 3. See [5] for more details.

Theorem 1.1 *[4][5] In space-time dimension $n = 4$ the endomorphisms ρ form a symmetric monoidal category C with duals. Moreover,*

$$C = Rep(G),$$

the representations of a compact group G, which is reconstructed. This G acts on a larger algebra $\mathcal{F}$ with fixed points $\mathcal{A} = \mathcal{F}^G$.

In this theorem, $\mathcal{A}$ is the algebra of observables while $\mathcal{F}$ is the bigger "field algebra" and G is the global internal symmetry group. Only the G-invariant combinations of the fields in $\mathcal{F}$ are physically observable. For example, is should be possible to identify quantities like a spinor field ψ^a as lying in $\mathcal{F}$ but only G-invariant combinations such as $\bar{\psi}\psi$ would be among the observables in $\mathcal{A}$. Such identification of $\mathcal{F}$ with physical fields remains to be proven in practice, but it can be expected. Also, the local endomorphisms describe super-selection sectors forming a general symmetric monoidal category i.e., there are isomorphisms $\Psi : \pi_1 \otimes \pi_2 \cong \pi_2 \otimes \pi_1$ giving an action of the symmetric group. However, in terms of $\mathcal{F}$ the fields can be chosen such that the action is that of bosonic or fermionic statistics.

By contrast in $n = 2$ or $n = 3$ we have for C a braided monoidal category[7][12]. This is a generalization for which $\Psi^2 \neq \mathrm{id}$. Briefly, a braided monoidal category is a collection of objects and morphisms, a functor $\otimes$, unit object $\underline{1}$, associativity isomorphisms Φ and quasisymmetry isomorphisms (or braiding) Ψ[10], see also [16, Sec. 7]. In the present case the objects are $\pi_1, \pi_2, \pi_3, \cdots$. The isomorphisms $\Phi_{\pi_1,\pi_2,\pi_3} : \pi_1 \otimes (\pi_2 \otimes \pi_3) \cong (\pi_1 \otimes \pi_2) \otimes \pi_3$ obey Maclane's well-known pentagon coherence condition[14]. This asserts that the different ways to bracket expressions can be consistently identified, i.e. we can suppress the brackets and Φ. The quasisymmetry isomorphisms $\Psi_{\pi_1,\pi_2} : \pi_1 \otimes \pi_2 \cong \pi_2 \otimes \pi_1$ obey the conditions (suppressing Φ)

$$\Psi_{\pi_1 \otimes \pi_2, \pi_3} = \Psi_{\pi_1,\pi_3} \circ \Psi_{\pi_2,\pi_3}, \quad \Psi_{\pi_1, \pi_2 \otimes \pi_3} = \Psi_{\pi_1,\pi_3} \circ \Psi_{\pi_1,\pi_2}, \quad \Psi_{\pi,\underline{1}} = \mathrm{id} = \Psi_{\underline{1},\pi}. \tag{1}$$

This asserts that Ψ gives an action of the braid group on tensor products of objects. The details were reviewed already in [15, Sec. 3] (in the context of quantum groups) so we do not do so again here. An example is the category of super-spaces where $\Psi = \pm 1\tau$ according to the degrees (τ is the usual twist map). More general examples where $\Psi^2 \neq \mathrm{id}$ will be described explicitly below.

Hence in $n = 2$ or $n = 3$ we cannot identify C with the representations of any group. Many authors have therefore conjectured as to what should be the object playing the role of G. For example, it is known that the representations of quantum groups (in the strict sense of quasitriangular Hopf algebras) form such categories. This was explained in [15, Sec. 3] and indeed quantum groups could furnish examples of generalized G. For a general algebraic quantum field however, there seems to be no theorem that the role of G has to be played by a quantum group. Rather, we have argued in [17][18] that in general the role of G should be played not by a quantum group, but something more general, which we call a braided group[18][20]. It is this notion that we wish to explain in the present paper.

A braided group A is a generalization of the ring of functions on a group or supergroup. In the familiar language of Hopf algebras, it is a braided-commutative Hopf algebra in a braided monoidal category (just as a supergroup is, for our purposes, a supercommutative Hopf algebra in the category of superspaces.) Note that because braided groups are commutative (albeit in a non-commutative category), they are geometric objects rather than non-commutative-geometric or "quantum" objects. This is in keeping with the situation in $n = 4$ where G is the classical symmetry group of the quantum theory. On the other hand every ordinary quantum group can be viewed by a procedure of transmutation[18] as an braided-commutative Hopf algebra in a braided category. Hence the point of view of braided groups is a very general one. It represents a shift from the philosophy of non-commutative geometry, now familiar in quantum probability, to the philosophy of super-geometry and its generalizations.

Theorem 1.2 *[19][18] Let C be a braided monoidal category with duals. Then there exists a braided group* $\mathrm{Aut}(C)$ *in a cocompletion of C and a functor $C \to$* $\mathrm{Rep}(\mathrm{Aut}(C))$ *compatible with the forgetful functor.*

Here Rep denotes $\mathrm{Aut}(C)$-comodules in C and indeed, $\mathrm{Aut}(C)$ can be characterized as the universal object with such properties[20]. The proof of the theorem is based on a Tannaka-Krein or Fourier Transform theorem well known in harmonic analysis (where a group is reconstructed from its representations), but now generalized to the braided setting. Some similar Hopf algebras have also been found in another context in [13].

Much work remains to be done in order to actually identify such $\mathrm{Aut}(C)$ as the physical "internal symmetries" of actual algebraic quantum field theories in $n = 2$, $n = 3$. On the other hand this need not deter us from going ahead to construct increment processes on purely mathematical examples. It is such mathematical examples of categories C and braided groups resulting from them that we shall describe in detail here. They are perhaps toy models of real physical systems along the lines above.

We turn in Section 2 to a more detailed explanation of what a braided group is. This is based on [18][20] but includes pedagogical details omitted there. In Section 3 we describe some examples of matrix braided groups, such as the braided group $BSL(2)$. This is based on [21]. Finally, in Section 4 we conclude with a brief discussion of what would be needed to actually develop stationary independent increment processed based on braided groups (or quantum braided groups), and give another example of a braided category which could be useful for this. The formalism that we propose is along the lines introduced for Hopf algebras and super-groups or super-Hopf algebras in [1]. Many results in the context of Hopf algebras have been obtained already in pioneering works such as [9][25][26][24] and their sequels. Moreover, it has become clear that there are still more general examples of processes, and some have definite braid-like aspects. See for example the talk of R. Speicher at this conference[3]. Hence it seems likely that at least some of these examples can be put into the general framework of processes built on braided groups, and perhaps given a systematic physical interpretation in terms of algebraic quantum field theory or otherwise. While such results will certainly not be achieved here, it is hoped that the present introduction to braided groups may at least be useful for experts in quantum probability wishing to undertake this line of development.

2 HOPF ALGEBRAS IN BRAIDED CATEGORIES

A braided group is a certain type of Hopf algebra in a braided category. We explain the latter notion first. In doing this, we give some of the pedagogical details omitted in the existing works by the author on this topic.

Recall first that a Hopf algebra in the ordinary sense (over a field, say $\mathbb{C}$) means (A,Δ,ϵ,S) where A is a unital algebra, $\Delta\cdot A\to A\otimes A$ (the coproduct) is coassociative in the sense $(\Delta\otimes\mathrm{id})\circ\Delta=(\mathrm{id}\otimes\Delta)\circ\Delta$ and an algebra homomorphism. Here $A\otimes A$ has the tensor product algebra structure. $\epsilon:A\to\mathbb{C}$ (the counit) obeys $(\epsilon\otimes\mathrm{id})\circ\Delta=\mathrm{id}=(\mathrm{id}\otimes\epsilon)\circ\Delta$. The antipode $S:A\to A$ obeys $\cdot\circ(S\otimes\mathrm{id})\circ\Delta=1_A\epsilon=\cdot\circ(\mathrm{id}\otimes S)\circ\Delta$ and plays the role of group inverse. For an introduction, see [15]. A well known example is $A=L^\infty(G)$, the functions on a discrete group (more generally, for locally compact groups, this is a Hopf-von Neumann or Kac algebra). The coproduct is

$$(\Delta f)(x,y)=f(xy)\tag{2}$$

for $f\in L^\infty(G)$ and $x,y\in G$. The right hand side defines an element of $L^\infty(G)\otimes L^\infty(G)$. Of course, examples of this type, corresponding to functions on groups, are commutative.

A super-Hopf algebra is similar except that $A=A_0\oplus A_1$ is a $\mathbb{Z}_2$-graded algebra (the product map respects the total degree) and Δ is an algebra homomorphism provided $A\otimes A$ is endowed with the super tensor product algebra structure,

$$(a\otimes b)(c\otimes d)=(-1)^{|b||c|}(ac\otimes bd).\tag{3}$$

Here $|a|$ is the degree of a homogeneous element a etc.

More generally, fix a braided monoidal category $\mathcal{C}$ with quasisymmetry Ψ. We suppress Φ. An algebra in $\mathcal{C}$ just means an object A of $\mathcal{C}$ and a morphism $A\otimes A\to A$, along with $\eta:\underline{1}\to A$ (the unit morphism). This is a slight abuse of notation (the term "monoid" would be more precise) – in practice our categories are k-linear over a field and we require all morphisms to respect this so that we really have algebras.

Proposition 2.1 *Let A be an algebra in $\mathcal{C}$. Then*

$$\cdot_{A\otimes A}:(A\otimes A)\otimes(A\otimes A)\xrightarrow{\Psi_{A,A}}(A\otimes A)\otimes(A\otimes A)\xrightarrow{\cdot\otimes\cdot}A\otimes A.$$

defines an associative product on $A\otimes A$.

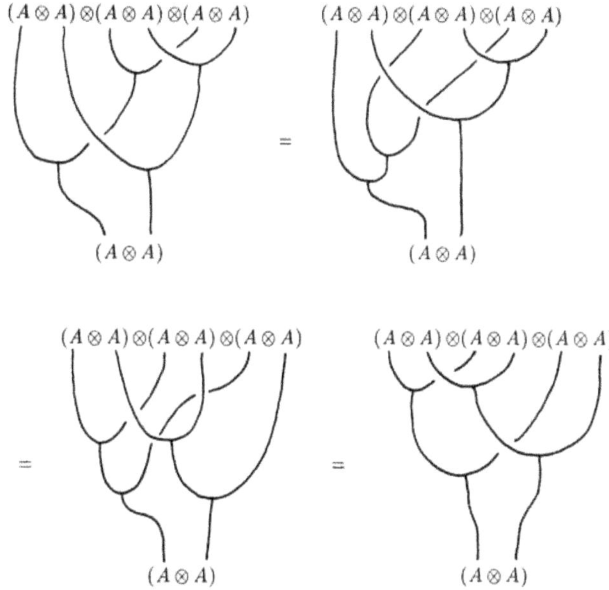

Figure 1: Proof of Proposition 2.1

Proof The proof is not entirely trivial, and is indicated in Figure 1. We have adopted here a diagrammatic notation essential to any serious work with braided categories, as for example in [19]. As explained in [15, Sec. 3], we write $\Psi = \times$ and $\Psi^{-1} = \times$. We will write all morphisms pointing downwards. The product is written as a $\cup$-vertex (and later on we will write the coproduct as a $\cap$-vertex). Functoriality of Ψ says that $\Psi_{X,Y}$ is well behaved as we change objects X, Y by morphisms in the category. In the diagrammatic notation it means that we can translate vertices (and other morphisms) in relation to the crossings Ψ, as long as no string is taken through another. The top left of Figure 1 shows the double product of $(A \otimes A) \otimes ((A \otimes A) \otimes (A \otimes A))$ in this notation. The first equality is by functoriality, the second by associativity of the product of A and the third by functoriality again. The last expression is the double product of $((A \otimes A) \otimes (A \otimes A)) \otimes (A \otimes A)$.

Definition 2.2 *A Hopf algebra in C is a unital algebra $(A, \cdot, \eta)$ in C equipped with a coproduct $\Delta : A \to A \otimes A$, counit $\epsilon : A \to \underline{1}$ and antipode $S : A \to A$, all now morphisms in C. Δ is a homomorphism provided $A \otimes A$ has the tensor product algebra structure obtained in Proposition 2.1. This axiom is shown Figure 2 (a). The other axioms are as usual for Hopf algebras, now written as morphisms in C.*

In concrete cases (such as our examples in later sections) one can write the product in $A \otimes A$ explicitly (with a slight abuse of notation) as

$$(a \otimes b)(c \otimes d) = a\Psi(b \otimes c)d. \tag{4}$$

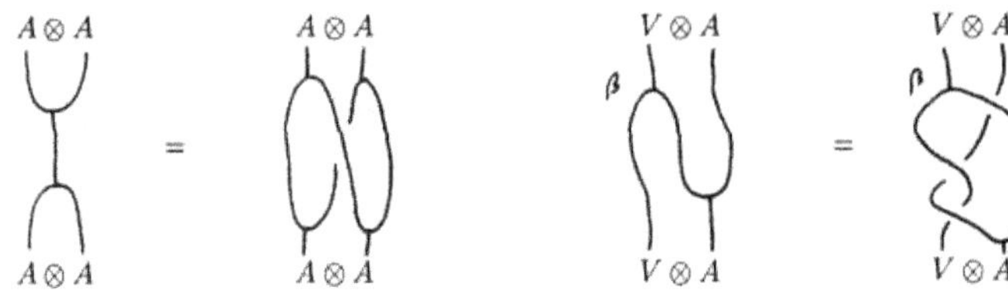

Figure 2: (a) Hopf algebra axiom. (b) Commutativity Axiom

We mean to first apply Ψ to $b\otimes c$ and then multiply on the left by a and on the right by d. Clearly the super tensor product algebra structure in (3) is an example where $\Psi(b\otimes c)=(-1)^{|b||c|}c\otimes b$.

The axioms of coassociativity, the counit, and the antipode do not directly involve Ψ, being just the usual ones (as morphisms). Nevertheless, the appearance of Ψ in the axiom for Δ means that they do involve Ψ in their various properties. For example,

Proposition 2.3 *Let (A,Δ,ϵ,S) be a Hopf algebra in a braided monoidal category. Then the antipode S obeys*

$$S\circ\cdot=\cdot\circ\Psi_{A,A}\circ(S\otimes S).$$

Proof This is to be expected since S has the role of group inverse. Its proof is straightforward and given in diagrammatic notation in [22].

We now introduce the necessary notion of "braided-commutativity" of a Hopf algebra in a braided monoidal category. The point is that when $\Psi^2\neq\mathrm{id}$ the naive definitions $\cdot=\cdot\circ\Psi_{A,A}$ or $\cdot=\cdot\circ\Psi^{-1}_{A,A}$ do not work. This is because these two candidates for an opposite product do not define Hopf algebra structures on A (with the same Δ) in the sense of Definition 2.2. This is also easily seen by means of the diagrammatic notation. We need a notion of an opposite Hopf algebra A^{op} (provided by an opposite product) before we can declare a braided-commutative one as a Hopf algebra for which $A=A^{\mathrm{op}}$.

The strategy of braided groups is to introduce a weaker notion of commutativity that is a property of objects on which the Hopf algebra acts rather than one of the Hopf algebra itself. Even for ordinary Hopf algebras one can view any Hopf algebra as commutative with respect to some class $\mathcal{O}$ of comodules. For details of comodules (which correspond for $A=L^\infty(G)$ to representations of G), see [15, Sec. 1]. Briefly, a (right) comodule (V,β_V) is a vector space V and a map $\beta_V:V\to V\otimes A$ such that $(\beta_V\otimes\mathrm{id})\circ\beta_V=(\mathrm{id}\otimes\Delta)\circ\beta_V$ and $(\mathrm{id}\otimes\epsilon)\circ\beta_V=\mathrm{id}$. For a comodule in a category $\mathcal{C}$, V should be an object of $\mathcal{C}$ and β_V a morphism. It should be thought of as a representation of the underlying group-like object of which A is the "ring of functions".

Definition 2.4 *[18][20] A Hopf algebra A in $\mathcal{C}$ is braided-commutative with respect to a class of (right) comodules $\mathcal{O}$ if*

$$(\mathrm{id}\otimes\cdot)\circ\beta_V=(\mathrm{id}\otimes\cdot)\circ Q_{V,A}\circ\Psi_{A,A}\circ\beta_V$$

on $V\otimes A$ for all right comodules β_V in $\mathcal{O}$. Here $Q_{V,A}=\Psi_{A,V}\circ\Psi_{V,A}$. The axiom is shown in Figure 2 (b).

This novel notion asserts that A behaves braided-commutatively as far as any representation in the class $\mathcal{O}$ is concerned. If $\Psi^2 = \mathrm{id}$ and $\cdot = \cdot \circ \Psi$ then this braided-commutativity condition holds for all comodules, but in general one cannot expect this as explained above. In the example $\mathrm{Aut}(\mathcal{C})$ explained in the introduction, $\mathcal{O}$ is the image of $\mathcal{C}$ under the functor $\mathcal{C} \to \mathrm{Rep}\,(\mathrm{Aut}(\mathcal{C}))$. Typically, as in the examples in Section 3, the braided-commutativity is reflected directly in the algebraic structure of A itself.

What should be the class $\mathcal{O}$ is in general for us to define. If A is not very commutative, $\mathcal{O}$ will not be very big. It can, however, always be taken to be closed under $\otimes$, as the next proposition shows.

Proposition 2.5 *Let A be a Hopf algebra in $\mathcal{C}$. Let $\mathcal{O}(A)$ denote the collection of all comodules of A for which A is braided-commutative in the sense of Definition 2.4. Then $\mathcal{O}(A)$ is a braided monoidal category. It can be called the category of* central comodules *of A.*

Proof The tensor product and quasisymmetry are given in the usual way as for groups (albeit in a dual form with comodules rather than modules): the role of the twist map for groups representations is played now by Ψ in $\mathcal{C}$, viewed now as an intertwiner in $\mathcal{O}(A)$. These results are proven in detail and in greater generality in [22, Sec. 3]. Explicitly, the tensor product of two comodules V, W is $V \otimes W$ with $\beta_{V\otimes W} = (\mathrm{id} \otimes \mathrm{id} \otimes \cdot) \circ \Psi_{A,W} \circ (\beta_V \otimes \beta_W)$ and Ψ can be seen, by diagrammatic methods, to be an intertwiner.

This completes our introduction to the central notions behind working with Hopf algebras and group-like objects in braided monoidal categories.

3 EXAMPLES OF MATRIX BRAIDED GROUPS

In this section we review some examples of matrix braided groups. In the setting of the Introduction, the internal symmetry group in 4 space-time dimensions is typically a matrix groups like $SU(n)$ (e.g. particle colour or isospin). So it should be braided group analogues of just such matrix groups that arise in 2 or 3 spacetime dimensions.

To explain our notation, recall first our treatment of ordinary groups G in terms of the algebra of functions. We saw in (2) that the group structure is expressed in terms of the coproduct Δ. For matrix groups the algebra of functions is generated by the tautological ones $u^i{}_j$ defined by $u^i{}_j(x) = x^i{}_j$ for any matrix $x \in G \subset M_n(\mathbb{C})$. According to (2) the coproduct of these is

$$(\Delta u^i{}_j)(x,y) = u^i{}_j(xy) = (xy)^i{}_j = \sum_k x^i{}_k y^k{}_j = \sum_k (u^i{}_k \otimes u^k{}_j)(x,y).$$

The counit and antipode can similarly be expressed in this way. Of course these $u^i{}_j$ are commutative as functions on G. Matrix braided groups are generated by similar $u^i{}_j$ with the same matrix form of Δ but now non-commutative. Instead, the $u^i{}_j$ are braided-commutative. This can be contrasted with matrix quantum groups as in [9][1]: the difference is that the generators live in a braided category and the apparent non-commutativity is of a very specific form, namely commutativity in that category. For example, if $\mathcal{C}$ is the category of super-spaces then a matrix braided group means a matrix super-group.

Theorem 3.1 *[21] There are braided group analogs of all the classical Lie groups.*

This is proven in [21] by borrowing some ideas from the theory of quantum groups. Indeed, it is known that all the classical Lie groups have quantum group analogs. The matrix versions of these were introduced in [6] by means of bialgebras $A(R)$ associated to $R \in M_n(\mathbb{C}) \otimes M_n(\mathbb{C})$ solving the quantum Yang-Baxter equations (QYBE). The latter is the equation $R_{12}R_{13}R_{23} = R_{23}R_{13}R_{12}$ in $M_n \otimes M_n \otimes M_n$, where $R_{12} = R \otimes \mathrm{id}$ etc. The construction was already reviewed in [15, Sec. 2]. Suitable R are known for all the classical Lie groups as well as ways to quotient the bialgebra by "determinant" type relations to obtain Hopf algebras with antipode. In fact, the construction works more generally for any invertible R obeying the QYBE for which R^{t_2} is invertible. Here ^{t_2} denotes transposition in the second M_n factor. Define the matrices Ψ, Ψ' in $M_{n^2} \otimes M_{n^2}$ by

$$\Psi^I{}_J{}^K{}_L = \sum_{a,b,c,d} R^{k_0}{}_a{}^d{}_{j_0} R^{-1}{}^a{}_{l_0}{}^{j_1}{}_b R^{l_1}{}_c{}^b{}_{i_1} \tilde{R}^c{}_{k_1}{}^{i_0}{}_d \tag{5}$$

$$\Psi'^I{}_J{}^K{}_L = \sum_{a,b,c,d} R^{-1}{}^d{}_{j_0}{}^{k_0}{}_a R^{j_1}{}_b{}^a{}_{l_0} R^{l_1}{}_c{}^b{}_{i_1} \tilde{R}^c{}_{k_1}{}^{i_0}{}_d. \tag{6}$$

Here $\tilde{R} = ((R^{t_2})^{-1})^{t_2}$ and $I = (i_0, i_1), J = (j_0, j_1)$ etc are multi-indices running from $(1,1), (1,2), \ldots, (n,n)$. Ψ is a matrix obeying the QYBE and Ψ' is a variant of Ψ. $\Psi' = \Psi$ in the ordinary symmetric case when R is such that $\Psi^2 = \mathrm{id}$.

Theorem 3.2 *[21] We define $B(R)$ as the associative algebra generated by 1 and n^2 indeterminates $\{u^i{}_j\}$ modulo the relations*

$$u^I u^K = \sum_{J,L} u^L u^J \Psi'^K{}_L{}^I{}_J \tag{7}$$

where $u^I = u^{i_0}{}_{i_1}$ etc. The action of the braid group on the $\{u^I\}$ (extending to an action on all of $B(R)$) is

$$\Psi(u^I \otimes u^K) = \sum_{J,L} u^L \otimes u^J \Psi^K{}_L{}^I{}_J. \tag{8}$$

With this action, the matrix coproduct and counit

$$\Delta u^i{}_j = \sum_k u^i{}_k \otimes u^k{}_j, \qquad \epsilon(u^i{}_j) = \delta^i{}_j \tag{9}$$

make $B(R)$ into a braided-commutative bialgebra in a braided category. For nice R we can quotient $B(R)$ further by suitable determinant-type relations to obtain a braided group

Example 3.3 *[21] Let* $R = \begin{pmatrix} q & 0 & 0 & 0 \\ 0 & 1 & q-q^{-1} & 0 \\ 0 & 0 & 1 & 0 \\ 0 & 0 & 0 & q \end{pmatrix}$. *This is the well-known solution of the QYBE associated to the Jones knot polynomial. Writing* $u = \begin{pmatrix} a & b \\ c & d \end{pmatrix}$, *the braided-commutativity relations (7) become*

$$ba = q^2 ab, \qquad ca = q^{-2} ac, \qquad da = ad, \qquad bc = cb + (1 - q^{-2})a(d - a) \tag{10}$$

$$db = bd + (1 - q^{-2})ab, \qquad cd = dc + (1 - q^{-2})ca \tag{11}$$

This describes a bialgebra $BM(2)$ in a braided category. The additional relation

$$ad - q^2 cb = 1 \tag{12}$$

gives the braided group $BSL(2)$. The quasisymmetry or "braiding" is given by (8). A sample of this is (the others are similar),

$$\begin{aligned}
\Psi(a\otimes a) &= a\otimes a+(1-q^2)b\otimes c & \Psi(c\otimes a) &= a\otimes c\\
\Psi(a\otimes b) &= b\otimes a & \Psi(c\otimes b) &= q^{-2}b\otimes c\\
\Psi(b\otimes a) &= a\otimes b+(1-q^2)b\otimes(d-a) & \Psi(d\otimes a) &= a\otimes d+(1-q^{-2})b\otimes c\\
\Psi(b\otimes b) &= q^2 b\otimes b & \Psi(d\otimes b) &= b\otimes d.
\end{aligned}$$

Note that this is *not* a Hopf algebra in the ordinary sense. The matrix coproduct (9) only extends consistently to products of generators if we give the generators braid statistics as shown. For example

$$\begin{aligned}
\Delta ba &= (\Delta b)(\Delta a) = (a\otimes b+b\otimes d)(a\otimes a+b\otimes c)\\
&= a\Psi(b\otimes a)a + a\Psi(b\otimes b)c + b\Psi(d\otimes a)a + b\Psi(d\otimes b)c\\
\Delta q^2 ab &= q^2(\Delta a)(\Delta b) = q^2(a\otimes a+b\otimes c)(a\otimes b+b\otimes d)\\
&= q^2 a\Psi(a\otimes a)b + q^2 a\Psi(a\otimes b)d + q^2 b\Psi(c\otimes a)b + q^2 b\Psi(c\otimes b)d.
\end{aligned}$$

That these expressions coincide can be seen after substituting $\Psi(b\otimes a)$ etc as shown and using the relations in $BM(2)$. That $\Psi^2\neq \mathrm{id}$ for generic $q\neq 1$ is evident from $\Psi(b\otimes b)=q^2 b\otimes b$, for example.

We see in this example that as $q\to 1$ the algebra becomes commutative and the braiding Ψ becomes trivial. We recover in this case the commutative algebra of functions on $SL(2)$ in the form explained above, while for generic q the algebra relations in this example are roughly as complicated as those for the well-known quantum group $SL_q(2)$[28]. However, the conceptual improvement is that whereas $SL_q(2)$ is non-commutative and therefore far from admitting a classical geometrical interpretation, the relations of $BSL(2)$ are simply those of braided-commutativity from (7) defining a braided 4-plane, $BM(2)$, plus one further quadratic constraint (the determinant condition (12)). This is exactly parallel to the description of $SU(2)$ as the space $S^3\subset \mathbb{R}^4$, or in our case $SL(2)\subset\mathbb{C}^4$ Thus working with braided groups rather than quantum groups means that our algebras are "commutative" (albeit in a braided sense) and hence retain their geometric interpretation as the "ring of functions" on a manifold (in the present case a complexified braided-S^3).

Example 3.4 *[21] Let* $R=\begin{pmatrix} q & 0 & 0 & 0\\ 0 & 1 & q-q^{-1} & 0\\ 0 & 0 & 1 & 0\\ 0 & 0 & 0 & -q^{-1}\end{pmatrix}$. *This is the solution of the QYBE known to be associated to the Alexander-Conway knot polynomial, see for example [11]. Writing* $u=\begin{pmatrix} a & b\\ c & d\end{pmatrix}$ *the braided-commutativity relations (7) become*

$$b^2=0,\qquad c^2=0,\qquad d-a\ \text{central},\tag{13}$$

$$ab=q^{-2}ba,\qquad ac=q^2 ca,\qquad bc=-q^2 cb+(1-q^2)(d-a)a.\tag{14}$$

This describes a bialgebra $BM_{1|1}$ in a braided category. The braiding is given by $d-a$ bosonic and

$$\begin{aligned} \Psi(a\otimes a) &= a\otimes a+(1-q^2)b\otimes c & \Psi(b\otimes b) &= -b\otimes b\\ \Psi(a\otimes b) &= b\otimes a & \Psi(b\otimes c) &= -c\otimes b-(1-q^2)(d-a)\otimes(d-a)\\ \Psi(b\otimes a) &= a\otimes b+(1-q^2)b\otimes(d-a) & \Psi(c\otimes b) &= -b\otimes c\\ \Psi(a\otimes c) &= c\otimes a+(1-q^2)(d-a)\otimes c & \Psi(c\otimes c) &= -c\otimes c\\ \Psi(c\otimes a) &= a\otimes c. \end{aligned}$$

We see in this example that a,d are almost bosonic while b,c are almost fermionic. For example, $\Psi(b\otimes b)=-b\otimes b$, $\Psi(c\otimes b)=-b\otimes c$ etc. This is reflected also in the algebra relations $b^2=0$ etc (because Ψ' is a variant of Ψ). Indeed, as $q\to 1$ this example becomes the algebra of super-matrices $M_{1|1}=\{\begin{pmatrix} a & b\\ c & d\end{pmatrix}\mid a,d$ bosonic, b,c fermionic$\}$. This demonstrates the sense in which braided groups and matrices really generalize super groups and matrices.

Finally in this section we want to describe a notation for working with braided groups and algebras that more closely resembles the notation used by physicists in describing fermions. In this notation, Ψ and $\otimes$ are suppressed. Instead, the exchange properties of independent particles are described by the *statistics relations* (such as commuting or anticommuting statistics). To be concrete we demonstrate the notation on $BSL(2)$. Firstly, distinguish elements of the two copies of $BSL(2)$ in $BSL(2)\otimes BSL(2)$ by labelling those in the second copy with a prime. Note that this is a property of where the element lies not of the element itself. Thus for any f in $BSL(2)$, we write $f\equiv f\otimes 1$ and $f'\equiv 1\otimes f$. Then from (4) we have

$$fg'=f\Psi(1\otimes 1)g=f\otimes g,\quad f'g=\Psi(f\otimes g). \tag{15}$$

The extension of Ψ from the generators of $BSL(2)$ to arbitrary elements, as well as the functoriality properties of Ψ and its braid properties (1) are now all encoded in the associativity of the algebra $BSL(2)\otimes BSL(2)$. This is just a concrete way to express the content of Proposition 2.1. The full statistics relations of $BSL(2)$ above are

$$a'a=aa'+(1-q^2)bc',\quad a'b=ba',\quad a'c=ca'+(1-q^2)(d-a)c',\quad a'd=da'+(1-q^{-2})bc'$$

$$b'a=ab'+(1-q^2)b(d'-a'),\quad b'c=q^{-2}cb'+(1+q^2)(1-q^{-2})^2bc'-(1-q^{-2})(d-a)(d'-a')$$

$$b'b=q^2bb',\quad b'd=db'+(1-q^{-2})b(d'-a'),\quad c'a=ac',\quad c'b=q^{-2}bc',\quad c'c=q^2cc',\quad c'd=dc'$$

$$d'a=ad'+(1-q^{-2})bc',\ d'b=bd',\ d'c=cd'+(1-q^{-2})(d-a)c',\ d'd=dd'-q^{-2}(1-q^{-2})bc'.$$

These statistics relations together with the algebra relations (10)-(12) of $BSL(2)$ (and an identical set for the primed variables) generate the associative algebra $BSL(2)\otimes BSL(2)$. This algebra contains all the information of the braided group. The notation can be extended to higher tensor powers by a labelling each copy of $BSL(2)$ by an integer for its place in the tensor power. Also in this notation, the coproduct takes the form

$$\Delta\begin{pmatrix} a & b\\ c & d\end{pmatrix}=\begin{pmatrix} a & b\\ c & d\end{pmatrix}\begin{pmatrix} a' & b'\\ c' & d'\end{pmatrix} \tag{16}$$

and similarly for higher coproducts. It means that we can work with matrix braided groups in many ways just as for ordinary matrix groups but now with braided-commutativity and braid statistics for their matrix entries.

4 TOWARDS BRAIDED INCREMENT PROCESSES

In this section we want to conclude with a (fairly obvious) proposal for braided increment processes. Our results here are limited to some modest category-theoretic considerations relevant to the general formulation. Of course, the abstract formulation has to be followed by some concrete examples and their realization on Fock spaces. We do not attempt this here but we do indicate some examples of the necessary data related to anyons, and briefly discuss the problems that remain.

The definition of a braided increment process that we propose is based on [1]. There a quantum independent increment process consists of $*$-algebra maps $j_{st} : B \to A$ where A is a (super)-$*$-algebra equipped with state $\phi : A \to \mathbb{C}$, and B is a (super)-bialgebra. The model is $A = L^\infty(\Omega)$, where Ω is a probability space and $B = L^\infty(G)$ where G is a locally compact group or semigroup, except that now A and B can be non-commutative and, if desired, $\mathbb{Z}_2$-graded. $(s,t) \in \mathbb{R}^2_+$ with $s \le t$. We consider now the problems that arise when we go beyond the category of super-spaces to consider A and B in a general braided monoidal category $\mathcal{C}$.

Lemma 4.1 *Let $\mathcal{C}$ be a monoidal category with unit object $\underline{1}$ and associativity Φ. Let $(A,\cdot,\eta)$ be an algebra (or simply monoid) in $\mathcal{C}$ and (B,Δ,ϵ) a coalgebra (or simply comonoid) in $\mathcal{C}$ as in Section 2. Then* $\mathrm{Mor}(B,A)$ *has an associative product $\star$ defined by*

$$j\star j' = \cdot\circ(j\otimes j')\circ\Delta, \qquad j,j'\in \mathrm{Mor}(B,A).$$

The unit in $\mathrm{Mor}(B,A)$ *is* $\eta\circ\epsilon : B\to A$.

Proof Given j,j',j'' in $\mathrm{Mor}(B,A)$ we compute

$$\begin{aligned} j\star(j'\star j'') &= \cdot\circ(\mathrm{id}\otimes\cdot)\circ(j\otimes(j'\otimes j''))\circ(\mathrm{id}\otimes\Delta)\circ\Delta \\ &= \cdot\circ(\mathrm{id}\otimes\cdot)\circ\Phi^{-1}_{A,A,A}\circ((j\otimes j')\otimes j'')\circ\Phi_{B,B,B}\circ(\mathrm{id}\otimes\Delta)\circ\Delta \\ &= \cdot\circ(\cdot\otimes\mathrm{id})\circ((j\otimes j')\otimes j'')\circ(\Delta\otimes\mathrm{id})\circ\Delta = (j\star j')\star j'' \end{aligned}$$

Here the first equality is the definition of $\star$ applied twice. The second is the functoriality of $\Phi\in\mathrm{Nat}(\otimes(\ \otimes\ ),(\ \otimes\ )\otimes)$[14]. The third is an expression of associativity of A and coassociativity of B in $\mathcal{C}$.

The associativity map Φ was suppressed in Section 2. After checking the last lemma, we continue to suppress writing it explicitly. We are now in a position to formulate our proposal.

Definition 4.2 *Let $\mathcal{C}$ be a braided monoidal category with unit $\underline{1}$ and quasisymmetry Ψ, A an algebra in $\mathcal{C}$ and B a bialgebra in $\mathcal{C}$. Let $\phi : A\to\underline{1}$ be a morphism in $\mathcal{C}$. A braided independent increment process over B is a collection of algebra homomorphisms $\{j_{st} : B\to A \mid (s,t)\in\mathbb{R}^2_+,\ s\le t\}$ in $\mathcal{C}$ such that*

(i) $j_{rs}\star j_{st} = j_{rt}$ for all $r<s<t$, $j_{tt}=\eta\circ\epsilon$ for all t,

(ii) $\phi\circ\cdot^{n-1}\circ(j_{t_1t_2}\otimes\cdots\otimes j_{t_nt_{n+1}}) = \phi\circ j_{t_1t_2}\otimes\cdots\otimes\phi\circ j_{t_nt_{n+1}}$ for all $t_1<\cdots<t_{n+1}$,

(iii) $\cdot\circ(j_{st}\otimes j_{s't'}) = \cdot\circ\Psi_{A,A}\circ(j_{st}\otimes j_{s't'})$ for all $t<s'$.

Here on the right in axiom (ii) we use the canonical isomorphisms $\underline{1}\otimes\underline{1}\cong\underline{1}$ etc with which the unit object in a monoidal category comes equipped[14]. $\cdot^{n-1}$ denotes the $(n-1)$-fold product

$A\otimes\cdots\otimes A\to A$. In axiom (iii) we have used that the set of (s,t) with $s\le t$ is ordered. This means that it is quite natural to impose (iii) only for half the cases as shown rather than for all disjoint $(s,t),(s',t')$ as in [1].

Lemma 4.3 *Axiom (iii) above is equivalent to*
$\cdot\circ(j_{st}\otimes j_{s't'})=\cdot\circ\Psi^{-1}_{A,A}\circ(j_{st}\otimes j_{s't'})$ *for all* $t'<s$.

Proof Assuming (iii) holds we compute for $t'<s$,

$$\cdot\circ(j_{st}\otimes j_{s't'})=\cdot\circ(j_{st}\otimes j_{s't'})\circ\Psi_{B,B}\circ\Psi^{-1}_{B,B}=\cdot\circ\Psi_{A,A}(j_{s't'}\otimes j_{st})\circ\Psi^{-1}_{B,B}=\cdot\circ(j_{s't'}\otimes j_{st})\circ\Psi^{-1}_{B,B}.$$

For the second equality we used functoriality of $\Psi\in \mathrm{Nat}(\otimes,\otimes^{\mathrm{op}})$. This was already used extensively in Section 2. Likewise, the last expression equals the right hand side in the lemma by functoriality of Ψ^{-1}. The above computation can also be done by diagrammatic means as in the proof of Proposition 2.1.

This axiom (iii) expresses a kind of partial commutativity of A. We have given a formulation based on the natural ordering of the (s,t). Some elements of this were also implicit in the talk of R. Speicher at this conference[3]. On the other hand, more sophisticated notions are certainly possible, as we saw for braided-commutative bialgebras in Definition 2.4. This is not particularly relevant for A but we could of course impose it for B.

If B is a braided-commutative bialgebra or a braided group, such as $BSL(2)$ in the previous section, then the process should be closer to having a geometrical interpretation as when $B=L^\infty(G)$. This was explained above for $BSL(2)$. On the other hand, as already noticed in [1], the notion of independent increment process is not limited to the case of B in any sense commutative. Thus B could be a quantum braided group rather than just a braided group. We describe now a family of examples of categories $\mathcal{C}$ and quantum braided groups B in them.

Example 4.4 *[23] The braided monoidal category* $\mathcal{C}=\mathcal{A}_n$ *of* anyonic spaces *can be described as follows. The objects are* $\mathbf{Z}_n$*-modules. The tensor product is the usual one for* $\mathbf{Z}_n$ *modules. The quasisymmetry is a non-standard one,*

$$\Psi_{V,W}(v\otimes w)=e^{\frac{2\pi i|v||w|}{n}}w\otimes v \tag{17}$$

on homogeneous elements of degree $|v|$, $|w|$. *Here the degree is defined by* $g\triangleright v=e^{\frac{2\pi i|v|}{n}}v$ *for the action of the generator* g *of* $\mathbf{Z}_n$.

This is well known to physicists in the context of anyons, e.g.[27][8]. The quantities $e^{\frac{2\pi i|v||w|}{n}}$ can be called fractional or anyonic statistics. The category was studied systematically in [23] where we showed that it can be identified with the category of representations of a certain quantum group with non-trivial $\mathcal{R}$. We also showed there that any ordinary quantum group containing a group-like element of order n can be turned into an anyonic one by a procedure of transmutation[23]. Likewise, any algebra A containing an element g of order n can be viewed as an anyonic algebra by making A into a $\mathbf{Z}_n$-module by the adjoint action. The following anyonic quantum group is obtained by transmutation. One could consider its use, or the use of its dual, for B in Definition 4.2.

Example 4.5 *[23]. Let* $n = 4r$ *and* $q = e^{\frac{8\pi i}{n}}$. *The anyonic quantum group* $B(\mathbf{Z}'_n, U_q(sl(2))')$ *in* $\mathcal{A}_n$ *has generators* g *of degree* 0 *and* E, F *of degrees* $|E| = 2$, $|F| = -2$ *and relations*

$$g^n = 1, \quad E^r = 0 = F^r, \qquad gEg^{-1} = e^{\frac{4\pi i}{n}}E, \quad gFg^{-1} = e^{-\frac{4\pi i}{n}}F, \qquad qEF - FE = \frac{g^4 - 1}{q - 1}.$$

The anyonic quantum group structure is

$$\underline{\Delta}E = E \otimes g^4 + 1 \otimes E, \quad \underline{\Delta}F = F \otimes 1 + 1 \otimes F, \qquad \underline{\epsilon}E = \underline{\epsilon}F = 0, \quad \underline{S}E = -Eg^{-4}, \quad \underline{S}F = -F$$

$$\underline{\Delta}^{\rm op}E = E \otimes 1 + 1 \otimes E, \quad \underline{\Delta}^{\rm op}F = F \otimes 1 + g^4 \otimes F, \qquad \underline{\mathcal{R}} = \sum_{m=0}^{r-1} E^m \otimes F^m \frac{(q-1)^{2m}}{(q^m - 1)\cdots(q-1)}$$

where the $m = 0$ *term is defined to be* $1 \otimes 1$. *The coproduct etc on* g *are as for the group algebra of* $\mathbf{Z}_n$. *Here* $\underline{\Delta}^{\rm op}$ *is a second coproduct structure and* $\underline{\mathcal{R}}$ *intertwines* $\underline{\Delta}$ *and* $\underline{\Delta}^{\rm op}$ *according to the axioms of an anyonic quantum group.*

The problem of actually realizing braided processes in the systematic form above remains an open one, even for ones based on such simple anyonic examples. A problem for the systematic theory is as follows. Firstly, recall that stationary processes are ones where $\phi \circ j_{st}$ depends only on $s - t$. Clearly there is no problem with this notion in the more general setting. In the usual formulation, under suitable conditions, such states give rise to GNS representations in terms of creation, differential second quantization and annihilation operators on a Fock space[1][25][26]. Here there is a problem. Firstly we need to introduce axioms for a $*$-structure on Hopf algebras or bialgebras in $\mathcal{C}$. This is not expected to pose any particular problem but remains to be spelled out in detail. We require of course that $\mathcal{C}$ is $\mathbb{C}$-linear. Next, from an abstract point of view the "simplest" procedure is simply to move all constructions systematically to the category $\mathcal{C}$. This means introducing inner products for objects in $\mathcal{C}$ and hence "braided inner product spaces" and ultimately doing braided analysis in $\mathcal{C}$ as a generalization of analysis on superspaces as in [2]. This is one direction for further work.

Alternatively, since every ordinary quantum group in the strict sense (with dual quasitriangular structure) can be turned onto a (quantum) braided group by a procedure of transmutation, one can expect that many braided increment processes could be obtained by transmutation of ones based on quantum groups. They could therefore be constructed along the lines of [1][25][26][3]. The proposal in this case is simply to interpret them more geometrically by shifting to the associated braided category.

For example, all of the braided groups in Section 3 above can be obtained by transmutation from ordinary quantum groups (with dual quasitriangular structure). The notion of dual quasitriangular structure needed here was already described in [15, Sec. 4] as a dual formulation of the axioms of Drinfeld. In particular $BSL(2)$ can be obtained from $SL_q(2)$ by transmutation. The latter, in the form of $SU_q(2)$ plays a central role in [3] so it seems reasonable to expect that the results there could be recast as a braided increment process based on $BSL(2)$ after including the $*$-structure. Although [3] is not given directly in the framework of [1] for ordinary bialgebras, it has features in common with it, as well as features in common with axiom (iii) in the braided Definition 4.2 above. The construction of braided processes by transmutation is a second direction for further work

Finally, from this point of view there should be yet more general braided process associated to (quantum) braided groups and braided categories that do *not* come in any way from quantum groups. Nevertheless they should still have realizations in terms of ordinary algebras and ordinary analysis in the context of algebraic quantum field theory. This setting was explained in the introduction and provides a third direction for further work.

ACKNOWLEDGEMENTS This is based on my talk at the conference. I want to thank the organizers K.R. Parthasarathy and K.B. Sinha for a most stimulating conference and for their very kind hospitality at the Indian Statistical Institute.

REFERENCES

[1] L. Accardi, M. Schürmann, and M. von Waldenfels. Quantum independent increment processes on superalgebras. *Math. Zeit.*, 198:451–477, 1988.

[2] F.A. Berezin. *Analysis with Anticommuting Variables.* Kluwer, Dordrecht, 1987.

[3] M. Bozejko and R. Speicher. An example of a generalized brownian motion, I, II. *Preprints*, 1990.

[4] S. Doplicher and J.E. Roberts. A new duality theory for compact groups. *Inv. Math.*, 98:157–218, 1989.

[5] S. Doplicher and J.E. Roberts. Why is there a field algebra with a compact gauge group describing the superselection structure in particle physics. *Comm. Math. Phys.*, 131:51–107, 1990.

[6] L.D. Faddeev, N.Yu. Reshetikhin, and L.A. Takhtajan. Quantization of Lie groups and Lie algebras. In M. Kashiwara and T. Kawai, editors, *Algebraic Analysis, Vol. I*, pages 129–139. Academic Press, 1988.

[7] K. Fredenhagen, K.H. Rehren, and B. Schroer. Superselection sectors with braid statistics and exchange algebras. *Comm. Math. Phys.*, 125:201–226, 1989.

[8] J. Fröhlich. Statistics of fields, the Yang-Baxter equation and the theory of knots and links. In *Proc. Cargèse (1987).*

[9] P. Glockner and W. von Waldenfels. The relations of the non-commutative coefficient algebra of the unitary group. In L. Accardi and W. von Waldenfels, editors, *Quantum Probability and Applications IV*, number 1396 in Lec. Notes Math., pages 182–220. Springer, 1989.

[10] A. Joyal and R. Street. Braided monoidal categories. Mathematics Reports 86008, Macquarie University, 1986.

[11] L. Kauffman. Knot polynomials and Yang-Baxter models. In *Proc. ICMP, Swansea.* Adam Hilger, 1989.

[12] R. Longo. Index of subfactors and statistics of quantum fields. *Comm. Math. Phys.*, 126:217, 1989.

[13] V.V. Lyubashenko. Tensor categories and RCFT I,II. *Preprint*, 1990.

[14] S. Mac Lane. *Categories for the Working Mathematician*. Springer, 1974. GTM vol. 5.

[15] S. Majid. Quantum groups and quantum probability. In L. Accardi et al, editor, *Quantum Probability and Applications, Trento, 1989.*

[16] S. Majid. Quasitriangular Hopf algebras and Yang-Baxter equations. *Int. J. Modern Physics A*, 5(1):1–91, 1990.

[17] S. Majid. Some physical applications of category theory. In C. Bartocci, U. Bruzzo, and R. Cianci, editors, *XIXth DGM, Rapallo (1990)*, volume 375 of *Lec. Notes in Phys.*, pages 131–142. Springer, 1991.

[18] S. Majid. Braided groups and algebraic quantum field theories, 1990. To appear in *Lett. Math. Phys.*

[19] S. Majid. Reconstruction theorems and rational conformal field theories, 1989. To appear in *Int. J. Mod. Phys.*

[20] S. Majid. Braided groups. *Preprint*, DAMTP/90-42, 1990.

[21] S. Majid. Examples of braided groups and braided matrices. *Preprint*, DAMTP/90-43, 1990.

[22] S. Majid. Transmutation theory and rank for qauntum braided groups. *Preprint*, DAMTP/91-10, 1991.

[23] S. Majid. Anyonic quantum groups. *Preprint*, DAMTP/91-16, 1991.

[24] M. Schürmann. White noise on involutive bialgebras. In *Quantum Probability and Applications, Trento, 1989.*

[25] M. Schürmann. A class of representations of involutive bialgebras. *Math. Proc. Camb. Phil. Soc.*, 107:149–175, 1989.

[26] M. Schürmann. Infiniteley divisible states on cocommutative bialgebras. *Heidel. math preprint*, 1989.

[27] F. Wilczek. *Fractional Statistics and Anyon Superconductivity*. World. Sci., 1990. Ed.

[28] S.L. Woronowicz. Twisted $SU(2)$-group, an example of a non-commutative differential calculus. *Publ. RIMS (Kyoto)*, 23:117–181, 1987.

Quantum Probability and Related Topics Vol. VIII (pp. 297–311)

A QUANTUM PROBABILITY THEORY FOR THE N-LEVEL ATOM

by

Lamberto Rondoni*

Center for Transport Theory and Mathematical Physics,

Virginia Polytechnic Institute and State University,

Blacksburg, VA 24061-0435 (U.S.A.)

rondoni @ vtcc1.bitnet

ABSTRACT

We define a class of discrete dynamical systems, called Quantum Boltzmann Maps, and we introduce an appropriate perturbative expansion of such maps, in order to construct a model for the dynamics of N-level atoms, $N \geq 2$, in a confined radiation field. In particular, we use the increase of the entropy under the maps to prove the convergence of the dynamics to the appropriate equilibrium positions, in the relevant phase spaces.

1. INTRODUCTION

In this paper, we construct a theory of N-level systems using a class of nonlinear dynamical systems, called "Quantum Boltzmann Maps" (QBM), which were first introduced in '87 by R.F. Streater [1]. Such maps represent a suitable mathematical tool for the description of certain nonequilibrium physical systems, as they approach their stationary states. It is known that the problem of convergence to equilibrium for different classes of physical systems is outstanding, and different theoretical approaches are under consideration. Among these, the method based on QBM belongs to a class of stochastic models which neglect the details of the microscopic interactions among particles, and such that irreversibility emerges as a consequence of the randomness which is put in the process by hand. This approach may not appear as satisfactory as a Hamiltonian theory would be, as the macroscopic behaviour of a physical system is not derived from its microscopic dynamics. Nonetheless, it has good physical justification, since it attempts to describe the effects of the myriad of unknown degrees of freedom, and of noise, which are always present in the real physical situations, and which are not likely to be included in a Hamiltonian formalism.

The study of N-level systems is precisely the subject of a vast literature. For instance, a very incomplete list of the research in the field includes [2 - 5], where the more mathematical aspects of the theory were investigated, and [6 - 10], where physical applications, with particular enphasis on the theory of LASER's, were considered. Our goal, here, is to show the versatility of the quantum probabilistic approach based on the QBM, in the construction of new models of physical interest. We will derive a description of N-level atoms in a finite radiation field, as an approximation of an appropriate QBM, and we will discuss some of its properties. In particular, our focus will be on the asymptotic behaviour of the dynamics. Our approach deals with discrete time models, differently from the most of the existing literature, as we think that discrete schemes are very interesting both from a mathematical and from a physical point of view, and thus worth being further developed [11]. However, our set up allows us to draw results also on the continuous time limit, as it will be explained in a future paper. This work differs from others in the literature also because the radiation field is assumed to be finite, that is to say it contains just a finite number of photons, and so its temperature is not constant. It follows that the resulting "isolated" system is properly described only by nonlinear processes, as explained by Streater in [12].

This paper is organized as follows. Section 2. summarizes some of the most important results on the QBM, from Streater's work [1,12]. In Section 3. a model for the two - level atom is constructed, and its properties are analyzed. Section 4. contains a generalization of the theory developed in Section 3. to the case of N-level systems.

2. THE QUANTUM BOLTZMANN MAP

In this section we briefly review for completeness some notions concerning the QBM, as defined by Streater [1,12]. This map is to be regarded as a model for a set of $N < \infty$ quantum harmonic oscillators which constitute a dilute gas whose particles seldom collide, so that memory of such events is lost before other collisions take place. Then, the Hamiltonian of the system is well approximated by the free Hamiltonian, and the interactions can be described by a scattering matrix computed as if there were an infinite time between successive collisions. This can be formalized as follows. Let $\mathcal{K}$ be the 1-particle space for a set of N harmonic oscillators with frequencies $\omega_1, ..., \omega_N$, and consider the symmetrized Fock space over $\mathcal{K}$: $\Gamma_s(\mathcal{K}) = \mathbb{C} \oplus \mathcal{K} \oplus (\mathcal{K} \otimes \mathcal{K})_s \oplus \cdots$, for bosons, or the antisymmetrized Fock space $\Gamma_a(\mathcal{K}) = \mathbb{C} \oplus \mathcal{K} \oplus \Lambda^2(\mathcal{K}) \oplus \cdots$, for fermions, where $\Lambda^k(\mathcal{K})$ is the subspace of antisymmetric tensors of $\otimes_1^k \mathcal{K}$. For bosonic systems, for instance, the algebra of observables is taken to be the set of bounded operators on $\Gamma_s(\mathcal{K})$, i.e. $\mathcal{A} = \mathcal{B}(\Gamma_s(\mathcal{K}))$, which is generated by bounded functions of the creation and annihilation operators a_j^* and a_j , $j = 1, ..., N$. Let Σ be the set of (normal) states on $\mathcal{A}$ –i.e. determined by a density operator ρ– and furnish $\mathcal{A}$ with the inner product $< A, B >= \mathrm{Tr}(A^* B)$ (for A and B of Hilbert-Schmidt class). The entropy of the state ρ is defined as $S(\rho) = -\mathrm{Tr}(\rho \log \rho)$, and the effect of the scattering among particles is represented by a doubly stochastic map T^*, adjoint of a doubly stochastic map on $\mathcal{A}$, which is also completely positive in all the practical cases. In particular, the doubly stochastic maps we deal with, in this paper, are obtained as mixtures of conjugations by unitaries, which also preserve the spectral projections, P_i say, of the Hamiltonian. Then, the average energy of the state ρ is preserved under the application of T^* to ρ. In fact, defining the Hamiltonian as $H_0 = \sum \omega_j a_j^* a_j = \sum E_i P_i$, we get

$$\mathrm{Tr}((T^*\rho)H_0) = \mathrm{Tr}(U^*\rho U H_0) = \mathrm{Tr}(\rho U H_0 U^*) = \mathrm{Tr}(\rho T H_0) = \mathrm{Tr}(\rho H_0) = E,$$

where U is a unitary operator, U^* is its adjoint, and T is defined by $TA = UAU^*$ for every A in $\mathcal{A}$. As discussed in Ref.s[1,12-14], a linear map like T^* cannot properly describe the time evolution of many systems of interest in Physics, and extra stochasticity must be introduced. Here, as we do not have a formalism of particles, we cannot resort to the classical Boltzmann's *stosszahlansatz*, in order to provide such extra stochasticity. However, by analogy, we can make use of the "quasi - free" projection, which replaces a state with finite first and second moments with the state whose third and higher moments vanish, and whose first and second moments equal those of the original state. Such a map was introduced long ago by Bogoliubov, and it has been considered also in more recent works by a number of authors; see Ref.s [15-17] for instance. It was proved by Streater [18] that the quasi free projection, which we denote by Q does not decrease the entropy, and that $S(Q\rho) > S(\rho)$ unless $Q\rho = \rho$.

Combining the actions of the maps T^* and Q we get a new map with many satisfactory features [1,12-14] which we call QBM:

DEFINITION. *A QBM is a map $\tau : \Sigma \to \Sigma$ defined by $\tau = QT^*$, where Q is the quasi - free projection on Σ, and T^* is a completely positive and doubly stochastic map on Σ.*

Thus, the dynamics determined by τ takes place on the subset of quasi - free states in Σ. It turns out that the limit of every convergent subsequence of $\{\tau^n \rho\}_1^\infty$ is a quasi - free state ρ_∞ such that $TT^*\rho_\infty = \rho_\infty$, [1].

3. THE TWO-LEVEL ATOM

In this section, we describe how a perturbation expansion of a QBM, τ, can be used to construct a theory for a dilute gas of 2-level atoms in a particular heath bath. Naturally, such a physical system has a large number of degrees of freedom, but we do not attempt to explicitly include all of them in our framework. Instead, our approach aims at providing a reduced description of the whole system, by predicting the values that the average populations of the atoms in the first and second energy levels –as well as the average population of the photons– will take at different instants of time. Note that our model may also be interpreted as a theory of a single atom in a radiation field. In order to do that, it suffices to normalize as appropriate the average populations, and view them as probabilities of finding the particle in a certain energy level.

Let ω_1 and ω_2 be the energies of the ground and of the excited states, respectively, and assume photons of energy $\omega = \omega_1 - \omega_2$ can be exchanged with a finite radiation field. Here, the terms energy and frequency are interchangebly used, as we take the multiplicative constant between the two quantities to be 1. Let a_i^* and a_i, $i = 1,2$ be the creation and annihilation operators of the atomic levels, and let b^* and b be those for the photonic field. The Hamiltonian of the system is $H_0 = \omega_1 N_1 + \omega_2 N_2 + \omega N$, where N_1, N_2, N are the number operators of level 1, level 2 and of the photons, and the space we must consider is $\mathcal{F}(\mathbb{C}^3)$, the Fock space over $\mathbb{C}^3$. The completely positive and doubly stochastic map that we adopt is defined by $T_\lambda A = S_\lambda A S_\lambda^*$, for every $A \in \mathcal{A} = \mathcal{B}(\mathcal{F}(\mathbb{C}^3))$, where

$$S_\lambda = \exp\{i\lambda(a_1^* b^* a_2 + a_2^* b a_1)\} = e^{i\lambda K}, \tag{3}$$

which is a unitary operator on $\mathcal{A}$, and K is defined by $K = a_1^* b^* a_2 + a_2^* b a_1$. We choose this form for S_λ because it intuitively describes the fact that one atom of level 2 is created when one atom of level 1 and one photon disappear by absorbtion, and viceversa. Moreover, we write T_λ to stress that there is a dependence on a real parameter, λ, which could also be a random variable on some sample space, so that averages with probability measures could also be considered. Something similar will be done, indeed, in the next section, in order to model the more general N-level atom. Clearly $[H_0, S_\lambda] = 0$, and thus the average energy turns out to be a constant of the motion under the action of $\tau = QT_\lambda^*$. For every ρ in Σ, let us introduce a linear functional W_ρ defined by $W_\rho(A) = \mathrm{Tr}(\rho A)$, for every A in $\mathcal{A}$. (This is what is more apropriately called a normal state). Then, if we define $n_i = W_\rho(N_i)$, for $i = 1,2$, and no index, we can write

$$E = \omega_1 n_1 + \omega_2 n_2 + \omega n = \text{constant}. \tag{4}$$

Moreover, if we consider the normal state relative to $\tau\rho$, $W_{\tau\rho}$, we get

$$n_i' = W_{\tau\rho}(N_i) = \mathrm{Tr}((QT_\lambda^*\rho)N_i) = \mathrm{Tr}((T_\lambda^*\rho)N_i) = W_\rho(T_\lambda N_i), \qquad i = 1, 2, \text{no index}, \quad (5)$$

because of the definition of the map Q. However, this should not mislead the reader, thinking that Q has no effect. In fact, Q shows up in the computation of higher order moments, setting to zero all the truncated functions of order 3 and more. It is Q that makes τ into a nonlinear dynamical system. If we now perform a perturbative expansion in the variable λ of the one point functions, we find that, to the second order, they are left unchanged by the map, provided that their initial values and the initial values of the "mixed" two - point functions –those involving other operators than N_1 , N_2 and N, like $W_\rho(a_1a_2)$, for instance– all vanish. As such moments do not correspond to physically measurable quantities, we can impose $W_\rho(a_1) = \dots = W_\rho(b^*) = 0$, without any harm, assuming that no off-diagonal long range order (ODLRO) occurs [19]. In certain cases, we may easily see that also the mixed second moments are left unchanged by the second order map, if the same conditions are satisfied. However, this may not be the case in the most general situation, therefore, in order to avoid unnecessary complications, we change the definition of the QBM introducing Pauli's "forget" map D . $\Sigma \to \Sigma$, which makes such mixed moments vanish, and which was proposed by Pauli in 1928. Thus, we take $\tau = DQT_\lambda^*$ as the new definition of a QBM, similarly to what was done in some example in Ref.[12], and the dynamics is now contained in the subset of diagonal states; but it should be noted that this doesn't alter the basic properties of τ described in Section 2. Moreover, we show that the resulting time evolution is still consistent with the appropriate quantum statistics. If we assume that the first and the mixed second moments all vanish at the initial time, we find that the perturbative expansion of the map defined by Eq.(5) produces the following expressions

$$\begin{aligned}
n_1' &= W_{\tau\rho}(N_1) = W_{DQ\rho}(T_\lambda N_1)\\
&= n_1 + W_{DQ\rho}\left(i\lambda[K, N_1] - \frac{1}{2}\lambda^2[K,[K,N_1]] + \dots\right) = n_1 + \lambda^2 G\\
n_2' &= W_{\tau\rho}(N_2) = W_{DQ\rho}(T_\lambda N_2)\\
&= n_2 + W_{DQ\rho}\left(i\lambda[K, N_2] - \frac{1}{2}\lambda^2[K,[K,N_2]] + \dots\right) = n_2 - \lambda^2 G\\
n' &= W_{\tau\rho}(N) = W_{DQ\rho}(T_\lambda N)\\
&= n + W_{DQ\rho}\left(\lambda i[K, N] - \frac{1}{2}\lambda^2[K,[K,N]] + \dots\right) = n + \lambda^2 G,
\end{aligned} \qquad (6)$$

where we have $[K, N_1] = [K, N] = -[K, N_2] = a_2^*ba_1 - a_1^*b^*a_2$, from which the equalities

$$[K,[K,\dots,[K,N_1]\dots]] = [K,[K,\dots,[K,N]\dots]] = -[K,[K,\dots,[K,N_2]\dots]],$$

for terms of the same order, follow. That's why Eq.(6) has a "$+G$" for n'_1 and n', while it has a "$-G$" for n'_2. Moreover, it is easy to see that the term G introduced in Eq.(6) depends on the order of the expansion, but it contains even terms only, so that the second order expansion, for instance, is indeed a third order one, similarly to what happens with the diffusion approximation of the classical Boltzmann equation (see Ref.[20] for instance). This is a consequence of the fact that the evaluation of the terms of odd order –those of the form $W_{DQ_\rho}([K,[K,...,[K,N_1]...]])$ with an odd number of K's– always involves either one-point functions, which are assumed to vanish, and/or t-point truncated functions, with t$\geq$ 3, which are set to zero by the application of Q. Also, we see that there are two independent conserved quantites, $n_1 + n_2 = c_1$ (the total number of atoms), and $n_2 + n = c_2$, so that the resulting dynamical system is 1-dimensional. It follows that the the total energy E, defined by Eq.(4), is conserved also under the approximate maps of any order.

Turning to the second order map (exact to third order), we find $[K,[K,N_2]] = 2\left(a_1 a_1^* b b^* a_2^* a_2 - a_1^* a_1 b^* b a_2 a_2^*\right)$, and, using the appropriate CCR or CAR relations, Eq.(6) reduces to

$$\left\{\begin{array}{l} n'_1 = c_1 - n'_2 \\ n'_2 = n_2 + \lambda^2\left[n_1(1 \pm n_2)n - (1 \pm n_1)n_2(1+n)\right] \\ n' = c_2 - n'_2, \end{array}\right. \tag{7}$$

where the "+" signs inside the brackets are for bosonic atoms, while the "−" signs are for the fermionic ones. Here, use was made of the fact that the mixed moments are initially vanishing, and so the action of the diagonalizing operator, D, turns to be necessary, to ensure that the form of Eq.(7) does not change at later times.

As the map defined by Eq.(7), τ_3 say, is only an approximation of τ, we don't know *a priori* whether it shares all the properties of the QBM. In fact, we are now going to prove that some care must be used, and that certain constraints must be satisfied for the approximate theory to be physically meaningful. The first aspect to check is whether τ_3 preserves the positivity of the average populations n_1 , n_2 and n, or not. To find out, we use the fact that the dynamics is 1-dimensional, and we parametrize it by n_2. Then, with a small abuse of notation, we can write

$$\tau_3(n_2) = n'_2 = n_2 + \lambda^2\left[3n_2^2 - (1+2c)n_2 + c_1 c_2\right] = n_2 + \lambda^2 G_b(n_2), \tag{8}$$

for bosonic atoms, where $c = c_1 + c_2$; from which it is easy to see that n'_2 is nonnegative if the initial populations are nonnegative and $\lambda^2 \leq 1/(1+2c) = \lambda_b^2$. Similarly, for fermionic atoms we get

$$\tau_3(n_2) = n'_2 = n_2 + \lambda^2\left[n_2^2 - (1+2c_2) + c_1 c_2\right] = n_2 + \lambda^2 G_f(n_2), \tag{9}$$

where $n'_2 \geq 0$ if $n_1, n_2 \in [0,1]$, $n \geq 0$, and $|\lambda| \leq |\lambda_f|$ with $\lambda_f^2 = 1/(1+2c_2)$. In both cases, the condition $|\lambda| \leq |\lambda_i|$, $i = b, f$, implies $\tau'_3(n_2) = (dn'_2/dn_2)|_{n_2} \geq 0$, if the initial conditions are in the appropriate ranges. Thus, we can write

$$0 \leq \tau_3(0) \leq \tau_3(n_2) \leq \tau_3(c/2) \leq \frac{c}{2} \ , \qquad \text{for} \quad n_2 \in [0, c/2], \tag{10}$$

for bosons, and

$$0 \leq \tau_3(0) \leq \tau_3(n_2) \leq \tau_3(1) \leq 1 \ , \qquad \text{for} \quad n_2 \in [0,1], \tag{11}$$

for fermions. That is to say, $I_b = [0, c/2]$ and $I_f = [0,1]$ are trivially compact and convex, invariant sets for τ_3, respectively for the bosonic and for the fermionic systems. Then, the continuity of τ_3 together with Brouwer fixed point theorem [21] ensures the existence of a fixed point for τ_3, in the appropriate invariant set. Moreover, the strict positivity of $\tau'_3(n_2)$ in the interior of I_b and I_f –provided I_b, I_f are not just one point– implies the uniqueness of such a fixed point, i.e.: there is a unique $\hat{n}_2 \in I_i$, $i = b, f$, such that $G_i(\hat{n}_2) = 0$. In fact, for $n_2 \in I_i$, it is easily seen that $G_i(n_2)$ is negative for $n_2 > \hat{n}_2$, and $G_i(n_2)$ is positive for $n_2 < \hat{n}_2$. This observation, coupled with Eq.s (10) or (11), and with the nonnegativity of τ'_3, allows us to conclude that $\hat{n}_2$ is a global attractor for the dynamics determined by τ_3 in I_i, and that the convergence is monotonic. Furthermore, using very simple inequalities, we can prove that there exists an invariant closed interval $J_i \subset I_i$ such that $\tau'_3(n_2) < 1$ for $n_2 \in J_i$, so that τ_3 is a contraction, and the convergence is exponential, in J_i. We find that J_f contains I_f, so τ_3 is a global contraction for fermionic systems, while J_b may or may not contain I_b, depending on the initial conditions n_1, n_2, n. With these results at hand, and using Eq.(7), it is possible to show that also the positivity of n_1 and n is preserved under the map τ_3, and that $n'_1 \leq 1$, for fermions, if $n_1, n_2 \in [0,1]$, $n \geq 0$. Also, performing some algebraic manipulation of the equations, we realize that the values of the three populations at the fixed point can be expressed as

$$\hat{n}_1 = \frac{1}{e^{\beta(\omega_1 - \mu)} - 1} \ , \quad \hat{n}_2 = \frac{1}{e^{\beta(\omega_2 - \mu)} - 1} \ , \quad \hat{n} = \frac{1}{e^{\beta\omega} - 1} \ , \tag{12}$$

for bosonic atoms, and

$$\hat{n}_1 = \frac{1}{e^{\beta(\omega_1 - \mu)} + 1} \ , \quad \hat{n}_2 = \frac{1}{e^{\beta(\omega_2 - \mu)} + 1} \ , \quad \hat{n} = \frac{1}{e^{\beta\omega} - 1} \ , \tag{13}$$

for fermionic atoms, where the inverse temperature, β, and the chemical potential, μ, are determined by the value of the energy, E in Eq.(4), and by the total number of atoms $c_1 = n_1 + n_2$. Summarizing, we we can state the following result.

LEMMA 5. *Let τ_3 be the map defined by Eq.(7). Assume the initial populations $P = (n_1, n_2, n)$ are nonnegative, and that $n_1, n_2 \leq 1$ in the fermionic case. Then, there exists a real number $\lambda' \geq |\lambda_i| > 0$, $i = b, f$, such that for $|\lambda| < \lambda'$ the following holds:*

a) *The iterations under τ_3 are positivity preserving and $\{\tau_3^k(P)\}_{k=1}^{\infty}$ is contained in a compact subset Q of the positive octant of $\mathbb{R}^3$.*

b) $\lim_{k\to\infty}\tau_3^k(P)=\hat{P}$, *where $\hat{P}$ contains the Plank distribution for the bosonic populations, and the Fermi distribution for the fermionic ones, relative to the initial values of the energy E and of the total number of atoms c_1 in the system. Moreover, the convergence is monotonic, and it is exponentially fast in a closed subset of Q.*

As a corollary of this lemma, we get a result on the increase of the entropy under the map τ_3. Note, that the entropy can be expressed in terms of n_1, n_2 and n, using the well known formulas for fermions [22] (and similarly for bosons), from which we get

$$S_b(n_1,n_2,n)=\sum_{i=1}^{2}[(n_i+1)\log(n_i+1)-n_i\log n_i]+(n+1)\log(n+1)-n\log n, \quad (14)$$

for bosonic atoms, and

$$S_f(n_1,n_2,n)=-\sum_{i=1}^{2}[n_i\log n_i+(1-n_i)\log(1-n_i)]+(n+1)\log(n+1)-n\log n, \quad (15)$$

in the fermionic case. If we express n_1 and n in terms of n_2, it is easy to see that

$$\frac{dS_b}{dn_2}=0 \quad \text{if and only if} \quad n_1(1+n_2)n=(n_1+1)n_2(n+1), \quad (16)$$

and

$$\frac{dS_f}{dn_2}=0 \quad \text{if and only if} \quad n_1(1-n_2)n=(1-n_1)n(1+n). \quad (17)$$

Moreover, (d^2S_i/dn_2^2) is negative (possibly equal to $-\infty$) in the whole domain of S_i, $i=b,f$. Therefore, the entropy of both the bosonic and the fermionic cases have a unique maximum, the unique fixed point $\hat{n}_2$ in I_i, and it is strictly convex. Using this fact, introducing the notation $\alpha_i(n_2)=(dS_i/dn_2)|_{n_2}$, and recalling that the iterations converge monotonically to the fixed point $\hat{n}_2$ (lemma 5.), we can write

$$\alpha_i(n_2)>\alpha_i(\tau_3(n_2))\geq 0 \quad \text{if} \quad n_2<\hat{n}_2, \quad (18)$$

because $n_2<\tau_3(n_2)$ in this case, and

$$\alpha_i(n_2)<\alpha_i(\tau_3(n_2))\leq 0 \quad \text{if} \quad n_2>\hat{n}_2, \quad (19)$$

as $n_2>\tau_3(n_2)$ in this second case. Then, integrating these expressions with respect to the variable n_2, we get the following result:

Corollary 6. *The entropy of the system (with two-level bosonic or fermionic atoms) is a strict Liapunov function for the map τ_3. Precisely,*

$$S_i(\tau(n_2))\geq S_i(n_2)+\lambda^2\,|\alpha_i(\tau_3(n_2))G_i(n_2)| \quad \text{for} \quad i=b,f, \quad (20)$$

where $\alpha_i(n_2)$ and $G_i(n_2)$ are different from zero unless n_2 equals $\hat{n}_2$, the unique fixed point in the domain of the map.

We complete the analysis of the two-level systems observing that the fixed points of τ_3 are stable with respect to small perturbations of the initial conditions. This is proved by the observation that the fixed points of τ_3 depend continuously on the initial conditions, and with an argument similar to the one for the proof of the stability of the fixed points of the classical Boltzmann maps examined in Ref.[14]. Let us turn, now, to the study of the N-level atom in a finite radiation field.

4. THE MULTILEVEL SYSTEM IN A FINITE RADIATION FIELD

The theory outlined in the previous section can be generalized to the case of bosonic or fermionic N-level systems in a number of different ways. The approach we adopt here is *to form convex combinations of the $(N^2 - N)/2$ two-level maps relative to all the pairs* of energy levels of the system. Let ω_i be the frequency of the i-th level, and $\omega_{ij} = \omega_j - \omega_i$, $\omega_j > \omega_i$, be the frequency of the photon γ_{ij} exchanged in a transition between the levels "i" and "j". Then, three different cases are possible. 1) $\omega_{ij} \neq \omega_{kl}$ if $(i,j) \neq (k,l)$. 2) $\omega_{ij} = \omega_{kl}$ for every two couples $(i,j),(k,l)$, such that direct transitions from i to j and from k to l are allowed. 3) $\omega_{ij} = \omega_{kl}$ for some of the couples (i,j) and (k,l) only, which is an intermediate situation between (1) and (2). The reason for considering these three different cases is technical, and has to do with the number of different transitions which are feeded with the same kind of photons. However, situations (2) and (3) can be treated with similar techniques to those of (1). Thus, we will concentrate on case (1) only. Note that these models correspond to systems whose transitions involve precisely one photon at a time. The case in which more photons are exchanged in one time step, in a transition from level i to level j, is more complicated and will be the subject of a future paper. The total number of degrees of freedom of our model is $d = N + (N^2 - N)/2 = (N^2 + N)/2$ but, as we will see, some of them are not independent from the others, and some others might be "inhibited", when certain transitions are forbidden.

The space we consider is $\mathcal{F}(\mathbb{C}^d)$, $\mathcal{A} = \mathcal{B}(\mathcal{F}(\mathbb{C}^d))$ is the algebra of observables, and we denote by Σ the set of density operators on $\mathcal{A}$. The Hamiltonian is given by

$$H_0 = \sum_{i=1}^{N} \omega_i N_i + \sum_{j=2}^{N} \sum_{i=1}^{j-1} \omega_{ij} N_{ij},$$

where N_i and N_{ij}, respectively, are the number operators of the atomic energy levels and of the photonic fields. Hereafter, the levels are ordered according to the growing values of the energy, so that in ω_{ij} the first index, i, is always less than the second one, j. A two-level map on Σ, which regards the energy levels i and j, is a map τ_{ij} that acts like the two-level maps we have studied in the previous section, on the subspace relative to the levels i, j, and to the photons γ_{ij} carrying the energy difference, and it acts as the identity on the rest of the space. Such a map can be written as $\tau_{ij}(\rho) = DQT_{\lambda_{ij}}(\rho)$, where $T_{\lambda_{ij}}$ is the adjoint action of $S_{\lambda_{ij}} = \exp(i\lambda_{ij}K_{ij}) \otimes I$, and $K_{ij} = a_i^* b_{ij}^* a_j + a_j^* b_{ij} a_i$, with obvious notation. Then, τ_{ij} takes Σ into itself, and any convex combination of maps of this kind, like

$$\tau = \sum_{j=2}^{N} \sum_{i=1}^{j-1} \nu_{ij} \tau_{ij} \ , \quad \nu_{ij} \geq 0 \ , \quad \sum_{i,j} \nu_{ij} = 1, \tag{21}$$

is another map of Σ into itself, because of the convexity of Σ. As the number of terms in the convex combination is finite, the perturbative second - order expansion of τ is just the convex combination, with same coefficients, of the second order expansions of each τ_{ij}. Note that the variable λ with respect to which the expansion is performed can be taken from the set $\{\lambda_{ij}\}$ of the coupling constants of each τ_{ij}, and all the remaining λ_{ij}'s can be expressed by $\lambda_{ij} = \lambda \cdot constant(i,j)$. If the constants are different from 1, a little care must be used, because the ranges over which the λ_{ij}'s can vary will be either enlarged or shrunk by a constant. Thus, for simplicity we assume $constant(i,j) = 1$ for all couples (i,j) with $i < j$. If we take the expectations of the N_i's and of the N_{ij}'s, and we expand to the second order, we end up with maps τ_{ij} and τ -with a small abuse of notation- of a convex subset, $\mathcal{Q}$, of $\mathbb{R}^d$ into itself, which can be written as

$$\begin{aligned} [\tau_{ij}(\mathbf{n})]_i &= n_i - \Delta_{ij}(\mathbf{n}) \\ [\tau_{ij}(\mathbf{n})]_j &= n_j + \Delta_{ij}(\mathbf{n}) \\ [\tau_{ij}(\mathbf{n})]_{ij} &= n_{ij} - \Delta_{ij}(\mathbf{n}) \\ [\tau_{ij}(\mathbf{n})]_k &= n_k \qquad \text{for} \quad k \neq i,j \\ [\tau_{ij}(\mathbf{n})]_k &= n_{kl} \qquad \text{for} \quad (k,l) \neq (i,j), \end{aligned} \tag{22}$$

for each two-level contribution, and

$$\left\{ \begin{array}{l} [\tau(\mathbf{n})]_k = n_k + \sum_{i=1}^{k-1} \nu_{ik} \Delta_{ik}(\mathbf{n}) - \sum_{i=k+1}^{N} \nu_{ki} \Delta_{ki}(\mathbf{n}) \\ [\tau(\mathbf{n})]_{kl} = n_{kl} - \lambda_{lk} \Delta_{lk}(\mathbf{n}), \end{array} \right. \tag{23}$$

for the total map τ. Here, $\mathbf{n} \in \mathcal{Q}$ is the vector whose components are the average populations n_i and n_{ij} , $[\tau_{ij}(\mathbf{n})]_k$ is the k-th element of $\tau_{ij}(\mathbf{n})$, and

$$\Delta_{ij}(\mathbf{n}) = \lambda_{ij}^2 \left[n_i(1 \pm n_j) n_{ij} - (1 \pm n_i) n_j (1 + n_{ij}) \right], \tag{24}$$

where the "+" signs in the brackets refer to bosonic atoms, while the "$-$" signs are for the fermionic ones. A first observation is that there is a conserved quantity for the dynamics, which is

$$L = \sum_{k=1}^{N} k n_k + \sum_{k=2}^{N} \sum_{l=1}^{k-1} (k-l) n_{lk}. \tag{25}$$

Then, the following holds.

Lemma 7. *Let the constants λ_{ij} relative to the 2-level maps τ_{ij}, which define τ in eq.(21), obey*

$$\lambda_{ij}^2 \leq \frac{1}{2L+1} \ . \tag{26}$$

Then, the map τ defined by Eq.(23) preserves the positivity of the populations of the energy levels and of the photons.

Proof: We know that each map τ_{ij} preserves the positivity of the populations of the energy levels as well as of the photons, provided that the relative constant $|\lambda_{ij}|$ is smaller than a positve constant $\lambda'_{ij} \geq 1/\sqrt{1+2c_{ij}}$, where $c_{ij} = c_1(i,j) + c_2(i,j) = n_i + 2n_j + n_{ij}$. Thus, a convex combination of maps like τ_{ij}, with small enough constants, is going to be positivity preserving. However, as we have seen in Section 3., the λ'_{ij}'s depend on the c_{ij}'s, which are determined by the initial conditions, and so they cannot be used in the present context, because of the action of more than one map on the energy levels i and j. On the other hand, each c_{ij} is less or equal than L, and positive, at the initial time. Therefore, if we let $\lambda_{ij}^2 \leq 1/(1+2L)$ for every i and j, all the constants of the two-level maps are in their positivity preserving ranges. As a consequence, after one time step, all the populations and the constants c_{ij} are nonnegative, and such that the new bounds on the λ_{ij}'s, for the positivity preserving property, still contain the range $[-1/\sqrt{1+2L}, 1/\sqrt{1+2L}]$. Thus, the total map τ preserves the positivity at all times. ∎

Note that the quantity L is just a special member of the family of conserved quantities given by

$$L(\mathbf{n}, \mathbf{y}) = \sum_k y_k n_k + \sum_{k>l} (y_k - y_l) n_{lk} \ , \tag{27}$$

where $\mathbf{n}$ is the initial condition and $\mathbf{y} = \{y_k\}$ is a vector in $\mathbb{R}^N$ Thus, there are at least N linearly independent conserved quantities for a given initial condition. In addition to these, we occasionally have other independent constants of motion –e.g. when some of the ν_{ij}'s vanish– and so we conclude that the dimension of the space where the dynamics takes place is at most $N(N-1)/2$.

We can now prove that every initial condition, $\mathbf{n} = (n_1, ..., n_N, n_{12}, n_{13}, ..., n_{N-1,N})$, is driven by the iterations of τ to a fixed point uniquely determined by $\mathbf{n}$ itself. The proof is based on a simple convexity argument, involving the uniqueness of the fixed point of τ in the convex set in which the iterations lie.

Theorem 8. *Consider an N-level system described by the map τ defined in Eq.(23). Let $\mathbf{n}$ be an initial condition with all nonnegative entries, and with $n_j \leq 1$ for fermionic atoms. Then, $\mathbf{n}$ determines a compact and convex set $\mathcal{Q}_\mathbf{n}$, through Eq.(25), which is invariant under τ. The dimension $d' \leq (N^2 - N)/2$ of $\mathcal{Q}_\mathbf{n}$ equals the number of positive coefficients ν_{ij} in the definition of τ. Moreover, there is a unique fixed point for τ in $\mathcal{Q}_\mathbf{n}$, $\hat{\mathbf{n}}$ say,* *to wich the iterations converge, i.e.* $\lim_{k\to\infty} \tau^k(\mathbf{n}) = \dot{\mathbf{n}}$. *If the atoms are fermionic, then* $[\tau^k(\mathbf{n})]_j \leq 1$ *for all* j*'s and* k*'s.*

PROOF: The fact the iterations will remain inside a convex set $\mathcal{Q}_\mathbf{n}$ follows from the fact that positivity is preserved and from Eq.(25). The dimension d' of the space $\mathbb{R}^{d'}$, where $\mathcal{Q}_\mathbf{n}$ is contained, is easily computed by the observation that each map τ_{ij} determines one direction in $\mathbb{R}^d$, in the sense that $\tau_{ij}(\mathbf{n})-\mathbf{n}$ is proportional to a given element, e_{ij} say, of a base of $\mathbb{R}^d$, for every $\mathbf{n} \in \mathbb{R}^d$. Thus, $\tau(\mathbf{n})- \mathbf{n}$ is a linear combination of a fixed sub-base, $\{e_{ij}\}$ of $\mathbb{R}^d$ for every $\mathbf{n} \in \mathbb{R}^d$. Then, we have $d' \leq$ (number of 2-level maps in τ). But each map τ_{ij} lives in a subspace which is not contained in the one determined by all the other maps, as the variable n_{ij} appears in the dynamics of τ_{ij} only. Thus $d' \geq$ (number of 2-level maps in τ). In particular, the map lives in a space of dimension $(N^2 - N)/2$, if all the coefficients ν_{ij} are positive.

The fermionic variables are confined inside $[0,1]$ simply because convex combinations of numbers in $[0,1]$ are numbers in $[0,1]$, and because each fermionic τ_{ij} takes $[0,1]$ into itself.

Concerning the fixed points of τ, Brouwer's fixed point theorem ensures the existence of at least one of such points in $\mathcal{Q}_\mathbf{n}$. Also, the entropy of the system, in a state with populations given by $\mathbf{n} = \sum_{i,j} \nu_{ij} P_{ij}$, obeys

$$S(\mathbf{n}) = S\left(\sum_{i,j} \nu_{ij} P_{ij}\right) \geq \sum_{i,j} \nu_{ij} S(P_{ij}), \tag{28}$$

and so $\mathbf{n}$ is a fixed point for τ if, and only if, it is fixed for each τ_{ij} separately. In fact, assume that $\mathbf{n}$ is not fixed for τ_{ij}, and write

$$S(\mathbf{n}) = \sum_k s(n_k) + \sum_{k,l} s(n_{kl}),$$

where $s(n_k)$ is the contribution to S of the k-th atomic degree of freedom, and $s(n_{kl})$ is the contribution of the photons γ_{kl}. Then, using inequality (20), we get

$$\begin{aligned} S(\mathbf{n}) &< S(\tau_{ij}(\mathbf{n})) = \\ &= \left[\sum_{\substack{k=1 \\ k\neq i,j}}^{N} s(n_k) + \sum_{(k,l)\neq(i,j)} s(n_{kl})\right] + s([\tau_{ij}(\mathbf{n})]_i) + s([\tau_{ij}(\mathbf{n})]_j) + s([\tau_{ij}(\mathbf{n})]_{ij}), \end{aligned} \tag{29}$$

and, letting $P_{ij} = \tau_{ij}(\mathbf{n})$ in Eq.(27), we obtain $S(\tau(\mathbf{n})) > S(\mathbf{n})$, i.e. $\mathbf{n}$ is not fixed for τ. Thus, S is a strict Liapunov function for τ and, being continuous, it is bounded in $\mathcal{Q}_\mathbf{n}$. It follows that the nonnegative quantity $F(\tau^k(\mathbf{n})) = S(\tau^{k+1}(\mathbf{n})) - S(\tau^k(\mathbf{n}))$ converges to zero as k tends to infinity and, because of equations (20) and (29), $\tau^k(\mathbf{n})$ must converge to the manifold of fixed points of τ, i.e. to the set of the joint fixed points in $Q_{\mathbf{n}}$ of all the τ_{ij}'s. The proof of the theorem will be completed if we can show that there is a unique $\hat{\mathbf{n}} \in Q_{\mathbf{n}}$ which is fixed for τ. Indeed, this is the case, as every fixed point of τ is a local maximum for S, and there can be no more than one point with that property in $Q_{\mathbf{n}}$. In fact, $\hat{\mathbf{n}}$ is a maximum for S along the direction of motion of τ_{ij}, for every (i,j), and the strict convexity of s implies that S is also strictly convex, thus making it impossible for two local maxima to coexist in $Q_{\mathbf{n}}$. ∎

We have thus shown that the discrete models defined by Eq.(23) admit a range for their parameters, within which convergence to an equilibrium point always occurs. From this, an appropriate limiting procedure similar to that described in Ref.[11] would show that the continuous time schemes associated with our QBM behave in the same way. This will be the subject of a future paper.

5. CONCLUSIONS

We have introduced the nonlinear dynamical systems called Quantum Boltzmann maps, and we have discussed some of their basic properties and an application to the theory of N-level systems. In particular, we have discussed the positivity preserving properties of the approximate theory, and the increase of the entropy under the relative map for the populations of the different energy levels and of the heat reservoirs. The global and the asymptotic behaviour of the system, under the approximate QBM has been investigated, and we have found that every initial condition is attracted to the uniquely determined fixed point.

ACKNOWLEDGEMENTS

The author wants to thank Prof. L. Accardi, Director of the "Centro Matematico V. Volterra", University of Rome II, and Prof. A. Frigerio for enlightening discussions. The author also thanks the anonymuos referees for pointing out several weaknesses of the first version of this work. Thanks are in order to Prof. R.F. Streater, for reading the manuscript and for constructive criticism. Support from GNFM-CNR is gratefully acknowledged.

REFERENCES

[1] R.F. Streater *A Boltzmann Map for Quantum Oscillators* J. Stat. Phys. **48**, 753 (1987)

[2] R. Dümcke *The Low Density Limit for an N-Level System Interacting with a Free Bose or Fermi Gas* Comm. Math. Phys. **97** 331 (1985)

[3] V. Gorini , A. Kossakowski and E.C.G. Sudarshan *Completely Positive Dynamical Semigroups of N-Level Systems* J. Math. Phys. **17**(5) 821 (1976)

[4] V. Gorini and A. Kosskowski *N-Level System in Contact with a Singular Reservoir* J. Math. Phys. **17**(7) 1298 (1976)

[5] A. Frigerio and V. Gorini *N-Level Systems in Contact with a Singular Reservoir. II* J. Math. Phys. **17**(12) 2123 (1976)

[6] M. H. Mittleman *Introduction to the Theory of Laser - Atom Interactions* Plenum Press, New York (1982)

[7] C. Cohen-Tannoudji *Light Shifts and Multiple Quantum Transitions* in *Physics of the One- and Two-Electron Atoms*, F. Bopp and H. Kleinpoppen Editors, Wiley Interscience, Amsterdam (1969)

[8] P.J. Ungar, D.S. Weiss, E. Riis and S. Chu *Optical Molasses and Multilevel Atoms: Theory* J. Opt. Soc. Am. B **6**(11) 2058 (1989)

[9] A. Aspect, E. Arimondo, R. Kaiser, N. Vansteenkiste and C. Cohen Tannoudji *Laser Cooling Below the one-photon Recoil Energy by Velocity - Selective Coherent Population Trapping: Theoretical Analysis* J. Opt. Soc. Am. B **6**(11) 2112 (1989)

[10] C.C. Gerry and R.F. Welch *Interactions of a Two-Level Atom with One Mode of Correlated Two-Mode Field States* J. Opt. Soc. Am. B **8**(4) 868 (1991)

[11] L. Rondoni *A Stochastic treatment of Reaction and Diffusion*, Ph.D. Thesis, Virginia Polytechnic Institute and State University, Blacksburg, Virginia, August 1991

[12] R.F. Streater *The Boltzmann Equation for Discrete Systems* in *Statistical Mechanics*, A. Solomon Editor, World Scientific (1988)

[13] R.F. Streater *Convergence of the Quantum Boltzmann Map* Comm. Math. Phys. **98**, 177 (1985)

[14] L. Rondoni *Nonlinear Boltzmann Maps in Classical and Quantum Probability* (in this volume)

[15] O.E. Lanford and D.W. Robinson *Approach to Equilibrium of Free Quantum Systems* Comm. Math. Phys. **24**, 193 (1972)

[16] E.H. Wichmann *Density Matrices Arising from Incomplete Measurements* J. Math. Phys. **4**, 884 (1963)

[17] A.G. Shuhov and Yu. M. Suhov *Ergodic Properties of Groups of Bogoliubov Transformation of CAR C^*-algebras* Ann. of Phys. **175**, 231 (1987)

[18] R.F. Streater *Entropy and the Central Limit Theorem in Quantum Mechanics* J. Phys. A. Math. Gen. **20**, 4321 (1987)

[19] L.E. Reichl *A Modern Course in Statistical Physics* University of Texas Press, Austin (1984)

[20] V. Boffi *Fisica del Reattore Nucleare* Vol. 1, Patron Editore, Bologna (1974)

[21] V. Hutson and J.S. Pym *Applications of Functional Analysis and Operator Theory* Academic Press, New York (1980)

[22] M. Fannes, Ann. Inst. H. Poincaré A, **28**, 187 (1978). See also N.M. Hugenholtz *Derivation of the Boltzmann Equation for a Fermi Gas* J. Stat. Phys. **32**(2) 231 (1983)

* Current address: Department of Theoretical Physics, School of Physics, The University of New South Wales, Kensington, NSW 2033, Australia

Quantum Probability and Related Topics Vol. VIII (pp. 313–328)

NONLINEAR BOLTZMANN MAPS IN CLASSICAL AND QUANTUM PROBABILITY

by

Lamberto Rondoni*

Center for Transport Theory and Mathematical Physics,
Virginia Polytechnic Institute and State University,
Blacksburg, VA 24061-0435 (U.S.A.)

ABSTRACT

We consider a class of discrete dynamical systems, called Boltzmann Maps, both in classical and quantum probability. We show that many physical models used in nonequilibrium statistical physics can be expressed in terms of Boltzmann Maps, and we discuss how new models can be constructed in this framework. In particular, the increase of the entropy under the Boltzmann Maps is used to prove the convergence of the dynamics to the appropriate equilibrium positions, in the relevant phase spaces.

1. INTRODUCTION

In this paper, we consider a class of nonlinear dynamical systems, called "Boltzmann Maps" (BM), which were first introduced in the 80's by R.F. Streater [1 - 3]. Such maps represent a suitable mathematical tool for the description of nonequilibrium physical systems, as they approach their stationary states. Several applications of the BM have already been investigated, with the result that certain known features of classical and non classical systems have been reproduced in the new framework, and new results have been derived for other systems. In particular, certain problems that arise in statistical, atomic and nuclear physics, in chemistry, and in biology have been studied in terms of BM. It is known that the problem of convergence to equilibrium for different classes of physical systems is outstanding, and different theoretical approaches are under consideration. Among these, the method based on BM belongs to a class of stochastic models which neglect the details of the microscopic interactions among particles, and such that irreversibility emerges as a consequence of the randomness which is put in the process by hand. This model may not appear as satisfactory as a Hamiltonian theory would be, as the macroscopic behaviour of a physical system is not derived from its microscopic dynamics. Nonetheless, it has good physical justification, since it attempts to describe the effects of the myriad of unknown degrees of freedom, and of noise, which are always present in the real physical systems, and which are not easily included in a Hamiltonian formalism. Moreover, we will show how some physical insight into the microscopic structure of certain processes can be drawn from the use of BM, and we will give an idea of their potentially very broad applicability. The aim of this paper is to familiarize the reader with the subject of BM, and to stimulate further research. Thus, the discussion will be rather informal and sketchy, as technical details can be found in the quoted references.

Our approach is not the most commonly used, and, in fact, the Hamiltonian formalism is more often found in the literature. In particular, we call the reader's attention to Alicki and Messer's work [25], where a model for a quantum Boltzmann equation is derived within a Hamiltonian formalism. Also, more recent results on mean-field dynamical semigroups have been presented by Duffield and Werner in Ref.[26].

2. THE CLASSICAL BOLTZMANN MAP

The concept of nonlinear "classical" Boltzmann maps (CBM) was first introduced by Stretaer in Ref.[1]. Here, we use the term classical to stress the fact that quantum mechanics doesn't play a role in this theory. The notion of CBM can be summarized as follows.

Let $\Omega = \{A_1, A_2, A_3, ...\}$ be a finite or countable set, to be regarded as our theory's sample space, and consider $\Omega^m = \Omega \times \cdots \times \Omega$, the cartesian product of m copies of Ω. Denote by $\mathcal{Q}$ the simplex of probability measures on Ω (real sequences $\{p_i\}$ in $l_1(\Omega)$, such that $p_i \geq 0$ and $\sum_i p_i = 1$), and let $\mathcal{Q}_m$ represent the set of probability measures on Ω^m (real sequences $\{p_{i_1 \dots i_m}\}$ in $l_1(\Omega^m)$, such that $\sum p_{i_1 \dots i_m} = 1$). Define the sampling map $\sigma : \mathcal{Q} \to \mathcal{Q}_m$, as $\sigma(P) = \otimes_1^m P$, which forms a probabilty in $\mathcal{Q}_m$ by taking the product of m copies of $P = (p_1, p_2, p_3, ...)$, a probability in $\mathcal{Q}$. We will interpret σ as the map that implements the Boltzmann's *Stosszahlansatz*, by noting that each component, p_i, of P could represent the probability of sorting one element, A_i, out of Ω, and thus each component of $\sigma(P)$ would be the probability of fishing out m independent elements before they interact.

Let $T : \mathcal{Q}_m \to \mathcal{Q}_m$ be a doubly stochastic map; i.e. T and its adjoint T^* are stochastic maps, which is to say: $\mathbf{a} \geq 0$ implies $T\mathbf{a} \geq 0$, $\|T\mathbf{a}\| = \|\mathbf{a}\|$ (l_1-norm) for all $\mathbf{a} \geq 0$, and the same holds if T is replaced by T^*. The matrix elements of T, under the inner product of l_2, will be interpreted as the probability that a set of m particles interact and produce another set of m particles.

Let $E : \mathcal{Q}_m \to \mathcal{Q}$ be the expectation onto the first factor, i.e. the summation over all but one index of an element of $\mathcal{Q}_m$. This takes probabilities on Ω^m to probabilities on Ω. Then we define a Boltzmann map of order m as follows.

DEFINITION 1. *A* **Boltzmann map** *of order m is a map of the form $\tau(P) = E(T(\sigma(P)))$, i.e.*

$$P \overset{\sigma}{\longmapsto} \otimes_1^m P \overset{T}{\longmapsto} T(\otimes_1^m P) \overset{E}{\longmapsto} P^* = \tau(P),$$

for every $P \in \mathcal{Q}$.

Just from the definition, one can easily deduce that

THEOREM 1. *(a) A Boltzmann map of order m maps $\mathcal{Q}$ into itself.*
(b) The entropy $S = -\sum_j p_j \log p_j$ of the system does not decrease under the action of τ; i.e. $S(\tau(P)) \geq S(P)$ for every $P \in \mathcal{Q}$.

These are desirable properties, for a theory that describes the approach to equilibrium of a physical system, although they are not shared by all the discrete models. In particular, the preservation of probabilities has a very important interpretation in terms of nonnegativity of densities of particles, or concentrations and so on. Furthermore, the choice of a doubly stochastic map for the description of the interaction of particles, is made necessary, in a sense, by a theorem of Alberti and Uhlmann [4], concerning "chaos - enhancement" due to linear operators. In fact, it is shown in Ref.[4] that the only chaos - enhancing linear operators are doubly stochastic, where the term chaos - enhancing is a generalization of the concept of entropy non-decrease.

Although it is possible to proceed with the analysis on countable sample spaces, we now turn to the special case of finite spaces, in order to focus on some applications of particular interest. Therefore, let us assume that Ω contains n elements only, then we can prove the following.

THEOREM 2. *If 1 is a simple eigenvalue of TT^* and $P \in \mathcal{Q}$, then $\tau^k(P) \to (1, 1, ..., 1)/n$ as $k \to \infty$.*

(Note that we are in a finite dimensional space, so TT^* is well defined).

The proof of Theorem 2. relies on the fact that the crucial step for the increase of the entropy determined by τ is the application of T, and it is based on the following fundamental result from Ref.[1]

THEOREM 3. *Let T be a bistochastic map on $\mathbb{R}^N$ and let $P = \{p_j\}_1^N$ be a semiprobability, i.e. $1 \geq p_j \geq 0$, $j = 1, ..., N$, and $\sum_j p_j \leq 1$. Let the eigenvalue 1 of TT^* be simple, with a gap Δ to the next largest eigenvalue. Let $q_l = \sum_k T_{kl}\ p_k$. Then*

$$\sum_k p_k \log p_k \geq \sum_j q_j \log q_j + \frac{1}{2}\Delta \|p - \hat{p}\|_2^2 \tag{1}$$

where $\hat{p} = \|P\|_1/N$, and $\|P\|_2 = \sqrt{|p_1|^2 + ... + |p_N|^2}$

Therefore, whenever TT^* has a spectral gap Δ, the entropy of the system increases a positive amount at every application of τ, unless P is the uniform distribution. It follows that the entropy converges to a finite limit as it is a continuous function on a compact domain. At the same time, the sequence $\{\tau^m P\}$ lies in the same compact set, and so it has a subsequence that converges to $\hat{P}$, say. This observation allows the author to prove that $S(\tau\hat{P}) = S(\hat{P})$, and that $\hat{P}$ is the limit of all the convergent subsequences of $\{\tau^m P\}$. Then $\hat{P}$ must be the uniform probability distribution, otherwise Theorem 3. would be violated.

Having introduced the CBM, we now take a look at its applications. We start with the observation that it is possible to find appropriate sample spaces and doubly stochastic maps, such that the resulting form of τ constitutes a discrete scheme for the time evolution of the concentrations of the different chemicals in a chemical reaction. For instance, consider the following reaction

$$n_1 A_1 + ... + n_N A_N = m_1 B_1 + ... + m_M B_M, \tag{2}$$

where the small letters are integers representing the stoichiometric coefficients of the reaction, and the capital letters are symbols of chemicals. Let us denote by p_i and by q_j, respectively, the concentrations of the A_i's and of the B_j's. Then, assuming that the forward and the backward reaction rates are equal, we have that the following system of ordinary differential equations describes the evolution of the concentrations.

$$\begin{aligned} \frac{dp_j}{dt} &= n_j \lambda(q_1^{m_1} ... q_M^{m_M} - p_1^{n_1} ... p_N^{n_N}), \qquad j = 1, ... N \\ \frac{dq_k}{dt} &= -m_k \lambda(q_1^{m_1} ... q_M^{m_M} - p_1^{n_1} ... p_N^{n_N}), \quad k = 1, ... M \end{aligned} \tag{3}$$

where $\lambda > 0$ is the rate constant. These generalize the equations of [5],[6]. If $n_1 + ... + n_N = m_1 + ... + m_M$ we say the system is "balanced". In that case, and normalizing the sum of the concentrations to 1, we can express the discrete form of Eq.(1)

$$\begin{aligned} p_j^* &= p_j + n_j \mu(q_1^{m_1} ... q_M^{m_M} - p_1^{n_1} ... p_N^{n_N}), \qquad j = 1, ... N \\ q_k^* &= q_k - m_k \mu(q_1^{m_1} ... q_M^{m_M} - p_1^{n_1} ... p_N^{n_N}), \quad k = 1, ... M \end{aligned} \tag{4}$$

as a Boltzmann map on the probability space relative to $\Omega = \{A_1, ..., A_N, B_1, ..., B_M\}$. In fact, our reaction describes a process in which a particular selection of n particles of kind "A" interact and disappear in order to produce another particular selection of n particles of kind "B", and viceversa. If the set of n reacting A-particles is different from a permutation of the n-tuple, $R_{1,...,1,...,N,...,N} = (A_1, ..., A_1, ..., A_N, ..., A_N)$, with n_1 particles of kind A_1, ... , n_N of kind A_N, then no reaction takes place, i.e. the reaction products are represented by a permutation of the reacting n-tuple itself. If, instead, the reacting n-tuple is $R_{1,...,1,...,N,...,N}$, then the reaction products form a permutation of $S_{1,...,1,...,M,...,M} = (B_1, ..., B_1, ..., B_M, ..., B_M)$, with m_1 particles of kind B_1, ..., m_M of kind B_M. A similar notation can be developed for a set of n reacting B-particles. Then, the following can be proven [7,8]

LEMMA 4. *Let R and S be generic n-tuples of elements of Ω, and assume that all the permutations, $\pi_i(R) \longmapsto \pi_j(S)$, are equally probable processes. Then Eq.(4) represents a CBM of order n, if and only if $\mu \in [0, L_B]$, where*

$$L_B = \min\left(\frac{(n-1)!}{n_1!\cdots n_N!}\ ,\ \frac{(n-1)!}{m_1!\cdots m_M!}\right).$$

In the case that Eq.(4) is a CBM, the form of the interaction operator is

$$T = T' \oplus M_1 \oplus ... \oplus M_k,$$

where T' is doubly stochastic, $M_i = (1/t_i)J_{t_i}$, and J_{t_i} is a $t_i \times t_i$ matrix whose entries are all equal to 1. Moreover, $T'(T')^*$ has a spectral gap $\Delta > 0$ if and only if $\mu \in (0, L_B)$.

The proof of this result is relatively complex in terms of algebraic manipulations of sets of equations, which invlove some combinatorial calculus. For the rest, the result is based on Perron - Frobenius theorem and on the observation that all the entries of T' are positive if and only if μ lies in $(0, L_B)$.

To make all these concepts clear, let us consider a simple example that shows the main general features: $\boxed{2A = B + C}$. In this case, the map τ acts on a probability space $\{P = (p_A, p_B, p_C)\} = \mathcal{Q} \subset \mathbb{R}^3$, and it transforms P into P^* acording to the following rules:

$$\begin{aligned} p_A^* &= p_A - \mu\left(p_A^2 - p_B p_C\right) = p_A - 2\mu D \\ p_B^* &= p_B + \mu D \\ p_C^* &= p_C + \mu D. \end{aligned}$$

The bistochastic matrix associated with this reaction is given by

$$\begin{array}{c} \\ AA \\ BC \\ CB \\ AB \\ BA \\ AC \\ CA \\ BB \\ CC \end{array} \begin{array}{c} \begin{array}{ccccccccc} AA & BC & CB & AB & BA & AC & CA & BB & CC \end{array} \\ \begin{pmatrix} 1-2\mu & \mu & \mu & 0 & 0 & 0 & 0 & 0 & 0 \\ \mu & \frac{1-\mu}{2} & \frac{1-\mu}{2} & 0 & 0 & 0 & 0 & 0 & 0 \\ \mu & \frac{1-\mu}{2} & \frac{1-\mu}{2} & 0 & 0 & 0 & 0 & 0 & 0 \\ 0 & 0 & 0 & .5 & .5 & 0 & 0 & 0 & 0 \\ 0 & 0 & 0 & .5 & .5 & 0 & 0 & 0 & 0 \\ 0 & 0 & 0 & 0 & 0 & .5 & .5 & 0 & 0 \\ 0 & 0 & 0 & 0 & 0 & .5 & .5 & 0 & 0 \\ 0 & 0 & 0 & 0 & 0 & 0 & 0 & 1 & 0 \\ 0 & 0 & 0 & 0 & 0 & 0 & 0 & 0 & 1 \end{pmatrix} \end{array} = T \qquad (5)$$

Here the capital letters have been used to better identify which process is represented by which matrix element, and each such matrix element $T_{ij;kl}$ can be interpeted as the probability that the process $(i,j) \mapsto (k,l)$ takes place. Clearly, T acts like the identity on a part of the vector that is given as input. In fact, T is applied to vectors of the form $P \otimes P$ only, and so the part relative to the components $p_{AB}, p_{BA}, p_{AC}, p_{CA}, p_{BB}$ and p_{CC} is left unchanged.

On the other hand, the invariant subspace relative to the components p_{AA}, p_{BC} and p_{CB} is "mixed" by the action of T. In fact, the submatrix $T'(T')^*$ of TT^*, which is relevant to such a subspace, has positive entries, and so it has a spectral gap. As a consequence, the application of τ increases the entropy until the input vector for T has $p_{AA} = p_{BC} = p_{CB}$, i.e. $p_A^2 = p_B p_C$, which is exactly the equation for the stationary points of τ. This, in general, is not enough to prove convergence to the fixed points of τ. In fact, τ has a smooth, connected and compact manifold of fixed points, and so the iterations could wander around, closer and closer to such a manifold, without converging to any one of its points. However, it is easily seen that single reactions give rise to one - dimensional dynamical systems, due to the presence of constants of motion. This observation, coupled with Brouwer's fixed point theorem, allows us to prove that, for every initial condition $P \in \mathcal{Q}$, there is a unique fixed point $\hat{P}$ to which the iterations can converge, and thus they do indeed converge. This result, although quite crude, is somewhat interesting, because the map τ is not a global contraction in $\mathcal{Q}$, neither it is locally contracting. Note that $\mathcal{Q}$ is a subset of $\mathbb{R}^K$ for some $K \in \mathbb{N}$, and so we prove that there is no norm or metric of $\mathbb{R}^K$ under which τ is contracting, simply by checking that τ is not a contraction in the $\|.\|_2$-norm (cf. theorem 88.1 in Ref.[24]). Nonetheless, knowing that the system is 1-dimensional, we can restrict our analysis to the appropriate subspace, and draw some more refined conclusions. Thus, parametrizing the motion with a unique variable, x say, we get the following result [7-9]

THEOREM 5. *For every reaction described by a CBM τ, and for every initial condition $P \in \mathcal{Q}$, there is a "line of reaction" $\Gamma_P \subset \mathcal{Q}$, such that the dynamics $\{\tau^j(P)\}_0^\infty$ is contained in Γ_P, and there is a unique fixed point $\hat{P} \in \Gamma_P$ to which the iterations converge provided that $\mu \in (0, L_B)$. Also, there is a $\mu_0 > 0$, such that the iterations converge in a monotonic way, if $0 < \mu \leq \mu_0$. Moreover, there exists an invariant compact subset I of Γ_P, which attracts the iterations, and there is a $\mu' > 0$, such that $\tau^j(P)$ is exponentially convergent to $\hat{P}$, if $P \in I$ and if $0 < \mu \leq \mu'$.*

The proof of this theorem is based on very standard arguments from the analysis of dynamical systems on the interval, and its consequence is this corollary [7-9]

COROLLARY 6. *(a) The entropy is a strict Liapunov function for τ. (b) All the fixed points in the interior of $\mathcal{Q}$ are stable.*

So far, we have considered single reactions only; but it is possible to construct "Generalized CBM", which share the same basic properties of the CBM. In particular, it is possible to take convex combinations of different CBM's –each one of which acts on the simplex relative to a particular sample space– by redefining every CBM on the probability space of the union of the different sample spaces. Doing this, we construct a new map which can be used to model complex chemical reactions, i.e. reactions which are combinations of many single reactions, and which better represent most of the "real life" ones [9]. Consider one of this generalized CBM, $\tau = \sum_i \lambda_i \tau_i$, where τ_i acts on $\mathcal{Q}_i$, and the λ_i's obey $\sum_i \lambda_i = 1$, $\lambda_i > 0$. Then, using the convexity properties of the entropy, we prove that a point P is fixed for τ if and only if it is fixed for each of the τ_i's separately. The analysis of this new kind of dynamical systems is necessarily more complex than that of the CBM, because the notation becomes a little heavier, and especially because the dynamics is not 1-dimensional any more. Therefore, we leave out the details, and we simply state the main result [8,10].

THEOREM 7. *Consider a set of N reactions, each described by a Boltzmann map τ_i of order n_i, $i = 1, ..., N$. Let τ be a convex combination of the τ_i's. Then*

(a) S is a strict Liapunov function for τ.

(b) Let $\mathcal{M}$ be the manifold of the fixed points of τ, and let $d(x, \mathcal{M}) = \min_{y \in \mathcal{M}} \|x - y\|_2$. Then, for every $P \in \mathcal{Q}$ and every $\epsilon > 0$, there exists a $k \in \mathbb{N}$ such that $d(\tau^j(P), \mathcal{M}) < \epsilon$ whenever $j \geq k$.

This theorem comes again from the convexity properties of the entropy S. Moreover, it is possible to show that every choice of the initial condition, P, in the set of probability measures over the union of the sample spaces of each single map, determines a convex subset in which the dynamics takes place, and such that, in its interior, there is a unique fixed point for τ. As we can also prove that the iterations cannot converge to the fixed points on the boundary of the set determined by P, we have got the result that also the systems of convex combinations converge to an equilibrium point.

It is now quite interesting to note that the discrete diffusion operator, which we get by taking the central difference approximation to the Laplacian in a finite dimensional space, is a CBM of order 1. Therefore, if we study a reaction in an m-dimensional container of finite volume, we can subdivide such a volume in an arbitrary, but finite, number of cells, and then consider the space inhomogeneus problem with the discrete Laplacian, as described by a generalized CBM. We cannot easily take the limit for the number of cells that goes to infinity. However, from a physical point of view the situation is satisfactory, as the space discretization of a finite volume can be made as fine as desired.

To make this clear, let us consider the example of a 1-dimensional medium made of L cells, in which a single non-autocatalyitc reaction takes place:

$$n_1 A_1 + \cdots + n_N A_N = m_1 A_{N+1} + \cdots + m_M A_{N+M}. \tag{6}$$

For simplicity, assume that the diffusion coefficients of each chemical do not vary from cell to cell, and call $d_i > 0$ the cell relative to the chemical A_i. Let p_{ij} be the concentration of A_i in the j-th cell, and assume that

$$\sum_{i=1}^{N+M} \sum_{j=1}^{L} p_{ij} = 1. \tag{7}$$

Then, the diffusion operator has the form $\delta = \delta_1 \oplus \ldots \oplus \delta_{N+M}$, where δ_i describes the diffusion of the chemical A_i. On the other hand, the full reaction operator takes the form

$$\tau_r = \sum_{i=1}^{L} \lambda_i \tau_i, \tag{8}$$

where τ_i describes (6) in the i-th cell, and so the combination of the reaction term with δ can be written as $\tau = \tau_r + (1-\lambda)\delta$, for $\lambda \in (0,1)$, if we have $\lambda_i > 0$ and $\sum_i \lambda_i = \lambda$. Then, given a $P \in \mathcal{Q}$, the simplex of probabilities on $\Omega = \{A_{1i}, \ldots, A_{N+M,i}\}_{i=1}^{L}$, the action of τ on P produces a new probability P^*, whose components p^*_{ji} can be computed as follows:

$$\begin{aligned}
p^*_{j1} &= p_{j1} - \lambda_1 \mu n_j \left(p_{11}^{n_1} \cdots p_{N1}^{n_N} - q_{11}^{m_1} \cdots q_{M1}^{m_M}\right) - (1-\lambda) d_j (p_{j1} - p_{j2}) \\
p^*_{ji} &= p_{ji} - \lambda_i \mu n_j \left(p_{1i}^{n_1} \cdots p_{Ni}^{n_N} - q_{1i}^{m_1} \cdots q_{Mi}^{m_M}\right) + (1-\lambda) d_j (p_{j,i-1} - 2p_{ji} - p_{j,i+1}), \\
p^*_{jL} &= p_{jL} - \lambda_L \mu n_j \left(p_{1L}^{n_1} \cdots p_{NL}^{n_N} - q_{1L}^{m_1} \cdots q_{ML}^{m_M}\right) + (1-\lambda) d_j (p_{j,L-1} - p_{jL}),
\end{aligned} \tag{9}$$

for $i = 2, \ldots, L-1$, for the chemicals on the left hand side of (6), and

$$\begin{aligned}
q^*_{j1} &= q_{j1} + \lambda_1 \mu m_j \left(p_{11}^{n_1} \cdots p_{N1}^{n_N} - q_{11}^{m_1} \cdots q_{M1}^{m_M}\right) - (1-\lambda) d_{N+j} (q_{j1} - q_{j2}) \\
q^*_{ji} &= q_{ji} + \lambda_i \mu m_j \left(p_{1i}^{n_1} \cdots p_{Ni}^{n_N} - q_{1i}^{m_1} \cdots q_{Mi}^{m_M}\right) + (1-\lambda) d_{N+j} (q_{j,i-1} - 2q_{ji} - q_{j,i+1}), \\
q^*_{jL} &= q_{jL} + \lambda_L \mu m_j \left(p_{1L}^{n_1} \cdots p_{NL}^{n_N} - q_{1L}^{m_1} \cdots q_{ML}^{m_M}\right) + (1-\lambda) d_{N+j} (q_{j,L-1} - q_{jL}),
\end{aligned} \tag{10}$$

for $i = 2, ..., L-1$, for the chemicals on the right, where we have used $q_{ji} = p_{N+j,i}$. Let us define now the total probability of the substance A_j by $p_j = \sum_{i=1}^{L} p_{ji}$. Then, we get

$$p_j^* = p_j - \mu n_j \sum_{i=1}^{L} \lambda_i D_i(P), \quad \text{and} \quad q_j^* = q_j + \mu m_j \sum_{i=1}^{L} \lambda_i D_i(P), \tag{11}$$

where $D_i(P) = p_{1i}^{n_1} \cdots p_{Ni}^{n_N} - q_{1i}^{m_1} \cdots q_{Mi}^{m_M}$, from which it follows that

$$n_j p_1 - n_1 p_j = K_j, \qquad j = 2, ..., N, \quad \text{and} \quad m_j p_1 + n_1 q_j = L_j, \qquad j = 1, ..., M, \tag{12}$$

are constants of motion. We can now find the fixed points of τ, given some initial condition $P(0) \in \mathcal{Q}$. Recall that $\tau(\hat{P}) = \hat{P}$ if and only if $\delta(\hat{P}) = \hat{P}$ and $\tau_i(\hat{P}) = \hat{P}$, for every i, because $\hat{P}$ is fixed for τ if and only if it is separately fixed for each of its components. Then, if we call $\hat{p}_{ji}$ the components of a fixed point $\hat{P}$ of τ, we must have $\hat{p}_{j1} = ... = \hat{p}_{jL} = \hat{p}_j / L$ for every $j = 1, ..., N+M$, and $D_i(\hat{P}) = 0$ for every i, i.e.:

$$D_i(\hat{P}) = \frac{1}{L^n}\left(\hat{p}_1^{n_1} \cdots \hat{p}_N^{n_N} - \hat{q}_1^{m_1} \cdots \hat{q}_M^{m_M}\right) = \frac{1}{L^n} D(\hat{P}) = 0, \tag{13}$$

where $n = \sum n_i = \sum m_i$, and $D(\hat{P})$ is the disequilibrium parameter of the stirred reaction, evaluated at the point whose components are the $\hat{p}_j$'s. We know from [7] that there exists a unique solution to (13) in $\mathcal{Q}$, given the particular choice of the constants of motion due to $P(0)$, and we conclude that for every $P(0) \in \mathcal{Q}$ there exists a unique $\hat{P} \in \mathcal{Q}$ such that $\tau^k(P(0)) \to \hat{P}$ as $k \to \infty$. Moreover, $\hat{P}$ is stable, as the fixed points of the stirred reaction are. Concerning the explicit form of the diffusion operator, δ, we observe that each one of its components, δ_i, has the form

$$\begin{array}{c} \\ 1 \\ 2 \\ 3 \\ 4 \\ \vdots \\ \vdots \\ L \end{array}
\begin{array}{c} \begin{array}{ccccccc} 1 & 2 & 3 & 4 & \cdots & L-1 & L \end{array} \\
\begin{pmatrix}
1-d_i & d_i & 0 & 0 & \cdots & 0 & 0 \\
d_i & 1-2d_i & d_i & 0 & \cdots & 0 & 0 \\
0 & d_i & 1-2d_i & d_i & \cdots & 0 & 0 \\
0 & 0 & d_i & 1-2d_i & \cdots & 0 & 0 \\
\vdots & \vdots & \vdots & \vdots & \ddots & \vdots & \vdots \\
0 & 0 & 0 & 0 & \cdots & 1-2d_i & d_i \\
0 & 0 & 0 & 0 & \cdots & \cdots d_i & 1-d_i
\end{pmatrix} \end{array}, \tag{14}$$

in a 1-dimensional medium with equal diffusion coefficients at every cell.

So far, we have seen how the study of chemical reactions, under certain hypothesis, can be cast into the form of a CBM. The results we have presented are not the only ones that can be derived in this framework, and we don't know of any other equally general treatment for discrete models. We should note that most of these results were known for the continuous time systems. However, we believe that the discrete systems are worth being investigated of their own merits, and not just because they approximate the continuous models. In fact, besides the physical motivation of this statement [8], there certainly is a strong mathematical motivation, which arises because of the use of discrete schemes in numerical analysis and especially because discrete models are substantially different mathematical objects from their continuous counterparts. In particular, properties like the positivity of the solutions, which are trivially verified for the systems of ordinary differential equations, are not shared by all the discrete schemes, so that only particular classes of maps, like the CBM's, can be used. It is also interesting to note that the known results for continuous systems, can be rederived in the framework provided by CBM's, if the rate constant μ is interpreted as $\lambda\triangle t$, where λ is the rate constant of Eq.(3). In fact, conditions can be given such that the discrete scheme converges to the solution of the continuous model when $\triangle t \to 0$, [9,10].

The study of chemical reactions is not the only application of the CBM that has been investigated. For instance, Streater has recently shown how the Ising model with Glauber dynamics, and cotransport through membranes can be treated via generalized CBM's of particular kind [11,12].

3. THE FIRST QUANTIZED BOLTZMANN MAP

The first quantized Boltzmann map (FQBM) was first introduced by Stretaer in Ref.[2]. It is a direct generalization to quantum mechanics of the CBM, which has the feature of treating quantum particles as distinguishable. However, this model is based on very clear physical intuitions, and many of the results drawn within its framework are used also in the second quantized formalism. Moreover, we may argue that Boltzmann's stosszahlansatz really makes sense only in a scheme with distinguishable particles. It is therefore worth taking a look at the FQBM as a discrete version of the quantum Boltzmann equation. The set up for the FQBM is this: $\mathcal{H}$ is a finite dimensional Hilbert space, with $< x, y >= \sum_i \overline{x_i}\, y_i$ and a state is obtained from a density operator ρ which is positive and has unit trace. If the set of linear operators is denoted by $\mathcal{A} = \mathcal{B}(\mathcal{H})$, and the set of density matrices by Σ, we can define a doubly stochastic map, $T : \mathcal{A} \to \mathcal{A}$, as in the previous section. Define the entropy of a state ρ as $S(\rho) = -\operatorname{Tr}(\rho \log \rho)$. Then, τ maps Σ into Σ, and $S(\tau\rho) \geq S(\rho)$, where equality holds if and only if $\tau(\rho) = \rho$. For a doubly stochastic map $T : \mathcal{A} \to \mathcal{A}$, a sharp estimate on the entropy increase under its action can be given. Furnishing $\mathcal{A}$ with the inner product $< A, B >= \operatorname{Tr}(A^* B)$, we get

THEOREM 8. *Let $P^\perp$ be the orthogonal projection of $\mathcal{B}$ onto $\operatorname{Ker}(I - TT^*)^\perp$, and assume that the spectrum $\sigma(TT^*)$ of TT^* is contained in $[0, 1 - \triangle] \cup \{1\}$, for some number*

$0 < \triangle < 1$. *Then the entropy obeys*

$$S(T^{*}\rho) - S(\rho) \geq \frac{\triangle}{2}\|P^{\perp}\rho\|^{2}, \tag{15}$$

where I is the identity map, and ρ is a positive operator of unit trace.

(Note that the domain of T^{*} is invariant under the action of T^{*} and it is contained in the domain of T. Thus TT^{*} is well defined). Let us introduce $\mathcal{K}$, a Hilbert space representing the one-particle space, and assume that our $\mathcal{H}$ is the tensor product $\mathcal{H} = \mathcal{K}\otimes\cdots\otimes\mathcal{K}$, which will be regarded as the n-particle space, if n is the number of factors. Then we define a FQBM, τ, as follows.

DEFINITION 2. *A FQBM is the composition of maps*

$$\rho \mapsto \otimes_1^n \rho \mapsto T(\otimes_1^n\rho) \mapsto \mathrm{Tr}_{2\ldots n}[T(\otimes_1^n\rho)] = \tau(\rho),$$

where T is doubly stochastic on $\mathcal{A}$ and commutes with the permutations of the factors in the tensor products, and $\mathrm{Tr}_{2\ldots n}$ is the trace over all but one of the n factors $\mathcal{K}$.

The assumption on the commutativity of T with permutations is technical, and has to do with the fact that the particles are treated as distinguishable, in this theory.

The physical picture behind the construction of the FQBM is that of a rarefied gas, in which the time between two collisions among a particle and other particles is very long, so that the single particle Hamiltonian is approximately equal to the free Hamiltonian. Let us assume that h_0, the single particle hamiltonian, have eigenvalues $e_1, e_2, ..., e_k$. We denote by H_0 the Hamiltonian for n independent particles of the gas. Then H_0 takes the form $H_0 = h_0 \otimes I \otimes\cdots\otimes I + ... + I\otimes I\otimes\cdots\otimes h_0$. Let us assume that H_0 has a spectral resolution $H_0 = \sum_j E_jP_j$, where the P_j's are the corresponding spectral projections, and the E_j's are the eigenvalues. Then we can prove that the FQBM, τ, preserves the average 1-particle energy, i.e. $\mathrm{Tr}((\tau\rho)h_0) = \mathrm{Tr}(\rho h_0)$, provided that the doubly stochastic map T preserves the spectral projections of H_0.

We now introduce the notion of ergodicity for doubly stochastic maps.

DEFINITION 3. *A doubly stochastic map which preserves the spectral projections, P_j, of H_0 is called ERGODIC if $TA = A = T^{*}A$ implies $A = \sum_j a_jP_j$, where a_j is a function of E_j for every j.*

Then, the following can be proved:

THEOREM 9. *Assume that $\{0, 1, 2, ...\}$ is the spectrum of h_0 on $\mathcal{K}$, and let the finite multiplicity $m(j)$ of the eigenvalue j obey $m(j) \leq cj^{r}$, with c and r fixed constants. Let $\mathcal{H} = \mathcal{K}\otimes\cdots\otimes\mathcal{K}$ and let $T : \mathcal{A} \to \mathcal{A}$ be doubly stochastic and commute with the* *permutations of tensor products. Assume T is ergodic on every eigenspace of H_0. Then $\tau^m(\rho)$ converges in the trace norm to a Gibbs state*

$$\rho_\infty = \frac{e^{-\beta h_0}}{Tr(e^{-\beta h_0})} \tag{16}$$

where β is determined by

$$Tr(\rho h_0) = \frac{Tr(h_0 e^{-\beta h_0})}{Tr(e^{-\beta h_0})} \quad . \tag{17}$$

Cearly, the condition on the multiplicity of the eigenvalues is always satisfied for a finite dimensional $\mathcal{K}$, but it should be noted that this theorem holds also if $\mathcal{K}$ is an infinite dimensional separable Hilbert space [2].

A possible application of the FQBM would be to investigate the properties of certain continuous time models, similarly to what was done with the CBM. For instance, Aratari and Frigerio [13], gave an explicit solution in terms of unitary dilations of the quantum Boltzmann equation introduced by Snider [14]. However, the form of the solution is complicated enough that it is difficult to analyze it in a direct way. This could be avoided, for instance, with the aid of the FQBM as, once again, such a map is nothing but a discrete form of the continuous system.

4. THE SECOND QUANTIZED BOLTZMANN MAP

The second quantized Boltzmann map (SQBM) was introduced by Streater in Ref.[3], as a model for a set of $M < \infty$ quantum harmonic oscillators. Let $\mathcal{K}$ be the 1-particle space for a set of M harmonic oscillators with frequencies $\omega_1, ..., \omega_M$, and consider the symmetrized Fock space over $\mathcal{K}$: $\Gamma_s(\mathcal{K}) = \mathbb{C} \oplus \mathcal{K} \oplus (\mathcal{K} \otimes \mathcal{K})_s \oplus \cdots$, for bosons, or the antisymmetrized Fock space $\Gamma_a(\mathcal{K}) = \mathbb{C} \oplus \mathcal{K} \oplus \Lambda^2(\mathcal{K}) \oplus \cdots$, for fermions, where $\Lambda^k(\mathcal{K})$ is the subspace of antisymmetric tensors of $\otimes_1^k \mathcal{K}$. For simplicity, we will mainly concentrate on bosonic systems. Therefore, the algebra of observables of the system is taken to be the set of bounded operators on $\Gamma_s(\mathcal{K})$, i.e. $\mathcal{A} = \mathcal{B}(\Gamma_s(\mathcal{K}))$, which is generated by the creation and annihilation operators a_j^* and a_j , $j = 1, ..., M$. Let Σ be the set of (normal) states on $\mathcal{A}$, and furnish $\mathcal{A}$ with the inner product $< A, B >= \mathrm{Tr}(A^* B)$. If we define the entropy of the state ρ as in the previous section, we can make use of the same results about the entropy gain under a doubly stochastic map. In particular, the doubly stochastic maps, T, we deal with, in this section, are obtained as mixtures of conjugations of unitaries, thus being completely positive, and we assume that all such unitaries commute with the spectral projections of the Hamiltonian. Then, the average energy of the state ρ is preserved under the application of T^* to ρ. In fact, defining the

Hamiltonian as $H_0 = \sum \omega_j a_j^* a_j = \sum E_i P_i$, we get

$$\mathrm{Tr}((T^*\rho)H_0) = \mathrm{Tr}(U^*\rho U H_0) = \mathrm{Tr}(\rho U H_0 U^*) = \mathrm{Tr}(\rho T H_0) = \mathrm{Tr}(\rho H_0),$$

where U is a unitary operator and U^* is its adjoint, and T is defined by $TA = UAU^*$ for every A in $\mathcal{A}$. We can also prove this result:

THEOREM 10. *Let $\mathcal{H}_j = P_j\mathcal{H}$, with $\mathcal{H} = \mathcal{F}(\mathbb{C}^N)$, the Fock space over $\mathbb{C}^N$. Then, the restriction T_j of T on $\mathcal{B}(\mathcal{H}_j)$ maps the identity operator I_j into itself, and*

$$S(T^*\rho) - S(\rho) \geq \frac{\triangle_j}{2}\left|\rho - \frac{1}{\dim \mathcal{H}_j} I_j\right|^2 , \quad \rho \in \mathcal{B}(\mathcal{H}_j), \tag{18}$$

where $\triangle_j$ is the spectral gap of $T_j T_j^*$. Then the iterations $(T^*)^n\rho$ for $\rho \in \mathcal{B}(\mathcal{H})$ converge to a density operator which is constant on each $\mathcal{B}(\mathcal{H}_j)$.

Therefore, a map like T^* cannot mix the different "energy shells" of a system, and the dynamics it induces can be interpreted as the evolution of a system of thermally isolated parts, each one of which reaches an independent microcanonical state. To achieve mixing of the different shells more stochasticity is needed than that provided by a doubly stochastic map. In the previous two sections, relative to the CBM and to the FQBM, this extra stochasticity is provided by the stosszahlansatz. Here, we do not have a formalism of particles any more, and so we cannot resort to such a tool. However, we can argue by analogy, and introduce a map Q which, like the stosszahlansatz, destroys the correlations by replacing a given state with the Gaussian state which has same first and second moments. Such a nonlinear map was already used in previous works [15 - 17], and is called "quasi- free" projection, as it replaces a state with finite first and second moments with the state whose third and higher moments vanish, and whose first and second moments equal those of the original state. It was proved by Streater [18] that such a map does not decrease the entropy, and that $S(Q\rho) > S(\rho)$ unless $Q\rho = \rho$.

DEFINITION 4. *A SQBM is a map $\tau : \Sigma \to \Sigma$ defined by $\tau = QT^*$, where Q is the quasi-free projection on Σ, and T^* is a completely positive and doubly stochastic map on Σ.*

Thus, the dynamics determined by τ takes place on the subset of quasi - free states in Σ. It turns out that the limit of every convergent subsequence of $\{\tau^n\rho\}_1^\infty$ is a quasi - free state ρ_∞ such that $TT^*\rho_\infty = \rho_\infty$.

DEFINITION 5. *We say that a state is ERGODICALLY MIXED if it can be expressed as $\rho = \sum \nu_j P_j$, where P_j are the spectral projections of H_0, and we say that the doubly stochastic map T is ERGODIC if the only fixed points of TT^* are ergodically mixed.*

Then the following theorem holds.

THEOREM 11. *Suppose the frequencies $\omega_1, ..., \omega_N$ are such that the only quasi-free, ergodically mixed states are canonical, i.e. of the form $\rho_\beta = e^{-\beta H_0} / Tr(e^{-\beta H_0})$ for some $\beta > 0$. Assume T^* conserves energy and is ergodic; then, for any state ρ of finite energy E, $\tau^n \rho \to \rho_\beta$, where β is determined by $Tr(\rho_\beta H_0) = E$.*

From this, it follows that

THEOREM 12. *Consider a bosonic system with Hamiltonian $H_0 = \sum_j \omega_j a_j^* a_j$. Let $\omega_1, ..., \omega_M$ be positive and mutually rational. Let T be ergodic and ρ a state of energy E. Then $\tau^n \rho$ converges to the canonical state of energy E.*

A similar result holds for fermionic systems. For some applications of this theory see Refs.[3] and [19]. In another paper [20] we show how a perturbation expansion of the map defined by τ can be used to construct a theory for the M-level atom.

Finally, we mention Streater and Koseki's works who have recently studied the problem of photon exchange with an infinite heat bath, deriving the relative F-theorems (F = free energy) both for the discrete, [11], and for the continuous, [22], cases. One application of the SQBM to nuclear physics can be found in [23].

5. CONCLUSIONS

We have introduced the nonlinear dynamical systems called Boltzmann maps, in classical and quantum probability, and we have discussed some of their basic properties and applications to physical problems. The stochastic treatment of the approach to equilibrium for isolated, closed and open systems have been outlined, thus showing the versatility of the BM's in describing a wide variety of physical phenomena. In particular, we have discussed how the asymptotic behaviour of the systems described by BM's can be inferred from the properties of the map, how the results about continuous systems can be recovered from the study of their discrete counterparts, and, in some cases, we have given a detailed analysis of the global dynamics.

ACKNOWLEDGEMENTS

The author wants to thank Prof. L. Accardi, Director of the "Centro Matematico V. Volterra", University of Rome II, for having suggested the writing of this paper and for the hospitality of the Center. The author is indebted to Prof. A. Frigerio, for enlightening discussions on the topic of this work, and to Prof. R.F. Streater, for many inspiring discussions and for critical reading of the manuscript. Thanks are in order to the anonymous referees for useful comments, and to Prof. P.F. Zweifel, Director of the "Center for Transport Theory and Mathematical Physics", Virginia Polytechnic Institute and State University, where this work has been partly carried out.

Partial support from GNFM-CNR is gratefully acknowledged.

REFERENCES

[1] R.F. Streater *Convergence of the Iterated Boltzmann Map* Publ. RIMS, Kyoto University, **20**, 913 (1984)

[2] R.F. Streater *Convergence of the Quantum Boltzmann Map* Comm. Math. Phys. **98**, 177 (1985)

[3] R.F. Streater *A Boltzmann Map for Quantum Oscillators* J.Stat. Phys. **48**, 753 (1987)

[4] P.M. Alberti and A. Uhlmann *Stochasticity and Partial Order* D. Reidel Pub. co., Dordrecht (1982)

[5] I. Prigogine *Evolution Criteria, Variational Properties and Fluctuations.* In *Nonequilibrium Thermodynamics, Variational Techniques and Stability* R.J. Donnelly, R. Herman and I. Prigogine (Eds.), University of Chicago Press (1965)

[6] B. Crell and A. Uhlmann *An Example of a Non-Linear Evolution Equation Showing "Chaos - Enhancement"* Letters Math. Phys. **3** 463 (1979)

[7] L. Rondoni and R.F. Streater *Chemical Reactions as Dynamical Systems on the Interval* J. Stat. Phys. **66**(5/6), 1557 (1992)

[8] L. Rondoni *A Stochastic Treatment of Reaction and Diffusion* Ph.D. Thesis, Virginia Tech, August 1991

[9] L. Rondoni *Autocatalytic Reactions as Dynamical Systems on the Interval* (submitted)

[10] L. Rondoni *Complex Chemical Reactions: a Probabilistic Approach* (to appear in NUc. Sci. Eng.)

[11] R.F. Streater *The F-Theorem for Stochastic models* (to appear in Annals of Physics)

[12] R.F. Streater *Stochastic Models of Cotransport* (to appear in Transport Theory and Stat. Phys.)

[13] A. Frigerio, C. Aratari *Unitary Dilation of a Nonlinear Boltzmann Equation* in *Quantum Probability and Applications IV*, L. Accardi and W. Waldenfels (Eds.), Sringer Verlag, New York (1989)

[14] R.F. Snider J. Chem. Phys. **32II** (1960) 1051

[15] O.E. Lanford and D.W. Robinson *Approach to Equilibrium of Free Quantum Systems* Comm. Math. Phys. **24**, 193 (1972)

[16] E.H. Wichmann *Density Matrices Arising from Incomplete Measurements* J. Math. Phys. **4**, 884 (1963)

[17] A.G. Shuhov and Yu. M. Suhov *Ergodic Properties of Groups of Bogoliubov Transformation of CAR C^*-algebras* Ann. of Phys. **175**, 231 (1987)

[18] R.F. Streater *Entropy and the Central Limit Theorem in Quantum Mechanics* J. Phys. A. Math. Gen. **20**, 4321 (1987)

[19] R.F. Streater *The Boltzmann Equation for Discrete Systems* in *Statistical Mechanics* A. Solomon (Ed.), World Scientific

[20] L. Rondoni *A Quantum Probability Theory for the N-Level Atom* (in this volume)

[21] N.M. Hugenholtz *Derivation of the Boltzmann Equation for a Fermi Gas* J. Stat. Phys. **32**(2), 231 (1983)

[22] S. Koseki, Ph.D. Thesis, King's College, London (1992)

[23] R.F. Streater, *Cold Fusion* King's College, preprint

[24] H.G. Heuser *Functional Analysis*, John Wiley and Sons, New York (1982)

[25] R. Alicki and J. Messer *Nonlinear Quantum Dynamical Semigroups for Many-Body Open Systems* J. Stat. Phys. **32**(2), 299 (1983)

[26] N.G. Duffield and R.F. Werner *Mean Field Dynamical Semigroups on C^*-Algebras*, DIAS-STP-90-13

* Current address: Department of Theoretical Physics, School of Physics, The University of New South Wales, Kensington, NSW 2033, Australia

Quantum Probability and Related Topics Vol. VIII (pp. 329–346)

WEAK COUPLING AND LOW DENSITY LIMITS IN TERMS OF SQUEEZED VECTORS

Sławomir Rudnicki, Sławomir Sadowski and Robert Alicki

Institute of Theoretical Physics and Astrophysics
University of Gdańsk
Wita Stwosza 57, PL-80-952 Gdańsk, Poland

Abstract

The weak coupling limit for a quantum system interacting with a free Bose or Fermi gas as well as the low density limit for the similar models with bilinear interactions are considered. In both cases the choice of collective squeezed vectors as reference states leads to the proper unitary limit dynamics driven by the quantum Brownian or Poisson processes respectively. The proof of the associated limit theorem is given for the case of weak coupling limit while the low density limit is treated in a separate paper.

1. INTRODUCTION

The investigation of limit theorems for quantum evolution initiated by Accardi, Lu and Frigerio [1, 2] and continued by the authors of the present paper also provide a rigorous description of the influence of large quantum reservoirs on small quantum systems in terms of *quantum noises*. The main idea is to consider the unitary dynamics of the composite system (small system + reservoir) in the interaction picture $\mathcal{U}_t^\lambda$ which depends on the constant $\lambda > 0$ describing the magnitude of a reservoir's influence on the system. The limit theorems state the convergence $\mathcal{U}_{t/\lambda^2}^\lambda \overset{\lambda\to 0}{\longrightarrow} \mathcal{U}_t$ in the sense of matrix elements constructed in terms of certain reference vectors from the Hilbert space of the reservoir. The limit dynamics $\mathcal{U}_t$ should be a unitary solution of the suitable quantum stochastic differential equation in the sense of Hudson and Parthasarathy [3, 4] driven by quantum Brownian or Poisson processes.

After the first successful implementation of this scheme for the case of weak coupling limit for a system linearly coupled to Bose field and coherent states as reference vectors [1] one has realized [5, 6] that the existence of unitary limit $\mathcal{U}_t$ depends essentially on the choice of reference vectors. Namely for the interactions bilinear in fields the use of coherent vectors leads to an unphysical limit dynamics [5].

The aim of this paper is to show that the *collective squeezed vectors* are proper reference vectors for bilinear interaction both for the weak coupling and low density limits. We give the proof of the limit theorem for the former case while the later is studied in details in ref. [6].

It is worthwhile to notice that the different and independent approach by Accardi and Lu who used *collective number vectors* as reference states in the case of weak coupling limit [7] as well as in low density limit [2] (in the Bose case only) gives the same physical solution, i.e. the shapes of the quantum stochastic differential equations describing the limit dynamics $\mathcal{U}_t$ are the same. Moreover it gives the same reduced dynamics for small system. Nevertheless the *collective squeezed vectors* have more clear physical interpretation, furthermore they allowed us to extend our results on the Fermi case too. The combined results suggest the following hypothesis: if for two different sets of reference states the limit dynamics are unitary then they are essentially the same. Finally one should mention that the limit theorems discussed here are essential extensions of the Markovian limit procedures [8, 9, 10] performed for the reduced dynamics of the open system. One assumed there as an initial state the tensor product of the fixed state of the reservoir and an arbitrary state of the small system while the description based on the notion of quantum noise admits more complicated and physically interesting initial conditions too. This formalism has found already applications in quantum optics and measurement theory [11, 12].

2. MAIN IDEA

In this chapter we present very briefly the main idea of the problem which is considered in this paper.

Let $\mathcal{H}_S$ and $\mathcal{H}_R$ denotes Hilbert spaces for the system S and the reservoir R respectively. Let $\mathcal{U}_t^\lambda$ denotes one-parameter family of unitary operators describing the time evolution of the whole system S + R in the interaction picture. Our aim is to find a certain set of vectors $\{\Psi\} \subset \mathcal{H}_R$ and a certain nonlinear averaging map $\Psi \longrightarrow P_\lambda \Psi$ and for each $P_\lambda \Psi$ to assign explicitly one vector $\Phi \in \Gamma_B(\mathcal{L}^2(\mathbf{R}, dt) \otimes \mathcal{H}_N)$ (for a certain Hilbert noise space $\mathcal{H}_N$) such that:

$$< u \otimes P_\lambda \Psi \quad , \quad \mathcal{U}^\lambda_{t/\lambda^2} \quad u' \otimes P_\lambda \Psi' > \quad \stackrel{\lambda \to 0}{\longrightarrow} \quad < u \otimes \Phi \quad , \quad \mathcal{U}_t \quad u' \otimes \Phi' > \quad , \tag{2.1}$$

where $u, u' \in \mathcal{H}_S$ and $\mathcal{U}_t$ is a solution of a certain *quantum stochastic differential equation* (QSDE).

It was already mentioned that in the limit $\lambda \longrightarrow 0$ the time evolution of the system and reservoir S + R is governed by QSDE. Those equations are driven by quantum Brownian motion or quantum Poisson process in a weak coupling limit or a low density limit respectively. Now we briefly present the ideas of these two types of quantum equations.

Let $\mathcal{U}_t$ is an operator on $\mathcal{H}_S \otimes \Gamma_B(\mathcal{L}^2(\mathbf{R}, dt) \otimes \mathcal{H}_N)$ where $\mathcal{H}_S$ and $\mathcal{H}_N$ are two Hilbert spaces and $\Gamma_B(\cdot)$ denotes boson-Fock space.

Definition 2.1 Let $\{D_\alpha\}$ be a family of bounded operators on $\mathcal{H}_S$ and $\{g_\alpha\}$ be a set of vectors $g_\alpha \in \mathcal{H}_N$. Then the following equation is called *QSDE driven by Brownian motion*:

$$\begin{aligned} d\mathcal{U}_t &= \sum_\alpha \{ \; D^*_\alpha \otimes dA_t(g_\alpha) \; - \; D_\alpha \otimes dA^*_t(g_\alpha) \; - \; D^*_\alpha D_\alpha \, \|g_\alpha\|^2_- \, dt \; \} \, \mathcal{U}_t \quad , \\ \mathcal{U}_0 &= \mathbf{1} \quad , \end{aligned} \tag{2.2}$$

where $A_t(g) \equiv A(\chi_{[0,t)} \otimes g)$ is an annihilation operator and the factor $\|g\|^2_-$ has the property: $2\Re(\| \cdot \|^2_-) = < \cdot, \cdot >$ (cf. (3.18), (3.26)). Solution of this equation exists [3] and one can easily check that it is **unitary**.

Definition 2.2 Let $a \in \mathcal{B}(\mathcal{H}_S)$, $b \in \mathcal{B}(\mathcal{H}_N)$ are bounded operators, $\eta \in \mathcal{H}_N$ and $\chi_{[0,t)}$ is a characteristic function of the interval $[0,t)$ which can be treated either as an operator on $\mathcal{L}^2(\mathbf{R},dt)$ or a function in $\mathcal{L}^2(\mathbf{R},dt)$. Then the *quantum Poisson noise* is a family of operators on $\mathcal{H}_S \otimes \Gamma_B(\mathcal{L}^2(\mathbf{R},dt)\otimes\mathcal{H}_N)$ defined as follows:

$$\begin{aligned} \mathcal{N}_t(a\otimes b,\eta) \;&=\; a\otimes A_t^*(b\eta) \;+\; a\otimes\Lambda_t(b)+ \\ &+\; a\otimes A_t(b^*\eta) \;+\; a<\eta, b\eta>t \quad , \end{aligned} \tag{2.3}$$

where $\Lambda_t(b) \equiv \Lambda(\chi_{[0,t)}\otimes b)$ is a preservation operator [3]. By linearity $\mathcal{N}_t$ can be extended on all bounded operators on $\mathcal{B}(\mathcal{H}_S)\otimes\mathcal{B}(\mathcal{H}_N)$.

The following equation is called *QSDE driven by quantum Poisson noise*:

$$\begin{aligned} d\mathcal{U}_t \;&=\; d\mathcal{N}_t(\mathbf{S}-\mathbf{1},\eta)\,\mathcal{U}_t \quad , \\ \mathcal{U}_0 \;&=\; \mathbf{1} \quad , \end{aligned} \tag{2.4}$$

where $\mathbf{S}$ is an operator on $\mathcal{B}(\mathcal{H}_S)\otimes\mathcal{B}(\mathcal{H}_N)$ Solution of this equation exists [13] and one can easily check that for unitary $\mathbf{S}$ (e.g. scattering matrix) U_t is also **unitary**.

3. WEAK COUPLING LIMIT

Let us assume that reservoir R consists of noninteracting bosons or fermions in a quasi-free equilibrium state $\omega_{\beta,z}$ (β is an inverse temperature, z is a fugacity). One-particle is described by a bounded from below Hamiltonian H_1 on the one-particle Hilbert space $\mathcal{H}_1$. $\mathcal{H}_S$ is a Hilbert space of the system S and H_R is a second quantized H_1 H_S is a Hamiltonian of the system S on the Hilbert space $\mathcal{H}_S$. Let us assume that the *interaction* between system S and reservoir R is given by:

$$V \;=\; \lambda\, D\otimes A(g^0)\,A(g^1) \;+\; \text{h.c.} \quad , \tag{3.1}$$

where $D \in \mathcal{B}(\mathcal{H}_S)$, $g^0, g^1 \in \mathcal{H}_1$ and $A(\cdot)$, $A^*(\cdot)$ satisfy CCR or CAR. We also assume rotating wave approximation:

$$e^{itH_S}\,D\,e^{-itH_S} \;=\; e^{-i\omega t}\,D \quad , \tag{3.2}$$

and

$$< g^0\,,\, e^{itH_1}\,g^1 >= 0 \quad \text{for} \quad \text{all} \quad t\in\mathbf{R} \quad , \tag{3.3}$$

i.e. g^0 and g^1 have disjoint energy spectra. The reservoir can be represented in terms of the Fock space [14]:

$$\Gamma(\mathcal{H}_1\oplus\mathcal{H}_1) \;=\; \Gamma(\mathcal{H}_1)\otimes\Gamma(\mathcal{H}_1) \tag{3.4}$$

and the correspondence:

$$A(f) \longrightarrow a(\sqrt{1\pm T}f) \;+\; b^*(J\sqrt{T}f) \;\equiv\; A_{\beta,z}(f) \quad , \tag{3.5}$$

$$\omega_{\beta,z}\big(A^{\#}(f_1)\cdots A^{\#}(f_n)\big) \;=< \Omega\,,\, A^{\#}_{\beta,z}(f_1)\cdots A^{\#}_{\beta,z}(f_n)\,\Omega > \quad , \tag{3.6}$$

where $A^{\#}$, $A^{\#}_{\beta,z}$ denote A, A^{*}, $A_{\beta,z}$, $A^{*}_{\beta,z}$ respectively, Ω is a vacuum state on $\Gamma(\mathcal{H}_1 \oplus \mathcal{H}_1)$. J is an involution operator commuting with H_1 (for each self-adjoint operator such J can be chosen). Here:

$$T : \mathcal{H}_1 \longrightarrow \mathcal{H}_1 \quad ,$$

$$T = z\,e^{-\beta H_1}\left(1 \mp z\,e^{-\beta H_1}\right)^{-1} \tag{3.7}$$

We have denoted by $a(f)$, $a^{*}(f)$ and $b(g)$, $b^{*}(g)$ the annihilation and creation operators on $\Gamma(\mathcal{H}_1 \oplus \mathcal{H}_1)$ labelled by vectors $f \oplus 0$ and $0 \oplus g$ respectively. In all formulae with double sign $\pm$ or $\mp$ the upper one corresponds to the Bose and the lower one to the Fermi statistics.

Now we define reference vectors. Let M_i $i = 1, 2$ are Hilbert-Schmidt operators on $\mathcal{H}_1$. Without lost of generality one can assume the symmetry property $M_i^{*} = \pm J M_i J$ Then the following operator exists and is unitary [15]:

$$\mathcal{B}(M_1, M_2) \equiv$$

$$\exp\left\{-\frac{1}{2}\sum_{i,j}\left[< h_i\,,\, M_1\, h_j > a^{*}(h_i)a^{*}(Jh_j) + < h_i\,,\, M_2\, h_j > b^{*}(h_i)b^{*}(Jh_j)\right] - \text{h.c.}\right\}, \tag{3.8}$$

where $\{h_i\}$ is an orthonormal basis in $\mathcal{H}_1$.

Using these operators one can define the *Bogoliubov transformation* :

$$\mathcal{B}(M_1, M_2)\, a(f)\, \mathcal{B}^{*}(M_1, M_2) = a\big(c(M_1)\, f\big) + a^{*}\big(J\, s(M_1)\, f\big) \quad , \tag{3.9}$$

$$\mathcal{B}(M_1, M_2)\, b(f)\, \mathcal{B}^{*}(M_1, M_2) = b\big(c(M_2)\, f\big) + b^{*}\big(J\, s(M_2)\, f\big) \quad , \tag{3.10}$$

where:

$$s(M) \equiv \sum_{n=0}^{\infty} (\pm 1)^n \,\frac{M^{*}\,(M\,M^{*})^n}{(2n+1)!} \quad , \tag{3.11}$$

$$c(M) \equiv \sum_{n=0}^{\infty} (\pm 1)^n \,\frac{(M\,M^{*})^n}{(2n)!} \tag{3.12}$$

Definition 3.1 The following vectors are called *squeezed vectors* :

$$\Psi(M_1, M_2) \equiv \mathcal{B}^{*}(M_1, M_2)\,\Omega \tag{3.13}$$

Using above properties one can easily check that the following state:

$$\omega_{M_1, M_2}\big(A^{\#}(f_1)\cdots A^{\#}(f_n)\big) \equiv < \Psi(M_1, M_2) \quad , \quad A^{\#}_{\beta,z}(f_1)\cdots A^{\#}_{\beta,z}(f_n)\; \Psi(M_1, M_2) >$$

is a quasi-free state being a perturbation of state $\omega_{\beta,z}$, hence it gives us motivation to choose squeezed vectors as reference vectors.

By analogy to refs. [1, 2, 5] we can define the averaging map P_λ :

$$P_\lambda \Psi(M_1,M_2) \equiv \Psi(M_1^\lambda,M_2^\lambda) \quad , \tag{3.14}$$

where

$$M_i^\lambda \equiv \lambda \int_{Q_i/\lambda^2}^{R_i/\lambda^2} dt\, S_t\, M_i\, S_t \qquad , \quad R_i\,,Q_i \in \mathbf{R} \tag{3.15}$$

and

$$S_t = e^{it(H_1+\omega/2)} \tag{3.16}$$

is a modified one-particle evolution in the reservoir; ω is defined in (3.2). One can easily check that:

$$e^{it(H_S+H_R)}\, V\, e^{-it(H_S+H_R)} = e^{it\tilde{H}_R}\, V\, e^{-it\tilde{H}_R} \quad , \tag{3.17}$$

where $\tilde{H}_R$ is a second quantization of the operator $H_1 + \omega/2$. It is obvious that both operators $H_S + H_R$ and $\tilde{H}_R$ give the same evolution in the interaction picture, so we shall use the second one. The vectors $P_\lambda\, \Psi(M_1,M_2)$ we shall name the *collective squeezed vectors.*

According to the scheme presented at the beginning of this chapter we should find a noise space $\mathcal{H}_N$ and then assign each $P_\lambda\Psi(M_1,M_2)$ with the one vector from $\Gamma_B(\mathcal{L}^2(\mathbf{R},dt)\otimes\mathcal{H}_N)$. In order to do this we introduce a subspace $\mathcal{L}$ which consists of operators satisfying the following condition:

$$(M\,,\,M') \equiv \int_{\mathbf{R}} \mathrm{Tr}(S_u\, M^*\, S_u\, M')du < \infty \quad \text{for all} \quad M,M' \in \mathcal{L} \quad . \tag{3.18}$$

Definition 3.2 The *noise space* $\mathcal{H}_N$ is equal to $\mathcal{H}'_N \oplus \mathcal{H}'_N$, where $\mathcal{H}'_N$ is a completion of the quotient of $\mathcal{L}$ with respect to $(\cdot,\cdot)$-norm defined by (3.18).

Now in order to formulate the final theorem we need a few technical assumption. Let $\mathcal{L}_0 \subset \mathcal{L}$ is a set of finite-range operators:

$$M = \sum_{i=1}^{N} |\, f_i >< g_j\,| \quad , \tag{3.19}$$

for $f_j\,,\,g_j \in \mathcal{K} \subset \mathcal{H}_1$, $\mathcal{K}$ is defined as follows:

$$\forall_{f,g\in\mathcal{K}} \qquad \int_{\mathbf{R}} |< f\,,\,S_u g >|\,du < \infty \qquad \text{and} \qquad \sqrt{T}\mathcal{K} \subset \mathcal{K} \quad , \quad \sqrt{1\pm T}\mathcal{K} \subset \mathcal{K} \quad . \tag{3.20}$$

We also assume that $g^0, g^1 \in \mathcal{K}$ (cf. (3.1)).

Let us introduce the notation: D^ϵ, $\epsilon = 0,1$ where $D^0 \equiv D$, $D^1 \equiv D^*$.

Theorem 3.3

With the notation as above for $M_i, M_i' \in \mathcal{L}_0$ $i = 1,2$ *(which satisfy some technical assumption described in Proposition 5.5) there exists* $t_0 > 0$ *such that for each* $u, u' \in \mathcal{H}_S$ *and for all* $t \in [0, t_0)$ *or* $t \in \mathbf{R}_+$ *in the Bose or Fermi case respectively and finite* $Q_i, Q_i', R_i, R_i' \in \mathbf{R}$ $i = 1,2$:

$$\lim_{\lambda \to 0} < u \otimes P_\lambda \Psi(M_1, M_2) \quad , \quad \mathcal{U}^\lambda_{t/\lambda^2} \quad u' \otimes P_\lambda \Psi(M_1', M_2') > \tag{3.21}$$

exists and is equal to

$$< u \otimes \Phi(\chi_{[Q_1,R_1]} M_1 \oplus \chi_{[-R_2,-Q_2]} M_2) \quad , \quad \mathcal{U}_t \quad u' \otimes \Phi(\chi_{[Q_1',R_1']} M_1' \oplus \chi_{[-R_2',-Q_2']} M_2') > \quad , \tag{3.22}$$

where $\mathcal{U}_t$ *is the unitary solution of QSDE:*

$$d\mathcal{U}_t = \sum_{\epsilon=0,1} \{ D^\epsilon \otimes dA_t^*(G^\epsilon) - D^{1-\epsilon} \otimes dA_t(G^\epsilon) - \|G^\epsilon\|_-^2 \, D^\epsilon D^{1-\epsilon} \otimes \mathbf{1} dt \} \mathcal{U}_t \quad , \tag{3.23}$$

$$\mathcal{U}_0 = \mathbf{1} \quad ,$$

where:

$\Phi(\cdot)$ *denotes a coherent state normalized to* 1 *in the Fock space* $\Gamma(\mathcal{L}^2(\mathbf{R}, dt) \otimes \mathcal{H}_N)$,

$$G^0 = -i\,(0 \oplus |J\sqrt{T}\, g^1 >< \sqrt{T}\, g^0|) \quad , \tag{3.24}$$

$$G^1 = -i\,(|\sqrt{1 \pm T}\, g^1 >< J\sqrt{1 \pm T}\, g^0| \oplus 0) \quad , \tag{3.25}$$

$$\|M \oplus M'\|_-^2 \equiv \|M\|_-^2 + \|M'\|_-^2 \qquad for \quad M, M' \in \mathcal{H}'_N \quad ,$$

$$\|M\|_-^2 \equiv \int_{-\infty}^0 dt \, \mathrm{Tr}\{S_t M^* S_t^* M\} \tag{3.26}$$

In the above theorem we have used the same symbols for elements of $\mathcal{L}_0 \oplus \mathcal{L}_0$ and $\mathcal{H}_N$ It is obvious that any element $\mathcal{L}_0 \oplus \mathcal{L}_0$ defines an element in the Hilbert space $\mathcal{H}_N$, so such notation should not be confusing.

For the proof see the chapter Main Calculus.

4. LOW DENSITY LIMIT

The low density limit was described in all details in ref. [6], so now we present only an outline of this problem.

We assume that the interaction between system S and reservoir R is given by:

$$V = D \otimes A^*(g^0) A(g^1) + D^* \otimes A^*(g^1) A(g^0) \quad , \tag{4.1}$$

where $A(\cdot)$ and $A^*(\cdot)$ satisfy CCR or CAR and $D \in \mathcal{B}(\mathcal{H}_S)$ $g^0, g^1 \in \mathcal{H}_1$. As in the weak coupling limit case, we also assume the rotating wave approximation (3.2) as well as that g_0 and g_1 have disjoint energy spectra (3.3). We use the same representation for this system as in the weak coupling limit case.

Let M is a Hilbert-Schmidt operator on $\mathcal{H}_1$ then the following unitary operator exists:

$$\mathcal{B}(M) = \exp\Big\{ -\sum_{i,j} < h_i, M\, h_j > a^*(h_i) b^*(Jh_j) - \text{h.c.} \Big\} \quad , \tag{4.2}$$

where $\{h_i\}$ is an orthonormal basis in $\mathcal{H}_1$ Such operators also define the *Bogoliubov transformation* (cf. [15]). By analogy to the previous case the following vectors we shall call *squeezed vectors*:

$$\Psi(M) \equiv \mathcal{B}^*(M)\Omega \quad , \tag{4.3}$$

where Ω is a vacuum state on $\Gamma(\mathcal{H}_1 \oplus \mathcal{H}_1)$ The states define by the above vectors:

$$\omega_M(\cdot) \equiv < \Psi(M) \, , \, \cdot \, \Psi(M) > \tag{4.4}$$

are quasi-free and gauge invariant and are perturbations of the equilibrium state $\omega_{\beta,z}$ (3.6).

The averaging map P_z is defined as follows:

$$P_z \Psi(M) = \Psi\Big(\sqrt{z} \int_{Q/z}^{R/z} dt\; S_t\, M\, S_t^* \Big) \quad . \tag{4.5}$$

One should notice that in the low density limit we choose as a magnitude of reservoir's influence on the system S the fugacity z (3.6) which is proportional (for small z) to the density of the reservoir. In the above definition S_t is a modified one-particle reservoir's evolution:

$$S_t \equiv e^{it(H_1 + \omega P_0)} \quad , \tag{4.6}$$

where P_0 is a projector commuting with $\mathcal{H}_1$ such that $P_0\, g^0 = g^0$ and $P_0\, g^1 = 0$. Existence of such projector is a consequence of the assumption about disjoint energy spectra of g^0 and g^1

Let $\mathcal{L}$ is a linear subspace of the space of Hilbert-Schmidt operators consists of operators which satisfy the following condition:

$$(M \, , \, M') \equiv \int_{\mathbf{R}} \text{Tr}(\, S_t\, M^*\, S_t^*\, M'\,) < \infty \quad \text{for all } M, M' \in \mathcal{L} \tag{4.7}$$

Definition 4.1 The *noise space* $\mathcal{H}_N$ we define as a completion of the quotient of $\mathcal{L}$ with respect to $(\cdot,\cdot)$-norm defined above.

The final theorem is proved for a certain subset of the set of squeezed vectors, indexed by a subset $\mathcal{L}_0$ of the space $\mathcal{L}$. $\mathcal{L}_0$. consists of finite-range operators:

$$M = \sum_{i=1}^{N} |f_i >< g_i| \quad f, g \in \mathcal{H}_1 \quad ,$$

where f, g satisfy the condition (3.20). We have to assume also that two families of t-functions parameterized by fugacity z:

$$|< \sqrt{T/z}\, f \,,\, S_t\, g >| \quad \text{and} \quad |< \sqrt{1 \pm T}\, f \,,\, S_t\, g >|$$

for all z small enough are majorized by a function which belongs to $\mathcal{L}^1(\mathbf{R})$ We should also assume that:

$$4\|D\| \int_{-\infty}^{0} dt\, |< g^\epsilon \,,\, S_t\, g^\epsilon >| \;<\; 1 \quad \text{for} \quad \epsilon = 0,1$$

In order to make the formulae in the following theorem simpler we introduce the notation:

$$\Pi(M) \;\equiv\; \int_{\mathbf{R}} du\; S_u\, M\, S_u^*$$

One can easily check that the next formulae in which the above notation is used are well defined.

Theorem 4.2
If $M, M' \in \mathcal{L}_0$ *and* t *satisfy the following condition:*

$$\begin{cases} C_1\, \|D\|\, t \,+\, C_2 \,<\, 1 & \textit{for bosons,} \\ C_3 \,<\, 1 & \textit{for fermions,} \end{cases} \tag{4.8}$$

where

$$\begin{aligned} C_1 \;=\; & \max_{\epsilon,\epsilon'=0,1}\; 4\|\Pi(M^* - M'^*)\, g^\epsilon\|\; \|\Pi(\xi \,+\, M'^*)\, g'^\epsilon\| \,+ \\ & +\; \max_{\epsilon,\epsilon'=0,1}\; 4\|\Pi(M^* - M'^*)\, \Pi(\xi \,+\, M'^*)\, g^\epsilon\|\; \|g'^\epsilon\| \,+ \\ & +\; \max_{\epsilon=0,1}\; 4\, \|\Pi(\xi \,+\, M'^*)\, g^\epsilon\|^2 \quad , \\ C_2 \;=\; & 4\|D\| \max_{\epsilon=0,1} \int_{-\infty}^{0} dt\, |< g^\epsilon \,,\, S_t\, g^\epsilon \;>| \quad , \\ C_3 \;=\; & \frac{5}{2}\, (C_1 + C_2) \quad , \end{aligned} \tag{4.9}$$

where $\xi \in \mathcal{H}_N$ *and is defined as follows:*

$$\xi \;=\; \sum_{\epsilon=0,1} |g_\beta^\epsilon >< g_\beta^\epsilon| \quad , \tag{4.10}$$

with:

$$g_\beta^\epsilon \;\equiv\; \int_{\mathrm{supp}\|g^\epsilon\|(E)} \frac{e^{-\beta E/4}}{\|g^\epsilon\|(E)}\; P(dE)\, g^\epsilon$$

where $P(dE)$ *is the spectral measure of* H_1 *and*

$$[\|g^\epsilon\|(E)]^2 = \frac{< g^\epsilon , P(dE) g^\epsilon >}{dE}$$

is a Radon-Nikodym derivative which exists due to (3.20). Then for all $u, u' \in \mathcal{H}_S$ *and finite* $Q, Q', R, R' \in \mathbf{R}$ *:*

$$\lim_{z \to 0} < u \otimes P_z \Psi(M) \quad , \quad \mathcal{U}^z_{t/z} \quad u' \otimes P_z \Psi(M') > \tag{4.11}$$

exists and is equal to

$$< u \otimes \Phi(\chi_{[Q,R]} M) \quad , \quad \mathcal{U}_t \quad u' \otimes \Phi(\chi_{[Q',R']} M') > \quad , \tag{4.12}$$

where $\mathcal{U}_t$ *is a unitary solution of QSDE driven by quantum Poisson noise:*

$$\begin{aligned} d\mathcal{U}_t &= d\mathcal{N}_t(\mathbf{S} - \mathbf{1}, \eta) \, \mathcal{U}_t \\ \mathcal{U}_0 &= \mathbf{1} \quad , \end{aligned} \tag{4.13}$$

where $\mathbf{S}$ *is a unitary operator on* $\mathcal{H}_S \otimes \mathcal{H}_N$ *associated with a scattering matrix* $\mathbf{S}_1$ *on* $\mathcal{H}_S \otimes \mathcal{H}_1$ *defined in terms of the operators*

$$\begin{cases} H_S \otimes 1 + 1 \otimes H_1 \quad , \\ H_S \otimes 1 + 1 \otimes H_1 + V_1 \qquad (V_1 = D \otimes |g^0 >< g^1| + \text{h.c.}) \, . \end{cases}$$

Namely, let $M, M' \in \mathcal{L}_0$ *(cf. (3.19)),* $u, u' \in \mathcal{H}_S$ *:*

$$< u \otimes M \quad , \quad \mathbf{S}\, u' \otimes M' >_{\mathcal{H}_S \otimes \mathcal{H}_N} \equiv \sum_{i=1}^{n} \sum_{j=1}^{m} \int_{\mathbf{R}} dt < g'_j , S_t g_i >< u \otimes S_t f_i , \mathbf{S}_1 u' \otimes f'_j > \, .$$

Φ *is a coherent state normalized to* 1 *in the Fock space* $\Gamma(\mathcal{L}^2(\mathbf{R}, dt) \otimes \mathcal{H}_N)$

5. MAIN CALCULUS

This chapter refers to the case of the weak coupling limit.

Lemma 5.1

Let $M_i, M'_i \in \mathcal{L}_0$, $i = 1,2$ *, then:*

$$\lim_{\lambda \to 0} < u \otimes P_\lambda \Psi(M_1, M_2) \quad , \quad u' \otimes P_\lambda \Psi(M'_1, M'_2) > =$$
$$= < u \otimes \Phi\big(\chi_{[Q_1,R_1]} M_1 \oplus \chi_{[-R_2,-Q_2]} M_2\big) \quad , \quad u' \otimes \Phi\big(\chi_{[Q'_1,R'_1]} M'_1 \oplus \chi_{[-R'_2,-Q'_2]} M'_2\big) > \, . \tag{5.1}$$

Proof One can check that (cf. ref. [15] ; §4 th. 3 and §5 th. 2 and discussion after it):

$$
\begin{aligned}
&< P_\lambda \Psi(M_1, M_2) \quad , \quad P_\lambda \Psi(M_1', M_2') > = \\
&= < \Omega \quad , \quad \mathcal{B}(M_1^\lambda, M_2^\lambda)\, \mathcal{B}^*(M_1'^\lambda, M_2'^\lambda)\, \Omega > = < \Omega \ , \ \tilde{\mathcal{B}}\, \Omega > = \\
&= \left[\det\left(c(M_1^\lambda, M_1'^\lambda)\, c^*(M_1^\lambda, M_1'^\lambda)\right) \ \det\left(c(M_2^\lambda, M_2'^\lambda)\, c^*(M_2^\lambda, M_2'^\lambda)\right)\right]^{\mp 1/2} \quad ,
\end{aligned} \tag{5.2}
$$

where $\tilde{\mathcal{B}}$ is a Bogoliubov transformation for which the corresponding functions $c(\cdot)$ and $s(\cdot)$ (cf. (3.9), (3.10)) are equal:

$$c(M_i^\lambda, M_i'^\lambda) = c(M_i^\lambda)\, c(M_i'^\lambda) \mp s(M_i^{\lambda *})\, s(M_i'^\lambda) \quad , \tag{5.3}$$

$$s(M_i^\lambda, M_i'^\lambda) = s(M_i^\lambda)\, c(M_i'^\lambda) - c(M_i^{\lambda *})\, s(M_i'^\lambda) \quad , \tag{5.4}$$

$\det(\cdot)$ denotes the Fredholm determinant (see eg. [16] or any other handbook on integral equations). One can easily check that:

$$
\begin{aligned}
&c(M_i^\lambda, M_i'^\lambda)\, c^*(M_i^\lambda, M_i'^\lambda) = \\
&= \exp\{\pm[M_i^\lambda M_i^{\lambda *} + M_i'^\lambda M_i'^{\lambda *} - M_i^\lambda M_i'^{\lambda *} - M_i'^\lambda M_i^{\lambda *}]\} + \delta(M_i^\lambda, M_i'^\lambda) \equiv \\
&\equiv e^{F_i^\lambda} + \delta_i^\lambda
\end{aligned} \tag{5.5}
$$

and F_i^λ , δ^λ are self-adjoint, trace-class operators. Hence:

$$\det\left(c(M_i^\lambda, M_i'^\lambda)\, c^*(M_i^\lambda, M_i'^\lambda)\right) = e^{\frac{1}{2}\mathrm{Tr}\, F_i^\lambda} \det\left[1 + e^{-\frac{1}{2}F_i^\lambda}\, \delta_i^\lambda\, e^{-\frac{1}{2}F_i^\lambda}\right] e^{\frac{1}{2}\mathrm{Tr}\, F_i^\lambda} \tag{5.6}$$

One can check, using the same method as in the Lemma 5.2 that:

$$\limsup_{\lambda \to 0} \|e^{-\frac{1}{2}F_i^\lambda}\| < \infty \quad ,$$

$$\lim_{\lambda \to 0} \mathrm{Tr}\, |\delta_i^\lambda| = 0 \quad .$$

Therefore there exists $\lambda_0 > 0$ such that for all $\lambda < \lambda_0$

$$0 \leq \mathrm{Tr}\, |e^{-\frac{1}{2}F_i^\lambda}\, \delta_i^\lambda\, e^{-\frac{1}{2}F_i^\lambda}| \leq \|e^{-\frac{1}{2}F_i^\lambda}\|^2\, \mathrm{Tr}\, |\delta_i^\lambda| \leq 1$$

Then the following inequalities are satisfied:

$$
\begin{aligned}
1 - \mathrm{Tr}\, |e^{-\frac{1}{2}F_i^\lambda}\, \delta_i^\lambda\, e^{-\frac{1}{2}F_i^\lambda}| &\leq |\det[1 + e^{-\frac{1}{2}F_i^\lambda}\, \delta_i^\lambda\, e^{-\frac{1}{2}F_i^\lambda}]| \leq \\
&\leq \exp\{\mathrm{Tr}\, |e^{-\frac{1}{2}F_i^\lambda}\, \delta_i^\lambda\, e^{-\frac{1}{2}F_i^\lambda}|\}
\end{aligned} \tag{5.7}
$$

So we have:

$$\lim_{\lambda \to 0} \det[1 + e^{-\frac{1}{2}F_i^\lambda}\, \delta_i^\lambda\, e^{-\frac{1}{2}F_i^\lambda}] = 1 \quad . \tag{5.8}$$

From the fact that M_i $i = 1, 2$ are symmetric or antisymmetric operators for the Bose or Fermi case respectively it results:

$$\mathrm{Tr}\,[M_i^\lambda \, M_i'^{\lambda *}] = \mathrm{Tr}\,[J\, M_i^{\lambda *}\, J\, J\, M_i'^\lambda\, J] = \mathrm{Tr}\,[M_i^{\lambda *}\, M_i'^\lambda] \tag{5.9}$$

Now after straightforward calculations using the method from Lemma 5.2, properties of coherent vectors and the shape of scalar product (3.18) we complete the proof.

■

Let us notice that the initial condition of the equation in the main Theorem 3.3 results from the above Lemma.

Let us introduce the notation

$$f_t \equiv S_t f \quad \text{for} \quad f \in \mathcal{H}_1 \quad , \tag{5.10}$$

$$\begin{aligned} T_1 &\equiv \sqrt{1 \pm T} \quad , \\ T_2 &\equiv \sqrt{T} \quad , \end{aligned} \tag{5.11}$$

$$V(t) = e^{it(H_S + H_R)}\, V\, e^{-it(H_S + H_R)}$$

Lemma 5.2

Let set $\mathcal{A} \subset \{1, \cdots, K\}$, $K < \infty$, $N = \mathrm{card}\,\mathcal{A}$ *and let* $\{t_k(M_k^\lambda)\}_{k=1}^K$ *be a series of operators such that:*

$$t_k(M_k^\lambda) = \begin{cases} J\, s(M_k^\lambda) & \text{for } k \in \mathcal{A} \quad , \\ c(M_k^\lambda) & \text{for } k \in \{1, \cdots, K\} \setminus \mathcal{A} \quad , \end{cases} \qquad M_k \in \mathcal{L}_0 \quad , \tag{5.12}$$

and $f, g \in \mathcal{K}$ *, then the following limit exists and is equal to:*

$$\lim_{\lambda \to 0} \lambda^{-N} < g_{s/\lambda^2 + t} \;,\; \prod_{k=1}^{K} t_k(M_k^\lambda)\, f_{s/\lambda^2 + t'} > =$$

$$= \prod_{p \in \mathcal{A}} \left(\chi_{[S_p, T_p]}(s)\right) \int_{\mathbf{R}^N} ds_1 \cdots ds_N < g_{t-t'} \;,\; \prod_{k \in \mathcal{A}} (\, J\, S_{s_k}\, M_k\, S_{s_k}\,) f > \tag{5.13}$$

If $\mathcal{A} = \emptyset$ *, then we mean the product* $\prod_{k \in \mathcal{A}} (\cdot)$ *as equal to* 1.

Proof The proof is exactly the same as Proof 7.1 in ref. [6].

■

Lemma 5.3

Let $M_i, M_i' \in \mathcal{L}_0$, $i = 1, 2$, $n \in \mathbf{N}$ *, then for all* $\Delta > 0$ *there exists* $\delta > 0$ *, such that for all* $\lambda < \delta$ *:*

$$\left| < u \otimes P_\lambda \Psi(M_1, M_2) \,,\, (-i)^n \lambda^n \int_0^{t/\lambda^2} ds_1 \cdots \int_0^{s_{n-1}} ds_n\, V(s_1) \cdots V(s_n)\, u' \otimes P_\lambda \Psi(M_1', M_2') > \right| \leq$$

$$\leq \begin{cases} C_0 \, (C_1 \, \|D\| \, t \, + \, C_2 \, + \, \Delta)^n & \textit{for bosons}, \\ C_0 \, e^{\|D\| \, t} \, (C_3 \, + \, \Delta)^n & \textit{for fermions}, \end{cases} \tag{5.14}$$

where:

$$C_0 \equiv \|u\| \, \|u'\| \quad , \tag{5.15}$$

$$C_1 \equiv (24)^2 \, \|D\| \max_{i=1,2} \max_{\epsilon=0,1} \Big\{ \Big|\big(M - M' \quad , \quad |T_i \, g^\epsilon >< J \, T_i \, g^{1-\epsilon}| \, \big)\Big| \quad ,$$

$$\Big|\big(M' \quad , \quad |T_i \, g^\epsilon >< J \, T_i \, g^{1-\epsilon}| \, \big)\Big| \quad , \quad \|T_i \, g^\epsilon\|^2 \Big\} \quad , \tag{5.16}$$

$$C_2 \equiv \int_{-\infty}^{0} ds \, f(s) \quad , \tag{5.17}$$

$$C_3 \equiv 2 \, C_1 \, + \, C_2 \quad , \tag{5.18}$$

$$f(s) \equiv 2 \cdot (24)^2 \, \|D\| \max_{i=1,2} \max_{\epsilon,\epsilon'=0,1} \; | < T_i \, g^\epsilon \quad , \quad T_i \, g_s^{\epsilon'} > | \quad . \tag{5.19}$$

Proof Because the proof of this lemma is long and rather technical we perform here an outline of it only.

The left-hand side of the equation (5.14) can be estimated by the sum of the following type:

$$\Big| \int_0^{t/\lambda^2} ds_1 \cdots \int_0^{s_{n-1}} ds_n < u \, , \, D^\# \cdots D^\# \, u' >$$

$$< \mathcal{B}^*(M_1^\lambda, M_2^\lambda) \, \Omega \quad , \quad c^\#(\cdot) \cdots c^\#(\cdot) \, \mathcal{B}^*(M_1'^\lambda, M_2'^\lambda) \, \Omega > \Big| \leq$$

$$\leq \|u\| \, \|u'\| \, \|D\|^n \, \Big| \int_0^{t/\lambda^2} ds_1 \cdots \int_0^{s_{n-1}} ds_n$$

$$< \mathcal{B}(M_1'^\lambda, M_2'^\lambda) \, \mathcal{B}^*(M_1^\lambda, M_2^\lambda) \, \Omega \quad , \quad \underbrace{\mathcal{B}(M_1'^\lambda, M_2'^\lambda) \, c^\#(\cdot) \cdots c^\#(\cdot) \, \mathcal{B}^*(M_1'^\lambda, M_2'^\lambda)} \, \Omega > \Big| \quad ,$$

where $c^\#(\cdot)$ denotes $a(\cdot)$, $a^*(\cdot)$, $b(\cdot)$ or $b^*(\cdot)$ operator. The proof can be done in the following steps:

1) We apply the Bogoliubov transformation given by the unitary operator $\mathcal{B}(M_1'^\lambda, M_2'^\lambda)$ (see the marked part above).
2) We bring our expression to the normally ordered form. Because an annihilation operator acting on the vacuum state gives zero, the final formula is equal to the sum of expressions:

$$< \mathcal{B}(M_1'^\lambda, M_2'^\lambda) \, \mathcal{B}^*(M_1^\lambda, M_2^\lambda) \, \Omega \quad , \quad c^*(\cdot) \cdots c^*(\cdot) \, \Omega >$$

   multiplied by the product of scalar products $< f, g >$ where $f, g \in \mathcal{H}_1$
3) We move creation operators $c^*(\cdot)$ to the left-hand side of the scalar product and we repeat the procedure described in steps 1) and 2). Now the Bogoliubov transformation is given by:

$$\mathcal{B}(M_1^\lambda, M_2^\lambda) \, \mathcal{B}^*(M_1'^\lambda, M_2'^\lambda)$$

4) The remaining terms:

$$< c^*(\cdot)\cdots c^*(\cdot)\,\Omega \quad , \quad \mathcal{B}(M_1^\lambda, M_2^\lambda)\,\mathcal{B}^*(M_1'^\lambda, M_2'^\lambda)\,\Omega > \tag{5.20}$$

we estimate by $\lambda^{3k}\, C^k\, k!$ or $\lambda^{3k}\, C^k$ ($2k$ is the number of creation operators) in the Bose or Fermi case respectively.

5) Now we have to make the most difficult step. We are going to apply the Pulè inequality [8] in the following form:

$$\int_0^{t/\lambda^2} ds_1 \cdots \int_0^{s_{n-1}} ds_n \sum_{\sigma\in S_m^1} \prod_{i=1}^{m} h(s_{q_{\sigma(j)}} - s_{p_j}) \leq$$

$$\leq \lambda^{2(m-n)} \frac{t^{n-m}}{(n-m)!} \left[\int_{-\infty}^{0} h(t)\, dt \right]^m ,$$

where $2 \leq q_1 < \cdots < q_m \leq n$, $1 \leq p_1 \leq \cdots \leq p_m \leq n-1$ and $p_j < q_j$, $j = 1, \cdots, m$. S_m^1 is a set of permutation σ of numbers $1, \cdots, m$ such that $p_i \leq q_{\sigma(i)}$, $i = 1, \cdots, m$, and h is a non-negative, integrable and symmetric function. In order to do this we first should notice that each scalar product (cf. step 2)) $| < f, g > |$, $f, g \in \mathcal{H}_1$ can be treated as a function of $s_i - s_j$ for certain i and j . Next we have to find as many as possible such functions which arguments $s_i - s_j$ satisfy the same condition as for arguments $s_{q_{\sigma(j)}} - s_{p_j}$ in the Pulè inequality for certain σ . All such functions we estimate using lemma 5.2 by one non-negative function $h \in \mathcal{L}^1(\mathbf{R})$. We must remember that such expression should be summed up over all permutations σ satisfying the same assumption as in the Pulè inequality. One can check that such summation really takes place in our expression. The sum over all remaining scalar products $< f', g' >$, $f', g' \in \mathcal{H}_1$ is equal to the sum of eight terms of the following shape:

$$| < \Omega \quad , \quad a^\#(\cdot)\cdots a^\#(\cdot)\,\Omega > |$$

where $a^\#(\cdot)$ denotes creation or annihilation operator. Using again Lemma 5.2 we estimate those terms by $\sqrt{k!}\, C^k\, \lambda^p$ or $C^k\, \lambda^p$ ($p > 0$ and C is a constant) in the Bose or Fermi case respectively. Now we apply Pulè inequality.

6) At the end we sum up all terms and we make simple estimations. The final result is of the form (5.14), where the constants depend on λ , but they tend to C_1 , C_2 or C_3 respectively in the limit $\lambda \longrightarrow 0$.

∎

Lemma 5.4
Let $M_i, M_i' \in \mathcal{L}_0$, $i = 1, 2$, $n \in \mathbf{N}$, *then:*

$$\lim_{\lambda\to 0} < u \otimes P_\lambda \Psi(M_1, M_2) \,,\, (-i)^n \lambda^n \int_0^{t/\lambda^2} ds_1 \cdots \int_0^{s_{n-1}} ds_n\, V(s_1)\cdots V(s_n)\, u' \otimes P_\lambda \Psi(M_1', M_2') >$$

exists.

Proof In the detailed proof of the Lemma 5.3 one can notice that the terms containing creation operators (5.20) are of order λ and vanish in the limit $\lambda \to 0$ Hence the non-zero contribution may give the pure scalar products of elements from the Hilbert space $\mathcal{H}_1$ only. The convergence of a very similar expressions has been shown by Accardi, Lu and Frigerio (ref. [1] theorem 5.1).

■

Proposition 5.5

If the constants C_1 , C_2 , C_3 *defined in Lemma 5.3 and time* t *satisfy the condition:*

$$\begin{cases} C_1\,\|D\|\,t + C_2 < 1 & \text{for bosons,} \\ C_3 < 1 & \text{for fermions,} \end{cases} \tag{5.21}$$

then the following limit exists:

$$\lim_{\lambda\to 0} < u \otimes P_\lambda \Psi(M_1, M_2) \quad , \quad \mathcal{U}^\lambda_{t/\lambda^2} \quad u' \otimes P_\lambda \Psi(M'_1, M'_2) > \quad . \tag{5.22}$$

Proof Using the Dyson expansion of $\mathcal{U}^\lambda_{t/\lambda^2}$ and Lemata 5.1, 5.3, 5.4 the proof is immediate.

■

Now we shall find a QSDE which describes the limit evolution of the system.

Let us introduce the notation:

$$< u, \mathcal{D}(t) > \equiv \lim_{\lambda\to 0} \; < u \otimes P_\lambda \Psi(M_1, M_2) \quad , \quad \mathcal{U}^\lambda_{t/\lambda^2}\, u' \otimes P_\lambda \Psi(M'_1, M'_2) > \quad , \tag{5.23}$$

$$\begin{aligned} V^0_R(t) \equiv\ & 1 \otimes \Big\{ a(T_1\, g^0_t)\, a(T_1\, g^1_t) + \\ & + a(T_1\, g^0_t)\, b\big(c(M^\lambda_2)\, J\, s(M^\lambda_2)\, T_2\, g^1_t\big) + b\big(c(M^\lambda_2)\, J\, s(M^\lambda_2)\, T_2\, g^0_t\big)\, a(T_1\, g^1_t)\Big\} \; , \end{aligned} \tag{5.24}$$

$$\begin{aligned} V^1_R(t) \equiv\ & 1 \otimes \Big\{ b(T_2\, g^1_t)\, b(T_2\, g^0_t) + \\ & + b(T_2\, g^1_t)\, a\big(c(M^\lambda_1)\, J\, s(M^\lambda_1)\, T_1\, g^0_t\big) + a\big(c(M^\lambda_1)\, J\, s(M^\lambda_1)\, T_1\, g^1_t\big)\, b(T_2\, g^0_t)\Big\} \end{aligned} \tag{5.25}$$

$$\begin{aligned} V^*_L(t) \equiv\ & D \otimes \mathcal{B}(M^\lambda_1, M^\lambda_2)\Big\{ a(T_1\, g^0_t)\, a(T_1\, g^1_t) + \\ & + a(T_1\, g^0_t)\, b^*(T_2\, g^1_t) - a^*\big(c(M^\lambda_1)\, J\, s(M^\lambda_1)\, T_1\, g^0_t\big)\, b^*(T_2\, g^1_t) + \\ & + b^*(T_2\, g^0_t)\, a(T_1\, g^1_t) - b^*(T_2\, g^0_t)\, a^*\big(c(M^\lambda_1)\, J\, s(M^\lambda_1)\, T_1\, g^1_t\big) \quad \Big\}\mathcal{B}^*(M^\lambda_1, M^\lambda_2) + \\ & + D^* \otimes \mathcal{B}(M^\lambda_1, M^\lambda_2)\Big\{ b(T_2\, g^1_t)\, b(T_2\, g^0_t) + \\ & + b(T_2\, g^1_t)\, a^*(T_1\, g^0_t) - b^*\big(c(M^\lambda_2) J s(M^\lambda_2)\, T_2\, g^1_t\big)\, a^*(T_1\, g^0_t) + \\ & + a^*(T_1\, g^1_t)\, b(T_2\, g^0_t) - a^*(T_1\, g^1_t)\, b^*\big(c(M^\lambda_2)\, J\, s(M^\lambda_2)\, T_2\, g^0_t\big) \quad \Big\}\mathcal{B}^*(M^\lambda_1, M^\lambda_2) \quad . \end{aligned} \tag{5.26}$$

After straightforward calculations and using (5.23)-(5.26) we get:

$$< u\,,\mathcal{D}(t) > \; - \; < u\,,\mathcal{D}(0) > =$$

$$= -\frac{i}{\lambda^2}\int_0^{t/\lambda^2} ds < u\otimes P_\lambda\Psi(M_1,M_2) \quad , \quad V(s)\,\mathcal{U}_s^\lambda\, u'\otimes P_\lambda\Psi(M_1',M_2') > =$$

$$= \lim_{\lambda\to 0}\frac{-i}{\lambda}\int_0^t ds\,\Big\{< 1\otimes\mathcal{B}^*(M_1^\lambda,M_2^\lambda)\,V_L^*(s/\lambda^2)\,u\otimes\Omega \quad , \quad \mathcal{U}_{s/\lambda^2}^\lambda \quad u'\otimes\mathcal{B}^*(M_1'^\lambda,M_2'^\lambda)\,\Omega > +$$

$$\sum_{\epsilon=0,1} < u\otimes\mathcal{B}^*(M_1^\lambda,M_2^\lambda)\,\Omega \quad , \quad D^\epsilon\otimes 1 \quad \mathcal{U}_{s/\lambda^2}^\lambda\, V_R^\epsilon(s/\lambda^2) \quad u'\otimes\mathcal{B}^*(M_1'^\lambda,M_2'^\lambda)\,\Omega > +$$

$$\sum_{\epsilon=0,1} < u\otimes\mathcal{B}^*(M_1^\lambda,M_2^\lambda)\,\Omega \quad , \quad D^\epsilon\otimes 1\,\left[V_R^\epsilon(s/\lambda^2)\,,\mathcal{U}_{s/\lambda^2}^\lambda\right] \quad u'\otimes\mathcal{B}^*(M_1'^\lambda,M_2'^\lambda)\,\Omega > \Big\} \equiv$$

$$\equiv \int_0^t ds\,\{I_1(s) + I_2(s) + I_3(s)\} \quad . \tag{5.27}$$

Lemma 5.6
With the same assumption as in Proposition 5.5

$$\begin{aligned} I_1(s) \;=\; & -i\,\chi_{[Q_1,R_1]}(s)\,\Big(M_1\oplus 0 \quad , \quad |\sqrt{1\pm T}\,g^1 >< J\sqrt{1\pm T}\,g^0|\oplus 0\Big) < D\,u\,,\mathcal{D}(s) > + \\ & -i\,\chi_{[-R_2,-Q_2]}(s)\,\Big(0\oplus M_2 \quad , \quad 0\oplus|J\sqrt{T}\,g^1 >< \sqrt{T}\,g^0|\Big) < D^*\,u\,,\mathcal{D}(s) > \;. \end{aligned} \tag{5.28}$$

Proof After straightforward calculations and using Lemma 5.2 the proof is immediate.

■

Lemma 5.7
With the same assumption as in Proposition 5.5

$$\begin{aligned} I_2(s) \;=\; & -\,i\chi_{[Q_1',R_1']}(s)\,\big(|\sqrt{1\pm T}\,g^1 >< J\sqrt{1\pm T}\,g^0|\oplus 0 \;,\; M_1'\oplus 0)\big) < D\,u\,,\mathcal{D}(s) > + \\ & -\,i\,\chi_{[-R_2',-Q_2']}(s)\,\Big(0\oplus|J\sqrt{T}\,g^1 >< \sqrt{T}\,g^0| \;,\; 0\oplus M_2'\Big) < D^*\,u\,,\mathcal{D}(s) > \end{aligned} \tag{5.29}$$

Proof The proof is similar to the proof of the Lemma 5.6.

■

Lemma 5.8
With the same assumption as in Proposition 5.5

$$\begin{aligned} I_3(s) \;=\; & -\,\|\,|\sqrt{1\pm T}\,g^1 >< J\sqrt{1\pm T}g^0|\,\oplus\, 0\,\|_-^2 \, < D\,D^*\,u\,,\mathcal{D}(s) > + \\ & -\,\|0\,\oplus\,|J\sqrt{T}\,g^1 >< \sqrt{T}\,g^0|\,\|_-^2 \, < D^*\,D\,u\,,\mathcal{D}(s) > \quad . \end{aligned} \tag{5.30}$$

Proof Expanding $\mathcal{U}^{\lambda}_{t/\lambda^2}$ in the Dyson series we have:

$$I_3(s) = \lim_{\lambda\to 0} -\frac{i}{2} \sum_{\epsilon=1,2} \sum_{n=1}^{\infty} (-i)^n \lambda^{n+1} \int_0^{s/\lambda^2} ds_1 \cdots \int_0^{s_{n-1}} ds_n$$

$$< u \otimes P_\lambda \Psi(M_1, M_2) \quad , \quad D^\epsilon \otimes \left[V_R^\epsilon(s/\lambda^2) \ , \ V(s_1)\cdots V(s_n)\right] u' \otimes P_\lambda \Psi(M_1', M_2') > \ . \tag{5.31}$$

It is obvious that:

$$\left[V_R^\epsilon(s/\lambda^2) \ , \ V(s_1)\cdots V(s_n)\right] = = \sum_{k=1}^{n} V(s_1)\cdots V(s_{k-1}) [V_R^\epsilon(s/\lambda^2) \ , \ V(s_k)] \, V(s_{k+1})\cdots V(s_n) \ . \tag{5.32}$$

An analogous method to Accardi, Frigerio and Lu [1] lead us to the conclusion that the non-zero contribution in (5.31) is given by the term with $k = 1$ in (5.32) only.

Let us consider the first term of $V_R^0(s/\lambda^2)$ (5.24) : $a(T_1\, g^0_{s/\lambda^2})\, a(T_1\, g^1_{s/\lambda^2})$ Using the condition (3.3) we obtain:

$$\lim_{\lambda\to 0} \frac{-i}{\lambda} < u \otimes \mathcal{B}^*(M_1^\lambda, M_2^\lambda)\Omega \ , \ D \otimes \left[a(T_1\, g^0_{s/\lambda^2})\, a(T_1\, g^1_{s/\lambda^2}) \ , \ \mathcal{U}^\lambda_{s/\lambda^2}\right] u' \otimes \mathcal{B}^*(M_1'^\lambda, M_2'^\lambda)\Omega >=$$

$$= \lim_{\lambda\to 0} - \int_0^{s/\lambda^2} ds_1 \ < u \otimes \mathcal{B}^*(M_1^\lambda, M_2^\lambda)\Omega \quad , \quad D\, D^* \otimes$$

$$\otimes \Big\{ < T_1\, g^0_{s/\lambda^2} \ , \ T_1\, g^0_{s_1} >< T_1\, g^1_{s/\lambda^2} \ , \ T_1\, g^1_{s_1} > \pm < T_1\, g^0_{s/\lambda^2} \ , \ T_1\, g^0_{s_1} > a^*(T_1\, g^1_{s/\lambda^2})\, a(T_1\, g^1_{s_1}) +$$

$$+ < T_1\, g^0_{s/\lambda^2} \ , \ T_1\, g^0_{s_1} > a(T_1\, g^1_{s/\lambda^2})\, b(T_1\, g^1_{s_1}) \pm < T_1\, g^1_{s/\lambda^2} \ , \ T_1\, g^1_{s_1} > a^*(T_1\, g^0_{s/\lambda^2})\, a(T_1\, g^0_{s_1}) +$$

$$+ < T_1\, g^1_{s/\lambda^2} \ , \ T_1\, g^1_{s_1} > a(T_1\, g^0_{s/\lambda^2})\, b(T_1\, g^0_{s_1}) \Big\} \ \mathcal{U}^\lambda_{s_1} \ \ u' \otimes \mathcal{B}^*(M_1'^\lambda, M_2'^\lambda)\,\Omega > \tag{5.33}$$

The terms of type $a^*(\cdot)\, a(\cdot)$ in the above formula vanish in the limit $\lambda \to 0$ after acting on the left-hand side of the scalar product. The terms of $a(\cdot)\, b(\cdot)$ - type we commute once again with $\mathcal{U}^\lambda_{s_1}$ and we see that they vanish in the limit $\lambda \to 0$ because of the same reasons as we discussed in case of formula (5.32). Therefore only the first term in the above formula has a non-zero contribution in the limit $\lambda \to 0$. Performing the change of variable $s_1 \longrightarrow s_1 - s/\lambda^2$ lead us to the expression:

$$\lim_{\lambda\to 0} - \int_{-s/\lambda^2}^{0} ds_1 \ < T_1\, g^0 \ , \ T_1\, g^0_{s_1} >< T_1\, g^1 \ , \ T_1\, g^1_{s_1} >$$

$$< D\, D^*\, u \otimes \mathcal{B}^*(M_1^\lambda, M_2^\lambda)\,\Omega \quad , \quad \mathcal{U}^\lambda_{s/\lambda^2+s_1} \ \ u' \otimes \mathcal{B}^*(M_1'^\lambda, M_2'^\lambda)\,\Omega >=$$

$$= - \int_{-\infty}^{0} ds_1 \ < T_1\, g^0 \ , \ T_1\, g^0_{s_1} >< T_1\, g^1 \ , \ T_1\, g^1_{s_1} > < D\, D^*\, u \ , \ \mathcal{D}(s) > \tag{5.34}$$

The rest term of $V_R^\epsilon(s/\lambda^2)$ can be calculated in the same way, what completes the proof.

■

Proof of the Theorem 3.3

The following formula results from the Lemata 5.1, 5.6, 5.7, and 5.8:

$$
\begin{aligned}
&< u\,, \mathcal{D}(t) > - < u\,, \mathcal{D}(0) > = \int_0^t ds \\
&\Big\{ - \| \, |T_1\, g^1 >< J\, T_1\, g^0| \,\oplus\, 0 \|^2_- \;< D\, D^*\, u\,, \mathcal{D}(s) > + \\
&\quad - i\, \chi_{[Q_1, R_1]}(s) \left(M_1 \,\oplus\, 0 \;\;,\;\; |T_1\, g^1 >< J\, T_1\, g^0| \,\oplus\, 0 \right) < D\, u\,, \mathcal{D}(s) > + \\
&\quad - i\, \chi_{[Q_1', R_1']}(s) \left(|T_1\, g^1 >< J\, T_1\, g^0| \,\oplus\, 0 \;\;,\;\; M_1' \,\oplus\, 0 \right) < D\, u\,, \mathcal{D}(s) > + \\
&\quad - \| 0 \,\oplus\, |J\, T_2\, g^1 >< T_2\, g^0| \, \|^2_- \;< D^*\, D\, u\,, \mathcal{D}(s) > + \\
&\quad - i\, \chi_{[-R_2, -Q_2]}(s) \left(0 \,\oplus\, M_2 \;\;,\;\; 0 \oplus |J\, T_2\, g^1 >< T_2\, g_0| \right) < D^*\, u\,, \mathcal{D}(s) > + \\
&\quad - i\, \chi_{[-R_2', -Q_2']}(s) \left(0 \,\oplus\, |J\, T_2\, g^1 >< T_2\, g_0| \;\;,\;\; 0 \oplus M_2' \right) < D^*\, u\,, \mathcal{D}(s) > \Big\} \;.
\end{aligned}
\tag{5.35}
$$

From the theory of quantum stochastic processes [3] it results that the following matrix element:

$$
< u \otimes \Phi\big(\chi_{[Q_1, R_1]}\, M_1 \oplus \chi_{[-R_2, -Q_2]}\, M_2\big) \quad , \quad \mathcal{U}_t \quad u' \otimes \Phi\big(\chi_{[Q_1', R_1']}\, M_1' \oplus \chi_{[-R_2', -Q_2']}\, M_2'\big) > \tag{5.36}
$$

satisfies the integral equation which has exactly the same form as in the formula (5.35), where $\mathcal{U}_t$ is the unitary solution of the QSDE (3.23). Using Lemma 5.1 we find the initial condition for $\mathcal{U}_t$. The vectors $M_i, M_i' \in \mathcal{H}_R$, $i = 1, 2$ in the formula (5.36) are natural inclusions into $\mathcal{H}_R$ of the vectors $M_i, M_i' \in \mathcal{H}_R'$, $i = 1, 2$ in the formula (5.35). Hence the proof of the theorem is completed.

■

Acknowledgement : This work is supported by The Ministry of Education Grants BW/5-400-4-048-1, PB 1436/2/91.

References

[1] L. Accardi, A. Frigerio, Y.G. Lu, *The weak coupling limit as a quantum functional central limit.* Commun. Math. Phys. **131**, 537–570 (1990).

[2] L. Accardi, Y.G. Lu, *The low density limit of quantum systems.* J. Phys. A Math. Gen. **24**, 3483–3512 (1991).

[3] R.L. Hudson, K.R. Parthasarathy, *Quantum Ito's formula and stochastic evolutions.* Commun. Math. Phys. **93**, 301–323 (1984).

[4] K.R. Parthasarathy, *Quantum Ito's formula* Rev. Math. Phys. **1**, 89–112 (1989).

[5] L. Accardi, Y.G. Lu, R. Alicki, A. Frigerio, *An invitation to the weak coupling and low density limits.* Quantum Probability vol. VI. World Scientific, Singapore (1991).

[6] S. Rudnicki, R. Alicki, S. Sadowski, *The low density limit in terms of collective squeezed vectors.* To appear in J. Math. Phys. (1992).

[7] Y.G. Lu, *The weak coupling limit for quadratic interactions.* Preprint (1990).

[8] J.V. Pulè, *The Bloch equation.* Commun. Math. Phys. **38**, 241–256 (1974).

[9] E.B. Davies, *Markovian master equations.* Commun. Math. Phys. **39**, 91–110 (1974).

[10] R. Dümke, *The low density limit for an N-level system interacting with a free Bose or Fermi gas.* Commun. Math. Phys. **97**, 331–359 (1985).

[11] A. Barchelli, *Quantum stochastic differential equations: an application to the electron shelving effect.* J. Phys. A **20** 6341–6355 (1987).

[12] P. Robinson and H. Maasen *Quantum Stochastic Calculus and the Dynamical Stark Effect* . Preprint (1990).

[13] A. Frigerio, H. Maasen, *Quantum Poisson processes and dilations of dynamical semigroups.* Prob. Th. Rel. Fields **83**, 489–508 (1989).

[14] O. Bratelli, D.W. Robinson, *Operator algebras and quantum statistical mechanics, vol. II* . Springer-Verlag, New York (1981).

[15] F.A. Berezin, *The method of second quantization.* Academic Press, New York (1966).

[16] F. Smithies, *Integral equations.* Cambridge University Press, New York (1958).

Quantum Probability and Related Topics Vol. VIII (pp. 347–352)

THE LATTICE OF ADMISSIBLE PARTITIONS

Roland Speicher

INSTITUT FÜR ANGEWANDTE MATHEMATIK
UNIVERSITÄT HEIDELBERG
IM NEUENHEIMER FELD 294
W-6900 HEIDELBERG
FEDERAL REPUBLIC OF GERMANY

Abstract. We introduce the notion of 'admissible' partitions, outline its importance for non-commutative probability theory and start the examination of the lattice of admissible partitions by calculating the corresponding Möbius function.

In the following we examine the lattice of admissible partitions, which determines the combinatorial structure of the 'free' convolution of Voiculescu [11]. After finishing this work we found out that our lattice of admissible partitions was considered before in literature, namely it appeared as 'lattice of non-crossing partitions' in a paper of G. Kreweras: Sur les partitions non croisees d'un cycle. Discr. Math. 1 (1972), 333-350. So our contribution should be regarded as a remind of the basic facts about the lattice of non-crossing partitions from a quantum probabilistic point of view. In particular, this point of view allows a proof of Theorem 2 which is much simpler than the one in the paper of Kreweras.

The combinatorial aspects of the convolution of probability measures are in a nice way connected with the lattice $\mathcal{L}_n = \mathcal{P}(1,\ldots,n)$ of all partitions of the set $\{1,\ldots,n\}$ (or better with the union of all such lattices for all n): For a measure μ with moments m_i and cumulants (semi-invariants) C_i ($i \in I\!N$) we can define on the lattice $\mathcal{L}_n$ the multiplicative functions φ_μ and Q_μ by ($\mathcal{V} = \{V_1,\ldots,V_s\} \in \mathcal{L}_n$)

$$\varphi_\mu(\mathcal{V}) := \prod_{i=1}^{s} \varphi_\mu(V_i) \qquad \text{and} \qquad Q_\mu(\mathcal{V}) := \prod_{i=1}^{s} Q_\mu(V_i),$$

where for some finite $V \subset I\!N$

$$\varphi_\mu(V) = \varphi_\mu(1_{\#V}) := m_{\#V} \qquad \text{and} \qquad Q_\mu(V) = Q_\mu(1_{\#V}) := C_{\#V}.$$

Then the connection between these two functions is given by

$$\varphi_\mu(\mathcal{V}_0) = \sum_{\mathcal{V} \le \mathcal{V}_0} Q_\mu(\mathcal{V})$$

This work has been supported by the Deutsche Forschungsgemeinschaft (SFB 123).
I thank Marek Bożejko for many interesting discussions.

and by Möbius inversion

$$Q_\mu(\mathcal{V}_0) = \sum_{\mathcal{V}\leq\mathcal{V}_0} \varphi_\mu(\mathcal{V})\mu(\mathcal{V},\mathcal{V}_0),$$

where $\mu(\cdot,\cdot)$ is the Möbius function of the lattice $\mathcal{L}_n$. The usual convolution of probability measures which are determined by their moments is easily described in terms of the Q-functions since

$$Q_{\mu_1\star\mu_2}(\mathbf{1}_n) = Q_{\mu_1}(\mathbf{1}_n) + Q_{\mu_2}(\mathbf{1}_n).$$

This extends multplicatively to all $\mathcal{V}$ and thus gives, by the above formula, the moments of $\mu_1 \star \mu_2$.

For the connection between cumulants and the lattice $\mathcal{L}_n$ of all partitions (and the extension of this to more general cumulants and lattices) and its unifying significance in statistics one should consult the papers of Speed (e.g. [5,6,7]).

In a non-comutative context, Voiculescu [10] introduced another notion of convolution, called 'free' convolution, which is in many respects quite analogous to the classical case and which has a lot of theoretical and practical relevance ([3,9,10, 11,12]). In [8] we showed that this form of convolution behaves from a combinatorial point of view in exactly the same way as the usual convolution, one only has to replace the lattice $\mathcal{L}_n$ of all partitions by the lattice $\mathcal{L}'_n = \mathcal{P}_a(1,\dots,n)$ of some special partitions, called 'admissible' partitions. (Of course, the notion of cumulants is changed in comparison with the classical case.)

Here we want to consider this interesting lattice $\mathcal{P}_a$ more closely. In particular, we want to calculate its Möbius function, which gives us the explicit formula for the 'free' cumulants $Q_\mu(\mathbf{1}_n)$ in terms of the moments $m_n = \varphi_\mu(\mathbf{1}_n)$.

One should also note that there is another subset of the lattice of all partitions, namely the lattice of interval partitions (which is isomorphic to the Boolean lattice of all subsets of a given set), which has gained some interest in the literature ([1,13]) and also gives rise to some form of convolution.

Although the lattice point of view shows in a nice way the combinatorial features of the involved forms of convolution, one should, however, keep in mind that it veils all positivity aspects. Thus from this point of view it is a mystery why the convolution of two positive functionals is positive again. That this is indeed the case for the above mentioned three lattices seems to be accidentially, and it would be an interesting task to determine all subsets of the lattice of all partitions having this feature.

Let us now start our examination of the lattice of admissible partitions.

DEFINITION: Let $\mathcal{V} = \{V_1,\dots,V_s\}$ be a partition of the ordered set S, i.e. the V_i are ordered and disjoint sets, whose union is S. Then $\mathcal{V}$ is called *admissible* if for all $i,j = 1,\dots,s$ with $V_i = (v_1,\dots,v_n)$ $(v_1 < \dots < v_n)$ and $V_j = (w_1,\dots,w_m)$ $(w_1 < \dots < w_m)$ we have

$$w_k < v_1 < w_{k+1} \Leftrightarrow w_k < v_n < w_{k+1} \qquad (k = 1,\dots,m-1).$$

We will denote the set of all admissible partitions of S by $\mathcal{P}_a(S)$.

We can reformulate the definition of 'admissible' in a recursive way: The partition $\mathcal{V} = \{V_1, \dots, V_s\}$ is admissible if at least one V_i is a segment of S, i.e. it contains exactly all points of S lying between two points, and $\mathcal{V}\backslash\{V_i\}$ is an admissible partition of $S\backslash V_i$.

In a more pictorial language: If we build bridges by connecting in S the numbers belonging to the same V_i, then a partition is admissible if it is possible to build the corresponding bridge in such a way that the lines do not cross.

EXAMPLE: E.g. $\{(1,3,5),(2),(4)\}$ and $\{(1,5),(2,4),(3)\}$ are admissible partitions of $\{1,2,3,4,5\}$, not admissible are $\{(1,3),(2,4,5)\}$ and $\{(1,4),(2),(3,5)\}$.

The V_i are called the blocks of a given partition $\mathcal{V} = \{V_1, \dots, V_s\}$, and the order structure on $\mathcal{P}_a(S)$ is defined as usual: $\mathcal{V}_1 \leq \mathcal{V}_2$ exactly if each block of $\mathcal{V}_1$ is contained in one block of $\mathcal{V}_2$. Of course, the lattice structure of $\mathcal{P}_a(S)$ depends only on the number of elements of S, i.e. we can restrict for each n to the consideration of some fixed S_n with $\#S_n = n$, let's say $S_n = (1, \dots, n)$. By $\mathbf{0}_n := \{(i) \mid i \in S_n\}$ and $\mathbf{1}_n := \{S_n\}$ we denote the minimal and the maximal element of $\mathcal{P}_a(S_n)$, consisting of n and one block, respectively.

One sees easily that the Möbius function $\mu(\cdot,\cdot)$ of $\mathcal{P}_a(S)$ is multiplicative in the following weak sense: If T is the disjoint union of T_1 and T_2 and $\mathcal{V}_i$, $\mathcal{W}_i$ are admissible partitions of T_i with $\mathcal{V}_i \leq \mathcal{W}_i$ $(i = 1,2)$, then

$$\mu(\mathcal{V}_1 \cup \mathcal{V}_2, \mathcal{W}_1 \cup \mathcal{W}_2) = \mu(\mathcal{V}_1, \mathcal{W}_1)\mu(\mathcal{V}_2, \mathcal{W}_2).$$

This shows that the determination of μ reduces to the calculation of $\mu(\mathcal{V}, \mathbf{1}_n)$ for all n and $\mathcal{V} \in \mathcal{P}_a(S_n)$.

One notices that for each $\mathcal{V} \in \mathcal{P}_a(S_n)$ there exists a unique set $\{k_1, \dots, k_r\}$ of integers $k_i \geq 2$ such that the segment $[\mathcal{V}, \mathbf{1}_n]$ is isomorphic to $\mathcal{P}_a(S_{k_1}) \times \dots \times \mathcal{P}_a(S_{k_r})$. We shall call this set the class of $\mathcal{V}$: class$(\mathcal{V}) := \{k_1, \dots, k_r\}$. The existence of this set can be seen as following: Let us consider $[\mathcal{V}, \mathbf{1}_n]$ with $\mathcal{V} = \{V_1, \dots, V_s\} \in \mathcal{P}_a(S_n)$ and put $T := (1, \dots, n)$. If possible, choose a V_i and two neighbouring elements $k < l$ in V_i, i.e. $V_i = (\dots k, l \dots)$, such that $k + 1 \neq l$. Put

$$T_0 := T\backslash[k+1, l-1], \qquad T_1 := \{k\} \cup [k+1, l-1].$$

Then

$$\mathcal{V}_0 := \mathcal{V} \cap T_0$$

is an admissible partition of T_0 and

$$\mathcal{V}_1 := \{k\} \cup (\mathcal{V} \cap [k+1, l-1])$$

is an admissible partition of T_1 and we have

$$[\mathcal{V}, \mathbf{1}_n] \cong [\mathcal{V}_0, \mathbf{1}_{T_0}] \times [\mathcal{V}_1, \mathbf{1}_{T_1}].$$

This decomposition reflects the admissible character of our partitions, which implies that the block V_i separates the points lying between k and l from the points lying outside this interval. If it should happen that one of the two factors is isomorphic to $\mathcal{P}_a(S_1)$, then we will forget this factor. Now we can repeat the above procedure (for each of the factors) until we end up with some $[\mathcal{W}, \mathbf{1}_m]$, where we do not find some V_i with the above properties any more. But in this case $[\mathcal{W}, \mathbf{1}_m] \cong \mathcal{P}_a(S_{|\mathcal{W}|})$. One should note that the resulting set $\{k_1, \dots, k_r\}$ is independent of the order of choosing the V_i or the neighbouring k, l.

Let us also give some examples for the notion of class:
$\text{class}(\{(1,3,4),(2)\}) = \{2\}$, $\text{class}(\{(1,6),(2,4,5),(3)\}) = \{2,2\}$,
$\text{class}(\{(1,6,11),(2,5),(3,4),(7,8),(9,10),(12),(13,14),(15)\}) = \{2,2,3,4\}$.
The class of $\mathcal{V}$ contains all information for the calculation of $\mu(\mathcal{V}, \mathbf{1}_n)$, because it allows to reduce $\mu(\mathcal{V}, \mathbf{1}_n)$ to the knowledge of some $\mu(\mathbf{0}_i, \mathbf{1}_i)$. But this can be given explicitely with the help of the Catalan numbers c_i, which may be defined as the number of lattice paths from $(0,-1)$ to $(i, i-1)$ lying entirely below the diagonal and which appear in a lot of contextes (see e.g. [2]). For example, $c_0 = 1$, $c_1 = 1$, $c_2 = 2$, $c_3 = 5$, $c_4 = 14$, $c_5 = 42$.

THEOREM 1. *1) Consider $\mathcal{V} \in \mathcal{P}_a(S_n)$ with class$(\mathcal{V}) = \{k_1, \dots, k_r\}$. Then*

$$\mu(\mathcal{V}, \mathbf{1}_n) = \prod_{j=1}^{r} \mu(\mathbf{0}_{k_j}, \mathbf{1}_{k_j}).$$

2) For all $k \in \mathbb{N}$ we have $\mu(\mathbf{0}_k, \mathbf{1}_k) = (-1)^{k-1} c_{k-1}$.

OUTLINE OF PROOF: The first part follows from the fact that the segment $[\mathcal{V}, \mathbf{1}_n]$ is isomorphic to $\mathcal{P}_a(S_{k_1}) \times \dots \times \mathcal{P}_a(S_{k_r})$.

For the second part one checks that the same argument as for the lattice of all partitions (c.f. [4], Proposition 3) gives

$$\mu(\mathbf{0}_n, \mathbf{1}_n) = -\sum_{i=1}^{n-1} \mu(\mathcal{V}_i, \mathbf{1}_n),$$

where $\mathcal{V}_i$ is the atom consisting of $n-2$ blocks which contain only one element and one block which contains the first and the $(i+1)$-th element of S_n. Since $\text{class}(\mathcal{V}_i) = \{n-i, i\}$ we have $\mu(\mathcal{V}_i, \mathbf{1}_n) = \mu(\mathbf{0}_{n-i}, \mathbf{1}_{n-i})\mu(\mathbf{0}_i, \mathbf{1}_i)$, hence

$$\mu(\mathbf{0}_n, \mathbf{1}_n) = -\sum_{i=1}^{n-1} \mu(\mathbf{0}_{n-i}, \mathbf{1}_{n-i})\mu(\mathbf{0}_i, \mathbf{1}_i).$$

With the definition $d_k = (-1)^k \mu(\mathbf{0}_{k+1}, \mathbf{1}_{k+1})$ this gives $d_{n-1} = \sum_{i=1}^{n-1} d_{n-i-1} d_{i-1}$, which is the recursion formula for the Catalan numbers c_k. Because of $d_0 = d_1 = 1$

we get the assertion $d_k = c_k$.

$\diamondsuit$

The connection between the lattice of admissible partitions and the Catalan numbers is even more intriguing. The next theorem shows that also the number of all admissible partitions of S_n, which corresponds in the case of the lattice of all partitions to the Bell number B_n, is given by c_n. Although this shows that the set of all paths below the diagonal and the set of all admissible partitions can be identified, one should note that the partial order on the partitions does not correspond to the canonical partial order on the paths. Thus, the language of paths does not seem to be the right way of thinking about our lattice structure.

In the next theorem we shall also consider such admissible partitions $\mathcal{V} = \{V_1, \dots, V_s\}$, where each block V_i contains exactly two elements. Let us denote the set of all admissible partitions of this form by $\mathcal{P}_a^2(S_n)$ (where, of course, n should be even).

THEOREM 2. *We have for all* $n \in \mathbb{N}$

$$\#\mathcal{P}_a(S_n) = \#\mathcal{P}_a^2(S_{2n}) = c_n.$$

OUTLINE OF PROOF: The essential part of the proof for the first equality is already contained in [8]. We only recall the idea: Let $\{e_i \mid i \in \mathbb{N}_0\}$ be an orthonormal basis of a separable Hilbert space and define the operators l with adjoint l^* and Λ by ($i \in \mathbb{N}_0$)

$$\begin{aligned} l^* e_i &= e_{i+1}, \\ l e_0 &= 0, \qquad l e_{i+1} = e_i, \\ \Lambda e_0 &= 0, \qquad \Lambda e_{i+1} = e_{i+1}. \end{aligned}$$

Then one checks that for all $n \in \mathbb{N}$

$$\begin{aligned} < e_0, (l + l^*)^{2n} e_0 > &= \#\mathcal{P}_a^2(S_{2n}) \\ < e_0, (\Lambda + l + l^* + 1)^n e_0 > &= \#\mathcal{P}_a(S_n). \end{aligned}$$

The equation $(l + l^*)^2 = \Lambda + ll + l^*l^* + 1$ and the fact that ll, l^*l^* are as good as l, l^* shows the first equality. (From a probabilistic point of view, the operators $l + l^*$ and $\Lambda + l + l^* + 1$ give a free Gaussian and a free Poisson distribution, respectively.)

For the second equality, let us count the elements of $\mathcal{P}_a^2(S_{2n})$ in the way that we first fix the first block of $\mathcal{V}$, let's say $V_1 = (1, i)$, and then sum over all possibilities for completing this block to an admissible $\mathcal{V} = \{V_1, \dots, V_n\}$. The nature of admissible partitions implies that V_1 separates the points $2, \dots, i-1$ from the points $i+1, \dots, 2n$ and thus $\mathcal{V} \in \mathcal{P}_a^2(1, \dots, 2n)$ is built up (if V_1 is fixed) from V_1, an element of $\mathcal{P}_a^2(2,\ldots,i-1)$, and an element of $\mathcal{P}_a^2(i+1,\ldots,2n)$. In particular, i has to be even, let's say $i=2k$. This implies

$$\#\mathcal{P}_a^2(S_{2n})=\sum_{k=1}^{n}\#\mathcal{P}_a^2(S_{2k-2})\cdot\#\mathcal{P}_a^2(S_{2n-2k}),$$

which is the *recursion formula* for the Catalan numbers.

◇

References

[1] G.C. Hegerfeld and S. Schulze, Noncommutative Cumulants for Stochastic Differential Equations and for Generalized Dyson Series, J. Stat. Phys. 51 (1988) 691-710

[2] P. Hilton and J. Pederson, Catalan Numbers, Their Generalization, and Their Uses, Math. Intelligencer 13, No. 2 (1991) 64-75

[3] F. Radulescu, The fundamental group of the von Neumann algebra of a free group with infinitely many generators, Preprint

[4] G.-C. Rota, On the Foundations of Combinatorial Theory. I Theory of Möbius Functions, Z. Wahrscheinlichkeitstheorie verw. Geb. 2 (1964) 340-368

[5] T.P. Speed, Cumulants and partition lattices, Austral. J. Statist. 25 (1983) 378-388

[6] T.P. Speed, Cumulants and partition lattices II: generalized k-statistics, J. Austral. Math. Soc. A 40 (1986) 34-53

[7] T.P. Speed, Cumulants and partition lattices III: multiply-indexed arrays, J. Austral. Math. Soc. A 40 (1986) 161-182

[8] R. Speicher, A New Example of 'Independence' and 'White Noise', Probab. Th. Rel. Fields 84 (1990) 141-159

[9] R. Speicher, Free convolution and the random sum of matrices, Preprint, Heidelberg, 1991

[10] D. Voiculescu, Symmetries of some reduced free product C^*-algebras, in: H. Araki, C.C. Moore, S. Stratila and D. Voiculescu, eds., Operator Algebras and their Connection with Topology and Ergodic Theory (Lect. Notes in Math., vol 1132, Springer, Heidelberg, 1985) 556-588

[11] D. Voiculescu, Addition of certain non-commuting random variables, J. Funct. Anal. 66 (1986) 323-346

[12] D. Voiculescu, Limit laws for random matrices and free products, Invent. math. 104 (1991) 201-220

[13] W. von Waldenfels, Integral Partitions and Pair Interactions, in: P.A. Meyer, ed., Séminaire de Probabilités IX (Lect. Notes in Math., vol 465, Springer, Heidelberg, 1975) 565-588

Quantum Probability and Related Topics Vol. VIII (pp. 353–369)

Phonion Limits and Macroscopic quasi-particle spectrum for the BCS-model

M. Broidioi, B. Momont [1] *and A. Verbeure*

K.U.Leuven

Instituut voor Theoretische Fysica

Celestijnenlaan 200 D

B-3001 LEUVEN (Belgium)

Abstract

On the basis of non-commutative central limit theorems, starting from a dynamical system in equilibrium $(\mathcal{A}, \alpha_t, \omega)$ the corresponding macroscopic system $(\mathcal{A}(\tilde{L}_k), \tilde{\alpha}_t, \tilde{\omega})$ of field fluctuations at momentum k is constructed. A mathematical mechanism for the appearance of a macroscopic quasi particle structure is developed in the BCS-model. Due to the spontaneous breaking of symmetry below the critical temperature the resulting dispersion relation shows an energy gap at zero momentum. The BCS-model is used as a toy model for explaining the general theory.

[1] Onderzoeker IIKW Belgium

1 Introduction

The role played by phon(i)ons in condensed matter physics is well known. Phon(i)ons mediate various interactions between quasi-particles and phon(i)ons interact with external fields, phon(i)ons carry an energy and a quasi-momentum. As essential ingredients we learned from the literature that phon(i)ons are linked to collective or macroscopic observables, and that their dynamics is induced by the microdynamics of the system.

Here we illustrate our ideas of how to obtain phonions from first principles in a quantum mechanical context.

Our general mathematical theory for the mechanism of the appearance of phonons is worked out for lattice systems in ref. [9].

We take the opportunity of these lecture notes to work out an example of the continuous BCS-model. This is a model for long range interactions. It is also completely soluble and in fact well known. So what is new in our treatment is the approach to the model illustrating the mechanism of the appearance of phonions as bona fide quasi-particles. It is also used to illustrate the phenomenon of coarse graining, i.e. at macroscopic level the number of degrees of freedom comes over as being much smaller than at the microscopic level.

For long range forces, the spontaneous breaking of a symmetry is accompanied with an energy gap ε_0 in the density spectrum at momentum $k = 0$. Recently it was proved that the energy gap ε_0 can be found back as a discrete point of the spectrum of zero-momentum fluctuations in models with very long range interactions [1][2][3]. In this paper we consider the continuous BCS-model [4] and extend the above results in the following sense.

We develop the theory of field fluctuations at momentum $k \neq 0$. We show that for any fixed $k \in \mathbb{R}^3$ these field fluctuations generate a finite dimensional Clifford algebra. The macroscopic dynamics leaves this algebra invariant and yields a discrete spectrum coinciding exactly with the quasi spectrum. However, the main result of this paper is the mathematical mechanism for the understanding of the quasi particle structure in solid state physics.

2 The equilibrium states of the BCS-model

In this section we determine the translation-invariant equilibrium states of the BCS-model. The model describes a dynamics of a system of Fermions (electrons) with two possible spin states labeled by the index $i = 1, 2$.

Therefore the algebra of observables, denoted by $\mathcal{A}$, is generated by the following set of smeared out creation and annihilation operators

$$C_i^+(f), C_i^-(f); f \in L^2(\mathbb{R}^3)$$

satisfying the usual anti-commutation relations (CAR)

$$\{C_i^-(f), C_j^+(g)\} = (f, g)\delta_{ij}$$

$$\{C_i^-(f), C_j^-(g)\} = 0$$

$$C_i^+(f)^* = C_i^-(f)$$

The formal $C_i^+(x)$ are related to these by the expression

$$C_i^+(f) = \int dx\, f(x) C_i^+(x).$$

The gauge groups $\{\gamma_j^\lambda | \lambda \in [0, 2\pi)\}_{j=1,2}$ act as automorphisms on $\mathcal{A}$ as follows

$$\gamma_j^\lambda(C_{j'}^-(f)) = C_{j'}^-(f)\, e^{i\delta_{jj'}\lambda} \tag{2.1}$$

The translation group $\{\tau_a | a \in \mathbb{R}^3\}$ acts as automorphisms τ_a on $\mathcal{A}$ in in the following way

$$\tau_a(C_j^-(f)) = C_j^-(f_a) \tag{2.2}$$

with $f_a(x) = f(x - a)$. The BCS-Hamiltonian is given by [4]

$$\begin{aligned} H_V &= \sum_{i=1}^{2} \frac{1}{2} \int_V dx\, \nabla C_i^+(x) \cdot \nabla C_i^-(x) \\ &- \frac{c}{V} \int_V dy\, S(y) \int_V dy'\, S(y') \int_V dx\, C_1^+(x) C_2^+(x+y) \int_V dx'\, C_2^-(x'+y') C_1^-(x') \end{aligned} \tag{2.3}$$

with S a real-valued function such that $\frac{\hat{S}(k)}{|k|} \in L^2(\mathbb{R}^3)$. Notice that the model-Hamiltonian is invariant under both gauge groups

$$\gamma_j^\lambda(H_V) = H_V \; ; \; j = 1, 2.$$

The model-Hamiltonian is also space translation invariant

$$\tau_a H_V \tau_{-a} = H_{V+a}.$$

As the algebra of observables is the separable CAR-algebra $\mathcal{A}$, which is space translation asymptotically abelian, the translation invariant equilibrium states ω can be uniquely decomposed into extremal invariant equilibrium states [5].

Now we determine the extremal space translation invariant states. One of the main properties of extremal space translation invariant states is that the average over the translations of a local observable exists i.e. for all $A \in \mathcal{A}$

$$\text{weak} - \lim_{V \to \infty} \frac{1}{V} \int_V da\, \tau_a(A) = \omega(A)\mathbb{1} \tag{2.4}$$

and for all $B \in \mathcal{A}$

$$\text{weak} - \lim_{V\to\infty} [\frac{1}{V}\int_V da\, \tau_a(A), B] = 0 \tag{2.5}$$

These averages are observables at infinity [6]. Let H_V be the BCS-model and ω an extremal space translation invariant state, then for all local $X \in \mathcal{A}$, using (2.4) and (2.5)

$$\begin{aligned}
&\lim_{V\to\infty} \omega(X^*[H_V, X]) \\
&= \lim_{V\to\infty} \{\omega(X^*[T_V, X]) - c\bar{b}\,\omega(X^*[\Delta_V, X]) \\
&\qquad -c\,b\,\omega(X^*[\Delta_V^*, X])\}
\end{aligned} \tag{2.6}$$

where

$$\Delta_V = \int_V dy\, S(y) \int_V dx\, \tau_x\left(C_2^-(y)C_1^-(0)\right)$$

and

$$b = \lim_{V\to\infty} \int_V dy\, S(y)\, \omega\left(\frac{1}{V}\int_V dx\, \tau_x\left(C_2^-(y)C_1^-(0)\right)\right)$$

These are the formal expressions for the rigorous ones: for any δ-function sequence $(f^n)_n$

$$\Delta_V = \text{weak} - \lim_{n\to\infty} \int_V dy\, S(y) \int_V dx\, \tau_x\left(C_2^-(f_y^n)C_1^-(f^n)\right).$$

The above computation yields that for each extremal space translation invariant state ω, there exists an effective Hamiltonian

$$H_V^\omega = T_V - c(b\Delta_V^* + \bar{b}\Delta_V) \tag{2.7}$$

such that for all local $X \in \mathcal{A}$

$$\begin{aligned}
&\text{weak} - \lim_{V\to\infty} [H_V, X] \\
&= \text{weak} - \lim_{V\to\infty} [H_V^\omega, X] \equiv \delta_\omega(X).
\end{aligned}$$

One computes explicitly

$$\delta_\omega(C_1^+(x)) = -\frac{\Delta}{2}C_1^+(x) - c\,\bar{b}\int dy\, S(y)C_2^-(x+y) \tag{2.8}$$

$$\delta_\omega(C_2^+(x)) = -\frac{\Delta}{2}C_2^+(x) + c\,\bar{b}\int dy\, S(y)C_1^-(x-y) \tag{2.9}$$

In order to determine the two point function and the gap-equation in a convenient way we introduce the CAR-two-component version of $\mathcal{A}$. For

$$f = (f_1, f_2) \in L = L^2(\mathbb{R}^3) \oplus L^2(\mathbb{R}^3)$$

define the creation operators $C^+(f)$, $f \in L$ by :

$$C^+(f) = \left(C_1^+(f_1), C_2^-(\overline{f_2})\right). \tag{2.10}$$

Then

$$\delta_\omega C^+(f) = C^+(h_\omega f) \tag{2.11}$$

where h_ω is the following operator ·

$$h_\omega = \begin{pmatrix} -\frac{\Delta}{2}\cdot & -c\,b\,S\,\bar{*}\cdot \\ -c\,\bar{b}\,S\,*\cdot & \frac{\Delta}{2}\cdot \end{pmatrix}$$

which can be defined as a self-adjoint operator on $L = L^2(\mathbb{R}^3) \oplus L^2(\mathbb{R}^3)$. The convolution products $*$ and $\bar{*}$ are defined as follows :

$$(S * f_1)(x) = \int dy\, S(x-y) f_1(y)$$

$$(S\,\bar{*}\, f_2)(x) = \int dy\, S(y-x) f_2(y).$$

Define the functions $c_\pm^k$ on $\mathbb{R}^3$, for all $k \in \mathbb{R}^3$, by

$$c_\pm^k(x) = \frac{1}{(2\pi)^{3/2}} \frac{1}{\sqrt{1 + |\mu_\pm(k)|^2}} \begin{pmatrix} 1 \\ \mu_\pm(k) \end{pmatrix} e^{ikx}$$

where

$$\mu_\pm(k) = \frac{k^2/2 \mp \sqrt{k^4/4 + c^2|b|^2(2\pi)^3|\hat{S}(k)|^2}}{cb(2\pi)^{3/2}\hat{S}(-k)}.$$

Then one computes that

$$h_\omega e^k_\pm = \varepsilon_\pm(k) e^k_\pm \tag{2.12}$$

with

$$\varepsilon_\pm(k) = \pm\sqrt{k^4/4 + c^2|b|^2(2\pi)^3|\hat{S}(k)|^2}$$

These are called the quasi particle energies at momentum k. Clearly the spectrum of h_ω is absolutely continuous, because using Fourier analysis, and the property $\mu_+(k)\overline{\mu_-(k)} = -1$ one checks that any $f \in L$ can be written as

$$f = \int dk \left((e^k_+, f)e^k_+ + (e^k_-, f)e^k_-\right). \tag{2.13}$$

The extremal space translation invariant KMS-states are determined in the following theorem.

Proposition 2.1. (KMS-solutions)

The extremal space translation invariant equilibrium states are the quasi-free states ω on $\mathcal{A}$ determined by the two-point function

$$\omega\left(C^+(f)C^-(g)\right) = \left(g, \frac{1}{1+e^{\beta h_\omega}} f\right) \tag{2.14}$$

where the value b in h_ω is determined as a solution of the gap-equation

$$b\left(1 - \int dk \frac{g|\hat{S}(k)|^2}{2\varepsilon_+(k)} \tanh\frac{\beta\varepsilon_+(k)}{2}\right) = 0. \tag{2.15}$$

proof the proof is completely analogous to the proof of theorem 2.2. in [7]. ∎

The gap-equation always admits a solution $b = 0$. This state is invariant for the gauge group automorphisms. In the low temperature region ($\beta > \beta_c$) there are solutions to the gap-equation with $b = \int dy S(y)\omega\left(C_2^-(y)C_1^-(0)\right) \neq 0$. Hence the broken symmetry in the BCS-model is the compact gauge group [4][8].

Introducing the Fourier transforms of the creation and annihilation operators

$$C_i^+(x) = \frac{1}{\sqrt{V}} \sum_{k \in \tilde{V}} e^{-ikx} b_i^+(k)$$

$$b_i^-(k) = b_i^+(k)^*$$

the order parameter $\bar{b}$ can be written as

$$\bar{b} = \frac{1}{(2\pi)^{3/2}}\omega\left(\int dk\, \hat{S}(k)\, b_1^+(-k) b_2^+(k)\right).$$

Hence there is a nonvanishing correlation between electron states with opposite spins and impulses in the superconducting phase $\bar{b} \neq 0$. One speaks of the appearance of Cooper pairs.

Notice also that the gap-equation fixes only the absolute value $|b|$, but not the phase. Therefore each non-zero solution $|b| \neq 0$ yields an infinite set of extremal invariant states parametrized by the phase of b.

The physical spectrum of the micro-dynamics in any state ω is as usual given by the spectrum of the Hamiltonian H_ω, implementing the time evolution in the GNS-representation of ω

$$e^{iH_\omega t}C^-(f)e^{-iH_\omega t} = e^{it\delta_\omega}C^-(f)$$

where δ_ω is given by formula (2.11) in terms of the one particle Hamiltonian h_ω.

If ω is an equilibrium state of the type described in propostion 2.1., it is necessarily time invariant. This implies that zero belongs to the spectrum of H_ω. Furthermore as any such state is quasi free and the spectrum of h_ω is absolutely continuous given by the set $]-\infty, +\infty[$, the spectrum of the physical Hamiltonian H_ω in the state ω is given by $\mathbb{R}$ and a discrete spectrum consisting of the single point zero.

3 Macroscopic quasi particle spectrum for the BCS-model

It is easily checked that the space translation invariant KMS-states of the BCS-model are: even, quasi free and L^1-clustering; i.e.

$$\int d\alpha_1 \ldots \int d\alpha_{n-1} \left|\omega_T^{(n)}\left(\tau_{\alpha_1} C^{\#_1}(f_1), \ldots, \tau_{\alpha_{n-1}} C^{\#_{n-1}}(f_{n-1}), C^{\#_n}(f_n)\right)\right| < \infty \qquad (3.16)$$

for all $n \geq 2$ and for all f_i having compact support in L, and where the $\omega_T^{(n)}$ are the truncated functions defined in the usual way [5] and $C^{\#_i} = C^+$ or C^-

From now on we take a space translation invariant equilibrium state of the BCS-model determined by (2.14) and (2.15) with $b \neq 0$ ($T < T_c$). In this state ω, we consider the local field fluctuations at momentum k in the volume Λ

$$F_{k,\Lambda}^{\#}(f) = \frac{1}{\sqrt{\Lambda}} \int_\Lambda d\alpha \, e^{\#ik\alpha} \tau_\alpha C^{\#}(f) \; ; \; f \in L.$$

We are interested in the macroscopic observables $\lim_{\Lambda \to \infty} F_{k,\Lambda}^{\#}(f)$, called the fluctuation observables of the Fermi fields $C^{\#}(f)$ at momentum k. The limit should be understood in the sense of a central limit.

Notice that one can also consider fluctuations of more complicated local operators, like higher order monomials in the creation and annihilation operators, e.g. energy density, particle density, orderparameter density, etc. [3]. Here we limit ourselves to the simplest situation of the field operators.

Because of the cluster condition (3.16) and the translation invariance of ω, one has the following theorem whose proof is now immediate.

Theorem 3.1. (Central limit theorem) Under the above conditions

$$\lim_{\Lambda\to\infty} \omega\left(F^{\#_1}_{k,\Lambda}(f_1)\dots F^{\#_{2n+1}}_{k,\Lambda}(f_{2n+1})\right) = 0$$

$$\lim_{\Lambda\to\infty} \omega\left(F^{\#_1}_{k,\Lambda}(f_1)\dots F^{\#_{2n}}_{k,\Lambda}(f_{2n})\right)$$

$$= \sum_{\pi} \epsilon_\pi \prod_{l=1}^{n} \int d\alpha\, e^{\#_{2l}ik\alpha}\, \omega\left(C^{\#_{2l-1}}(f_{\pi_{2l-1}})\tau_\alpha C^{\#_{2l}}(f_{\pi_{2l}})\right)$$

where the sum is over all permutations

$$\pi = \begin{pmatrix} 1 & \dots & 2n \\ \pi_1 & \dots & \pi_{2n} \end{pmatrix} \quad \text{with } \pi_{2l-1} < \pi_{2l}.$$

■

It is obvious that different micro observables can lead to the same macroscopic fluctuation observable, e.g. $C^+(f)$ and $e^{-ik\alpha}\tau_\alpha C^+(f)$. This is the property of coarse graining. As in [9] and [10,p. 68] we define an equivalence relation on the microscopic algebra with the property that all the micro observables in one equivalence class lead to the same macroscopic fluctuation. We define on $\mathcal{A}$ the equivalence relation $\sim_k$ by:

$$C^{\#}(f) \sim_k C^{\#}(g) \qquad f, g \in L \equiv L^2(\mathbb{R}^3) \oplus L^2(\mathbb{R}^3)$$

$$\Leftrightarrow \lim_{\Lambda\to\infty} \omega\left[\left(F^{\#}_{k,\Lambda}(f-g)\right)^* F^{\#}_{k,\Lambda}(f-g)\right] = 0.$$

It follows from a straightforward computation that e.g.

$$\begin{aligned}
&\lim_{\Lambda\to\infty} \omega\left(F^{+}_{k,\Lambda}(f-g)F^{-}_{k,\Lambda}(f-g)\right)\\
&= \int d\alpha\, e^{-ik\alpha}\omega\left(C^{+}(f-g)\tau_\alpha C^{-}(f-g)\right)\\
&= \frac{(2\pi)^3}{1+e^{\beta\epsilon_+(k)}}\frac{1}{1+|\mu_+(k)|^2}\left|\left(\hat{f}_1(k)-\hat{g}_1(k)\right)+\overline{\mu_+(k)}\left(\hat{f}_2(k)-\hat{g}_2(k)\right)\right|^2\\
&+\frac{(2\pi)^3}{e^{1+\beta\epsilon_-(k)}}\frac{1}{1+|\mu_-(k)|^2}\left|\left(\hat{f}_1(k)-\hat{g}_1(k)\right)+\overline{\mu_-(k)}\left(\hat{f}_2(k)-\hat{g}_2(k)\right)\right|^2
\end{aligned}$$

So we have the following

Lemma 3.2.

$$C^{-}(f)\sim_k C^{-}(g) \Leftrightarrow \begin{cases} \hat{f}_1(k)=\hat{g}_1(k)\\ \hat{f}_2(k)=\hat{g}_2(k)\end{cases}$$

It is clear that $\sim_k$ is indeed an equivalence relation. ∎

It is possible to construct a CAR-algebra of macroscopic field fluctuations. We first consider the limit of the anti-commutators of the local field fluctuations

Lemma 3.3.

$$\begin{aligned}
&\lim_{\Lambda}\omega\{F^{-}_{k,\Lambda}(f),F^{+}_{k,\Lambda}(g)\} = \int d\alpha\, e^{ik\alpha}\,(f,g_\alpha)\\
&= (2\pi)^3\left(\overline{\hat{f}_1(k)}\hat{g}_1(k)+\overline{\hat{f}_2(k)}\hat{g}_2(k)\right)\\
&\equiv (f,g)^k_{\sim}
\end{aligned}$$

∎

We define the algebra of Fermion field fluctuations of the system at momentum k as the CAR-algebra $\mathcal{A}(\tilde{L}_k)$ generated by the new Fermion creation and annihilation operators

$$F^{\#}_k([f])\ ;\ [f]\in\tilde{L}_k = L/_{\sim_k}\ \cong\ \mathbb{C}^2$$

satisfying the CAR:

$$\{F_k^-([f]), F_k^+([g])\} = (f,g)_\sim^k$$

$$\{F_k^-([f]), F_k^-([g])\} = 0.$$

The central limit theorem fixes a representation of this algebra $\mathcal{A}(\tilde{L}_k)$. This becomes clear in the following theorem.

Theorem 3.4. There exists a quasi free state $\tilde{\omega}$ on $\mathcal{A}(\tilde{L}_k)$ defined by

$$\tilde{\omega}\left(F_k^{\#_1}([f_1])\dots F_k^{\#_n}([f_n])\right) = \tag{3.17}$$

$$\lim_\Lambda \omega\left(F_{k,\Lambda}^{\#_1}(f_1)\dots F_{k,\Lambda}^{\#_n}(f_n)\right) \tag{3.18}$$

with two point function

$$\tilde{\omega}\left(F_k^{\#_1}([f_1])F_k^{\#_2}([f_2])\right)$$
$$= \int d\alpha\, e^{\#_2 ik\alpha}\,\omega\left(C^{\#_1}(f_1)\tau_\alpha C^{\#_2}(f_2)\right)$$

for all $[f_1],[f_2] \in \tilde{L}$. The microscopic time evolution $\alpha_t = \exp it\delta_\omega$ in the state ω induces a dynamics $\tilde{\alpha}_t$ on $\mathcal{A}(\tilde{L}_k)$ in the macrostate $\tilde{\omega}$ through $\tilde{\alpha}_t F_k^\#([f]) = F_k^\#(e^{it\tilde{h}_k}[f])$ leaving the state $\tilde{\omega}$ invariant and where $\tilde{h}_k$ is determined by $\tilde{h}_k[f] \equiv [h_\omega f]$

proof Determine the linear functional $\tilde{\omega}$ on $\mathcal{A}(\tilde{L}_k)$ by formula (3.17). It follows from proposition 2.1 and theorem 3.1 that the functional is quasi-free with the right two-point function. The positivity of $\tilde{\omega}$ follows from the positivity of the microstate ω. Using (2.14) one has

$$0 \le \omega\left(C^+(f)C^-(f)\right) \le (f,f)\ ;\ f \in L$$

Since the functional is gauge-invariant in the two-component version a necessary and sufficient condition for the positivity is [5, vol.II, p.44]

$$0 \leq \tilde{\omega}\left(F_k^+([f])F_k^-([f])\right) \leq ||f||^2$$

Using the translation invariance of ω one has

$$\begin{aligned} 0 &\leq \tilde{\omega}(F_k^+([f])F_k^-([f])) = \int d\alpha\, e^{-ik\alpha}\omega(C^+(f)\tau_\alpha C^-(f)) \\ &\leq \int d\alpha\, e^{-ik\alpha}(f, f_\alpha) = (f,f)_\sim^k \end{aligned}$$

proving the positivity of the quasi-free functional. The existence of the dynamics on $\mathcal{A}(\tilde{L}_k)$ induced by α_t follows from the fact that

$$\begin{aligned} &(e^{ith_\omega}f, e^{ith_\omega}g)_\sim^k = (f,g)_\sim^k \ ; \ f,g \in L \\ \text{i.e.} \quad &(e^{it\tilde{h}_k}[f], e^{it\tilde{h}_k}[g]) = ([f],[g]) \ ; \ [f],[g] \in \tilde{L}_k. \end{aligned}$$

This is an immediate consequence of the translation invariance of the dynamics : $\tau_\alpha\alpha_t = \alpha_t\tau_\alpha$. It is well known that each unitary $e^{it\tilde{h}_k}$ map on $\tilde{L}_k$ yields a *-automorphism $\tilde{\alpha}_t$ of the corresponding CAR-algebra $\mathcal{A}(\tilde{L}_k)$ via the formula

$$\tilde{\alpha}_t F_k^\#([f]) = F_k^\#(e^{it\tilde{h}_k}[f])$$

Again the translation invariance of α_t and the time invariance of ω : $\omega\alpha_t = \omega$ yield:

$$\begin{aligned} &\tilde{\omega}\left(\tilde{\alpha}_t\left(F_k^+([f])F_k^-([g])\right)\right) \\ &= \int d\alpha\, e^{-ik\alpha}\omega\left(\alpha_t(C^+(f))\tau_\alpha C^-(g)\right) \\ &= \int d\alpha\, e^{-ik\alpha}\omega\left(C^+(f)\tau_\alpha C^-(g)\right) \\ &= \tilde{\omega}\left(F_k^+([f])F_k^-([g])\right) \end{aligned}$$

implying the $\tilde{\alpha}_t$-invariance of $\tilde{\omega}$. ■

We obtained a new dynamical system, namely that of the Fermion field fluctuations of the BCS-model at momentum k. It is given by the triplet $\left(\mathcal{A}(\tilde{L}_k), \tilde{\omega}, \tilde{\alpha}_t\right)$ with $\tilde{\omega}\tilde{\alpha}_t = \tilde{\omega}$; $\mathcal{A}(\tilde{L}_k)$ is the CAR-algebra of the macroscopic field fluctuations, $\tilde{\omega}$ is the macro state induced by the micro equilibrium state ω as given by theorem 2.1., $\tilde{\alpha}_t$ is the macro evolution induced by the micro evolution α_t.

The central limit theorem and therorem 3.4. suggest the identification

$$F_k^{\#}([f]) = \lim_{\Lambda} F_{k,\Lambda}^{\#}(f) \; ; \; f \in L$$

i.e. the limit of the Fermion field fluctuations can be identified to a new and well defined Fermion field representation determined by the quasi free state $\tilde{\omega}$. Furthermore the $\tilde{\alpha}_t$-time invariance of the state $\tilde{\omega}$, yields as a consequence of the GNS-theorem, that the macro dynamics $\tilde{\alpha}_t$ has a Hamiltonian representation in the state $\tilde{\omega}$:

$$\tilde{\alpha}_t F_k^{\#}([f]) = e^{it\tilde{H}_k} F_k^{\#}([f]) e^{-it\tilde{H}_k}$$

Where

$$\tilde{H}_k = \frac{k^2}{2}\left(F_k^+\begin{pmatrix}1\\0\end{pmatrix}F_k^-\begin{pmatrix}1\\0\end{pmatrix} + F_k^-\begin{pmatrix}0\\1\end{pmatrix}F_k^+\begin{pmatrix}0\\1\end{pmatrix}\right)$$

$$-c(2\pi)^{3/2}\left(\bar{b}\hat{S}(k)F_k^+\begin{pmatrix}0\\1\end{pmatrix}F_k^-\begin{pmatrix}1\\0\end{pmatrix} + b\hat{S}(-k)F_k^+\begin{pmatrix}1\\0\end{pmatrix}F_k^-\begin{pmatrix}0\\1\end{pmatrix}\right)$$

What has to be observed on the level of the dynamics of the fluctuations is the spectrum of the macroscopic Hamiltonian $\tilde{H}_k$. Our main result of this section is the determination of the spectrum of $\tilde{H}_k$.

Theorem 3.5. The spectrum of $\tilde{H}_k$ is completely discrete and given by the set

$$\{\pm\sqrt{k^4/4 + c^2|b|^2|\hat{S}(k)|^2(2\pi)^3}\}$$

proof

$$\begin{aligned} \tilde{\alpha}_t F_k^{\#}([f]) &\equiv F_k^{\#}(e^{it\tilde{h}}[f]) \\ &\equiv F_k^{\#}([e^{ith_\omega} f]) \end{aligned}$$

or equivalently we determine $\tilde{\delta}$, the infinitesimal generator of $(\tilde{\alpha}_t)_{t\in\mathbb{R}}$

$$\begin{aligned} \tilde{\delta} F_k^{\#}([f]) &\equiv \# F_k^{\#}(\tilde{h}_k[f]) \\ &\equiv \# F_k^{\#}([h_\omega f]) \end{aligned}$$

We already showed that the equivalence class $[g]$ is characterized by the two complex numbers $\hat{g}_1(k)$ and $\hat{g}_2(k)$. Applying this to the equivalence class $[h_\omega f]$, we find for $\tilde{h}_k$ the following matrix:

$$\tilde{h}_k = \begin{pmatrix} k^2/2 & -cb\hat{S}(-k)(2\pi)^{3/2} \\ -c\bar{b}\hat{S}(k)(2\pi)^{3/2} & -k^2/2 \end{pmatrix}$$

with eigenvalues

$$\epsilon_k^{\pm} = \pm\sqrt{k^4/4 + c^2|b|^2|\hat{S}(k)|^2(2\pi)^3}.$$

The corresponding normalized eigenvectors are

$$e_k^{\pm} = \frac{1}{(2\pi)^{3/2}} \frac{1}{\sqrt{1 + |\mu_k^{\pm}|^2}} \begin{pmatrix} 1 \\ \mu_k^{\pm} \end{pmatrix}$$

with

$$\mu_k^{\pm} = \frac{k^2/2 - \epsilon_k^{\pm}}{cb(2\pi)^{3/2}\hat{S}(-k)}.$$

For $k \neq 0$ we obtain a result analogous to the case $k = 0$ worked out for the Overhauser-model [7]. Remark that on the level of the fluctuations the different $k \neq 0$ modes correspond to discrete eigenvalues, and hence macroscopically give rise to bound states, macroscopic quasi-states. Notice that

$$\lim_{k \to 0} \varepsilon_k = \varepsilon_0 \neq 0 \text{ if } \hat{S}(k = 0) \neq 0$$

corresponding to a macroscopic quasi paricle of zero momentum. This is the situation of an energy gap, due to the Cooper pairing.

The $k \neq 0$ field fluctuations can be worked out rigorously in an analogous way for the Overhauser-model, and in principle for any mean field model.

We showed that on the level of the field fluctuations in equilibrium a new dynamical system of macroscopic Fermions with a discrete spectrum appears. As such this might be understood as the mathematically rigorous understanding of the so-called selfconsistent field of macroscopic quasi-particle structure.

At the basis of our work is a central limit theorem, our theory should also give an understanding of this phenomenon for more realistic systems (i.e. beyond mean field systems), at least in the situations that there is enough clustering; e.g. at high enough temperatures.

References

1. D. Goderis, A. Verbeure, P. Vets; Il Nuovo Cimento 106 B, 375 (1991).
2. A. Verbeure; J. Math. Phys. 32, 689 (1991).
3. M. Broidioi, A. Verbeure; Plasmon frequency for a spin-density wave model; Preprint KUL-TF-91/18; to appear in Helv. Phys. Acta 1991.
4. R. Haag; Il Nuovo Cimento 19, 154 (1961).
5. O. Bratelli, D.W. Robinson; Operator Algebras and Quantum Statistical Mechanics, Springer, Berlin, 1979, 1981, vol. I, II.
6. O. Lanford, D. Ruelle; Commun. Math. Phys. 13, 194 (1969).
7. M. Broidioi, B. Nachtergaele, A. Verbeure; The Overhauser model: equilibrium fluctuation dynamics; Preprint KUL-TF-90/28; to appear in J. Math. Phys. 1991.
8. P.C. Hohenberg; Phys. Rev. 158,383 (1967).
9. D. Goderis, A. Verbeure, P. Vets; Comm. Math. Phys. 128, 533 (1990).
10. P. Vets; phD. thesis: Non-commutative central limits and fluctuations (1990).

H. Maassen:	Free Probability and Stochastic Calculus	Mathematics Institute University of Nijmegen Toernooiveld 6525 ED Nijmegen The Netherlands
K. R. Parthasarathy:	Connections with Classical Probability	Indian Statistical Institute 7, S. J. S. Sansanwal Mrg New Delhi 110016 India
D. Petz:	Operator Theoretical Methods	Hungarian Acad. of Science Mathematical Institute H-1364 Budapest, PF. 127 Hungary
K. B. Sinha:	Stochastic Differential Equations	Indian Statistical Institute 7, S. J. S. Sansanwal Mrg New Delhi 110016 India
W. von Waldenfels:	Algebraic Probability Theory	Institut fur Angewandte Mathematik Univ. Heidelberg SFB 123 Im Neuenbeimer feld 294 6900 Heidelberg 1 Germany

QP-PQ

www.ingramcontent.com/pod-product-compliance
Lightning Source LLC
LaVergne TN
LVHW010538300926
849162LV00002B/166